S0-BNY-082

The Math Advantage
Daily Practice for FCAT
Student Workbook

Harcourt Brace & Company

ORLANDO   ATLANTA   AUSTIN   BOSTON   SAN FRANCISCO   CHICAGO   DALLAS   NEW YORK
TORONTO   LONDON

Copyright © by Harcourt Brace & Company

All rights reserved. No part of this publication may be reproduced or transmitted in any form or by any means, electronic or mechanical, including photocopy, recording, or any information storage and retrieval system, without permission in writing from the publisher.

Teachers using MATH ADVANTAGE may photocopy complete pages in sufficient quantities for classroom use only and not for resale.

HARCOURT BRACE and Quill Design is a registered trademark and MATH ADVANTAGE is a trademark of Harcourt Brace & Company.

Grateful acknowledgment is made to the Florida Department of Education for permission to reprint mathematics rubrics language and mathematics reference sheet.

Printed in the United States of America

ISBN 0-15-310677-8

11 12 13 14 15   073   06 05 04 03 02

Contents

Harcourt Brace School Publishers

Name _____

1 Angela and Tom set up a race course for their bicycles. The course goes around their neighborhood. It begins and ends at Angela's house.

Multiple Choice
Test Taking Tips

Which unit is commonly used to measure distances on streets?

Which is a reasonable distance for a bicycle race?

Ⓐ 2 inches

Ⓑ 2 feet

Ⓒ 2 yards

Ⓓ 2 miles

2 Terri built a pattern with groups of counters. It looked like this.

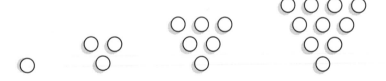

Gridded Response
Test Taking Tips

What pattern do you see?

REMINDER: Did you mark your answer in the grid?

If this pattern continues, how many coins will be in her next group?

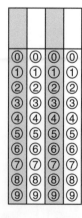

Grade 4 • Harcourt Brace School Publishers

Name _____

3 Cassie saw that her score on a video game was 4053. When she looked again, the 5 had changed to an 8. By how much had her score changed?

Explain your reasoning.

_____

_____

_____

_____

_____

Test Taking Tips

What do you know about place value that will help you solve this problem and explain your reasoning?

---

4 Sort these shapes into two groups.

On the lines below, draw the groups. Explain your sorting rules.

_____

_____

_____

_____

Test Taking Tips

Do all the shapes have straight lines? Are they all closed?

Grade 4 • Harcourt Brace School Publishers

Name _____

5 Mario is taking the bus to his aunt's house. The bus only takes exact change. The fare is 75¢. Mario has these coins.

Think • Solve • Explain
Long Answer

Test Taking Tips

What coins does Mario have?

What does "exact change" mean?

Grade 4 • Harcourt Brace School Publishers

Name _____

5 Use Mario's coins.
Show two ways to make exact change for 75¢.

Test Taking Tips

How can you check to see if your answers are right?

What are the fewest coins that Mario can use for bus fare? Explain how you decided.

_____

_____

_____

_____

_____

_____

_____

_____

_____

Grade 4 • Harcourt Brace School Publishers

Name _____

1 When Simon finished cleaning his room, his building blocks were piled neatly in the corner as shown below.

How many blocks are in his pile?

- Ⓐ 10 blocks
- Ⓑ 16 blocks
- Ⓒ 17 blocks
- Ⓓ 24 blocks

Test Taking Tips

Are there parts of the picture that you cannot see?

How can making a model help solve the problem?

---

2 Li-Ping and her brother Sam were playing video games. Li-Ping scored 2,398 points. Sam said, "Guess what my score is. It is 1,000 points more than yours."

How many points did Sam have?

Test Taking Tips

How can you use place value to help solve the problem?

REMINDER: Did you mark your answer in the grid?

3 Philip and Carlota drew these shapes for a geometry art project.

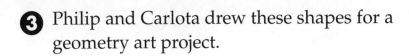

Are their shapes congruent? How do you know? Explain your reasoning.

_____

_____

_____

_____

Think • Solve • Explain
Short Answer

Test Taking Tips

What does congruent mean?

What is always true about congruent figures?

---

4 On Monday, March 1, three students came to Karate class. On Tuesday, March 2, six students came. On Wednesday, March 3, nine students came. This pattern continued all week. How many students came on Saturday, March 6?

Mon. Mar. 1	Tues. Mar. 2	Wed. Mar. 3	Thurs. Mar. 4	Fri. Mar. 5	Sat. Mar. 6

Describe a pattern in your chart.

_____

_____

Think • Solve • Explain
Short Answer

Test Taking Tips

How can you complete the chart to find the pattern?

Grade 4 • Harcourt Brace School Publishers

Name _____

5 Look at the shape of Wendy's gingerbread cookie house. The roof is a triangle. The sides of the triangle have the same measure. The triangle has a perimeter of 18 inches. Wendy is going to outline her house with ribbon. How much ribbon will she need to go all the way around?

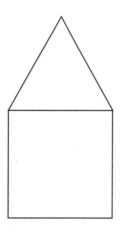

Remember that perimeter is the distance around an object.

Solve
Long Answer
Think • Explain
Test Taking Tips

How long is one side of the triangle? How do you know?

Grade 4 • Harcourt Brace School Publishers

Name _____

5 Explain how you solved the problem. Use words and pictures.

Think • Solve • Explain

Long Answer

Test Taking Tips

How can you check to see if your answer is right?

_____

_____

_____

_____

_____

_____

_____

Grade 4 • Harcourt Brace School Publishers

Name _____

1 Mrs. Castillo has four daughters.

Our Birthdays

Andrea	June 8, 1994
Josefina	May 23, 1992
Martina	July 27, 1990
Teresa	March 13, 1996

Who is the oldest child?

Ⓐ   Andrea

Ⓑ   Josefina

Ⓒ   Martina

Ⓓ   Teresa

Test Taking Tips

Which of these years came before all the other years?

---

2 What number is missing in the following number sentence?

$$3 + (4 + 7) = (3 + 4) + \underline{\hspace{1cm}}$$

Test Taking Tips

What property of addition does the number sentence show?

REMINDER: Did you mark your answer in the grid?

Name _____

3 Judy drew some stairs on dot paper. She decided to make more stairs going the other way. She will flip her design. Draw the flip of Judy's design.

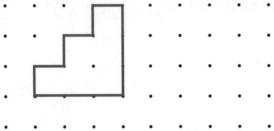

On the lines below explain why the two designs are symmetrical.

_____

_____

Think • Solve • Explain
Short Answer

Test Taking Tips

What does symmetrical mean?

What is true of all symmetrical figures?

---

4 Jonathan went into the store to get a gallon of milk for his mother.

He gave the cashier $5 to pay for the milk. He received $0.57 in change. Use ESTIMATION to help Jonathan explain to the cashier how much change he should have received.

_____

_____

Think • Solve • Explain
Short Answer

Test Taking Tips

What is the price for one gallon of milk?

How can you use rounding to estimate the correct change?

Grade 4 • Harcourt Brace School Publishers

Name _____

5 William, Thomas, and Beth were doing a set of 12 homework problems. At the end of 30 minutes, William had three-fourths finished, Thomas had 6 problems done, and Beth had two-thirds finished.

Draw pictures to show the number of problems each student had finished.

Test Taking Tips

Think • Solve • Explain
Long Answer

How can you use the picture to show how many problems William had completed?

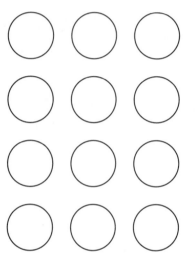

Grade 4 • Harcourt Brace School Publishers

Name _____

5 Who had the most problems finished?

_____

Explain how you solved the problem.
Use pictures, words, and numbers.

How can you check that your answers are correct?

William's work:

Thomas's work:

Beth's work:

Grade 4 • Harcourt Brace School Publishers

Name _____

1 Enrico's favorite T.V. program comes on at 5:00 P.M. His sister's favorite program is on $2\frac{1}{2}$ hours later and lasts 30 minutes. At what time does his sister's favorite program end?

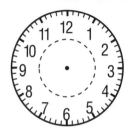 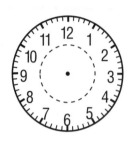 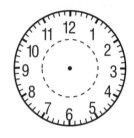

Ⓐ 5:30 P.M.

Ⓑ 7:30 P.M.

Ⓒ 8:00 P.M.

Ⓓ 8:30 P.M.

Test Taking Tips

How does showing each time help solve the problem?

---

2 How many lines of symmetry are in this hexagon?

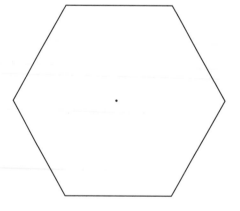

Test Taking Tips

What math words do you need to know?

REMINDER: Did you mark your answer in the grid?

Grade 4 • Harcourt Brace School Publishers

Name _____

3 Eric is the manager of the Must See Video Store. He wants to buy 30 new videos for the store. He surveyed his customers to learn their favorite type of movie.

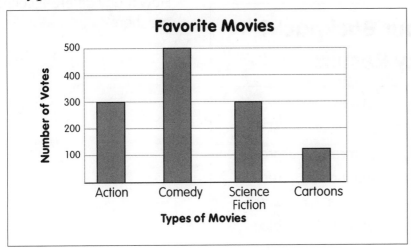

What kinds of movies should he buy?

_____

Explain how you used the data from the survey to make your decision.

_____

_____

Test Taking Tips

What does each bar of the graph show?

4 Sabrina says that there are exactly three ways to make 45 cents without using any pennies. Is she correct?

_____

Explain your reasoning. List all the ways you can find.

_____

_____

Test Taking Tips

How will making a chart help you know if you have found all the ways?

Grade 4 • Harcourt Brace School Publishers

THE MATH ADVANTAGE • Daily Practice for FCAT • Week 4   **17**

Name _____

5 Mr. Toma's class took a survey to find out the most common color of the backpacks students carry. Here are the results.

Test Taking Tips

How many different colors are there?

What Color Is Your Backpack?
Class Survey Results

Stephanie	black		Alyssa	brown
Taoran	green		Angelo	black
Matei	black		Kamran	green
Rachel	blue		Alanna	red
Rebecca	blue		Nikolas	black
Monika	blue		Linnea	blue
Corey	black		Sophie	blue
Justin	brown		Jeremiah	black
Kurtis	green		William	green
Adelina	blue		Patricia	red
Lana	red		Tanisha	red

Organize the results in a table.

Class Survey Results

Color of Backpack	Number of Students

Name _____

5 Make a bar graph that shows the results of the survey. Be sure to

- give your graph a title
- use a scale
- tell what each bar shows
- correctly show the data

Write two statements that compare the data on backpack colors.

Test Taking Tips

How many different colors do the data show?

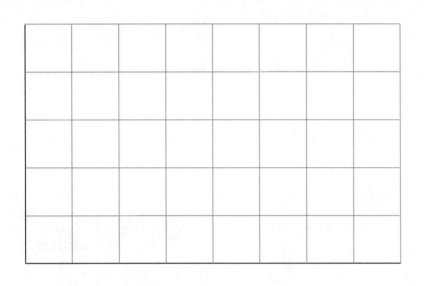

_____

_____

_____

_____

Grade 4 • Harcourt Brace School Publishers

Name _____

1 Bob made $28 mowing grass over the weekend. He gave half to his mother. He spent $4.78 for lunch.

How much money did Bob have left?

Ⓐ   Ten dollars and thirty-two cents

Ⓑ   Nine dollars and twenty-two cents

Ⓒ   Sixteen dollars

Ⓓ   Four dollars and seventy-eight cents

Test Taking Tips

How much money did Bob have when he went to lunch?

---

2 Samantha has laid out 30 cookies. Half are chocolate chip. How many cookies are NOT chocolate chip?

⓪	⓪	⓪	⓪
①	①	①	①
②	②	②	②
③	③	③	③
④	④	④	④
⑤	⑤	⑤	⑤
⑥	⑥	⑥	⑥
⑦	⑦	⑦	⑦
⑧	⑧	⑧	⑧
⑨	⑨	⑨	⑨

Test Taking Tips

How can drawing a picture help you solve the problem?

REMINDER: Did you mark your answer in the grid?

Grade 4 • Harcourt Brace School Publishers

3 Alonzo and Janet are playing a game. If the spinner lands on a 2-digit number, Alonzo gets five points. If the spinner lands on a 1-digit number, Janet gets five points. The purpose of the game is to win 50 points.

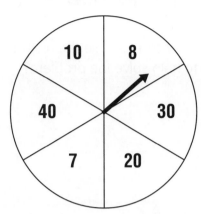

Who is more likely to win the game?

_____

On the lines below, explain how you decided.

_____

_____

_____

Test Taking Tips

How many spaces on the spinner have a 2-digit number?

How many spaces on the spinner have a 1-digit number?

4 Which measure is the best ESTIMATE to describe the length of the salamander?

Circle the best estimate.

     3 inches   3 miles   3 pounds

On the lines below, explain how you decided.

_____

_____

Test Taking Tips

How do you use each unit of measure?

Grade 4 • Harcourt Brace School Publishers

Name _____

5 Jolene is setting up a store. She has 36 erasers. She wants to charge 10¢ for each eraser. She is trying to decide how many erasers to put into each package.

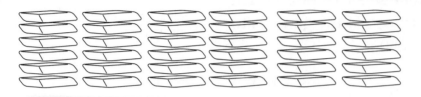

Help Jolene find three different ways to package 36 erasers with no leftovers. Record your ideas on the chart.

Test Taking Tips

How can you use the picture to help solve the problem?

Idea	Number of Erasers in 1 Package	Number of Packages	Price per Package

Grade 4 • Harcourt Brace School Publishers

5 Explain how you decided how to make packages with no leftovers. Use pictures, words and numbers.

Test Taking Tips

How can you make a list to solve the problem?

Grade 4 • Harcourt Brace School Publishers

Name _____

1 Tonya and Brittany were setting up a race course for their pet hamsters. They want to see how far the hamsters can run in 10 seconds.

Which is the most reasonable unit to measure the distance the hamsters run in 10 seconds?

Ⓐ miles

Ⓑ square feet

Ⓒ tons

Ⓓ feet

Test Taking Tips

Which choices do NOT seem reasonable?

---

2 Kendra and Marco drew a map of their neighborhood. How many blocks long is the perimeter of the park?

Remember: Perimeter is the distance around a shape.

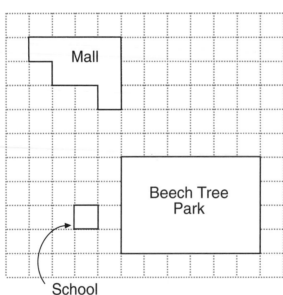

Mall

Beech Tree Park

School

⓪	⓪	⓪	⓪
①	①	①	①
②	②	②	②
③	③	③	③
④	④	④	④
⑤	⑤	⑤	⑤
⑥	⑥	⑥	⑥
⑦	⑦	⑦	⑦
⑧	⑧	⑧	⑧
⑨	⑨	⑨	⑨

Test Taking Tips

How can you use the map to find the length and width of the park?

REMINDER: Did you mark your answer in the grid?

Grade 4 • Harcourt Brace School Publishers

3 Fran is buying treats for a party. There will be four people to share the treats. She wants everyone to have the same number of treats. She can buy a package of 16, 26, or 30.

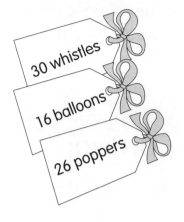

30 whistles

16 balloons

26 poppers

Which package should she buy?

_____

Explain how you decided.

_____

_____

Test Taking Tips

Think • Solve • Explain
Short Answer

What operation will you use to solve the problem?

---

4 Tanisha and Joel collect baseball cards. Tanisha has 569 cards. Joel has 596 cards. Who has more cards?

Show each number on the number line.

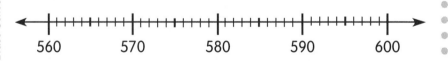

560   570   580   590   600

Write a number sentence that compares their card collections.

_____

Test Taking Tips

Think • Solve • Explain
Short Answer

How can a number line help you decide who has more cards?

Grade 4 • Harcourt Brace School Publishers

Name _____

5 Adam bought two snacks at the school carnival. He paid $0.06 in sales tax. When Adam gave the cashier $1.00, he received $0.20 in change.

What did Adam buy?

Test Taking Tips

How can working backward help you solve the problem?

Snacks at the Carnival

$0.29

$0.49

$0.25

$0.80

$0.45

$0.65

Name _____

5 What did Adam buy?

_____

_____

On the lines below, explain your thinking. Use numbers and words.

_____

_____

_____

_____

Test Taking Tips

How can estimation help you check that your answers are reasonable?

Grade 4 • Harcourt Brace School Publishers

Name _____

1 There is a new student in Victor's class. What are the chances that this student's birthday is in a month that starts with the letter J?

January	April	July	October
February	May	August	November
March	June	September	December

  Ⓐ   1 out of 3

  Ⓑ   3 out of 1

  Ⓒ   3 out of 3

  Ⓓ   3 out of 12

Test Taking Tips

What information do you need in order to solve the problem?

---

2 Tomiwa and his team are writing a report about recycling. They found out that their city recycled one thousand two hundred three pounds of newspaper in September and nine hundred fifty-nine pounds in October.

How many pounds of newspaper were recycled in those two months?

Test Taking Tips

How can writing the data in standard form help solve the problem?

REMINDER: Did you mark your answer in the grid?

Grade 4 • Harcourt Brace School Publishers

Name _____

3 Louis wants to measure the amount of water in his swimming pool.

Which unit of measurement should he use, cups, pints, quarts, or gallons?

_____

Explain how you decided.

_____

_____

_____

Think • Solve • Explain
Short Answer

Test Taking Tips

Which measure holds more, a cup, a pint, a quart, or a gallon?

4 Ricky has a riddle for his friend Marcello. There are two numbers. Their difference is 24. Their sum is 60. What are the two numbers?

_____

On the lines below, explain how you decided.

_____

_____

_____

Think • Solve • Explain
Short Answer

Test Taking Tips

How can you use a table to organize your guess and checks for the two numbers?

Grade 4 • Harcourt Brace School Publishers

5 Look at the quilt square Miranda designed. How many different congruent patch pieces make up the design?

Daily Practice
FCAT

Test Taking Tips

Think • Solve • Explain
Long Answer

What math word do you need to know to solve the problem?

What is always true about congruent figures?

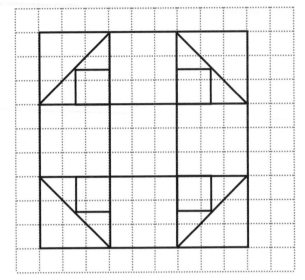

Draw each one.

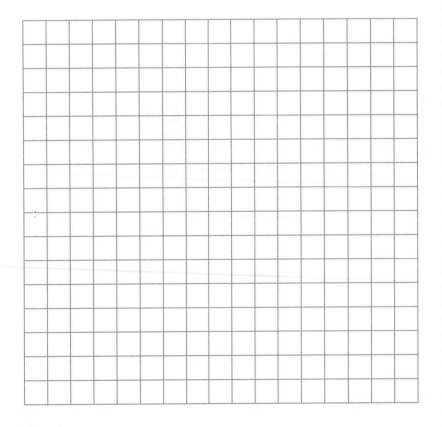

Name _____

5 Name each figure that you found.

_____

_____

_____

_____

How do you know which figures are congruent?
Use pictures and words to explain your
reasoning.

_____

_____

_____

_____

_____

_____

_____

_____

Test
Taking
Tips

How can making a list of the
shapes and sizes help you
solve the problem?

Name _____

1 Last month a music concert was held at an auditorium with 4,000 seats. Half the auditorium was closed for repairs. The rest of the auditorium was full. The seating area is divided into sections. Each section seats 500 people.

How many SECTIONS were used?

Ⓐ 4

Ⓑ 80

Ⓒ 250

Ⓓ 2,000

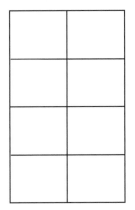

Test Taking Tips

How can you use the picture to solve the problem?

---

2 Meg and Ramirez made a value chart for the letters of the alphabet.

The word HAT is worth $0.29. How much is the word WHALE worth?

Letter	Value
A	1¢
B	2¢
C	3¢
D	4¢
E	5¢
F	6¢
G	7¢
H	8¢
I	9¢
J	10¢
K	11¢
L	12¢
M	13¢

Letter	Value
N	14¢
O	15¢
P	16¢
Q	17¢
R	18¢
S	19¢
T	20¢
U	21¢
V	22¢
W	23¢
X	24¢
Y	25¢
Z	26¢

Test Taking Tips

What information in the chart do you need to solve the problem?

REMINDER: Did you mark your answer in the grid?

Grade 4 • Harcourt Brace School Publishers

Name _____

3 Sam is going to visit his grandmother. He arrives at the airport at 2:00 P.M. His flight is supposed to leave at 3:10 P.M. The flight attendant announces that the departure will be delayed by 30 minutes.

How long will Sam have to wait at the airport?

_____

On the lines below, explain your reasoning.

_____

_____

Test Taking Tips

At what time should Sam's flight leave now?

---

4 Which of these pictures show a slide ONLY? Circle your answer.

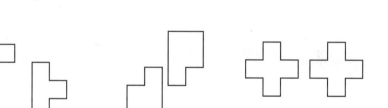

Figure 1   Figure 2   Figure 3

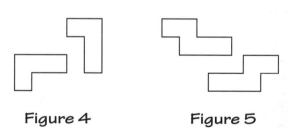

Figure 4   Figure 5

On the lines below, explain your reasoning.

_____

_____

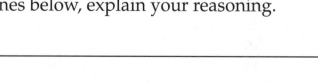

Test Taking Tips

What is the difference between a slide and a turn?

Name _____

5 Adelina and Amad are working as a team to find the word with the most points. They are using this chart.

Test Taking Tips

Which letters have the most points?

Points per Letter

A	B	C	D	E	F	G	H	I	J	K	L	M
1	2	3	4	5	6	7	8	9	10	11	12	13

N	O	P	Q	R	S	T	U	V	W	X	Y	Z
14	15	16	17	18	19	20	21	22	23	24	25	26

Find a word that is worth more points than "roadrunner."

Letter	Value		Letter	Value
R	18			
O	15			
A	1			
D	4			
R	18			
U	21			
N	14			
N	14			
E	5			
R	+18			
Total	128		Total	

Daily Practice
FCAT

5 One team said, "A longer word is always worth more than a shorter word." Do you agree?

_____

Explain your thinking. Give examples of words to prove your opinion.

_____

_____

_____

_____

_____

_____

Test Taking Tips

How can you check that your explanation is clear and correct?

Grade 4 • Harcourt Brace School Publishers

Name _____

1 Which of the following shapes DOES NOT have a line of symmetry?

   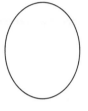

Ⓐ   Ⓑ   Ⓒ   Ⓓ

Test Taking Tips

How can you use a mirror to test for a line of symmetry?

---

2 Maria made a weekly calendar of her activities.

Maria's Weekly Schedule		
Mon.	dance class	4:00 – 5:15
Tues.	soccer	5:00 – 6:30
Wed.	dance class	4:00 – 5:15
Thurs.		
Fri.	babysit	6:00 – 8:15
Sat.	dance class	12:15 – 1:30

How many minutes does she spend at dance class each week?

Reminder: 1 hour = 60 minutes

Test Taking Tips

What do you need to find out first to solve the problem?

REMINDER: Did you mark your answer in the grid?

Grade 4 • Harcourt Brace School Publishers

3 Carlos is balancing wooden blocks. He notices that two square blocks will balance one triangle. He also notices that one triangle and one circle will balance five square blocks.

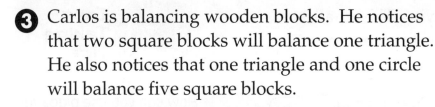

How many square blocks must Carlos use to balance one circle? Explain your thinking.

_____

_____

_____

Test Taking Tips

What information does the first scale give you that can help solve the problem?

How can Carlos balance a circle WITHOUT the triangle?

---

4 Joanie and Phil are playing a game like Tic Tac Toe. Three markers in a row win. Phil is using X's. It's Phil's turn. Where must he put his marker to keep from losing the game?

(___ , ___)

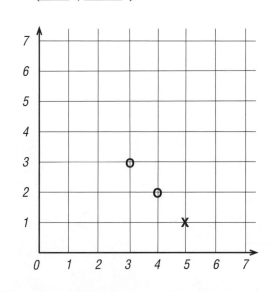

Test Taking Tips

How do you name a point on a grid?

Which number do you name first, the number across or the number up?

Grade 4 • Harcourt Brace School Publishers

Name _____

5 Mrs. Munoz's class is playing The Smallest Number Game. Arlette's team drew these number cards.

Rafael's team drew these cards.

What is the smallest number that each team can make?

Arlette _____

Rafael _____

Explain how you decided.

_____

_____

_____

_____

_____

Test Taking Tips

Solve • Explain • Think • Long Answer

How can you use place value to solve the problem?

Name _____

5 In the Greatest Sum Game, can the two teams together make a SUM that is greater than 100,000?

_____

Explain how you decided.

_____

_____

_____

_____

_____

_____

_____

_____

_____

How can you use ESTIMATION to solve the problem?

Grade 4 • Harcourt Brace School Publishers

Name _____

1 Look at Figure 1.

Which of the figures below show Figure 1 flipped to the right?

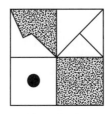

Figure 1

Multiple Choice
A
B
C
D
Test Taking Tips

How do you move a shape when you flip it?

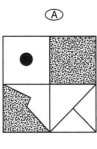

Ⓐ

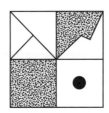

Ⓑ

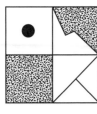

Ⓒ

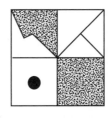

Ⓓ

2 There are 40 ounces of juice in Jar A. ESTIMATE about how many ounces of juice are likely to be in Jar B.

Jar A

Jar B

Gridded Response
Test Taking Tips

How much fuller is Jar B than Jar A?

REMINDER: Did you mark your answer in the grid?

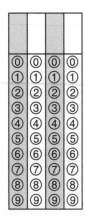

Grade 4 • Harcourt Brace School Publishers

3 Jamie has started a pattern using tiles.

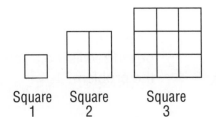

Square 1   Square 2   Square 3   Square 4   Square 5

How many tiles will Jamie need to make the fifth square in his pattern?

_____

Explain how you decided.

_____

_____

Test Taking Tips

How many tiles are in one row of each square?

4 Allen took $100.00 on his shopping trip. He bought his mom a new dress for $59.57. He bought his sister a new baseball cap for $16.19. He bought his brother a toy truck for $18.92. These prices INCLUDE sales tax.

How much change did Allen receive?

_____

Show your work below.

Test Taking Tips

What two steps could be used to solve this problem?

Name _____

5 Jacob is deciding what to wear for picture day at school. Here are his choices.

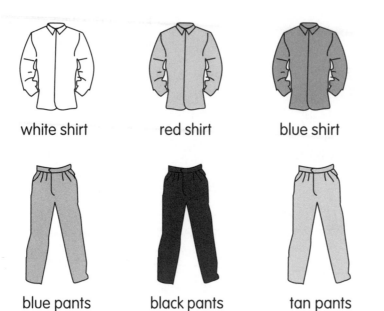

white shirt   red shirt   blue shirt

blue pants   black pants   tan pants

He is going to try them in different ways.

How can making an organized list help you solve the problem?

Name _____

5 How many choices does Jacob have?

_____

What are they?

_____

_____

_____

_____

_____

_____

_____

_____

_____

Explain how you know that you have found all the possible outfits he can make.

_____

_____

_____

_____

Grade 4 • Harcourt Brace School Publishers

Name _____

1 Fran noticed that there were nine commercials in the half-hour TV program that she watched. About how many commercials would she see if she watched TV for eight hours?

Ⓐ Fewer than 50

Ⓑ Between 50 and 90

Ⓒ Between 90 and 125

Ⓓ More than 125

Test Taking Tips

About how many commercials will there be in each full hour of programming?

---

2 On July 14, the hot dog vendor expects to sell twice as many hot dogs as she sold on June 9. How many hot dogs does she expect to sell on July 14?

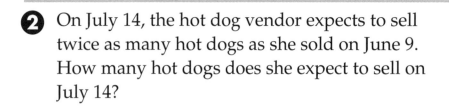

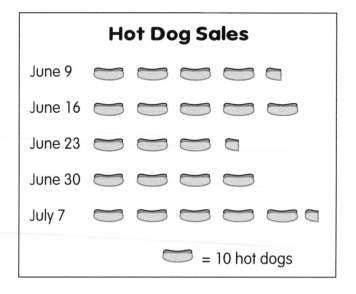

Hot Dog Sales

June 9

June 16

June 23

June 30

July 7

= 10 hot dogs

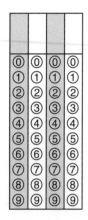

Test Taking Tips

What information in the pictograph will help you solve the problem?

REMINDER: Did you mark your answer in the grid?

Grade 4 • Harcourt Brace School Publishers

3 Angelo was waiting to order his pizza. He noticed a pattern on the ordering chart on the wall.

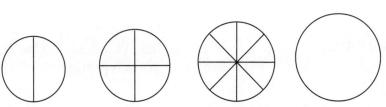

Mini　　　　Small　　　　Medium　　　　Large

If the pattern continues, how many slices will there be in the Large Size Pizza?

_____

On the lines below, explain how you decided.

_____

_____

Test Taking Tips

Think • Solve • Explain • Short Answer

How does the number of slices in each pizza compare to the number of slices in the next size?

---

4 Sarah's class is selling pencils for a fund raising project. They sold 142 pencils for 20¢ each.

Are their total sales closer to $10, $20, $30, or $40?

_____

Explain how you decided.

_____

_____

Test Taking Tips

Think • Solve • Explain • Short Answer

How can ESTIMATING help you solve the problem?

Grade 4 • Harcourt Brace School Publishers

5 Mariel and John are making up a geometry game. They are using these shapes in their game. The object of the game is to guess the sorting rule.

Here is how they sorted the shapes.

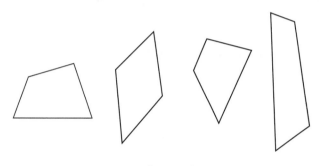

Group I

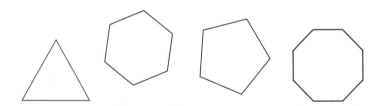

Group II

Test Taking Tips

What geometry words will help you describe the shapes in each group?

Name _____

5 Write three statements that describe the shapes in each group.

Group I

_____

_____

_____

_____

_____

How can you check that your answer is clear and complete?

Group II

_____

_____

_____

_____

_____

Daily Practice
FCAT

1

```
7
6  •
5
4           •
3
2
1     •     •
   0  1  2  3  4  5  6  7
```

William is drawing a W on grid paper. He has used ordered pairs to show these points.

(1, 6), (2, 1), (3, 4), (4, 1)

Multiple Choice
(A)
(B)
(C)
(D)

Test Taking Tips

Which number shows the position horizontally or straight across?

Which number shows the position vertically or straight up?

Where should his last point be?

Ⓐ (4, 6)

Ⓑ (5, 6)

Ⓒ (6, 5)

Ⓓ (6,6)

---

2 Kristina's mother is making a birthday cake in the shape of an octagon. She will put three small candles on each corner of the cake.

How many candles will she put on the cake?

Gridded Response

Test Taking Tips

What math words do you need to know to solve the problem?

REMINDER: Did you mark your answer in the grid?

Grade 4 • Harcourt Brace School Publishers

Daily Practice
FCAT

3 Julian's class took a survey of peanuts in a snack pack. They made a graph of their findings.

About how many peanuts would you expect to find in a similar snack pack of peanuts?

_____

Number of Peanuts in a Pack

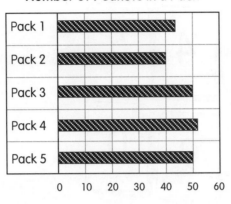

On the lines below, explain how you decided.

_____

_____

Test Taking Tips

Think • Solve • Explain
Short Answer

What is the least number of peanuts counted in a snack pack?

What is the greatest number of peanuts found in a snack pack?

---

4 Your team wins a point when the pointer lands on red.

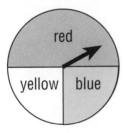

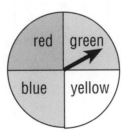

Which spinner would give your team a better chance of winning?

_____

On the lines below, explain how you decided.

_____

_____

Test Taking Tips

Think • Solve • Explain
Short Answer

What fraction of each spinner is red?

Grade 4 • Harcourt Brace School Publishers

Name _____

5 Shayla and Brian are playing a math game. They are using these cards.

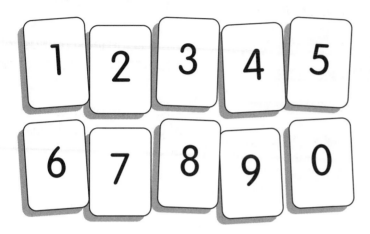

Test Taking Tips

Think • Solve • Explain
Long Answer

How can you use place value to solve the problem?

How can you arrange 6 of these cards in the addends to find the largest possible sum?

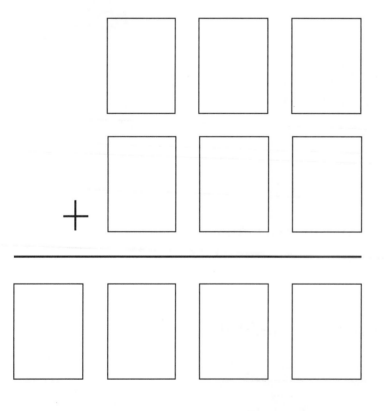

Grade 4 • Harcourt Brace School Publishers

Name _____

5 What strategy did you use to make the largest sum?

On the lines below, explain how you decided.

_____

_____

_____

_____

_____

_____

_____

_____

_____

How can you use place value to explain how to make the greatest sum?

Grade 4 • Harcourt Brace School Publishers

Name _____

1 In LaKeisha's city there are four tall buildings. The height of each building is shown below.

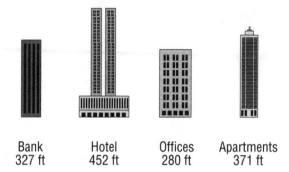

Bank	Hotel	Offices	Apartments
327 ft	452 ft	280 ft	371 ft

What is the difference in feet between the tallest building and the shortest building?

Ⓐ 91 feet

Ⓑ 125 feet

Ⓒ 172 feet

Ⓓ 732 feet

Test Taking Tips

What operation will you use to find the difference?

---

2 How many sides are on seven hexagons?

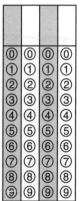

Test Taking Tips

What math word do you need to know to solve the problem?

How many sides does one hexagon have?

REMINDER: Did you mark your answer in the grid?

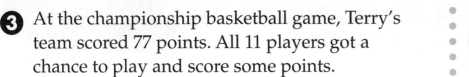

Name _____

3 At the championship basketball game, Terry's team scored 77 points. All 11 players got a chance to play and score some points.

Points Scored	
Mark	2
Luke	8
Josh	3
Juan	12
Terry	10
Lucas	4
Valentine	6
Jamal	12
Hector	10
Tony	6
Daniel	7

Which score represents the median (middle) score?

_____

Explain how you decided.

_____

_____

_____

_____

_____

_____

Test Taking Tips

Think • Solve • Explain
Short Answer

How can you organize the numbers to find the median score?

---

4 Jessica is greeting relatives at the airport. The schedule shows arrival times. Jessica will meet her grandmother from Miami at the gate. How much longer will they have to wait for her cousin, who is coming from Atlanta?

_____

Explain how you decided.

Flight Arrivals	
Arriving From	Arrival Time
Atlanta	5:35 P.M.
Boston	2:10 P.M.
Chicago	8:20 P.M.
Miami	1:45 P.M.

_____

_____

Test Taking Tips

Think • Solve • Explain
Short Answer

When will her grandmother arrive from Miami?

When will her cousin arrive from Atlanta?

How can you find the difference between those times?

Grade 4 • Harcourt Brace School Publishers

Name _____

5 Mr. Romero's class and Ms. Caldwell's class are going on a picnic. They need to pack box lunches for their classes.

Test Taking Tips

Each lunch box measures 8 in. x 8 in. x 8 in.

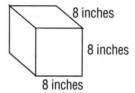

8 inches
8 inches
8 inches

How many boxes fit in the first layer of the large container?

They will pack the lunch boxes in a large container like the one drawn below.

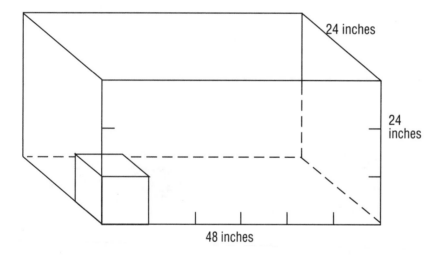

24 inches

24 inches

48 inches

How many boxes can fit in the lowest layer of the box?

_____

Draw a picture of that layer.

Grade 4 • Harcourt Brace School Publishers

Name _____

5 How many little boxes will fit in the large
container?

_____

Use pictures and diagrams to help explain your
thinking.

Test Taking Tips

Think • Solve • Explain
Long Answer

How can you use a picture to
understand the problem?

Name _____

1 Which of the following figures shows a right triangle inside a quadrilateral?

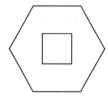

Ⓐ   Ⓑ

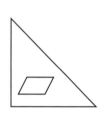

Ⓒ   Ⓓ

Multiple Choice
Ⓐ Ⓑ Ⓒ Ⓓ
Test Taking Tips

What is always true about right triangles?

What is a quadrilateral?

2 The faces of a number cube are numbered from 1 to 6.

After you toss the cube, you'll look at the number on top.

Which of the following outcomes is possible?

0   4   8   10   12

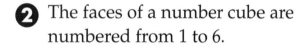

Gridded Response
Test Taking Tips

How can making a list of possible outcomes help you solve the problem?

REMINDER: Did you mark your answer in the grid?

Grade 4 • Harcourt Brace School Publishers

③ Jamal noticed a pattern in these numbers.

6, 12, 18, 24, 30,…

What would be the next three numbers in the pattern?

_____

Describe the pattern that helps you predict the next numbers.

_____

_____

_____

_____

Test Taking Tips

How is each number related to the number that comes before it?

---

④ Which graph is more likely to show the heights of a group of fourth graders?

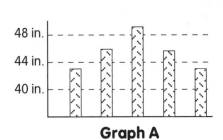

Graph A   **Graph B**

Explain how you decided.

_____

_____

_____

Test Taking Tips

What do you notice about the height of students in your class?

Grade 4 • Harcourt Brace School Publishers

Name _____

5 Use the grid below. Start at the point (6, 4).
Then show three more points to form a square
when the points are connected.

Test Taking Tips

Think • Solve • Explain Long Answer

How do you know where the point (6,4) is located?

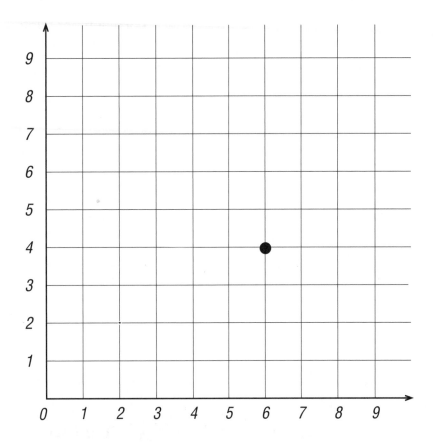

Write the ordered pair for each point on your
square.

(6, 4)

_____

_____

_____

Grade 4 • Harcourt Brace School Publishers

Name _____

5 Explain how to name ordered pairs in a grid. Use words and numbers.

_____

_____

_____

_____

_____

On the lines below, explain how you know that your shape is a square.

_____

_____

_____

_____

_____

_____

_____

_____

How can you check that your answer is clear and complete?

Grade 4 • Harcourt Brace School Publishers

Name _____

1 Jesse harvested pints of strawberries from the garden. How many more pints did Jesse harvest in 1994 than in 1997?

Jesse's Strawberry Crops

1997	🍓 🍓 🍓
1996	🍓 🍓 🍓 🍓
1995	🍓 🍓 🍓
1994	🍓 🍓 🍓 🍓
1993	🍓 🍓

Each 🍓 = 4 pints

Ⓐ 2 pints

Ⓑ 4 pints

Ⓒ 6 pints

Ⓓ 8 pints

Multiple Choice **Test Taking Tips**

How many pints of strawberries did Jesse harvest each year?

---

2 Sandy won a $15,000 shopping spree to an electronics store. She bought all the items on the list.

What Sandy Bought

2 big-screen TV sets	$2,150 each
1 stereo system	$1,750
2 computer systems	$3,580 each

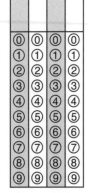

ESTIMATE to the nearest $1,000 how much money Sandy has left.

Gridded Response **Test Taking Tips**

How much did each TV set cost? How many did she buy?

How can rounding help you solve the problem?

REMINDER: Did you mark your answer in the grid?

Grade 4 • Harcourt Brace School Publishers

3 Terrell had these coins. She bought a soda for 60¢ and crackers for 65¢.

How much money does she have left over?

_____

On the lines below, explain how you decided.

_____

_____

_____

Solve
Think · Explain
Short Answer
Test Taking Tips

What do you need to find out first?

---

4 A large window is made up of four square panes. The perimeter of each pane is 40 inches. There is a piece of wood 1 inch wide around each pane.

What is the perimeter of the large window?

_____

_____

Remember: Perimeter is the distance around the outside edge of an object.

Solve
Think · Explain
Short Answer

Test Taking Tips

How can you label parts of the picture to help solve the problem?

Grade 4 • Harcourt Brace School Publishers

Name _____

5 Joey has a new puppy. His sister, Jenna, has a big dog.

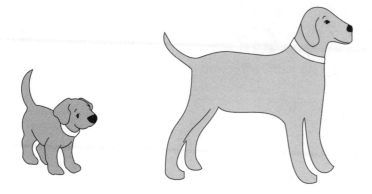

Jenna's dog weighs eight times as much as the puppy. Both pets together weigh 54 pounds.

How much does Joey's puppy weigh?

Solve • Explain • Think
Long Answer

Test Taking Tips

How can you make a model to show the information?

Grade 4 • Harcourt Brace School Publishers

Name _____

5 Explain how you solved the problem.

Use words, pictures and numbers.

_____

_____

_____

_____

_____

_____

Test Taking Tips

How can you check that your answer is clear and complete?

Grade 4 • Harcourt Brace School Publishers

Name _____

1 Annie went to the state fair to see the pumpkin contest. Her dad said, "The winning pumpkin weighs about four times as much as you do."

Which measure below is the most reasonable ESTIMATE of the weight of the winning pumpkin?

- Ⓐ 2,000 pounds
- Ⓑ 1,200 pounds
- Ⓒ 300 pounds
- Ⓓ 75 pounds

Test Taking Tips

How can you estimate the weight of a fourth grader?

---

2 Use the information in the diagram below. What is the perimeter of the Sports Center building?

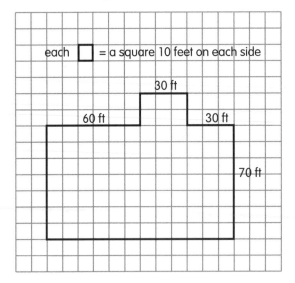

each ☐ = a square 10 feet on each side

30 ft

60 ft   30 ft

70 ft

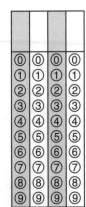

Remember: Perimeter is the distance around a shape.

Test Taking Tips

Which measures do you need in order to solve the problem?

Grade 4 • Harcourt Brace School Publishers

Name _____

3 Mrs. Jacoby's class and Mr. Ralston's class are going on a field trip. There are 27 students in Mrs. Jacoby's class and 31 students in Mr. Ralston's class. The two teachers and five other adults are also going. The bus can carry 65 passengers. Is there enough room on the bus?

_____

On the lines below, explain how you decided.

_____

_____

_____

Test Taking Tips

How many people are going on the field trip?

---

4 Roberta and Salina went to the beach and drew a rectangle in the sand. Then they measured two sides using Roberta's sand shovel.

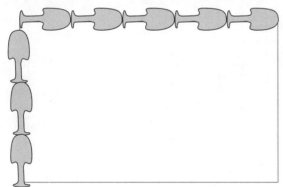

One side was 3 shovels long. Another side was 5 shovels long.

What was the perimeter of the rectangle? _____ Explain your thinking.

_____

_____

Test Taking Tips

How can labeling the diagram help you solve the problem?

Grade 4 • Harcourt Brace School Publishers

Name _____

5 Burt has 24 feet of fencing to build a cage for his new pet rabbit. The cage can be a square or a rectangle. Use the grid paper to draw 4 different ways that Burt can build the cage using all 24 feet of fencing.

Test Taking Tips

How can you check that your answer is clear and complete?

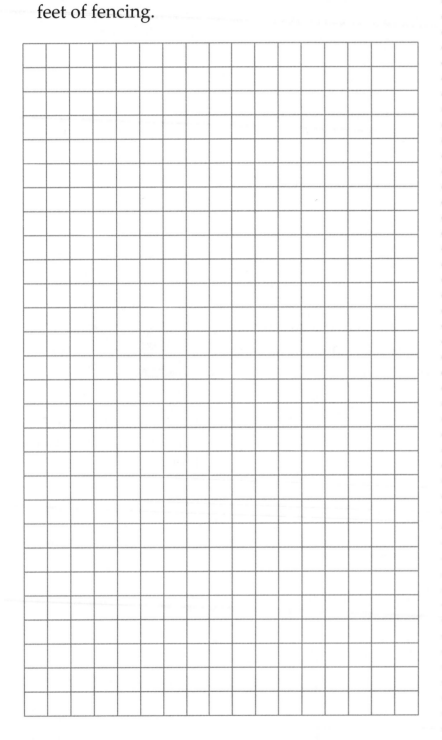

Name _____

5 Record the length and width of each possible cage. Then check that you have used all the fencing each time. Find the area of each cage.

Test Taking Tips

How can you check that each possiblility uses all of the fencing?

Length	Width	Fence Needed (feet)	Area (square feet)

What are the length and width of the cage that will give his rabbit the most room?

length: _____

width: _____

On the lines below, explain how you decided.

_____

_____

_____

_____

_____

Grade 4 • Harcourt Brace School Publishers

Name _____

1 Alexandro went to the mall with $25 to spend. He spent $11.41 for a tee-shirt, $7.93 for a book, and $3.15 for a snack. About how much does he have left?

Multiple Choice
Test Taking Tips

How can you use estimation to choose a reasonable answer?

  Ⓐ about $10

  Ⓑ about $5

  Ⓒ about $4

  Ⓓ about $3

$11.41

$3.15

Jokes

$7.93

2 Billy's mom wrote a check for fifty-three dollars and twelve cents.

Gridded Response
Test Taking Tips

How can you write the words as numbers?

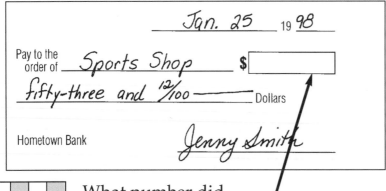

Jan. 25   19 98

Pay to the order of _Sports Shop_ $ [ ]

fifty-three and ¹²/₁₀₀ ————— Dollars

Hometown Bank   _Jenny Smith_

What number did she write in the box?

Mark your answer in the grid.

Grade 4 • Harcourt Brace School Publishers

Daily Practice
FCAT

3 Georgia's class has 26 students. On Saint Patrick's Day, they had a party. Everyone either wore a green shirt or wore a green hat. Some students did both. 17 students wore a green shirt. 21 students wore green hats. How many students wearing green shirts were also wearing green hats?

Use this number line to help find the answer.

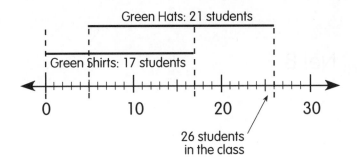

Test Taking Tips

What does the overlap of the two line segments represent?

---

4 Write three different measurements you can use to describe a person.

_____

_____

_____

Then explain what tool you would use for each measurement.

_____

_____

_____

Test Taking Tips

What information does each kind of measurement give you?

Name _____

5 Jodi is making cardboard cubes. Which of these patterns can she fold to make a cube?

Write OK by each net that will fold to make a cube. Write NO if it will not fold to make a cube.

Test Taking Tips

Which can be folded to look like a box with a top and bottom?

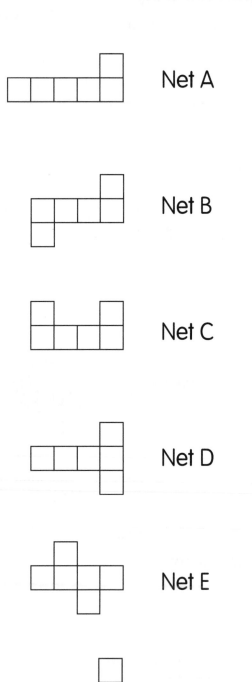

Net A

Net B

Net C

Net D

Net E

Net F

Grade 4 • Harcourt Brace School Publishers

Name _____

5 For each net that will not be a cube, explain why it would not work.

How many faces are on a cube?

_____

_____

_____

_____

_____

_____

_____

_____

_____

_____

_____

Daily Practice
FCAT

1 Margo put 5 dimes in an empty jar on Monday morning. Each day she added the same number of dimes as the previous day, plus two more.

How much money did she have in the jar on Friday evening?

Monday	Tuesday	Wednesday	Thursday	Friday
5 dimes 50¢				

Ⓐ $1.30

Ⓑ $3.20

Ⓒ $4.50

Ⓓ $45.00

Test Taking Tips

How will making a table help you record all the information you need?

---

2 How many sides do eight octagons have?

Test Taking Tips

What do you need to know about an octagon to solve this problem?

Grade 4 • Harcourt Brace School Publishers

3 Fred works at an ice cream shop. He sells vanilla, chocolate, and strawberry ice cream. He can make double-scoop cones using one flavor or two flavors. He can make several different double-scoop cones. If he uses chocolate and vanilla, for example, he can put the chocolate on the top, or on the bottom.

How many different double-scoop cones can he make?

_____

Use pictures or diagrams to help explain your answer.

Test Taking Tips

How can you keep a record of all the combinations?

---

4 For each gallon of gasoline Marisol's mom puts in her car, the car can travel about 26 miles. The tank holds 12 gallons.

Can Marisol's family drive 400 miles on one full tank?

_____

Explain how you decided. Use the lines below.

_____

_____

_____

Test Taking Tips

How can you use estimation to solve the problem?

Grade 4 • Harcourt Brace School Publishers

Name _____

5 Savannah built the following three designs from tiles.

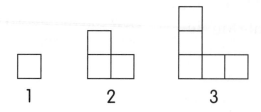

1   2   3

Test Taking Tips

How many tiles does she need for each design?

Follow her pattern and draw her fourth design.

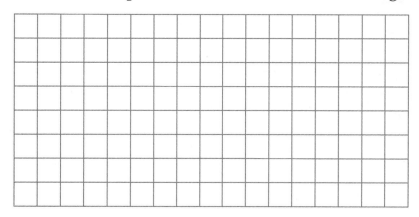

Draw her tenth design.

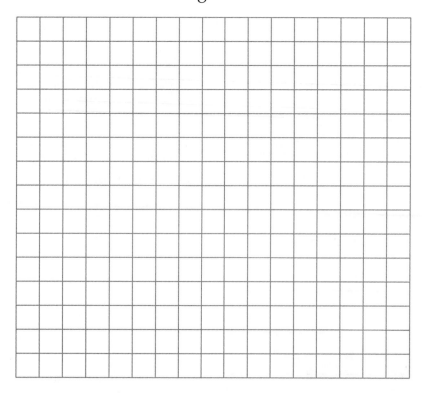

Grade 4 • Harcourt Brace School Publishers

Name _____

5 What pattern is she using to make each design?

_____

_____

_____

Explain in words and drawings.

_____

_____

_____

_____

How can you check that your answer is correct and that your explanation is clear?

Grade 4 • Harcourt Brace School Publishers

1 The fourth graders are making a book for their principal that shows a card made by each student. There are 138 students participating. A card is pasted on each side of a sheet of paper.

How many sheets of paper will they need to make the book?

Ⓐ   49 sheets of paper

Ⓑ   57 sheets of paper

Ⓒ   69 sheets of paper

Ⓓ   77 sheets of paper

Multiple Choice

Test Taking Tips

What operation can you use to solve the problem?

2 Mrs. Jones is thinking of a mystery number. She calls the number *m*. Starting with *m*, she adds 4. Then she multiplies her answer by 5. Next, she divides her new answer by 3. Then she subtracts 8. Her final answer is 7. What is her mystery number?

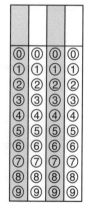

Gridded Response

Test Taking Tips

How can working backward help you solve the problem?

Grade 4 • Harcourt Brace School Publishers

Name _____

3 Lucas took 5 spelling tests this grading period. He used a calculator to average his scores.

Test #1	91
Test #2	85
Test #3	94
Test #4	92
Test #5	88

Is his solution reasonable?

_____

Explain how you decided.

_____

_____

Test Taking Tips

How do you find an average?

---

4 Rachel is on a swim team. She practices five days a week. The pool is 25 meters long. She swims 50 lengths at each practice. How many meters does she swim each week?

_____

Show how you found out.

_____

_____

_____

_____

Test Taking Tips

What operation can you use to solve the problem?

Grade 4 • Harcourt Brace School Publishers

5 Mr. Boromeo is planting a rectangular garden. He wants it to have an area of EXACTLY 48 square feet.

Draw three different rectangles that have an area of 48 square feet.

Think • Solve • Explain
Long Answer

Test Taking Tips

How does drawing a diagram for each rectangle help you know that the rectangles have an area of 48 square feet?

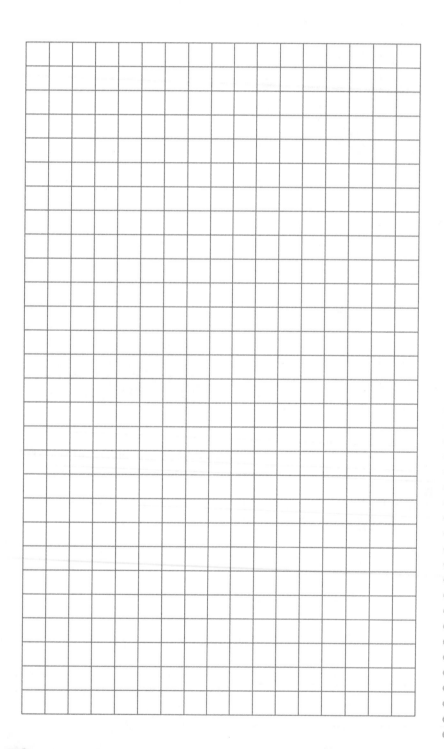

Grade 4 • Harcourt Brace School Publishers

Name _____

5 Make a table. Show the length and width of each rectangle that you found.

Length	Width	Fence Needed (feet)	Area (square feet)

How can making a table help you solve this problem?

What is the least amount of fencing he can use to enclose a rectangle with an area of 48 square feet?

_____

Use words and pictures to explain your answer.

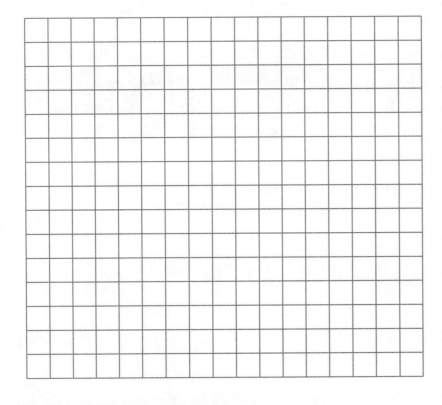

Grade 4 • Harcourt Brace School Publishers

Name _____

1 Richard drew this design.

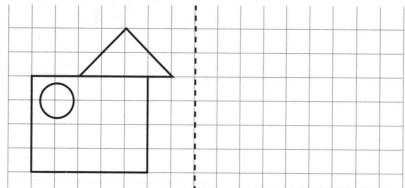

Then he folded his paper on the dotted line and traced the design. When he unfolds the paper, what does the tracing look like?

Ⓐ   Ⓑ   Ⓒ   Ⓓ

Test Taking Tips

Where will the triangle be in the tracing?

---

2 Alexander lives in a city where it snows. In January it snowed 24 inches. In February it snowed 18 inches and in March it snowed 18 inches. In April 8 inches of snow fell.

What was the average snowfall in inches for these months?

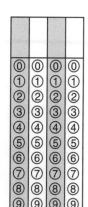

Test Taking Tips

How do you find an average?

Grade 4 • Harcourt Brace School Publishers

3 Six friends are on a tennis team. Each player will play each of the other players only once. How many tennis games will they play?

Show how you solved the problem. Explain your thinking.

Solve • Think • Explain • Short Answer
Test Taking Tips

How can a chart or picture help you solve the problem?

_____

_____

_____

_____

4 Leo has a sister, Melanie. Melanie is twice as old as Leo. If you add their ages together you get 27.

How old is Leo? _____

How old is Melanie? _____

Show your work or explain how you solved the problem.

Solve • Think • Explain • Short Answer
Test Taking Tips

How can you use a model to solve the problem?

_____

_____

_____

_____

Grade 4 • Harcourt Brace School Publishers

5 Samuel saw a bus full of children bringing their dogs to a dog show. He counted 42 legs in all. How many children and how many dogs could be on the bus?

Write some guesses in the chart. You may find more than one correct answer!

Test Taking Tips

How can you use guess and check to solve the problem?

Why should you guess fewer than 21 children?

Children		Dogs		Total: Children + Dogs	
How many children?	How many legs?	How many dogs?	How many legs?	How many legs total?	Comments

Grade 4 • Harcourt Brace School Publishers

Name _____

5 What did you learn by studying your guess and check chart? Explain your thinking on the lines below.

_____

_____

_____

_____

_____

Test Taking Tips

How can you check that your answer is clear and complete?

Name _____

1 Raoul likes to keep a weather watch. One February morning, the temperature was 4° C. Later that day, the temperature rose to 23° C.

How many degrees Celsius did the temperature rise?

50
40
30
20
10
0
−10

Ⓐ 4°C

Ⓑ 19°C

Ⓒ 23°C

Ⓓ 27°C

Multiple Choice

Test Taking Tips

How can using the scale on the thermometer help you solve the problem?

---

2 Three people drove together to Tennessee. Murray drove five hours. Malcolm drove three hours, and Xavier drove two hours. They averaged sixty miles per hour.

How many miles did they travel?

Gridded Response

Test Taking Tips

How many hours did their trip last?

How will this information help you solve the problem?

84

Grade 4 • Harcourt Brace School Publishers

Name _____

3 On Michael's basketball team, the height of each player's jump is measured in inches. Each player's best jump is shown in the chart.

How High Can You Jump?	
Michael	42 inches
Patrick	27 inches
Charles	37 inches
Angelo	31 inches
Taoran	40 inches

What is the **median** of the players' best jumps?

_____

Explain how to find the median.

_____

_____

Test Taking Tips

How can you organize the numbers to help you find the median?

---

4 Kendra made this design with pattern blocks.

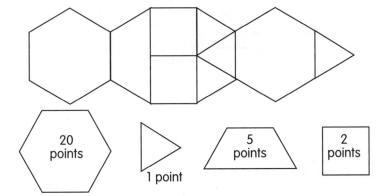

How many points is her design worth? Explain how you found your answer.

_____

_____

Test Taking Tips

How many points is each shape worth?

How many of each shape did she use?

Grade 4 • Harcourt Brace School Publishers

Daily Practice
FCAT

5 Jodi is buying supplies to start a window garden. She has $5 to spend.

Solve • Explain
Think
Long Answer
Test Taking Tips

How does rounding decimals to the nearest tenth help solve the problem?

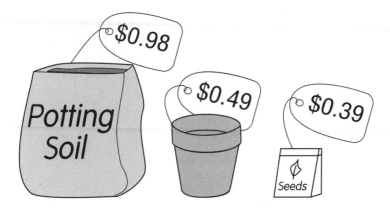

$0.98

Potting Soil

$0.49

$0.39

Seeds

One bag of soil fills 3 clay pots.

Make a shopping list of the items she can buy on her trip to the garden center. Round the prices shown on the tags to the nearest 10 cents and include rounded prices for the items on your shopping list.

_____

_____

_____

_____

_____

_____

5 She wants to have 5 clay pots full of plants. Can she afford to buy 5 clay pots, enough soil for the pots, and seeds?

_____

Remember, one bag of soil fills 3 clay pots.

Explain on the lines below.

_____

_____

_____

_____

_____

Test Taking Tips

How can you check that your answer is clear and complete?

Grade 4 • Harcourt Brace School Publishers

Name _____

1 Which of the following is NOT equal to 3?

⍐ 3.00

⍑ 12 ÷ 4

⍒ $\frac{3}{9}$

⍓ 1 x 3

Test Taking Tips

How does thinking about equivalent forms of a number help you solve the problem?

2 Leah and her brother are saving to buy a computer game that costs $49.95. They saved dimes in their piggy bank. They put the dimes into four rolls. Each roll holds 50 dimes.

How many more rolls of dimes do they need in order to buy the computer game?

Test Taking Tips

What do you need to find out first?

⓪	⓪	⓪	⓪
①	①	①	①
②	②	②	②
③	③	③	③
④	④	④	④
⑤	⑤	⑤	⑤
⑥	⑥	⑥	⑥
⑦	⑦	⑦	⑦
⑧	⑧	⑧	⑧
⑨	⑨	⑨	⑨

Name _____

3 Mr. Clark's class conducted an experiment to see how many raisins there are in a snack-size box. One group made this graph of their data:

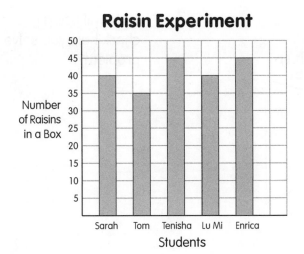

Raisin Experiment

What is the AVERAGE number of raisins per box? _____

Show how you solved the problem.

_____

_____

Test Taking Tips

Think • Solve • Explain
Short Answer

What clues about the average number of raisins in a box can you get from the bars on the bar graph?

4 What is the probability of spinning a lion?

_____

Explain how you decided.

_____

_____

_____

Test Taking Tips

Think • Solve • Explain
Short Answer

What are the possible outcomes?

What fractional part of the spinner shows a lion?

Grade 4 • Harcourt Brace School Publishers

Name _____

5 Alice spent $28 on new notebooks and organizer pockets. Notebooks cost $3 and organizer pockets cost $2. Alice bought 12 items.

Test Taking Tips

How will using guess and check help you solve the problem?

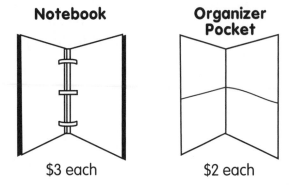

Notebook

$3 each

Organizer Pocket

$2 each

How many of each did she buy?

Use the chart to record your guesses.

	Notebooks ($3 each)		Organizer Pockets ($2 each)		Total Expense
	How many?	How much money?	How many?	How much money?	Notebooks + Organizers
Guess #1					
Guess #2					
Guess #3					
Guess #4					
Guess #5					
Guess #6					

Grade 4 • Harcourt Brace School Publishers

Name _____

5 How many of each did she buy?

_____

Explain how you can learn from a wrong guess to make a better one the next time.

_____

_____

_____

_____

_____

Grade 4 • Harcourt Brace School Publishers

Daily Practice

FCAT

1 Sam said, "We ate five sixths of the pizza." Which picture can be used to show the amount of pizza left?

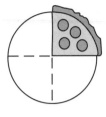

Ⓐ

Ⓑ

Ⓒ

Ⓓ

Multiple Choice

Ⓐ Ⓑ Ⓒ Ⓓ

Test Taking Tips

How many slices could have been in the pizza?

---

2 In music class there are four more boys than girls. If there are 28 students in all, how many are boys?

Gridded Response

Test Taking Tips

How can you use guess and check to solve the problem?

Grade 4 • Harcourt Brace School Publishers

Name _____

3 Circle the polygons.

A

B

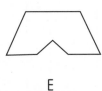

C

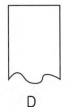

D

E

Explain how you know.

_____

_____

_____

What math word do you need to know to solve the problem?

---

4 Jasmin's kindergarten class has three boxes of crayons at each activity table in the classroom. There are four activity tables.

Half of the boxes hold 15 crayons each. The rest hold 20 crayons each. How many crayons are in the classroom?

Explain how you found out.

_____

_____

_____

_____

Test
Taking
Tips

What do you need to figure out first?

Grade 4 • Harcourt Brace School Publishers

Name _____

5 Kim, Cathy, and Jenny drove together to Florida for a vacation. Kim drove for 3 hours. Cathy drove for 5 hours. Jenny drove the rest of the time.

Together, Kim and Cathy drove twice as many hours as Jenny. How many hours did Jenny drive? Draw a diagram to help solve the problem.

Test Taking Tips

How can you draw a picture or make a diagram to show the information?

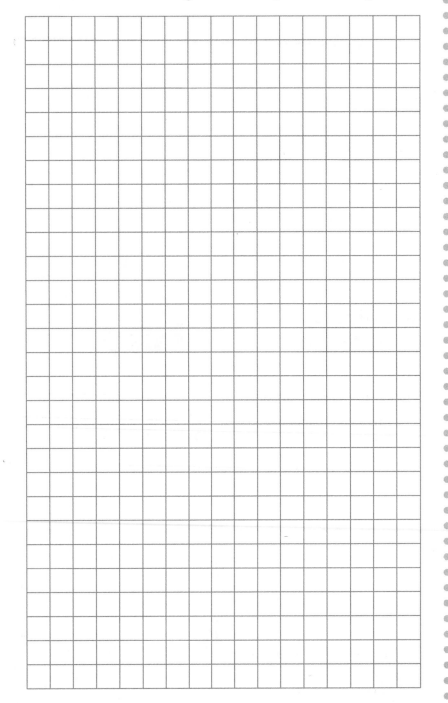

Grade 4 • Harcourt Brace School Publishers

Name _____

5 How long did Jenny drive?

_____

Explain how you solved the problem. Use pictures, numbers, and words.

_____

_____

_____

_____

_____

How can you use what you know about fractions to help solve the problem?

How can you check to see if your answer is right?

Name _____

1 Megan measured the dimensions of her room. She is going to buy a paper border to put all the way around where the wall meets the ceiling.

Which could she use to find out how much border to buy?

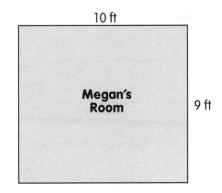

10 ft

Megan's Room

9 ft

Ⓐ 9 ft x 10 ft

Ⓑ 9 ft + 10 ft

Ⓒ 10 ft ÷ 9 ft

Ⓓ 9 ft + 10 ft + 9 ft + 10 ft

Multiple Choice

Test Taking Tips

How can you use the information in the diagram to help you choose the correct answer?

2 Jana noticed that the temperature at 4 P.M. was 65 °F. During the night, the temperature was 50 °F.

How many degrees did the temperature drop?

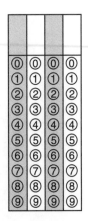

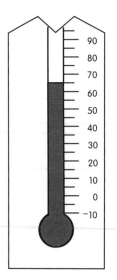

Gridded Response

Test Taking Tips

How can you use the scale on the thermometer to solve the problem?

Grade 4 • Harcourt Brace School Publishers

Name _____

3 Timmy's family spent the weekend at the beach. Timmy recorded the temperature each day at noon. On Friday, the temperature was 78° F. On Saturday, it was 82° F. On Sunday, the temperature was 83° F.

What was the average noon temperature?

_____

Explain how you found your answer.

_____

_____

_____

_____

 **Test Taking Tips**

How do you find an average?

---

4 Ms. Schultz's class conducted a survey to find out which hand students use to write. Here are their results.

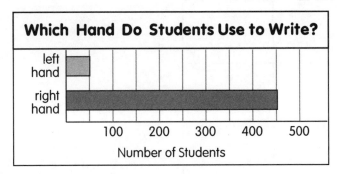

Which Hand Do Students Use to Write?

left hand

right hand

100   200   300   400   500
Number of Students

What fraction of the people surveyed are left-handed? _____

Explain how you decided.

_____

_____

 **Test Taking Tips**

How many students were surveyed?

How many of the surveyed students are left-handed?

Grade 4 • Harcourt Brace School Publishers

THE MATH ADVANTAGE • Daily Practice for FCAT • Week 24

Name _____

5 The students in Mrs. Hunter's class are making their own trail mix for a hiking trip. Each bag will contain exactly 12 ounces of mix.

Each mix must contain three or more ingredients. Use the chart to show 3 different trail mixes the students could make.

Test Taking Tips

Why can't you put 12 ounces of peanuts in a mix?

Ingredient	Mix #1	Mix #2	Mix #3
peanuts			
chocolate chips			
sunflower seeds			
raisins			
banana chips			
date pieces			
other:			
TOTAL AMOUNT	12 ounces	12 ounces	12 ounces

Grade 4 • Harcourt Brace School Publishers

5 The students are making a special bag of trail mix for Mrs. Hunter, the teacher who will take them hiking. This mix will have 5 ingredients and weigh one pound, or 16 ounces.

Use the ingredients listed on page 98 to make a one-pound mix for the teacher.

Test Taking Tips

Why can't you put 4 oz of each ingredient in this mix for the teacher?

Ingredient	Amount in Ounces
#1	
#2	
#3	
#4	
#5	
TOTAL AMOUNT	

Grade 4 • Harcourt Brace School Publishers

Name _____

1 Jeff went to the store to buy ten bottles of water.

2 for $1.89

About how much did he pay?

Ⓐ about $2

Ⓑ about $10

Ⓒ about $19

Ⓓ about $22

Test Taking Tips

How can you use estimation to choose a reasonable answer?

2 The track team finished a race. The runners recorded their times.

Tim	23.03 seconds
Bill	23.30 seconds
Joe	23.27 seconds
Pete	23.72 seconds

Which time was the fastest?

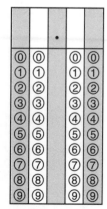

Test Taking Tips

How can ordering the decimals from least to greatest help you solve the problem?

Grade 4 • Harcourt Brace School Publishers

Name _____

3 Jon and Alexis are sharing a bowl of pretzels. Jon ate $\frac{3}{10}$ of the pretzels. Alexis ate 0.20 of the pretzels.

Who ate more pretzels?

_____

Explain how you decided. Write an expression using < or >.

_____

_____

_____

_____

Solve
Think • Explain
Short Answer

Test Taking Tips

How does writing an equivalent decimal for the fraction help you solve the problem?

---

4 Maria likes to go far on inline skates. She skated 1 kilometer her first week. Each week, she skated 1 more kilometer than the week before. She did this for 7 weeks.

What was the total number of kilometers she skated?

_____

Solve
Think • Explain
Short Answer

Test Taking Tips

How can you organize the information into a table to find the total?

Grade 4 • Harcourt Brace School Publishers

Name _____

5 In the 1996 Olympic Games in Atlanta, Georgia, the United States and Germany had the fastest times in the Women's Swimming Relay.

Test Taking Tips

How many seconds are in a minute?

How can you find the total time for the U.S. team?

Germany's team of four swimmers finished in 3 minutes, 57 seconds.

Here are the times for swimmers on the U.S. team:

1st swimmer	48.20 seconds
2nd swimmer	48.36 seconds
3rd swimmer	49.46 seconds
4th swimmer	47.98 seconds

Which team was faster? _____

By how many seconds? _____

Name _____

5 Write step-by-step directions to explain how to solve the problem.

_____

_____

_____

_____

_____

Did you write clear directions so that someone else could solve the problem in the same way?

Name _____

1 Which are the names of the faces on this three-dimensional figure?

- Ⓐ triangles and squares
- Ⓑ rectangles and squares
- Ⓒ triangles and hexagons
- Ⓓ triangles and rectangles

Test Taking Tips

What math words do you need to know to solve the problem?

2 Marvella and her dad were making a swing. They bought two lengths of heavy rope that were each 3.5 meters long. The rope cost $3.29 per meter. How much did they pay for the rope?

Test Taking Tips

How much rope did they buy?

$\$$

⓪	⓪		⓪	⓪
①	①		①	①
②	②		②	②
③	③		③	③
④	④		④	④
⑤	⑤		⑤	⑤
⑥	⑥		⑥	⑥
⑦	⑦		⑦	⑦
⑧	⑧		⑧	⑧
⑨	⑨		⑨	⑨

Grade 4 • Harcourt Brace School Publishers

Name _____

3 Colin's stepfather is making a new board for Colin's electric train village.

What is the perimeter of Colin's train board?

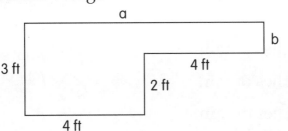

_____

Remember: Perimeter is the distance around a shape.

Explain how you decided.

_____

_____

Solve • Explain • Think

Short Answer

Test Taking Tips

How can you use the drawing to find the missing measures?

---

4 Lloyd has a coupon worth $1.00 to spend at a store. He will have to pay the sales tax in cash.

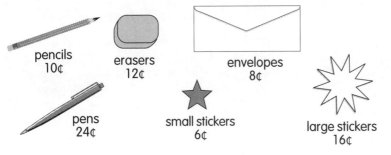

pencils 10¢   erasers 12¢   envelopes 8¢

pens 24¢   small stickers 6¢   large stickers 16¢

What can Lloyd buy that will use up his coupon? Make a table to show two different ways he can use the whole coupon.

Solve • Explain • Think

Short Answer

Test Taking Tips

How will you set up your table to help solve the problem?

You can write 10¢ as $0.10. How will writing all the prices of the items with a dollar sign and decimal help you know when you have spent $1.00?

Name _____

5 Sasha's class put a rain gauge outside their classroom for five months. They collected the following information:

September	7 inches of rain
October	8 inches of rain
November	7 inches of rain
December	3 inches of rain
January	2 inches of rain

Use this information to make a bar graph.

Check to see that your graph has

- a title
- an appropriate scale
- label for each bar
- accurate data

Test Taking Tips

What scale can you use for your bar graph?

Grade 4 • Harcourt Brace School Publishers

Name _____

5

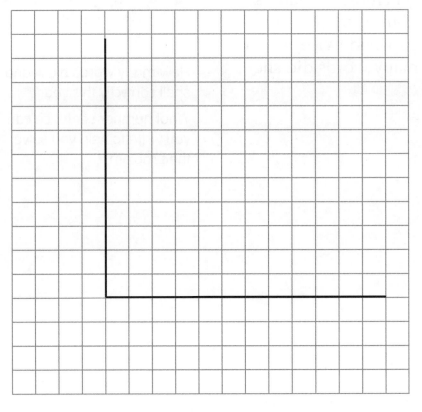

Test
Taking
Tips

How much rainfall will be
represented by each line on
your graph?

Write a question you can answer using the
information in the graph.

_____

_____

_____

_____

Grade 4 • Harcourt Brace School Publishers

1 Alana entered the county spelling bee. She spelled 47 words correctly before she made a mistake. If she had spelled three more words correctly, she would have spelled twice as many words as last year. How many words did she spell correctly last year?

Ⓐ 25

Ⓑ 27

Ⓒ 32

Ⓓ 35

Test Taking Tips

How many words did Alana spell correctly this year?

What number sentence can you write to help you solve the problem?

2 Michelle and Heather are measuring the volume of a new aquarium for their classroom. Michelle poured in 83 cups of water. Heather poured in 77 cups.

How many QUARTS of water did the girls put into the aquarium?

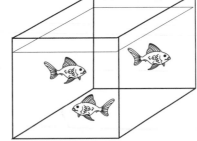

Remember, 4 cups = 1 quart.

Test Taking Tips

How can you change cups to quarts?

Grade 4 • Harcourt Brace School Publishers

Name _____

3 The music teacher at school is encouraging children to practice playing their instruments.

Sauri, Thomas, and Dan each begin practicing at 7 o'clock.

Sauri practiced half an hour a day for 5 days.

Thomas practiced one third of an hour a day for 10 days.

Dan practiced one fourth of an hour a day for 12 days.

Who spent the most time practicing? _____

Explain how you found your answer.

_____

_____

Test Taking Tips

How can you use the clocks to find the number of minutes in each fraction of an hour?

4 Angelo and Justin were painting a mural. Justin needed $\frac{2}{6}$ cup of purple paint. "Just put in $\frac{1}{3}$ cup of blue paint and $\frac{1}{3}$ cup of red paint. That will give you exactly $\frac{2}{6}$ cup of purple paint," Angelo said.

Do you think Angelo is right? _____

Explain your thinking.

_____

_____

_____

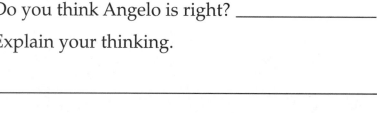

Test Taking Tips

What is the correct way to add fractions?

Grade 4 • Harcourt Brace School Publishers

Name _____

5 Benjie is helping his little sister build with blocks. The blue blocks are all the same length. They can fit 24 blocks end to end down the length of the hall.

Benjie measures one blue block and finds that it is 6 inches long.

How long is the hallway?

_____ inches

_____ feet

_____ yards

Use pictures or diagrams to help explain your thinking.

Test Taking Tips

How many inches are in one foot?

How many feet are in one yard?

Daily Practice
FCAT

5 Benjie got some new large blocks for his birthday. Each block is 2 feet long.

How many of these blocks will fit in the hall placed end to end? _____

Draw a picture and use numbers to explain.

Test Taking Tips

When you change feet to yards, will you have fewer or more yards than feet?

Grade 4 • Harcourt Brace School Publishers

Name _____

1 Janet is putting away her little brother's blocks. There are 48 blocks in all. She wants to put them on a shelf with the same number of blocks in each layer. Which of the following is NOT a possible way to arrange the blocks?

Ⓐ 4 wide, 4 high, and 3 deep

Ⓑ 12 wide, 2 high and 2 deep

Ⓒ 16 wide, 1 high and 3 deep

Ⓓ 9 wide, 2 high and 2 deep

Multiple Choice Test Taking Tips

How many blocks are in each answer choice?

---

2 The sign in the elevator shows that it cannot carry more than 2,000 pounds. Twelve people are waiting for the elevator. Their average weight is 160 pounds. How many more pounds can the elevator carry?

Gridded Response Test Taking Tips

What do you need to find first?

Grade 4 • Harcourt Brace School Publishers

Name _____

3 Paula surveyed the girls in her school. One third of the girls are wearing shirts with buttons and half of the girls are wearing tee-shirts. The rest of the girls are wearing dresses.

Are there more girls wearing shirts with buttons or tee-shirts?

Explain how you decided.

How can you compare the two fractions?

_____

_____

_____

_____

4 Tony's aunt served pizza for his birthday party. Each person at the party ate two pieces of pizza. All the pieces were the same size. Here is what was left at the end of the party.

How many people were at the party?

Explain your thinking.

Test Taking Tips

What information is given in the picture?

_____

_____

_____

Grade 4 • Harcourt Brace School Publishers

Daily Practice
FCAT

5 LaShondra is helping her brother make a float at the football parade. They are making a model of a rocket because the team is called the Elm Street School Rockets. This is what they want their rocket to look like.

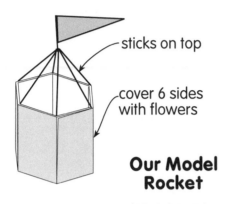

sticks on top

cover 6 sides with flowers

Our Model Rocket

Test Taking Tips

Think • Solve • Explain
Long Answer

How can you use the pattern for the rocket to help solve the problem?

They have made this pattern for the cardboard pieces they need for the rocket.

side of rocket	side of rocket	side of rocket	side of rocket	side of rocket	side of rocket

1 ft

1 ft

flag

Grade 4 • Harcourt Brace School Publishers

Name _____

5 Find the area of one side of the rocket. How many square feet?

_____

Find the total area of all the rocket sides. How many square feet?

_____

Estimate the area of the flag. How many square feet?

_____

How much cardboard will be left over after they cut out the rocket and flag?

_____

Explain how you found the area of the rocket, the flag, and the extra cardboard.

_____

_____

_____

_____

How can you check to see if your answers are right?

Grade 4 • Harcourt Brace School Publishers

Name _____

1 Which shape is Jorge least likely to spin?

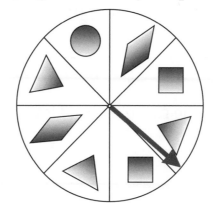

Ⓐ

Ⓑ

Ⓒ

Ⓓ

2 Megan glued a lace border around a picture of her cat. How much lace did she need?

20 inches

15 inches

Which expression can you use to solve the problem?

Ⓐ   15 x 20

Ⓑ   20 ÷ 15

Ⓒ   20 + 15

Ⓓ   15 + 20 + 15 + 20

Grade 4 • Harcourt Brace School Publishers

3 Which tool should Glenn use to measure the milk that he needs to make the pancakes in the recipe?

Ⓐ

Ⓑ

Ⓒ

Ⓓ

Pancakes

1 c pancake mix

1 egg

$\frac{3}{4}$ c milk

1 tsp vanilla

4  Sandor's mother bought him a shirt for $29.50 and a jacket for $60. She paid with a $100 bill. How much change should she receive?

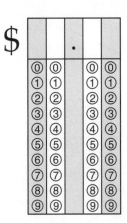

Grade 4 • Harcourt Brace School Publishers

Name _____

5 Janie wants to buy balloons. The ones she wants come in packages of eight. How many packages should she buy if she needs 48 balloons?

Balloons

Number of Packages	1	2	3	4	5	6	7	8
Number of Balloons	8	16						

6 Nick's aunt brought five trays of muffins to the bake sale. Each tray had nine muffins. By noon, they had sold 20 muffins. How many muffins were left to sell?

Name _____

7 Which figures have a line of symmetry?
Write yes or no below each figure.
Then draw the line of symmetry if there is one.

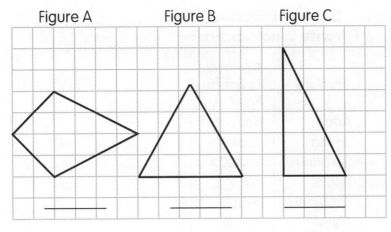

Figure A   Figure B   Figure C

On the lines below, explain how you decided.

_____

_____

8  Anna wants to buy a set of pastels for her art projects. The pastels cost $49.95 plus tax. Anna has saved $3 each week for seven weeks. How much has she saved?

Write a number sentence and solve. Then explain how you know your answer is correct.

_____

_____

_____

_____

_____

Name _____

9 Miss Jones's fourth-grade class took a survey of students' favorite lunch box fruit. The results are shown in the list below.

**Survey of Student
Favorite Lunchbox Fruit**

Student	Favorite Lunchbox Fruit
Arturo	banana
Li Ping	apple
Susanna	apple
Tyrone	orange
Shalvindra	peach
Myron	peach
Janine	peach
Sherell	orange
Megan	banana
Nick	apple
Simon	apple
Stephanie	apple
Daesun	orange
Beth	raisins
Jon	orange
Willie	apple
Anna	apple
Marison	banana
Taleah	apple
Charles	raisins

Complete a table that shows the data.

Favorite Lunchbox Fruit

Fruit	Votes by Tally	Number
Banana		
Apple		
Orange		
Peach		
Raisins		

Grade 4 • Harcourt Brace School Publishers

9   In the grid below, make a bar graph to show the results of the survey. Be sure that your graph has

a title

a label for each bar

a scale

accurate information

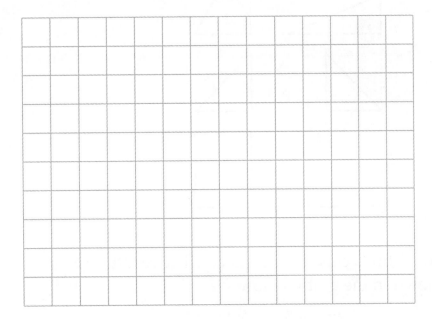

On the lines below, write one conclusion about Favorite Lunchbox Fruits. Explain how the graph supports your conclusion.

_____

_____

_____

_____

_____

_____

Grade 4 • Harcourt Brace School Publishers

Name _____

10 Which pair of figures is the same size and shape?

Ⓐ
Ⓑ
Ⓒ
Ⓓ

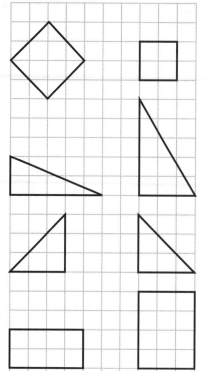

11 Mrs. Chin's class took a survey. The results are shown in the graph below.

Favorite Take-Out Food

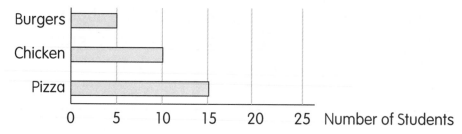

Which statement is NOT supported by the graph?

Ⓐ   Fifteen people prefer pizza.

Ⓑ   More people prefer chicken to burgers.

Ⓒ   Five out of 30 people prefer burgers.

Ⓓ   Most girls voted for pizza.

Grade 4 • Harcourt Brace School Publishers

Name _____

12 Tallie needs $\frac{3}{-}$ cup of sugar to make a cake. Which picture shows the amount of sugar she needs?

 Ⓐ

 Ⓑ

 Ⓒ

 Ⓓ

13 What temperature is shown on the thermometer?

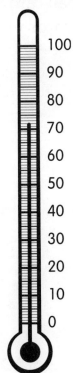

Grade 4 • Harcourt Brace School Publishers

14 Myra, Tricia, Eleanor, and Rusty formed a reading club. On Saturday, Myra read from 3:00 P.M. to 4:00 P.M. Tricia read from 3:00 P.M. to 3:45 P.M.. Eleanor read from 3:00 P.M. to 4:10 P.M.. Rusty read from 3:00 P.M. to 3:35 P.M.. How many minutes did they read in all?

Remember: 1 hour = 60 minutes

15 Rachel's brother said, "I have three thousand, nine hundred seven baseball cards in my collection." Write the number of cards in his collection in standard form.

STOP

Name _____

Notes

Problems that I answered correctly.

_____

_____

_____

_____

_____

_____

Problems that I did not understand.

_____

_____

_____

_____

_____

Vocabulary that I need to learn.

_____

_____

_____

_____

Grade 4 • Harcourt Brace School Publishers

Name _____

1

The fourth graders took a pet survey. The pictograph shows the number of students by class in the fourth grade who own pets. Which class has the most students who own pets?

Who Owns Pets?

Class	Number of Students
Mrs. Jenson	🐱 🐱 🐱
Mr. Chen	🐱 🐱
Ms. Clarke	🐱 🐱 🐱 🐱
Ms. Morales	🐱 🐱 🐱 🐱

Key: 🐱 = 5 students.

Ⓐ Mrs. Jenson

Ⓑ Mr. Chen

Ⓒ Ms. Clarke

Ⓓ Ms. Morales

2

LaToya and James each have some marbles. They keep them in jars. Compare the number of marbles that James and LaToya have. Which expression below shows the comparison of James's marbles to LaToya's marbles?

Ⓐ $6 > 9$

Ⓑ $9 < 6$

Ⓒ $9 = 6$

Ⓓ $6 < 9$

LaToya's Jar

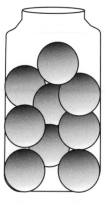

James's Jar

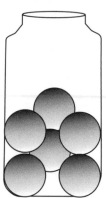

Grade 4 • Harcourt Brace School Publishers

3

Akhmed recorded the temperature outside his house at noon each day from August 10 to August 20. On which days was the temperature below 80°?

Ⓐ August 11,12, and 13

Ⓑ August 13, 15, and 16

Ⓒ August 13, 16, and 19

Ⓓ August 15, 16, and 17

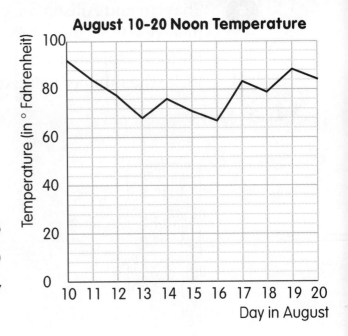

August 10–20 Noon Temperature

4

Greg is adding beans to a jar by small scoops. Look at the pictures below showing the jar after each scoop.

If Greg keeps adding beans by the scoop, how many beans will there be after the fifth scoop?

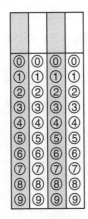

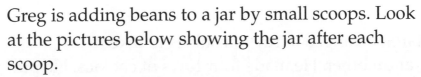

Greg's Jar

After 1st scoop   After 2nd scoop   After 3rd scoop

Grade 4 • Harcourt Brace School Publishers

5 Shaylan and Allison are collecting buttons. They have one thousand, three hundred seventy-four buttons. How many buttons are in their collection? Write the number using digits.

6 Aaron arranged a tray of cookies to take to his grandmother. He made four rows of cookies. He put 12 cookies in each row. How many cookies did he take to his grandmother?

Grade 4 • Harcourt Brace School Publishers

7

Joey and his dad went on a five-day camping trip. They wanted to fill up a jar with interesting pebbles to put in their turtle's aquarium. The picture shows how many pebbles they collected the first day. ESTIMATE how many days it will take to fill the jar if they collect about the same number every day. On the lines below, explain your estimation strategy.

Joey's Jar

After 1st day

_____

_____

_____

8

Sabrina drew the following picture for her friend, Andrew. She asked him to find all the squares he could. How many can you find? On the lines below, tell how many squares you found and write how you know you have found them all.

_____

_____

_____

_____

_____

Grade 4 • Harcourt Brace School Publishers

Name _____

9 The McCoys are making plans to build a house. They have a lot that measures 60 feet wide and 100 feet deep.

They have made a drawing of the house they want to build.

**The McCoys' Property
and House Plan**

60 feet

Grass

House

100 feet

Garage

Driveway

10 feet

Street

10 feet

Name _____

9

Find the area of the house in square feet.

_____

_____

_____

_____

Find the area of the garage and the driveway in square feet.

_____

_____

_____

_____

The McCoys will plant grass from the front wall of their house to the back of the lot. How many square feet of grass will they plant?

On the lines below, explain how you solved the problem.

_____

_____

_____

_____

_____

_____

Grade 4 • ©Harcourt Brace School Publishers

Name _____

10 Darlene's class took a survey. They counted the buttons they were wearing. The list shows their findings. What is the mode of the data?

Remember: The mode is the number that is listed most often in a set of data.

- Ⓐ 1
- Ⓑ 3
- Ⓒ 4
- Ⓓ 6

How Many Buttons Do We Have?

Student	Number of Buttons
Darlene	4
Sarah	5
Micah	1
Sander	0
Terry	4
Henry	3
Jamal	1
Thomas	4
Jared	4
Shelly	6

11  What is the area of the square in Tracy's design?

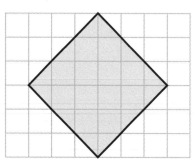

- Ⓐ 12 square units
- Ⓑ 14 square units
- Ⓒ 16 square units
- Ⓓ 18 square units

132

Grade 4 • Harcourt Brace School Publishers

12

Shirin made a map of her neighborhood on a grid.

Which ordered pair names the location of the school?

Ⓐ   (2, 3)

Ⓑ   (1, 6)

Ⓒ   (6, 2)

Ⓓ   (3, 2)

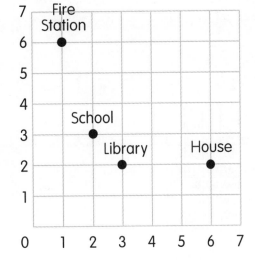

13

A fourth grader should get about 9 hours of sleep every night. How many hours should a fourth grader sleep every week?

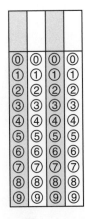

Grade 4 • Harcourt Brace School Publishers

14 Cliff made a pattern of numbers.

1, 2, 4, 7, 11, 16, 22, 29, 37, 46, ____?____

Write the next number in Cliff's pattern.

15 Deanna's class is making fabric flags. Each flag uses a 12-inch length of fabric. If there are 24 students in Deanna's class, how many feet of fabric will the class need to buy? 1 foot = 12 inches

STOP

Notes

Problems that I answered correctly.

_____

_____

_____

_____

_____

_____

_____

Problems that I did not understand.

_____

_____

_____

_____

Vocabulary that I need to learn.

_____

_____

_____

_____

1

For a class project, Thomas drew a map of his town. Look at his map and decide which answer below tells the best way to measure the actual distance from the school to the fire station.

Ⓐ   inches

Ⓑ   miles

Ⓒ   feet

Ⓓ   millimeters

2

Roberta and her classmates collected data about students' families. Their findings are shown in the Venn diagram below. Each "x" stands for a student in their class. How many students have both brothers and sisters?

Ⓐ   8

Ⓑ   4

Ⓒ   5

Ⓓ   12

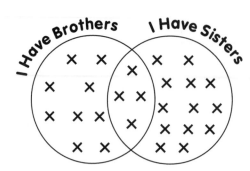

3

Sarah's grandmother is starting a small garden outside her apartment, shown below. About how many feet of wood edging will she need to fit around the perimeter of her garden?

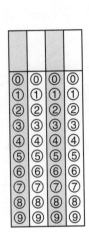

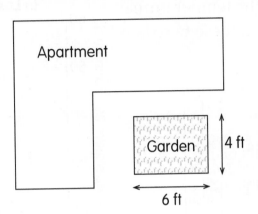

4

Adam is 3 years older than Nick. If Nick is 15 years old, which equation below will give you Adam's age?

Ⓐ   15 + 3 = ?

Ⓑ   15 − 3 = ?

Ⓒ   15 ÷ 3 = ?

Ⓓ   15 x 3 = ?

Name _____

5 Rosie's class kept records of the daytime high temperatures in their hometown for 5 days in December. They made a line graph of their findings. For two days in a row, the temperature was the same. What was the temperature on these two days?

Daytime High Temperatures in Last Week of December

6 Richard and John started a chess club. There were two people at the first meeting, 4 people at the second meeting, and 8 people at the third meeting. If this pattern continues, how many people will be at the fifth meeting?

138

Grade 4 • Harcourt Brace School Publishers

7

Lana is making a bead necklace. She is following a pattern. Her unfinished necklace looks like this.

O o o O o o O

She will use 34 beads in all. How many large beads will Lana need if she continues the pattern? On the lines below, explain how you solved the problem.

_____

_____

_____

_____

8

The students in Jeremy's fourth grade class had a contest to see who could fill a jar of beans first. Jeremy collected 50 beans. Sharon collected 100 beans. Xavier collected 125 beans. ESTIMATE how full Sharon's jar is. ESTIMATE how full Xavier's jar is compared to Jeremy's jar.

Jeremy's Jar   **Sharon's Jar**   **Xavier's Jar**

On the lines below, explain your reasoning.

_____

_____

_____

_____

_____

Grade 4 • Harcourt Brace School Publishers

9

Timothy surveyed his classmates about their favorite outdoor activities. Each student voted for one favorite activity. His data is shown on the list.

**Survey of Students' Favorite
Outdoor Activities**

Activity	Number of Students
soccer	7
camping	2
riding bicycles	14
skating	5

On the answer sheet, make a bar graph to show the favorite outdoor activities of the students in Timothy's survey.

Then write two statements that compare the data.

Name _____

9 Make a bar graph to show the favorite outdoor activities of the students in Timothy's survey. Be sure to

give your graph an appropriate title

label each bar

choose an appropriate scale

accurately graph the data

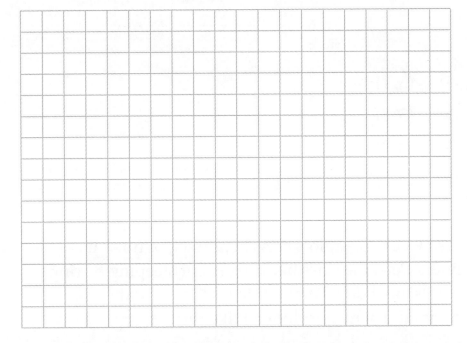

Use the information in your bar graph to write two statements that compare the data.

1. _____

_____

_____

2. _____

_____

_____

Grade 4 • Harcourt Brace School Publishers

Name _____

10 Ben plotted points A and B on the grid below. Which of the ordered pairs below should Ben graph to make a right triangle?

Ⓐ   (6, 1)

Ⓑ   (3, 8)

Ⓒ   (2, 2)

Ⓓ   (1, 5)

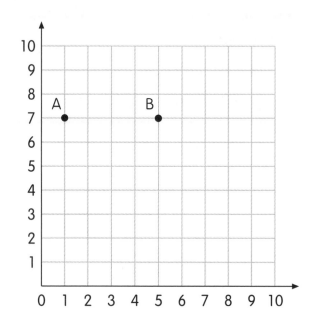

11 Last week, Georgio's class kept track of how many hours Pete, the class hamster, spent sleeping during each school day. They wrote each day's data as a decimal. Their chart looked like this:

Write the total number of hours Pete spent sleeping during school time last week.

How Long Pete Slept

Day	Number of Hours
Monday	3.5
Tuesday	4
Wednesday	2.5
Thursday	3.5
Friday	4.5

Grade 4 • Harcourt Brace School Publishers

12 Multiple Choice
A B C D

Amber made this design. Patrick rotated her design. Which picture below shows the design after it was rotated?

Ⓐ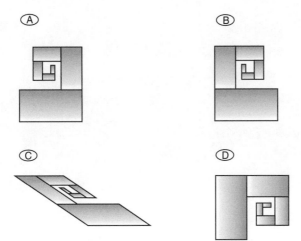

Ⓑ

Ⓒ

Ⓓ

13 Gridded Response

A bullfrog can jump 18 inches in one jump. How far can a bullfrog travel in 4 jumps if he keeps up the same rate?

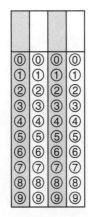

Grade 4 • Harcourt Brace School Publishers

14 LaShondra was dismissed from school at 2:30. Immediately after school ended, LaShondra spent 15 minutes helping her teacher. Then she spent 45 minutes tutoring another student. After this, she played on the playground for half an hour. After playing, she started to walk home. What time was this?

Ⓐ   2:30

Ⓑ   3:00

Ⓒ   3:45

Ⓓ   4:00

15  Troy has a piece of rope that measures three feet. Stephanie has a piece of rope that measures three yards. How many feet of rope do they have together?

Remember: 1 yard = 3 feet

STOP

STANDARD	CHAPTER
Professional History	
Understanding and Mastery...	
Historical roots of human services	2, 5, 8, 9, 10, 11, 12
Creation of human services profession	2, 5, 8, 9, 10, 11, 12
Historical and current legislation affecting services delivery	2, 5, 8, 9, 10, 11, 12
How public and private attitudes influence legislation and the interpretation of policies related to human services	2, 5, 8, 9, 14, 15
Differences between systems of governance and economics	2, 14, 15
Exposure to a spectrum of political ideologies	1, 2, 15
Skills to analyze and interpret historical data application in advocacy and social changes	1, 2, 15
Human Systems	
Understanding and Mastery...	
Theories of human development	1, 2, 3, 4, 5, 6
How small groups are utilized, theories of group dynamics, and group facilitation skills	4
Changing family structures and roles	4, 5, 6, 7, 12
Organizational structures of communities	2, 4, 5, 6, 7, 13, 14, 15
An understanding of capacities, limitations, and resiliency of human systems	1, 4, 13, 14, 15
Emphasis on context and the role of diversity in determining and meeting human needs	1, 2, 3, 4, 5, 6, 7, 8, 9, 10, 11, 12, 13, 14, 15
Processes to effect social change through advocacy (e.g., community development, community and grassroots organizing, local and global activism)	1, 2, 8, 9, 13, 14, 15
Processes to analyze, interpret, and effect policies and laws at local, state, and national levels	2, 4, 5, 6, 7, 13, 14, 15
Human Services Delivery Systems	
Understanding and Mastery...	
Range and characteristics of human services delivery systems and organizations	1, 4, 5, 6, 7, 8, 9, 10, 11, 12, 13, 14,
Range of populations served and needs addressed by human services	1, 2, 5, 6, 7, 8, 9, 10, 11, 12, 13, 14, 15
Major models used to conceptualize and integrate prevention, maintenance, intervention, rehabilitation, and healthy functioning	1, 2, 4, 5, 6, 7, 8, 10, 11, 12, 13, 14
Economic and social class systems including systemic causes of poverty	1, 2, 9, 14, 15
Political and ideological aspects of human services	2, 4, 5, 6, 7, 13, 14, 15
International and global influences on services delivery	1, 2, 4, 5, 6, 7, 13, 14, 15
Skills to effect and influence social policy	1, 2, 4, 5, 6, 7, 13, 14, 15

Adapted from the October 2010 Revised CSHSE National Standards

STANDARD | CHAPTER

Information Management

Understanding and Mastery...

Obtain information through interviewing, active listening, consultation with others, library or other research, and the observation of clients and systems

Recording, organizing, and assessing the relevance, adequacy, accuracy, and validity of information provided by others

Compiling, synthesizing, and categorizing information

Disseminating routine and critical information to clients, colleagues or other members of the related services system that is provided in written or oral form and in a timely manner

Maintaining client confidentiality and appropriate use of client data

Using technology for word processing, sending email, and locating and evaluating information

Performing elementary community-needs assessment

Conducting basic program evaluation

Utilizing research findings and other information for community education and public relations and using technology to create and manage spreadsheets and databases

Planning & Evaluating

Understanding and Mastery...

Analysis and assessment of the needs of clients or client groups

Skills to develop goals, and design and implement a plan of action

Skills to evaluate the outcomes of the plan and the impact on the client or client group

Program design, implementation, and evaluation

Interventions & Direct Services

Understanding and Mastery...

Theory and knowledge bases of prevention, intervention, and maintenance strategies to achieve maximum autonomy and functioning

Skills to facilitate appropriate direct services and interventions related to specific client or client group goals

Knowledge and skill development in: case management, intake interviewing, individual counseling, group facilitation and counseling, location and use of appropriate resources and referrals, use of consultation

STANDARD	CHAPTER

Interpersonal Communication

Understanding and Mastery...

Clarifying expectations	
Dealing effectively with conflict	
Establishing rapport with clients	
Developing and sustaining behaviors that are congruent with the values and ethics of the profession	

Administration

Understanding and Mastery...

Managing organizations through leadership and strategic planning	
Supervision and human resource management	
Planning and evaluating programs, services, and operational functions	
Developing budgets and monitoring expenditures	
Grant and contract negotiation	
Legal/regulatory issues and risk management	
Managing professional development of staff	
Recruiting and managing volunteers	
Constituency building and other advocacy techniques such as lobbying, grassroots movements, and community development and organizing	

Client-Related Values & Attitudes

Understanding and Mastery...

The least intrusive intervention in the least restrictive environment	
Client self-determination	
Confidentiality of information	
The worth and uniqueness of individuals including: ethnicity, culture, gender, sexual orientation, and other expressions of diversity	
Belief that individuals, services systems, and society change	
Interdisciplinary team approaches to problem solving	
Appropriate professional boundaries	
Integration of the ethical standards outlined by the National Organization for Human Services and Council for Standards in Human Service Education	

Self-Development

Understanding and Mastery...

Conscious use of self	
Clarification of personal and professional values	
Awareness of diversity	
Strategies for self-care	
Reflection on professional self (e.g., journaling, development of a portfolio, project demonstrating competency)	

THIRD EDITION

Introduction to Human Services

Through the Eyes of Practice Settings

Michelle E. Martin
Dominican University

Boston Columbus Indianapolis New York San Francisco Upper Saddle River
Amsterdam Cape Town Dubai London Madrid Milan Munich Paris Montréal
Toronto Delhi Mexico City São Paulo Sydney Hong Kong Seoul Singapore Taipei Tokyo

Editorial Director: Craig Campanella
Editor in Chief: Ashley Dodge
Editorial Product Manager: Carly Czech
Editorial Assistant: Nicole Suddeth
Vice President/Director of Marketing: Brandy Dawson
Executive Marketing Manager: Kelly May
Marketing Coordinator: Courtney Stewart
Senior Digital Media Editor: Paul DeLuca

Project Manager: Pat Brown
Image credit: oldmonk/Shutterstock
Editorial Production and Composition Service:
    Sudip Sinha/PreMediaGlobal
Interior Design: Joyce Weston Design
Creative Director: Jayne Conte
Cover Designer: Jodi Notowitz
Printer/Binder: RRD Crawfordsville

Credits and acknowledgments borrowed from other sources and reproduced, with permission, in this textbook appear on appropriate page within text.

If you purchased this book within the United States or Canada you should be aware that it has been imported without the approval of the Publisher or the Author.

Copyright © 2014, 2011, 2007 by Pearson Education, Inc. All rights reserved. Printed in the United States of America. This publication is protected by Copyright and permission should be obtained from the publisher prior to any prohibited reproduction, storage in a retrieval system, or transmission in any form or by any means, electronic, mechanical, photocopying, recording, or likewise. To obtain permission(s) to use material from this work, please submit a written request to Pearson Education, Inc., Permissions Department, One Lake Street, Upper Saddle River, New Jersey 07458 or you may fax your request to 201-236-3290.

Many of the designations by manufacturers and seller to distinguish their products are claimed as trademarks. Where those designations appear in this book, and the publisher was aware of a trademark claim, the designations have been printed in initial caps or all caps.

10 9 8 7 6 5 4 3 2 1

ISBN-10: 0-205-91478-0
ISBN-13: 978-0-205-91478-4

Contents

13. Faith-Based Agencies   305

14. Violence, Victim Advocacy, and Corrections   336

Preface

The third edition of *Introduction to Human Services: Through the Eyes of Practice Settings* includes many important additions. When I reflect back on all of the changes that have occurred since I began writing the first edition, I am in awe. Never could I have imagined the various tragedies that would unfold in the last decade! An agonizingly long war in the Middle East; a globalized economic crisis as we have not seen in decades; political and religious polarization that threatens to further fragment the social, political, and economic landscape in the United States; and "culture wars" that have pitted "social conservatives," including those on the religious right against social progressives, including many social advocates. But there were so many good things that happened as well—the first African American president was elected to office in the United States, and sexual orientation was included in hate crimes legislation, followed by increasing momentum gained in the marriage equity movement. We've also seen a dramatic increase in the effects of globalization fueled at least in part by the globalization of communication technologies. Do you want to start a social movement? Create a Facebook page and mobilize thousands of people globally, creating social awareness through the posting of status updates, online news articles, blogs, and YouTube videos!

What you'll notice throughout the third edition of this book is an exploration of all of these events, their precursors, and some of their consequences. You'll also notice a reflection of the effects of our ever-shrinking world—what we call globalization. I have updated all chapters with regard to research, terminology, and applicable legislation. In particular, I have made significant changes in Chapter 1 where I've included some exciting information about the continued growth of the human services profession, including information on the new certification process for human service professionals. Because of the continued professional development within the human services field, I have reduced the material focusing on related fields, such as the social work profession, and increased the focus on the human services profession. In Chapter 2 I explored numerous changes in social welfare legislation and policies that took effect under the Obama administration, including discussions on increasing rights afforded to the LGBTQ population, challenges facing migrant populations and the poor, and the most recent information on the healthcare debate. In Chapters 3 and 4 I have enhanced the focus on the human services profession. In Chapter 5 I included a section on the history of child labor, making a connection between this dark part of U.S. history and current patterns of abuse of vulnerable children in the United States, and around the world. I also explored recent changes in child welfare legislation. In chapters 6 through 12 I have updated the research and theories, and in chapter 13 I have increased interfaith content. In Chapter 14 I've added content on batterers intervention services, including information on the efficacy of these programs. In Chapter 15 I've added content on viewing global social problems from a human rights framework, as well as very important content on refugees, genocide, and other at-risk populations. Overall I hope I have captured the most recent trends, research, and contemporary issues on a local and global level that are important to human service professionals.

I would like to thank several people who helped make this edition possible. First, and foremost, I would like to thank my family—my son Xander, who was only 9 when I started writing this book, and is now 17. I'd also like to thank my two surrogate Rwandan daughters, Elodie Shami and Annabella Uwineza, who have shared my life, my home, and my family for the last three years. My aunt Jeri Serpico has always been my rock. My dear friend Karen Acevedo was a constant support for me throughout the writing of this edition. I would like to thank my colleagues at Dominican University's Graduate School of Social Work—Kim Kick, Myrna McNitt, Leticia Villarreal Sosa, and Charlie Stoops—for their professional insights and perspectives; they helped to sharpen my thinking. I would like to thank Asma Yousef with Islamic Relief USA for her insights on the Muslim faith. Finally I'd like to thank my social work students who sharpen my mind, and give me new ways to think about this wonderful profession.

CHAPTER I

Introduction to the Human Services Profession

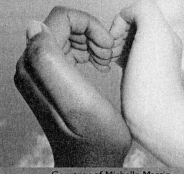

Courtesy of Michelle Martin

Purpose, Preparation, Practice, and Theoretical Orientations

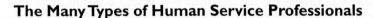

Learning Objectives

- Identify and describe the varied reasons why people may need human services intervention
- Describe the various ways one can enter the field of human services, and the various types of careers within the human services profession
- Identify the most common degree and licensure requirements associated with the human services profession
- Describe the new human services certification process developed by the Council for Standards in Human Service Education
- Identify and describe the most common theoretical frameworks used in the human services discipline

The Many Types of Human Service Professionals

Sara works for a hospice agency and spends one hour twice a week with Steven, who has been diagnosed with terminal cancer of the liver. He has been told he has approximately six months to live. He has been estranged from his adult daughter for four years, and Sara is helping him develop a plan for reunification. Sara helps Steve deal with his terminal diagnosis by helping him talk through his feelings about being sick and dying. Steve talks a lot about his fear of being in pain and his overwhelming feeling of regret for many of the choices he has made in his life. Sara listens and also helps Steve develop a plan for saying all the things he needs to say before he dies. During their last meeting, Sara helped Steve write a list of what he would like to say to his daughter, his ex-wife, and other family members. Sara is also helping Steve make important end-of-life decisions, including planning his own funeral. Sara and Steve will continue to meet until his death, and if possible, she will be with him and his family when he passes away.

Gary works for a public middle school and meets with six seventh graders every Monday to talk about their feelings. Gary helps them learn better ways to explore feelings of anger and frustration. During their meetings, they sometimes do fun things like play basketball, and sometimes they play a board game where they each take turns picking a "self-disclosure" card and answering a personal question. Gary uses the game to enter into discussions about healthy ways of coping with feelings, particularly anger. He also uses the game to get to know the students in a more personal manner, so that they will open up to him more. Gary spends one session per month to discuss their progress in their classes. The goal for the group is to help the students learn how to better control their anger and to develop more prosocial behavior, such as empathy and respect for others.

Cynthia works for her county's district attorney's office and has spent every day this past week in criminal court with Kelly, a victim of felony home invasion, aggravated kidnapping, and aggravated battery. Cynthia provides Kelly with both counseling and advocacy. Kelly was in her kitchen one morning feeding her baby when a man charged through her back door. The offender was recently released from state prison, had just robbed a gas station, and was running from the police in a stolen car. He ran from home to home until he found an unlocked door and entered it, surprising Kelly. Kelly immediately started screaming but stopped when he pulled a gun out and held it to her baby's head. During the next hour the defendant threatened both Kelly and her infant son's life and at one point even threatened to sexually assault Kelly. The offender became enraged and hit Kelly several times when she couldn't find any cash in her home. The police arrested him when he was attempting to force Kelly to drive him to an ATM to obtain money. Cynthia keeps Kelly apprised of all court proceedings and accompanies her to court, if Kelly chooses to assert her right to attend the proceedings. She also accompanies Kelly during all police interviews and helps her prepare for testifying. During these hearings, as well as during numerous telephone conversations, Cynthia helps Kelly understand and deal with her feelings, including her recent experience of imagining the violent incident again and again, her intense fear of being alone, and her guilt that she had not locked her door. Lately, Kelly has been experiencing an increasing amount of crying and unrelenting sadness, so Cynthia has referred her to a licensed counselor, as well as to a support group for Kelly and her husband.

Frank works for county social services, child welfare division, and is working with Lisa, who recently had her three young children removed from her home for physical and emotional neglect. Frank has arranged for Lisa to have parenting classes and individual counseling so that she can learn how to better manage her frustrations with her children. He has also arranged to have her admitted to a drug rehabilitation program to help her with her addictions to alcohol and cocaine. Frank and Lisa meet once a week to talk about her progress. He also monitors her weekly visitation with her children. Frank is required to attend court once per month to update the judge of Lisa's progress on her parenting plan. Successful completion of this plan will enable Lisa to regain custody of her children. Frank will continue to monitor her progress, as well as the progress of the children, who are in foster care placement.

Allison is currently lobbying several legislators in support of a bill that would increase funding for child abuse prevention and treatment. As the social policy advocate for a local grassroots organization, Allison is responsible for writing position statements and contacting local lawmakers to educate them on the importance of legislation aimed at reducing child abuse. Allison also writes grants for federal and private funding of the organization's various child advocacy programs.

What do all these professionals have in common? They are all human service professionals working within the interdisciplinary field of human or social services, each possessing a broad range of skills and having a wide range of responsibilities related to their roles in helping people overcome a variety of social problems. The National Organization for Human Services (NOHS) defines the human services profession as follows: "The Human Services profession is one which promotes improved service delivery systems by addressing not only the quality of direct services, but by also seeking to improve accessibility, accountability, and coordination among professionals and agencies in service delivery." *Human services* is a broad term covering a number of careers, but all have one thing in common—helping people meet their basic physical and emotional needs that for whatever reason cannot be met without outside assistance. The human services field can include a variety of job titles, including social worker, caseworker, program coordinator, outreach counselor, crisis counselor, and victim advocate, to name just a few.

Why Is Human Services Needed?

All human beings have basic needs, such as the need for food, health, shelter, and safety. People also have social needs, such as the need for interpersonal connectedness and love, and psychological needs, such as the need to deal with the trauma of past abuse, or even the psychological ramifications of disasters such as a hurricane or house fire. People who are fortunate have several ways to get their needs met. Social and psychological needs can be met by family, friends, and places of worship. Needs related to food, shelter, and other more complicated needs such as healthcare can be met through employment, education, and family.

But some people in society are unable to meet even their most basic needs either because they do not have a supportive family or because they have no family at all. They may have no friends or have friends who are either unsupportive or unable to provide help. They may have no social support network of any kind, having no faith community, and no supportive neighbors, perhaps due to apartment living or the fact that many communities within the United States tend to be far more transient now than in prior generations. They may lack the skills or education to gain sufficient employment; thus, they may not have health insurance or earn a good wage. Perhaps they've spent the majority of their lives dealing with an abusive and chaotic childhood and are now suffering from the manifestation of that experience in the form of psychological problems and substance abuse and, thus, cannot focus on meeting their basic needs until they are able to deal with the trauma they had been forced to endure.

Some people, particularly those who have good support systems, may falsely believe that anyone who cannot meet their most basic needs of shelter, food, healthcare, and emotional needs must be doing something wrong. This belief is incorrect because numerous barriers exist that keep people from meeting their own needs, some of which might be related to their own behavior, but more often, the reasons why people cannot meet their needs are quite complicated and often lie in dynamics beyond their control. Thus while some people who are fortunate enough to have great families, wonderfully supportive friends, the benefit of a good education, not faced racial oppression or social exclusion, and no significant history of abuse or loss may be self-sufficient in meeting their own needs. This does not mean that others who find themselves in situations where they cannot meet their own needs are doing something wrong. Human service agencies come into the picture when people find themselves confronting barriers to getting their needs met and their own resources for overcoming these obstacles are insufficient. Some of these barriers include the following:

- Lack of family (or supportive family)
- Lack of a healthy support system of friends
- Mental illness
- Poverty
- Social exclusion (due to racial discrimination for instance)
- Racism
- Oppression (e.g., racial, gender, age)
- Trauma
- Natural disasters
- Lack of education
- Lack of employment skills
- Unemployment
- Economic recession
- Physical and/or intellectual disability

A tremendous amount of controversy surrounds how best to help people meet their basic needs, and various philosophies exist regarding what types of services truly help those in need and which services may seem to help initially but may actually create more problems down the road, such as the theory that public assistance creates dependence. For instance, most people have heard the old proverb, "Give a man a fish and he will eat for a day. Teach a man to fish and he will eat for a lifetime." One goal of the human services profession is to teach people to fish. This means that human service professionals are committed to helping people develop the necessary skills to become self-sufficient and function at their optimal levels, personally and within society. Thus although an agency may pay a family's rent for a few months when they are in a crisis, human service professionals will then work with the family members to remove any barriers that may be keeping them from meeting their housing needs in the future, such as substance abuse disorders, a lack of education or vocational skills, health problems, mental illness, or gaining self-advocacy skills necessary for combating prejudice and discrimination in the workplace.

> **Human service professionals are committed to helping people develop the necessary skills to become self-sufficient and function at their optimal levels, personally and within society.**

In addition to a commitment to working with a broad range of populations, including high-needs and disenfranchised populations, and providing them with the necessary resources to get their basic needs met, human service professionals are also committed to working on a *macro* or societal level to remove barriers to optimal functioning that affect large groups of people. By advocating for changes in laws and various policies, human service professionals contributed to making great strides in reducing prejudice and discrimination related to one's race, gender, sexual orientation, socioeconomic status

(SES), or any one of a number of characterizations that might marginalize someone within society.

Human service professionals continue to work on all social fronts so that every member of society has an equivalent opportunity for happiness and self-sufficiency. The chief goal of the human service professional is to support individuals as well as communities function at their maximum potential, overcoming personal and social barriers as effectively as possible in the major domains of living.

Human Service Professionals: Educational Requirements and Professional Standards

Each year numerous caring individuals will decide to enter the field of human services and will embark on the confusing journey of trying to determine what level of education is required for specific employment positions, when and where a license is required, and even what degree is required. There are no easy answers to these questions, because the human services profession is a broad one encompassing many different professions, including human service generalist, mental health counselor, psychologist, social worker, and perhaps even psychiatrist, all of whom are considered human service professionals if they work in a human service agency working in some manner with marginalized, disenfranchised, or other individuals who are in some way experiencing problems related to various social or systemic issues within society.

Another area of confusion relates to the educational and licensing requirements needed to work in the human services field. Determining what educational degree to earn, the level of education required, and what professional license is needed depends in large part on variables such as specific state and federal legislation (particularly for highly regulated fields, such as in the educational and healthcare sectors), industry-specific standards, and even agency preference or need. To make matters even more confusing, these variables can vary dramatically from one state to the next; thus, a job that one can do in one state with an Associate of Arts (AA) degree may require a Master of Social Work (MSW) degree and a clinical license in another state. In addition, many individuals may work in the same capacity at a human service agency with two different degrees.

According to the NOHS website, a "human service professional" is

> [a] generic term for people who hold professional and paraprofessional jobs in such diverse settings as group homes and halfway houses; correctional, mental retardation, and community mental health centers; family, child, and youth service agencies, and programs concerned with alcoholism, drug abuse, family violence, and aging. Depending on the employment setting and the kinds of clients served there, job titles and duties vary a great deal. (National Organization for Human Services, 2009, para.11)

Within this text, I use the title *human service professional* to refer to all professionals working within the human services field, but if I use the term *social worker,*

Human Services Delivery Systems

Understanding and Mastery of Human Services Delivery Systems: Range of populations served and needs addressed by human services

Critical Thinking Question: Human service professionals often—but not always—work with the most disadvantaged members of society. What are some roles in which they serve the most vulnerable populations? What are some roles in which they might serve more affluent clients?

then I am referring to the legal definition and professional distinction of a licensed social worker, indicating either a Bachelor of Social Work (BSW) or an MSW level of education. Also, I use the term *human service agency,* but this term is often used synonymously in other literature with *social service agency.* One reason for the dramatic variation in educational and licensing requirements is that the human services field is a growing profession, and with the evolution of professionalization comes increasing practice regulations. Yet, issues such as the stance of legislators in a particular state regarding practice requirements, the need for human service professionals within the community, or even whether the community is rural or urban can affect educational and licensing requirements for a particular position within the human services profession (Gumpert & Saltman, 1998).

Some human service agencies are subject to federal or state governmental licensing requirements, such as the healthcare industry (hospitals, hospices, home healthcare), government child welfare agencies, and public schools, and as such may be required to hire a professional with an advanced degree in any of the social science fields, or a particular professional education requirement might be specified. For instance, in many states, school social workers must have an MSW degree and educational credentials in school social work, and school counselors must have a master's degree in educational counseling.

There is still considerable variability among state licensing bodies in terms of how professional terms such as *counselor, social worker,* and *related field* are defined. For instance, most states require hospice social workers to be licensed social workers, thus requiring either a BSW or an MSW degree. But in Illinois, for instance, the Hospice Program Licensing Act provides that a hospice agency can also employ bereavement counselors who have a bachelor's degree in counseling, psychology, or social work with one year of counseling experience. Some states require child welfare workers to be licensed social workers with an MSW, whereas other states require child welfare workers to have a master's degree in any related field (i.e., psychology, human services, sociology). In states where there is a significant need for bilingual social workers, such as California, educational requirements may be lowered if the individual is bilingual and has commensurate counseling and/or case management experience.

Keeping such variability within specific human services fields in mind, as well as differences among state licensing requirements, Table 1.1 shows a very general breakdown of degrees in the mental health field, their possible corresponding licenses, as well as what careers these professionals might be able to pursue, depending on individual state licensing requirements.

Human Service Education and Licensure

The Council for Standards in Human Service Education (CSHSE) was established in 1979 for the purposes of guiding and directing human service education and training programs. This organization has developed national standards for the curriculum and subject area competencies in human service degree programs and serves as the accreditation body for colleges and universities offering degrees in the growing human services discipline at the associate's, bachelor's, and master's levels.

Table 1.1	Multiple Discipline Degree Requirements		
Degree	**Academic Area/Major**	**License/Credential**	**Possible Careers**
BA/BS	Human Services	BS-BCP	Caseworker, youth worker, residential counselor, behavioral management aide, case management aide, alcohol counselor, adult day care worker, drug abuse counselor, life skills instructor, social service aide, probation officer, child advocate, gerontology aide, juvenile court liaison, group home worker, child abuse worker, crisis intervention counselor, community organizer, social work assistant, psychological aide
BA/BS	Psychology, Sociology	N/A	Same as above, depends on state requirements
BSW	Social Work (program accredited by CSWE)	Basic licensing (LSW) depends on state	Same as above, depends on state requirements
MA/MS	Counseling Psychology	LCP (Licensed Clinical Professional—on graduation)	Private practice, some governmental and social service agencies
30–60 credit hours		LCPC (Licensed Clinical Professional Counselor—~3,000 postgrad supervised hours)	
MSW	Social Work (program accredited by CSWE)	LSW (on graduation)	Private practice, all governmental and social service agencies (some requiring licensure)
60 credit hours		LCSW (Licensed Clinical Social Worker—~3,200 postgrad supervised hours)	
PsyD 120 credit hours	Doctor of Psychology	PSY# (Licensed Clinical Psychologist—~3,500 postgrad supervised hours)	Private practice, many governmental and social service agencies, teaching in some higher education institutions
PhD (Psychology)	Doctor of Philosophy in Psychology	PSY# (~3,500 postgrad supervised hours)	Private practice, many governmental and social service agencies, teaching in higher education institutions
120 credit hours			

The CSHSE requires that curriculum in a human services program cover the following standard content areas: *knowledge* of the human services field through the understanding of relevant *theory, skills, and values* of the profession; *history* of the profession; *human systems; scope* of the human services profession; standard clinical *interventions;* common *planning and evaluation* methods; and information on *self-development.* The curriculum must also meet the minimum requirements for *field experience* in a human service agency, as well as appropriate *supervision.*

The term *human services* is new compared to the title *social work* or *mental health* counselor, and grew in popularity partly in response to the narrowing of the definition and increasing professionalization of the social work profession. For instance, in the early 1900s many of those who worked in the social work field were called social workers; yet, as the social work field continued to professionalize, the title of social worker eventually became reserved for those professionals who had either an undergraduate or a graduate degree in social work from a program accredited by the Council on Social Work Education (CSWE), the accrediting body responsible for the accreditation of social work educational programs in the United States.

There is a wide variation between states with regard to what types of degrees are required; education levels required; what careers require licensing, certifications, or credentials as well as the variation in titles used to identify social workers, human service professionals, and counselors (Rittner & Wodarski, 1999). In many states, the human services profession is still largely unregulated, but this is quickly changing for several reasons, including the fact that many third-payer insurance companies will not reimburse for services unless rendered by a licensed mental health provider (Beaucar, 2000).

> In many states the human services profession is still largely unregulated, but this is quickly changing.

In 2010, the CSHSE and the NOHS in collaboration with Center for Credentialing & Education took a significant step toward the continuing professionalization of the human services profession by developing a voluntary professional certification called the Human Services Board Certified Practitioner (HS-BCP) (2009 was a "grandfather" year that allowed human service practitioners to apply for the certificate without taking the national exam). In order to take the national certification exam, applicants must have earned at least a "technical certificate" in the human services discipline from a regionally accredited college or university and completed the required amount of postgraduate supervised hours in the human services field. The number of required hours worked in the human services field ranges based upon the level of education earned, from 7,500 hours required for those applicants with a technical certificate, 4,500 hours required for those applicants with an associate degree, 3,000 hours for those applicants with a bachelor's degree, and 1,500 hours for those applicants with a master's degree. Applicants who have earned degrees in other than a CSHSE-approved program, such as in counseling, social work, psychology, marriage and family therapy, or criminal justice, must complete coursework in several different content areas related to human services, such as "ethics in the helping professions," "interviewing and intervention skills," "social problems," "social welfare/public policy," and "case management." The implementation of the HS-BCP certification has moved both the discipline and the profession of human services toward increased professional identity and recognition within the larger area of helping professions (for more information on the HS-BCP certification, go to http://www.nationalhumanservices.org/certification).

Duties and Functions of a Human Service Professional

Despite the broad range of skills and responsibilities involved in human services, most human services positions have certain work-related activities in common. The NOHS describes the general functions and competencies of the human service professional on its website located at www.nationalhumanservices.org. These include the following:

1. Understanding the nature of human systems: individual, group, organization, community and society, and their major interactions. All workers will have preparation which helps them to understand human development, group dynamics, organizational structure, how communities are organized, how national policy is set, and how social systems interact in producing human problems.
2. Understanding the conditions which promote or limit optimal functioning and classes of deviations from desired functioning in the major human systems. Workers will have understanding of the major models of causation that are concerned with both the promotion of healthy functioning and with treatment rehabilitation. This includes medically oriented, socially oriented, psychologically-behavioral oriented, and educationally oriented models.
3. Skill in identifying and selecting interventions which promote growth and goal attainment. The worker will be able to conduct a competent problem analysis and to select those strategies, services, or interventions that are appropriate to helping clients attain a desired outcome. Interventions may include assistance, referral, advocacy, or direct counseling.
4. Skill in planning, implementing, and evaluating interventions. The worker will be able to design a plan of action for an identified problem and implement the plan in a systematic way. This requires an understanding of problems analysis, decision-analysis, and design of work plans. This generic skill can be used with all social systems and adapted for use with individual clients or organizations. Skill in evaluating the interventions is essential.
5. Consistent behavior in selecting interventions which are congruent with the values of one's self, clients, the employing organization, and the human services profession. This cluster requires awareness of one's own value orientation, an understanding of organizational values as expressed in the mandate or goal statement of the organization, human service ethics, and an appreciation of the client's values, life style and goals.
6. Process skills which are required to plan and implement services. This cluster is based on the assumption that the worker uses himself as the main tool for responding to service needs. The worker must be skillful in verbal and oral communication, interpersonal relationships, and other related personal skills, such as self-discipline and time management. It requires that the worker be interested in and motivated to conduct the role that he has agreed to fulfill and to apply himself to all aspects of the work that the role requires.

How Do Human Service Professionals Practice?

Since human beings have walked this planet, people have been trying to figure out what makes them "tick." If we were to construct a historical time line, we would see that each era tends to embrace a particular philosophy regarding the psychological nature of humans. Were we created in the image of God? Are we inherently good? Are personal problems a product of social oppression, or are individuals responsible for their lot in life? Do we have various levels of consciousness with feelings outside our awareness, motivating us to behave in certain ways? What will make us happy? What leads to our emotional demise? These questions are often left to philosophers and more recently to psychologists, but they also relate very much to human services practice because the view of humankind held by human service professionals will undoubtedly influence how they both view and help their clients.

One of the most common questions human service professionals are asked in a job interview is about their "theoretical orientation." I recall having a professor in my graduate program who cautioned that when we were asked that question to make sure we never said we were "eclectic" because this was a clear indication to any employer that we had no idea what theoretical orientation we embraced. Essentially what this question is addressing is what theoretical orientation the human service professional operates from as a foundation. In any mental health clinic, one practitioner might counsel from a psychoanalytic perspective, another from a humanistic perspective, and yet another from a cognitive-behavioral perspective. The theoretical orientation of mental health professionals will serve as a sort of lens through which they view their clients. Depending on the theory, a human service professional's theoretical orientation may include certain *underlying assumptions* about human behavior (e.g., what motivates humans to behave in certain ways), *descriptive aspects* (e.g., common experiences of women in middle adulthood), as well as *prescriptive aspects,* defining adaptive versus maladaptive behaviors (e.g., is it normal for children to experience separation anxiety in the toddler years? Is adolescent rebellion a normal developmental stage?).

Most theoretical orientations will also extend into the clinical realm by outlining ways to help people become emotionally healthy, based on some presumption of what caused them to become emotionally unhealthy in the first place. For instance, if a practitioner embraces a psychoanalytic perspective that holds to the assumption that early childhood experiences influence adult motivation to behave in certain manners, then the counseling will likely focus on the client's childhood. If the practitioner embraces a cognitive-behavioral approach, the focus of counseling will likely be on how the client frames and interprets the various occurrences in his or her life.

Theoretical Frameworks Used in Human Services

When considering all the various theories of human behavior, it is essential to remember that culture and history affect what is considered healthy thinking and behavior. Common criticism of many major psychological theories is that they are often based on mores common in Western cultures in developed countries and are not necessarily representative or reflective of individuals living in developing or non-Western cultures. For instance, is it appropriate to apply Freud's psychoanalytic theory of human behavior, which was developed from his work with higher society women in the Victorian era, to individuals of the Masai tribe in Africa? Or, is it appropriate to use a theory of human behavior developed during peacetime when working with those who grew up in a time of war? Any theory of human behavior one considers using in relation to understanding the behavior of clients should include a framework addressing many systems, such as culture, historical era, ethnicity, and gender, as well as other systems within which the individual operates. In other words, it is imperative that the human service professional consider environmental elements that may be a part of the client's life as a part of any evaluation and assessment.

> **It is imperative that the human service professional consider environmental elements that may be a part of the client's life as a part of any evaluation and assessment.**

Consider this example:

> A woman in her forties is feeling rather depressed. She spends her first counseling session describing her fears of her children being killed. She explains how she is so afraid of bullets coming through her walls that she doesn't allow her children to watch television in the living room. She never allows her children to play outside and worries incessantly when they are at school. She admits that she has not slept well in weeks, and she has difficulty feeling anything other than sadness and despair.

Would you consider this woman paranoid? Correctly assessing her does not depend solely on her thinking patterns and behavior, but on the *context* of her thinking patterns and behavior, including the various elements of her environment. If this woman lived in an extremely safe, gate-guarded community where no crimes had been reported in 20 years, then an assessment of some form of paranoia might be appropriate. But what if she lived in a high-crime neighborhood, where "drive-by" shootings were a daily event? What if you learned that her neighbor's children were recently shot and killed while watching television in the living room? Her thinking patterns and behavior do not seem as bizarre when considered within the context or systems in which she is operating.

Human service professionals are often referred to as "generalists," implying that their knowledge base is broad and varied. This does not mean that they do not have areas of specialization; in fact, in the last 100 years human service professionals have increasingly ventured into practice areas previously reserved for social workers, psychologists, and professional counselors (Rullo, 2001). But many believe that in order to be most effective, human service professionals must be competent in working with a broad range of individuals and a broad range of issues, using a wide range of interventions. A conceptual framework that is most commonly associated with human service generalist practice is one that views *clients* in the context of their *environment*, specifically focusing on the transaction or relationship between the two.

Human Systems

Understanding and Mastery of Human Systems: Emphasis on context and the role of diversity in determining and meeting human needs.

Critical Thinking Question: Human service professionals are generalists, drawing on a wide range of knowledge, skills, and theoretical perspectives in order to best serve their clients. How might this broad array of tools help a professional to effectively serve clients from diverse cultural and/or socioeconomic backgrounds?

Several theories capture this conceptual framework, and virtually all are derived from general systems theory, which is based on the premise that various elements in an environment interact with each other, and this interaction (or transaction) has an impact on all elements involved. This has certain implications for the hard sciences such as ecology and physics, but when applied to the social environment, its implications involve the dynamic and interactive relationship between environmental elements, such as one's family, friends, neighborhood, church, culture, ethnicity, and gender, on the thoughts, attitudes, and behavior of the individual. Thus, if someone asked you who you were, you might describe yourself as a female, who is a college student, married, with two high school–aged children, who attends church on a regular basis. You might further describe yourself as having come from an Italian family with nine brothers and sisters and as a Catholic.

On further questioning you might explain that your parents are older and you have been attempting to help them find alternate housing that can help them with their extensive medical needs. You might describe the current problems you're having with your teenage daughter, who was recently caught "ditching" school by the truancy officer. Whether you realize it or not, you have shared that you are interacting with the following environments (often called ecosystems): family, friendships, neighborhood, Italian-American culture, church, gender, marriage covenant, adolescence, the medical community, the school system, and the criminal justice system.

Your interaction with each of these systems is influenced by both your expectations of these systems and their expectations of you. For instance, what is expected of you as a college student? What is expected of you as a woman? As a wife? As a Catholic? What about the expectations of you as a married woman who is Catholic? What about the expectations of your family? As you attempt to focus on your academic studies, do these various systems offer stress or support? If you went to counseling, would it be helpful for the practitioner to understand what it means to be one of nine children from a Catholic, Italian-American family?

This focus on transactional exchange is what distinguishes the field of human services from other fields such as psychology and psychiatry, although recently, systems theory has gained increasing attention in these latter disciplines as well. Several theories have been developed to describe the reciprocal relationship between individuals and their environment. The most common are *Ecological Systems Theory*, *Person-in-Environment* (PIE), and *Eco-Systems Theory*.

BRONFENBRENNER'S ECOLOGICAL SYSTEMS THEORY   Urie Bronfenbrenner (1979) developed the Ecological Systems Theory. In his theory, Bronfenbrenner categorized an individual's environment into four expanding spheres, all with increasing levels of intimate interaction with the individual. The Microsystem includes the individual and his family, the Mesosystem (or Mezzosystem) includes entities such as one's neighborhood and school, the Exosystem includes entities such as the state government, and the Macrosystem would include the culture at large. Figure 1.1 illustrates the various systems and describes the nature of interaction with the individual. Again, it is important to remember that the primary principle of Bronfenbrenner's theory is that individuals

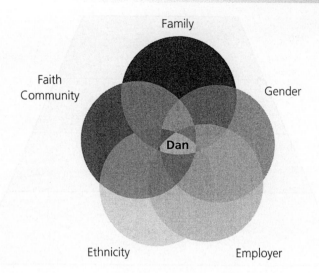

Family

Faith
Community

Gender

Dan

Ethnicity

Employer

FIGURE 1.1

Example of Common
Eco-Systems with the
Person in the Middle

can best be understood when seen in the context of their relationship with the various systems in their lives. Understanding the nature of these reciprocal relationships will aid in understanding the individual.

PERSON-IN-ENVIRONMENT　Another theory that is similar in nature to Ecological Systems Theory is referred to as "Person-in-Environment," or PIE. The premise of this theory is quite similar to Bronfenbrenner's theory, as it encourages seeing individuals within the context of their environment, both on a micro and macro levels (i.e., intra and interpersonal relationships and family dynamics) and on a macro (or societal) level (i.e., the individual is an African American, who lives in an urban community with significant cultural oppression).

ECO-SYSTEMS THEORY　Similar to Bronfenbrenner's theory, in Eco-Systems Theory, the various environmental systems are represented by overlapping concentric circles indicating the reciprocal exchange between a person and environmental system. Although there is no official recognition of varying levels of systems (from micro to macro), the basic concept is very similar, and most who embrace this theory understand that there are varying levels of systems, all interacting and thus impacting the person in various ways. It is up to the human service professional to strive to understand the transactional and reciprocal nature of these various systems (Meyer, 1988).

　　It is important to note that these theories do not presume that individuals are necessarily aware of the various systems they operate within, even if they are actively interacting with them. In fact,

Human Systems

Understanding and Mastery of Human Systems: Theories of human development

Critical Thinking Question: The field of human services focuses on the individual within the context of her environment. How might this perspective lead a human service professional to respond to a client differently than would, say, a psychiatrist who focuses on childhood trauma as the root of adult dysfunction?

FIGURE 1.2
Maslow's Hierarchy of
Needs

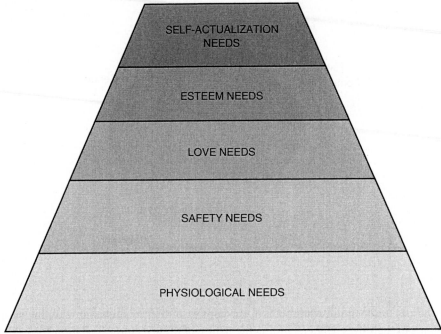

Maslow, Abraham H.; Frager, Robert D.; Fadiman, James, *Motivation and Personality*, 3rd Ed., ©1987.
Reprinted and Electronically reproduced by permission of Pearson Education, Upper Saddle River,
New Jersey

effective human service professionals will help their clients increase their personal awareness of the existence of these systems and how they are currently operating within them
(i.e., nature of reciprocity). It is through this awareness that clients increase their level of
empowerment within their environment and consequently in all aspects of their life.

MASLOW'S HIERARCHY OF NEEDS   Another effective model for understanding how many people are motivated to get their needs met was developed by
Abraham Maslow. Maslow (1954) created a model focusing on needs motivation. As
Figure 1.2 illustrates, Maslow believed that people are motivated to get their most
basic physiological needs met first (such as the need for food and oxygen) before
they attempt to meet their safety needs (such as the security we find in the stability
of our relationships with family and friends). According to Maslow, most people
would find it difficult to focus on higher-level needs related to self-esteem or self-
actualization when their most basic needs are not being met. Consider people you
may know who suffer from low self-esteem and then consider how they might react
if a war suddenly broke out and their community was under siege. Maslow's theory
suggests that thoughts of low self-esteem would quickly take a back seat as worries
about mere survival took hold. Maslow's Hierarchy of Needs can assist human service professionals in helping clients by recognizing a client's need to prioritize more
pressing needs over others.

Understanding Human Services through a Look at Practice Settings

It is important to remember that the nature of intervention is completely dependent on the specific practice setting where the human service professional is providing direct service. Thus, how clients are helped to improve their personal and social functioning will look very differently depending on whether services are provided in a school setting, a hospice, or a county social service agency. Human service professionals practice in numerous settings, some of which include schools, hospitals, advocacy organizations, faith-based agencies, government agencies, hospices, prisons, and police departments, as well as in private practice.

It would be difficult to present an exhaustive list of categories of practice settings due to the broad and often very general nature of this career. Practice settings could be categorized based on the social issue (i.e., domestic violence, homelessness), target population (i.e., older adults, the chronically mentally ill), or the area of specialty (i.e., grief and loss, marriage and family). Regardless of how we choose to categorize the various fields within human services, it is imperative that the nature of this career be examined and explored through the lens of practice settings in some respect to truly understand both the career opportunities available to human service professionals and the functions they perform within these various settings.

> Human service professionals practice in numerous settings, some of which include schools, hospitals, advocacy organizations, faith-based agencies, government agencies, hospices, prisons, and police departments, as well as in private practice.

Some of these practice settings include (but are not necessarily limited to) medical facilities, including hospitals and hospices; schools; geriatric facilities, including assisted-living facilities; victim advocacy agencies, including domestic violence, sexual assault, and victim–witness assistance departments; child and family service agencies, including adoption agencies and child protective service agencies; services for the homeless, including shelters and the government housing authority; mental health centers; faith-based agencies; and social advocacy organizations, such as human rights agencies and policy groups.

Regardless of the manner in which practice settings are categorized, there is bound to be some overlap because one area of practice could conceivably be included within another field, and some practice settings could also be considered an area of specialization. For instance, there are Christian hospices (medical social work and faith-based practice), some human service professionals work with both victims of domestic violence (victim advocacy) and batterers (forensic human services), and adoption is sometimes considered a practice setting unto itself and sometimes included under the umbrella of child welfare.

For the purposes of this text, the roles, skills, and functions of human service professionals will be explored in the context of particular practice settings, as well as areas of specialization within the human services field—general enough to cover as many functions

Human Services Delivery Systems

Understanding and Mastery of Human Services Delivery Systems: Range and characteristics of human services delivery systems and organizations

Critical Thinking Question: Human service professionals work in a wide variety of settings, including hospitals, schools, the legal system, child advocacy agencies, and mental health clinics, to name just a few. In what settings have you come into contact with human service professionals so far in your life?

and settings as possible within the field of human services, but narrow enough to be descriptively meaningful. The role of the human service professional will be examined by exploring the history of the practice setting, the range of clients, the clinical issues most commonly encountered, mode of service delivery, case management, and most common generalist intervention strategies within the following practice settings and areas of specializations: child welfare, adolescents, geriatric and aging, mental health, housing, healthcare and hospice, substance abuse, schools, faith-based agencies, violence, victim advocacy and corrections, and macro practice, including international human rights work.

The following questions will test your knowledge of the content found within this chapter.

1. The following are reasons why people may need to utilize human services:
   a. Mental Illness
   b. Racism
   c. Trauma
   d. All of the above

2. According to the chapter, someone is considered to be working in the human services field if he is working
   a. in the occupational and/or speech therapy fields
   b. with marginalized, disenfranchised, or other individuals who are in some way experiencing problems related to various social or systemic issues within society
   c. with marginalized, disenfranchised, or other individuals who are in some way experiencing problems related to various personal or pathological issues within oneself
   d. None of the above

3. According to the National Organization for Human Services, the human services profession is one which promotes _____ not only by addressing the quality of direct services, but by also seeking to improve _____ among professionals and agencies in service delivery.
   a. a healthy lifestyle/collaboration
   b. societal structures/accessibility and collaboration
   c. improved service delivery systems/accessibility, accountability, and coordination
   d. None of the above

4. The Human Services Board Certified Practitioner (HS-BCP) is a
   a. voluntary national professional certification
   b. license that allows paraprofessionals to work in schools and hospitals
   c. name for the accreditation of human services educational programs
   d. national professional certification required by insurance companies for payment reimbursement

5. The foundational theoretical approaches to the human services discipline include
   a. Person-in-Environment
   b. Bronfenbrenner's Ecological Systems Theory
   c. Eco-systems Theory
   d. All of the above

6. In Maslow's Hierarchy of Needs, a person would first need to meet her _____ needs, before meeting her _____ needs.
   a. higher level/lower level
   b. central level/lower level
   c. internal/external
   d. lower level/higher level

7. Compare and constrast the human services field with the social work and psychology disciplines.

8. Describe the basic tenets of Bronfenbrenner's Ecological Sytems Theory and provide an example of how this theory applies in the human services discipline.

Internet Resources

American Counseling Association: http://www.counseling.org

Council for Accreditation of Counseling & Related Educational Programs: http://www.cacrep.org

Council for Standards in Human Service Education: http://www.cshse.org

Human Services Career Network: http://www.hscareers.com

National Organization for Human Services: http://www.nationalhumanservices.org

References

Beaucar, K. O. (2000). Licensing a mixed bag in '99. *NASW News, 45*(2), 9.

Bronfenbrenner, U. (1979). *The ecology of human development: Experiments by nature and design.* Cambridge, MA: Harvard University Press.

Gumpert, J., & Saltman, J. E. (1998). Social group work practice in rural areas: The practitioners speak. *Social Work with Groups, 21*(3), 19–34.

Maslow, A. (1954). *Motivation and personality.* New York: Harper.

Meyer, C. H. (1988). The eco-systems perspective. In R. A. Dorfman (Ed.), *Paradigms of clinical social work* (pp. 275–294). Philadelphia: Brunner/Mazel, Inc.

National Organization for Human Services. (n.d.). What is human services? Retrieved from http://www.nationalhumanservices.org/what-is-human-services

Rullo, D. (2001). The profession of social work. *Research on Social Work Practice, 11*(2), 210–216.

Rittner, B., & Wodarski, J. S. (1999). Differential uses for BSW and MSW educated social workers in child welfare services. *Children & Youth Services Review, 21*(3), 217–238.

CHAPTER 2

History and Evolution of Social Welfare Policy

· ·

Effect on Human Services

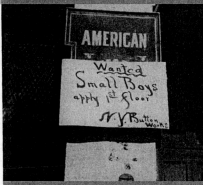

Everett Collection/SuperStock

The practice of helping others in need can be traced back to ancient times, but the human services profession in its current context has historic roots dating back to at least the late 1800s. The development of the social welfare system in the United States was very much influenced by England's social welfare system; therefore, it is important to understand the evolution of how the poor were treated in England to truly understand how social welfare policy has developed within the United States.

The Feudal System of the Middle Ages

A good place to begin this examination would be in England's Middle Ages (the 11th century), where a system called *feudalism* prevailed as England's primary manner of caring for the poor. Under this elitist system, privileged and wealthy landowners would parcel off small sections of their land, which would then be farmed by peasants or serfs. Many policy experts frame the feudal system not only as an effective method for controlling poverty, but also as a governmentally imposed form of slavery or servitude, because individuals became serfs through both racial and economic discrimination and were commonly born into serfdom with little hope of ever escaping. Serfs were considered the legal property of their landowner, or "lord"; thus although lords were required to provide for the care and support of serfs in exchange for farming their land, the lords had complete control over their serfs and could sell them or give them away as they deemed fit (Stephenson, 1943; Trattner, 1998).

Despite the seeming harshness of this system, it did provide insurance against many of the social hazards associated with being poor, and it was complemented by the prevailing attitude toward the poor during this time period, which was based on the notion that there was no shame in poverty. In fact, the commonly held societal more during medieval

Learning Objectives
- Recognize how economic, religious, and social policies influence society's perception of the poor
- Recognize the historic role of people of color in the development of the human services profession
- Understand the impact of the Great Depression and the context in which New Deal social welfare programs were created
- Compare and contrast contemporary socio-political perspectives
- Identify and develop methods of identifying and addressing bias in perceptions of disenfranchised populations, including bias based on race, socioeconomic status (SES), gender, sexual orientation, and age

times was that poverty within society was unavoidable, and the poor were a necessary component of society, in that it gave an opportunity for the rich to show their grace and goodwill through the giving of alms to those less fortunate than themselves. The poor were also necessary because without them there would be no servants.

This attitude was influenced by religious teachings, particularly teachings within the Judeo-Christian tradition, and was reinforced by church authorities, who shouldered the primary responsibility, within a governmental capacity, of administering relief to those unable to support themselves. Thus, caring for the poor was perceived as a noble duty that rested on the shoulders of all those who were able-bodied. Almost in the same way that evil was required to highlight good, poverty was likewise necessary to highlight charity and goodwill as required by God.

A policy of charity is not limited to Judeo-Christian faiths though; in fact, most religions include charity as requirements of faith. For instance, in Islam followers are required to contribute a fifth of their income to the poor (Qur'an 8:41), and believers also practice regular charity (Qur'an 2:43) and care for the orphans (Qur'an 2:177). In the tradition of Buddhism (more correctly referred to as a philosophy than a religion because of the lack of deity) suffering and giving are foundational to understanding the meaning of life.

The Middle Ages was a time when there was no separation of church and state; therefore, the church and government were one and the same. Poor relief was handled on a local level, with Catholic bishops administering aid through local parishes, which were supported by mandatory taxes or compulsory tithing. Much of the reason for the relative success of this system was due to the absence of many of the issues with which contemporary society must contend. Populations were not nearly as transient as they are now; in fact, residency requirements were strictly enforced, thus many of the poor were known within the community and had perhaps been former contributing members who had fallen on hard times. The concept of community as family was easier to envision when communities were small and completely governed by the church (Trattner, 1998).

Poor Laws of England

Many economic and environmental conditions led to the eventual phasing out of the traditional feudal system in the mid-14th century to the mid-16th century (1350 through 1550), including several natural disasters such as massive crop failures as well as the bubonic plague; mass urbanization spawned by the wool industry; as well as the Industrial Revolution in general. The increased demand for factory wage labor in the cities ultimately led to droves of individuals moving to the city to work in factories, and this trend, coupled with the decline of the feudal system as well as the diminishing influence of the church with its complex and effective framework of charitable provision, led to the need for a complete overhaul of the social welfare system in England. Thus, although this vast wave of urbanization led to freedom of serfdom for the poorest in England, it also generated a vacuum in how poverty was managed, creating the necessity for the development of England's earliest poor laws (Trattner, 1998).

Although these social changes were gradual, they led to a dramatic shift in not only how poverty was managed, but also how poverty was perceived. It is always easier to have a gracious attitude and extend a helping hand to someone we know, but such graciousness becomes more challenging when the poor are no longer extended family and longtime neighbors, whose personal circumstances are well known, but rather are nameless, faceless strangers living en masse, often from different countries, speaking different languages, and behaving in very different manners (Martin, 2010; Trattner, 1998).

The increasingly impersonal nature of poor care as well as the complexity of life in the city ultimately led to the belief that the incorporation of punitive measures into relief policy was needed to control what was becoming a true social ill: begging, vagrancy, and increased crime in the cities. In response, England passed several relief laws during the mid-1500s through the early 1600s, which set forth guidelines for dealing with the poor. England's relief act of 1536 placed responsibility for dealing with the poor at the local level and reflected a complete intolerance of idleness. Local law enforcement scoured the cities in search of beggars and vagrants, and once found, a determination was made as to whether they were true victims of poverty (the worthy poor) or legally defined vagrants (the unworthy poor). Legislative guidelines typically stipulated that only pregnant women, individuals who were extremely ill and unable to work, or any person over the age of 60 were considered justifiably poor, and thus they were treated more leniently, including given governmental authorization to beg (typically in the form of a letter of authorization), or they were given other forms of sustenance. If a person was found to be able-bodied and unemployed, they were determined to be vagrant, punishable by whippings, naked parading through the streets, being returned to the town of birth, or incarceration. Repeat offenders were often subjected to having an ear cut off or even death (Beier, 1974; Birtles, 1999).

Clearly, there was no sympathy to be had for individuals, male or female, who were deemed capable of working but found themselves without a job or any means of support, and little consideration was given to economic difficulties or what is now termed the *cycle of poverty*. Also, little sympathy was extended to children, particularly adolescents who were found begging, and district officials often took these children into custody, placing them into an apprenticeship program, which was later considered to be little different from child slavery. Thus, vagrancy was handled as a criminal matter, and the local authorities provided sustenance only for those truly unable to work (Trattner, 1998).

The Elizabethan Poor Laws***

The earlier English poor laws laid the foundation for the Elizabethan Poor Laws of 1601, which in turn acted as a foundation for American social welfare policy. Thus, rather than viewing the Elizabethan Poor Laws of 1601 as a single act, it is more appropriate to view it as an evolution, and the more final in a series of previous acts. The Elizabethan Poor Laws of 1601 served to set the stage for poor relief for several centuries and is still considered foundational in contemporary social welfare policy in both England and America. This act established three driving principles as the foundation for social legislation, including the belief that the primary responsibility for provision lay with one's family, that poor relief should be handled at the local level, and finally, that individuals should not be allowed to move to a new community if unable to provide for themselves financially.

Charity included both *indoor* and *outdoor* relief, with the former referring to assistance provided in almshouses and other institutionalized settings and the latter referring to services provided in the home environment of the person in need and might involve the delivery of food baskets or medicine.

It was quite common for community members to bring charges against others if it could be proven that they had moved into the district within the last 40 days and had no means to support themselves. Such individuals would be charged as vagrants by the local officials and returned to their home districts. The underlying notion was that local parishes didn't mind supporting those individuals who had fallen on hard times after years of paying taxes, but they didn't want to be forced to support strangers who came to their district for the sole purpose of receiving aid. Elements of these residency requirements can be found among current U.S. welfare policy; in fact, most welfare reform bills today contain residency requirement language.

The Elizabethan Poor Laws of 1601 were then an organized merging of England's earlier, sometimes conflicting and erratic, social welfare legislation, which not only brought order and organization to England's poor laws, but also served as the foundation for such legislation in colonial America. Life in colonial America not only offered tremendous opportunity, but also presented significant hardship related to life on the frontier. Many immigrants were quite poor to begin with, and the long and difficult ocean voyage to the New World often left them unprepared for the rigors of life in America. Thus, even though colonial America offered many opportunities not available in the Old World, such as land ownership and numerous employment opportunities, many of the social ills plaguing new immigrants in their homeland followed them to America.

English colonization of North America began around the 16th century and continued throughout the 17th century. Because there was no existing infrastructure in the original 13 colonies (such as religious monasteries or other social welfare programs), poor relief consisted primarily of mutual kindness, family support, and distant help from the motherland. Self-sufficiency was a must, and life was not easy on the frontier. But as the population increased within the colonies, the need arose for a more organized form of relief, and it makes sense that the colonies would rely on the English Poor Laws.

Although the prevailing assumption among many is that the United States was founded based on a desire to be completely different than England, in reality, the overriding reasons for the American Revolution, although certainly complex, were based more on the desire for independence, rather than solely on the desire for a completely different governmental structure. This presumption is evident in the development of many of the social customs, governmental infrastructures, and legislation, including the social welfare policy of the American colonies. Thus, the colonies adopted not only the social welfare legislation of England, but much of the perceptions of and attitudes about the poor and indigent as well.

The practice of human services is wholly influenced by social welfare policy, and to be truly effective in helping the poor and indigent, it is essential that all human service providers gain a level of social and cultural objectivity so that they can more fully understand both how social welfare policy and legislation has evolved over the years and how the complex relationship between such social welfare policy and legislation and the current prevailing attitudes toward the poor influence one another. It would be naïve to assume that any current trends in how the poor are perceived and treated developed in a vacuum; thus, a general understanding of the roots of current social welfare legislation, policy, and attitudinal trends is essential to any practicing human service professional.

A general understanding of the roots of current social welfare legislation, policy, and attitudinal trends is essential to any practicing human service professional.

The Protestant Reformation and Social Darwinism

Despite popular contention that economic policy practice is evidence-based, objective, and free of ideological bias, significant evidence exists indicating that both historic and current economic policy practice is solidly interwoven with moral philosophy, reflecting the cultural mores of the times as well as of the particular society (Hausman & McPherson, 2006). Social policy, particularly policy addressing the social welfare of its citizenry, commonly reflects particular philosophical movements and themes. In the mid-19th century several philosophical movements existed that attempted to address problems in the social world, particularly problems of social inequity and poverty.

In his book *The Protestant Ethic and the Spirit of Capitalism,* Max Weber described in detail the vast influence of John Calvin's theory of Predestination, an integral aspect of the Protestant Reformation and Puritan theology in the mid-16th century, on European and American society. According to Weber, Calvin asserted that God perceived all humans as sinful and wholly undeserving of salvation, yet God in his infinite wisdom and providence determined who would go to heaven and who would be condemned to hell, based solely upon his all-knowing determination of what action would best glorify himself. Human action in an attempt to secure salvation thus was futile since one's eternal fate rested not upon human goodness (which according to Calvin would always fall short of the perfection of God), but solely upon God's mysterious desire (Weber, 1905/1958).

Although Calvin rejected the notion that one could determine the state of one's salvation from any outward signs, Weber notes that determining the "state of grace" of oneself and others became an integral part of Reformed doctrine in part because a considerable amount of social functioning depended upon society's ability to separate the "elect" from the condemned. For instance, only God's faithful were allowed to become members of the church, receive communion, and enjoy other benefits of salvation (such as societal respect).

In time particular behaviors and conditions became certain indicators—or signs—of one's eternal fate. Most notably among these behaviors were hard work and good moral conduct. The high value placed upon hard work, what Weber referred to as the Protestant ethic, is reflective of Calvin's belief that one was called to a particular vocation and should work tirelessly as a sign of faithfulness. Thus, according to Weber, individuals did not need to endure a lifetime of questioning their salvation; rather, the commitment to a strong work ethic was "the best possible means of attaining this self-assurance. This and this alone would drive away religious doubt and give assurance of one's state of grace" (Weber, 1905/1958, pp. 77–78).

A life lived in pursuit of purity and denial of worldly pleasures, what Weber referred to as Puritan asceticism, also became an indicator of one's "state of grace" because, according to Calvin and Reformed theology, only members of the elect were capable of manifesting such a state of sanctified holiness. Thus, material success in response to hard work and high moral conduct became the universally accepted signs among mainstream (i.e., respectable) society of those predestined for eternal salvation (Hudson & Coukos, 2005; Weber, 1905/1958).

The influence of the Protestant ethic and Calvin's doctrine of predestination on society as a whole, and specifically upon society's cultural mores related to poverty, and the poor were significant, extending beyond that of the religious community (Kim, 1977). With hard work, good moral conduct, and material success serving as the best signs of election to salvation, it did not take long for poverty and presumed immoral behavior (because it was presumed that only the elect had the spiritual fortitude to behave morally) to become a clear indication of one's condemnation (Chunn & Gavigan, 2004; Gettleman, 1963; Hudson & Coukos, 2005; Kim, 1977; Schram, Fordingy, & Sossz, 2008; Tropman, 1986; Weber, 1905/1958).

Social Darwinism was another social philosophy that significantly influenced how poverty and disadvantage were perceived and treated within the American social welfare system. Social Darwinism involved the application of Charles Darwin's theory of natural selection to the human social world. Darwin's theory, developed in the mid-19th century, was based upon the belief that environmental competition—a process called *natural selection,* ensured that only the strongest and most fit organisms would survive (allowing the biologically fragile to perish), thus guaranteeing successful survival of a species (Darwin, 1859/2009). Darwin's theory was focused primarily upon the biological fitness of animals and plant life; yet, he did apply his theory to humans as well, providing natural-istic explanations for various phenomena in human social life. Weikart (1998) describes written discussions with contemporaries where Darwin espoused a belief that humans were subject to natural law and that economic competition was a necessary component of natural selection in the human species. In fact, Darwin even went so far as to argue that socioeconomic inequality was primarily due to biological inequality, thus it could not be avoided intimating that those in society who suffered poverty and other forms of misfortune were merely victims of their own biological inferiority; therefore, their demise was necessary in order for the survival of society as a whole (Weikart, 1998).

Thiel, another social Darwinist, argued that not only was the struggle for survival within society unavoidable, it was desirable, asserting that competition for economic resources should be maximized in order to weed out the weaker members of society, thus allowing the biologically (and mentally) superior to prevail. Thiel (1868, as cited in Weikart, 1998) cautioned against most forms of government intervention designed to lift individuals out of poverty and misfortune, or create social equality, asserting that giving the weak an opportunity to survive could actually pose a threat to society. In defending inequality within human society, Darwin and his colleague Thomas Henry Huxley "advocated social structures that allowed the more talented to advance and the less competent to sink. They advocated economic inequality and the accumulation of wealth as necessary for the progress of humanity" (Weikart, 1998, p. 27).

One of the most influential social Darwinists was Herbert Spencer, an English philosopher who actually preceded Darwin in applying concepts of natural selection to the social world. Spencer coined the term *survival of the fittest* (a term often incorrectly attributed to Darwin) in reference to the importance of human competitiveness for limited resources in securing the survival of the "fittest" members of society. Spencer was a fierce opponent of any form of government intervention or charity on behalf of the poor and disadvantaged, arguing that such interventions would interfere with the natural order, thus threatening society as a whole (Hofstadter, 1992). Although Spencer's theory of social superiority was developed in advance of Darwin's theory, his followers relied upon Darwin's theory of natural selection for scientific validity of social Darwinism.

The fatalistic nature of social Darwinism and the Protestant ethic became deeply imbedded in both American religious and secular culture and were used to justify a laissez-faire approach to charity and social welfare throughout most of the 19th and 20th centuries (Duncan & Moore, 2003; Hofstadter, 1992). Although the specific tenets of these ideologies may have softened over the years, the significance of hard work, good fortune, material success, and living a socially acceptable life have remained associated with a collective sense of entitlement to special favor and privilege in life, whereas poverty and disadvantage have remained associated with weak character, laziness, and questionable behavior. Standing back then and leaving the poor and disadvantaged to their own devices was perceived as nothing more than complying with God's (or nature's) grand plan (Duncan & Moore, 2003).

The popularity of social Darwinism and the Protestant ethic in American culture was related, at least in part, to the American cultural more of rugged individualism and self-sufficiency. Whereas traditional Catholicism focused on the transformation of the community and the giver by being blessed through the act of giving, the Protestant ethic and social Darwinism focused on the individual who was transformed (behaviorally) by the act of receiving (Duncan & Moore, 2003). With the focus of charity placed upon the one in need, the dilemma faced by the state and charity providers was determining who deserved help and who did not (Chunn & Gavigan, 2004; Duncan & Moore, 2003; Gettleman, 1963; Hudson & Coukos, 2005; Kim, 1977; Schram et al., 2008; Tropman, 1986; Weber, 1905/1958). This dilemma led to the practice of categorizing the poor as "worthy" or "unworthy" based upon the perceived cause of their impoverishment and misfortune and presumed likelihood of behavioral change in response to charity. Yet, with many asserting that providing charity to the poor would only serve to increase their immorality and dependence, even the worthy poor experienced difficulty in obtaining material assistance (Chunn & Gavigan, 2004; Gettleman, 1963; Weber, 1905/1958).

Professional History

Understanding and Mastery of Professional History: How public and private attitudes influence legislation and the interpretation of policies related to human services

Critical Thinking Question: The influences of the Protestant ethic and social Darwinism are unmistakable in the history of U.S. social welfare policy. In what ways are the concepts of work and morality, survival of the fittest, and the "worthy" and "unworthy" poor reflected in current policies and in the attitudes of the U.S. public today?

These ideological themes of moral deficiency of the poor and the belief that giving material support to the poor would only serve to increase their immoral nature, laziness, and dependency have been reflected in the policy perspectives of the American social welfare system at some level throughout U.S. history (Chunn & Gavigan, 2004; Duncan & Moore, 2003; Gettleman, 1963; Hudson & Coukos, 2005; Kim, 1977; Schram et al., 2008; Tropman, 1986).

Charity Organization Societies

The Charity Organization Society (COS), often considered the genesis of the social services, marked one of the first organized efforts within the United States to provide charity to the poor. The COS movement started in about 1870 in response to frustration with the current welfare system that was less of a system and more of a disorganized and often chaotic practice of almsgiving. The COS movement itself was started by a pastor, Rev. S. Humphreys Gurteen, who believed that it was the duty of good Christians everywhere to provide an organized and systematic way of addressing the plight of the poor. Gurteen and his colleagues strongly believed that the indiscriminate giving of alms by many of the relief agencies of that time encouraged fraud and abuse, which in turn encouraged laziness on the part of those who were beneficiaries of relief.

The COS philosophy was built on the concept of voluntary coordination, in which various charities worked within a larger network-coordinating services delivered to the local community. The first COS was created in New York in 1877, and the concept quickly spread to large cities across the nation. Soon, most large cities had at least one COS serving the community, acting as an umbrella organization for smaller agencies and churches offering charity services to the community. The COSs practiced what was called *scientific charity*, which embraced social Darwinist philosophies of "intelligent" giving and embraced the notion that charity should work with natural selection, not against it (Gettleman, 1963). A primary motivation of the COS movement was to coordinate charity efforts by serving as an umbrella organization for the myriad of independent and private charities, thus maximizing the best use of material relief (Schlabach, 1969). Outdoor relief, such as cash assistance or indiscriminate giving, was highly discouraged and actually considered evil based upon the long-standing belief that such assistance encouraged dependence and laziness, while discouraging self-sufficiency, which ultimately led to increased poverty (Gettleman, 1963; Kusmer, 1973).

In this respect, those involved in the COS movement embraced the concepts of the unworthy and worthy poor, and it was their goal to determine which category aid recipients fell into and then prescribe what each recipient actually needed—material aid for those who would not abuse it and other services for those who would. To accomplish this goal, the COSs employed "friendly visitors," an early version of caseworkers, who visited the homes of aid applicants and attempted to diagnose the reason for their poverty and, if possible, develop a "case plan" to authentically alleviate their suffering (Trattner, 1998).

A social hierarchy was reflected in the philosophical motivation of the COS leaders, often the community's most wealthy members, who agreed to provide charity to the poor dependent upon the poor remembering "his place of inferiority" (Gettleman, 1963, p. 319). Yet, even the deserving poor did not escape the demands of the Protestant ethic or the fatalism of social Darwinism, both of which were deeply imbedded in COS culture. These philosophical values were clearly reflected in a speech given by Josephine Shaw Lowell, a leader in the COS movement, at a charity conference held in 1895, where she stated "Even the widow with little children, if she finds that everything is made easy for her, may lose her energy, may even, by being relieved of anxiety for them, lose her love for the children" (1895 as cited in Gettleman, 1963, p. 323). The unworthy poor were often provided with indoor relief almshouses only and, according to COS leaders, should be allowed to perish according to natural selection. Many in the COS movement argued that to provide charity to those destined to perish was immoral and unkind because it just served to prolong their suffering to no good end for either the poor or society (Gettleman, 1963).

Mary Richmond, the general secretary of the Baltimore COS, is often associated with the COS movement because of her passion for social advocacy and social reform. Richmond believed that charities could employ both good economics and compassionate giving at the same time. Richmond became well known for increasing public awareness of the COS movement and for her fund-raising efforts. Richmond's compassion for the poor was the likely result of her own experience with poverty. Orphaned at the age of 2 and later left by an aunt to fend for herself in New York when she was only 17 years old, Richmond no doubt understood the social components of poverty, and how devastating it could be to one's life.

Richmond was responsible for developing the early conception of casework, having written several books and articles on the service delivery model. As a result, the concept of the friendly visitor grew, as did the debate about material relief continued, with many arguing that the best opportunity to truly effect change in those suffering from poverty was through the services of the friendly visitor who could help identify and address any barriers to self-sufficiency (Kusmer, 1973).

Despite the general success of the COS movement and the difficult task of basically cleaning up the social welfare system in the post–Civil War climate, the COS philosophy was tinged by the Reformation theology that anyone who worked hard enough would be blessed and could rise from the depths of poverty. This sentiment added to the general sense of rugged individualism often worn as a badge of strength by many U.S. citizens. But it was naïve to presume that poverty was primarily caused by individual failure and that material relief would lead to moral decline. The country was about to learn a very hard lesson during the Depression era, one that immigrants and ethnic minorities had known for years—that sometimes conditions exist that are beyond an individual's control and that create immovable barriers to self-sufficiency, leading to poverty and complete destitution.

Jane Addams and the Settlement House Movement

Not all social welfare movements within the United States reflected these harsh philosophical approaches though. Jane Addams, an advocate for social reform, was responsible for beginning the U.S. settlement house movement in the late 1800s. Addams's social action efforts reflected a far more compassionate approach to poverty alleviation and social inequity. Addams started the Hull-House Settlement house in Chicago as an alternative to the more religiously oriented charity organizations, which she perceived as "heartless and overly concerned with efficiency and rooting out of fraud" (Schneiderhan, 2008, p. 3). Addams used a relational model of poverty alleviation based upon the belief that the problems of poverty and disadvantage resulted from problems within society, not idleness and moral deficiency (Lundblad, 1995). Addams's social action efforts reflected a far more compassionate approach to poverty alleviation and social inequity. Addams advocated for changes within the social structure of society that created barriers to lateral contribution of all members of society, which she viewed as an essential aspect of a democracy (Hamington, 2005; Martin, 2012). In fact, the opening of the first settlement house in the United States was considered the beginning of one of the most significant social movements in U.S. history (Commager, 1961, as cited in Lundblad, 1995).

Addams was born in Cedarville, Illinois, in 1860. She was raised in an upper-class home where higher education and philanthropy were highly valued. Addams greatly admired her father, who encouraged her to pursue an education at a time when women were primarily encouraged to pursue only marriage and motherhood. She graduated from Rockford Female Seminary in 1881, the same year her father died. After her father's death, Addams entered Woman's Medical College in Pennsylvania, but dropped out because of chronic illness. Addams had become quite passionate about the plight of immigrants in the United States, but due to her poor health and the societal limits placed on women during that era, she did not believe that she had a role in social advocacy.

The United States experienced another significant wave of immigration in the 19th and early 20th centuries (between 1860 and 1910), with 23 million people emigrating from Europe, including Eastern Europe. Many of these immigrants were from non-English-speaking countries, such as Italy, Poland, Russia, and Serbia, and thus did not speak English, and were very poor. Unable to obtain work in the skilled labor force, many immigrants were forced to live in subhuman conditions, crammed together with several other families in deplorable tenements in large urban areas. New York's Lower East Side had 330,000 inhabitants per square mile (Trattner, 1998). With no labor laws for protection, racial discrimination and a variety of employment abuses were common, including extremely low wages, unsafe working conditions, and child labor. Poor families, particularly non-English-speaking families, had little recourse, and their mere survival depended on their coerced cooperation.

Addams was aware of these conditions because of her father's political involvement, but she was not sure how to respond. Despondent after her father's death and her failure in medical school, as well as over her chronic medical problems, Addams took an extended trip with friends to Europe, where among other activities she visited Toynbee Hall, England's response to poverty and other social problems. Toynbee Hall was a settlement house, which was essentially a neighborhood welfare institution in an urban slum area, where trained workers endeavored to improve social conditions, particularly by providing community services and promoting neighborly cooperation.

This concept was revolutionary, in that in its attempt to improve conditions through the promotion of social and economic reform, it actually called for the settlement house workers to reside in the home alongside the immigrant families they helped. In addition to providing a safe, clean home, settlement houses also provided comprehensive care, such as assistance with food, healthcare, English language lessons, child care, and general advocacy. The settlement house movement was different from the traditional charity organizations, in that it had as its goal the mission of no longer distinguishing between the worthy and unworthy poor.

> **The settlement house movement was different from the traditional charity organizations, in that it had as its goal the mission of no longer distinguishing between the worthy and unworthy poor.**

Addams returned home convinced that it was her duty to do something similar in the States, and with the donation of a building in Chicago, the Hull House became America's first settlement house in 1889. Addams and her colleagues lived in the settlement house, in the middle of what was considered a "bad" neighborhood in Chicago, offering services targeting the underlying causes of poverty such as unfair labor practices, the exploitation of non-English-speaking immigrants, and child labor. Services ranged from child care to education classes. The Hull House became the social center for all activities in the neighborhood and even offered residents an opportunity to socialize in the resident's café.

Addams's influence of American social policy was significant, in that it represented a shift away from the fatalistic and metaphysical philosophies of Calvinism and social Darwinism, marking recognition of the need for social change within society in order to remove barriers to upward mobility and optimal functioning (Martin, 2012). Addams and her counterparts were committed to viewing all individuals equally, to be treated with respect and dignity. Addams clearly saw societal conditions and the hardship of immigration as the primary cause of poverty, not necessarily one's own moral failing. Focus was placed on making changes in the community, and social inequality was perceived as the manifestation of exploitation, with social egalitarianism perceived as not just desirable but achievable (Lundblad, 1995, Martin, 2012).

The settlement house movement radically transformed not only how the poor were cared for, but also how they were *perceived* by the majority population. Now, immigrants had a safe place to live, a voice to advocate for them, and a way to better integrate into American society, so that their dream of obtaining a better life for themselves and their children could actually be realized. Addams also lobbied tirelessly for the passage of child labor laws and other legislation that would protect the working-class poor, who were often exploited in factories with sweatshop conditions. She also worked alongside Ida B. Wells, an African American reformer, confronting racial inequality in the United States, such as the extrajudicial lynching of black men.

Abolishing the Sweating System poster. Jane Addams Memorial Collection, Hull House Association records, HHA negative 33, Special Collections, University Library, University of Illinois at Chicago

Although there are no working settlement houses today, the prevailing concept espoused by this model involves recognition of the need for a holistic approach to poverty alleviation that encompasses challenges to social structures, and not just a focus on individual behavioral management. Elements of this concept can still be seen in the current U.S. social welfare system, as well as its current mental healthcare system (see Chapter 6); yet, unfortunately there would be far more future challenges to any philosophical approach to poverty alleviation that considers social inequality as a core reason for poverty, rather than personal moral failing. Thus, despite the overall success of the settlement house movement and the particular success of Addams with regard to achieving social reform in a variety of arenas, the influences of Calvinism, particularly the Protestant Ethic and social Darwinism remained strong, experiencing cyclical decline only during difficult economic times or civil unrest (as experienced in the 1960s).

The New Deal and the Social Security Act of 1935

In 1929 the stock market crashed, leading to a series of economic crises such as the United States had never before experienced. For the first time in modern U.S. history, large segments of the middle-class population were unemployed, and within a very short time thousands of people who had once enjoyed secure lives were without jobs and soon without homes and food as well. This served as a wake-up call for social reformers, many of whom had abandoned their earlier commitment to social activism. In response, many within the human service fields started pushing President Hoover to develop the country's first federal system of social welfare.

Hoover was resistant, though, fearing that federal social welfare would create dependency and displace private and local charities. He wanted to allow time for democracy and capitalism to self-correct before intervening with broad entitlement programs. But much of the country, many of whom were literally starving, apparently did not agree, and in 1933, Hoover lost his bid for reelection, and Franklin D. Roosevelt was elected as president. Roosevelt immediately set about to create dramatic changes in federal policy with regard to social welfare, promising a *New Deal* to the country, where a minimum standard of living was seen as a right, not a privilege.

Within his first 100 days in office, Roosevelt passed 13 acts including the Civil Works Administration (sometimes referred to as the CWA), which provided over a million temporary jobs to the unemployed; the Federal Emergency Relief Act, which provided direct aid and food to the unemployed; and the Civilian Conservation Core (CCC), which put thousands of young men aged 18 to 25 to work in reforestation and other conservation programs. Yet, as progressive as Roosevelt was, and as compassionate as the country had become due to the realization that poverty could strike anyone, racism was still rampant, as illustrated by Roosevelt placing a 10 percent limit on the enrollment of black men in the CCC program (Trattner, 1998).

By far the most famous of all programs in the *New Deal* and *Great Society* programs were those created in response to the Social Security Act of 1935, which among other things created old age assistance, unemployment compensation, aid to dependent mothers and children, and aid to the blind and disabled. In total Roosevelt created 15 federal programs as a part of the New Deal, some of which remain and some of which were dismantled once the crisis of the Depression subsided. Although some claim that the New Deal was not good for the country in the long run, it did pull the country out of the Depression, and it provided relief for millions of Americans who may have literally starved had the federal government not stepped in when it did. Programs such as the Federal Deposit Insurance Corporation (FDIC), which provided insurance for deposits, helped to instill a sense of confidence in the banking system once again, and the development of the Securities and Exchange Committee (SEC), which regulates the stock market, helped to ensure that a crash similar to the one in 1929 would be unlikely to occur again. In later times though, the dismantling of some post-Depression financial regulations would contribute to yet another devastating economic downturn—perhaps not as severe as the Great Depression, but more serious and long-lasting than any other recession experienced in the U.S. post-Depression era, particularly because of its global consequences.

Professional History

Understanding and Mastery of Professional History: Historical and current legislation affecting services delivery

Critical Thinking Question: The New Deal created a number of social welfare programs, many of which are still in place. Who benefits from these programs? How have they contributed to a shift in societal attitudes about basic economic rights?

Influences of African American Social Workers

A review of the historical elements influencing the development of the human services field would be remiss if the influences of African Americans reformers, particularly African American women in the last part of the 19th century, weren't explored. Black activists had a significant influence on the development of social justice and human services, particularly in the South, filling the vacuum left by a racist society that often created barriers to service in the black community in earlier eras.

Ida B. Wells was an African American reformer and social activist whose campaign against racial oppression and inequity laid the foundation for the civil rights movement of the 1960s. Wells was born in 1862 to parents who were slaves in rural Mississippi, and although her parents were ultimately freed, Wells's life was never free from the crushing effects of severe racial prejudice and discrimination. Wells as orphaned at the age of 16, and went on to raise her five younger siblings. This experience not only forced Wells to grow up quickly, but also seemed to serve as a springboard for her subsequent advocacy against racial injustice. In Wells's early advocacy career, she was the owner of a black newspaper (the only one of its kind) called *Free Speech,* where she consistently wrote about matters of racial oppression and inequity, including the vast amount of socially sanctioned crimes committed against blacks (Hamington, 2005).

The indiscriminate lynching of black men was prevalent in the South during Wells's lifetime, and was an issue that Wells became quite passionate about. Black men were commonly perceived as a threat on many levels, and there was virtually no protection of their personal, political, or social rights. The black man's reputation of an angry rapist was endemic in white society, and many speeches were given and articles written by white community members (including clergy) about this "growing problem." Davidson (2008) references an article published in the mainstream newspaper in the South, the

Commercial, entitled *More Rapes More Lynchings,* which cites the black man's penchant for raping white women, stating:

> The generation of Negroes which have grown up since the war have lost in large measure the traditional and wholesome awe of the white race which kept the Negroes in subjection . . . There is no longer a restraint upon the brute passion of the Negro . . . The facts of the crime appear to appeal more to the Negro's lustful imagination than the facts of the punishment do to his fears. He sets aside all fear of death in any form when opportunity is found for the gratification of his bestial desires. (p. 154)

Wells wrote extensively on the subject of the "myth of the angry black man," and the myth that all black men raped white women (a common excuse used to justify the lynching of black men) (Hamington, 2005). She challenged the growing sentiment in white communities that black men, as a race, were growing more aggressive and "lustful," raping white women, accusations often used as a justification for lynching, and prompted in part by the increasing number of biracial couples. The response to Wells's articles was swift and harsh. A group of white men surrounded her newspaper building with the intention of lynching her, but when they could not find her they burned down her business (Davidson, 2008).

Although this act essentially stopped her newspaper career, what this act of revenge really did was to motivate Wells even further. After the burning of her newspaper Wells left the South and moved to Chicago where she continued to wage a fierce anti-lynching campaign, often coordinating efforts with Jane Addams. She wrote numerous books and articles on racial inequality, challenging socially entrenched notions that all black men were angry and violent sexual predators (Hamington, 2005). Wells and Addams worked as colleagues, coordinating their social justice advocacy efforts fighting for civil rights. Together, they ran the *Chicago Association for the Advancement of Colored People,* and worked collectively on a variety of projects, including fighting against racial segregation in schools (Martin, 2012).

Many other key African American social welfare reformers made significant advances in the human services field, particularly with regard to confronting the disenfranchisement and marginalization of African Americans within U.S. society. In this absence of mainstream human services within this population, African American social welfare reformers operated as a tight community, developing close relationships with each other, even though many of these women were spread across the United States. Because racism excluded African Americans from receiving many services, including educational opportunities and health services, many early social welfare reformers focused on these two areas, developing "Negro" schools and healthcare facilities. One such reformer was Modjeska Simkins, who developed healthcare programs for the black community focusing on everything from infant mortality to tuberculosis. Another creative example of human services in the face of extreme opposition was the work of the black sorority Alpha Kappa Alpha, whose members were determined to provide healthcare services to sharecroppers in Mississippi. When the white community refused to rent them office space, they offered the health services from cars (Gordon, 1991).

Other black women who significantly influenced social welfare reform include Anna Cooper, who pushed for increased educational opportunities for blacks; and Jane Hunter, who formed the first black Young Women's Christian Association (YWCA) (Gordon, 1991). Although often unreported and undervalued, African American social welfare reformers not only assisted their own communities but helped the broader community as well by modeling the power of networking and relentlessly pursuing social justice for all, particularly for those who are the subject of social oppression and discrimination.

Gay Rights: From Marriage Equality to "Don't Ask Don't Tell" Repeal

Ethnic minorities, women, and immigrants are not the only groups in U.S. society to be used as scapegoats. The gay community, typically referred to as the LGBTQ (lesbian, gay, bisexual, transgendered, and questioning and/or queer), has long been a marginalized group in the United States (as well as in most countries in the world). Members of the gay community are often victims of horrific hate crimes, often solely because of their sexual orientation. For years this community has been excluded from many of the social welfare laws designed to protect disenfranchised and socially excluded groups. Yet, in the last three decades, several LGBTQ advocacy organizations, such as the Gay & Lesbian Alliance Against Defamation (GLAAD), have become increasingly vocal about the right of the LGBTQ community to live openly, and enjoy the same rights and protections as heterosexuals without fear of reprisal. Specific issues GLAAD has advocated for include the right to be included as a specially protected group in hate crimes legislation, the right to legal marriage (often referred to as marriage equality), and the right to serve openly in the military.

Despite strong opposition from social conservative groups, the LGBTQ community has experienced recent success in response to their efforts. In 2009 President Obama signed into law the Matthew Shepard and James Byrd, Jr. Hate Crimes Prevention Act, which expanded existing hate crime legislation to include crimes committed against individuals based upon perceived gender, sexual orientation, and gender identity. Marriage equality—the right of same-sex couples to get legally married—is currently a battle fought on both a federal and state level. In 1996 the Defense of Marriage Act was passed, which defines marriage on a federal level as a union between one man and one woman. Yet, some states have passed laws legalizing same-sex marriage, including Massachusetts and Iowa. Arguments for same-sex marriage are typically based upon rights of equality. The GLAAD website lists several protections that marriage offers that are currently unavailable to the LGBTQ population in same-sex partnerships. These include

- automatic inheritance
- child custody/parenting/adoption rights
- hospital visitation
- medical decision-making power
- standing to sue for wrongful death of a spouse
- divorce protections

- spousal/child support
- access to family insurance policies
- exemption from property tax upon death of a spouse
- immunity from being forced to testify against one's spouse
- domestic violence protections, and more (GLAAD, 2010, p. 7).

Arguments against same-sex marriages are often based upon religious values that hold homosexuality as sinful and unnatural, and define traditional marriage as being between a man and a woman. There also appears to be a general fear that the normalization of homosexuality will lead to the lowering of moral standards in a variety of respects throughout society. Yet, advocates of same-sex marriage confront religious arguments by citing research that disputes allegation that same-sex marriage will somehow dilute "traditional" marriage or harm children. They also cite the increasing acceptance among U.S. citizens of same-sex marriage and of homosexuality in general (according to a series of Gallup polls, in 2009, 63 percent of the U.S. population surveyed stated that they believed that same-sex couples should be able to marry or have a legal civil union, compared to 55 percent in 2004).

Another area of success for the LGBTQ population involves the right to serve in the U.S. military openly. Historically, gays and lesbians were systematically discharged from the military if their homosexuality was discovered. In December 1993, in response to mounting pressure to change this policy, the Clinton administration compromised by implementing *Don't Ask Don't Tell* (DADT), an official policy of the U.S. government that prohibited the military from discriminating against homosexual military personnel as long as they kept their sexual orientation as secret. In other words, military personnel could no longer investigate one's sexual orientation, but if a member of the military admitted to being a homosexual, he or she could legally be discharged from the military. DADT was repealed by Congress in December 2010 pending review by military leadership who were to determine the effect on military readiness, but in July 2011 a federal court of appeals ruling barred further enforcement of the policy, and it was officially repealed by President Obama in September 2011. In May 2012 President Obama officially declared his support for marriage equality, citing his daughters' friends with same-sex parents, and his recognition that he could not defend a position that would prohibit them from having the same right to legally marry as heterosexual parents.

These achievements by the LGBTQ population seem to illustrate a movement toward greater acceptance of what some call "alternative lifestyles"; yet, there remains considerable resistance to the inclusion of homosexuality into mainstream America, particularly among social and religious conservatives.

Welfare Reform and the Emergence of Neoliberal Economic Policies

A resurgence of earlier negative sentiments toward the poor and their plight began in the mid-1970s, peaking in the 1990s, perhaps in response to increased economic prosperity within mainstream America. This increased negative attitude toward the poor is reflected in several studies and national public opinion surveys that reflected the general belief that the poor were to blame for their "lot in life." For instance, a national survey conducted in 1975 found that the majority of those living in the United States attributed poverty to personal failures, such as having a poor work ethic, poor money management skills, a lack of any special talent that might translate into a positive contribution to society, and low personal moral values. Those questioned ranked social forces, such as racism, poor schools, and the lack of sufficient employment, the lowest of all possible causes of poverty (Feagin, 1975).

> **A national survey conducted in 1975 found that the majority of those living in the United States attributed poverty to personal failures, such as having a poor work ethic, poor money management skills ... and low personal moral values.**

Ronald Reagan capitalized on this negative sentiment toward the poor during the 1976 presidential campaign when he based his platform in large part on welfare reform. In several of Reagan's speeches he cited the story of the woman from the South Side of Chicago who was finally arrested after committing egregious welfare fraud:

> She has eighty names, thirty addresses, twelve Social Security cards and is collecting veteran's benefits on four non-existing deceased husbands. And she is collecting Social Security on her cards. She's got Medicaid, getting food stamps, and she is collecting welfare under each of her names. (Zucchino, 1999, p. 65)

While Reagan never mentioned the woman's race, the context of the story as well as the reference to the South Side of Chicago (a primarily black community) made it clear that he was referring to an African American woman on welfare—thus matching the common stereotype of welfare users (and abusers) (Krugman, 2007). And with that, the enduring myth of the *Welfare Queen* was born.

Journalist David Zucchino attempted to debunk the myth of the welfare queen in his expose on the reality of being a mother on welfare, but stated in his book *The Myth of the Welfare Queen* that the image of the African American woman who drove a Cadillac while collecting welfare illegally from numerous false identities was so imbedded in American culture it was impossible to debunk the myth, even though the facts do not back up the myth (Zucchino, 1999). Krugman (2007) also cites how politicians have used the myth of the welfare queen in order to reduce sympathy for the poor and gain public support for welfare cuts ever since, arguing that while covert, such images clearly play on negative racial stereotypes. They also play on the common belief in the United States that those who receive welfare benefits are poor due to immoral behavior and a lack of motivation to work.

More recent surveys conducted in the mid-1990s revealed an increase in the tendency to blame the poor for their poverty (Weaver, Shapiro, & Jacobs, 1995), even though a considerable body of research points to social and structural issues as the primary cause of poverty, such as shortages in affordable housing, recent shifts to a technologically based society requiring a significant increase in educational requirements, longstanding institutionalized oppression and discrimination against certain racial and ethnic groups, and a general increase in the complexity of life (Martin, 2012; Wright, 2000).

The general public's perception of social welfare programs seems to be based in large part on this negative bias against the poor, and the misguided belief that the poor were lazy, immoral, and dependent. In several studies during the 1980s and 1990s those surveyed claimed support for the general idea of helping the poor, but when asked about specific programs or policies, most became critical of governmental policies, specific welfare programs, and welfare recipients in general. In fact, a 1987 national study found that 74 percent of those surveyed believed that most welfare recipients were dishonest and collected more benefits than they deserved (Kluegal, 1987).

A new conservative political movement during this time period was born at least in part out of this increasingly negative attitude toward the poor and social programs designed to alleviate poverty, beginning during the Reagan administration in the 1980s and ultimately leading to both Republican and Democratic support for drastic welfare reform. Focus once again shifted from social and structural causes of poverty to personal ones with a renewal of punitive social welfare policies reflecting the paternalistic ideologies of the past (Schram et al., 2008).

Political discourse in the mid-1990s reflected what are often referred to as economic *neoliberal philosophies,* a political movement embraced by most political conservatives, espousing a belief that capitalism and the free market economy were far better solutions to many social conditions, including poverty, than government programs. Advocates of neoliberalism pushed for social programs to be privatized based upon the belief that getting social welfare out of the hands of government and into the hands of private enterprise, where market forces could work their magic, would increase efficiency and lower costs. Yet, research consistently showed that social welfare services did not lend themselves well to free market theory due to the complexity of client issues, as well as unknown outcomes, lack of competition, and other dynamics that makes social welfare services so unique (Nelson, 1992; Van Slyke, 2003).

In 1994, during the U.S. Congressional campaign, the Republican Party released a document entitled *The New Contract with America,* which represented "a plan that would reform welfare and, along with it, the behavior of the poor" (Hudson & Coukos, 2005, p. 2). *The New Contract with America,* introduced just a few weeks prior to the 1994 Congressional election, Clinton's first mid-term election, was signed by all but two of the Republican members of the House of Representatives, as well as all of the party's Congressional candidates. In addition to a renewed commitment to smaller government and lower taxes, the contract also pledged a complete overhaul of the welfare system to root out fraud and increase the poor's commitment to employment and self-sufficiency.

Hudson and Coukos (2005) note the similarities between this political movement in the mid-1990s and the one just 100 years before, arguing that Protestant ethic theology served as the driving force behind both. Take for instance the common arguments for welfare reform (policies that reduce and restrict social welfare programs and services), which have often been predicated upon the beliefs that (1) hardship is often the result of laziness; (2) providing assistance will increase laziness (and thus dependence), hence increasing hardship, not decreasing it; and (3) those in need often receive services at the expense of the working population (all of which were sentiments cited during the COS era). A 1995 article in *Time Magazine* entitled "100 Days of Attitude" captured this "us versus them" dynamic fostered in the debate on welfare reform in the mid-1990s. In his article, Stacks (1995) described how the country was 'up-in-arms' over public assistance program, and this outrage spread quickly through the country. The House held hearings on the state of public welfare in the country in response to the uproar. One of the most inflammatory speeches heard on the House floor was when John Mica compared public assistance users to alligators, arguing that 'if you treat the alligator like a pet or a child, it will become dependent.'

Such perspectives negate the complexity of economic disadvantage often experienced by vulnerable and marginalized populations, and categorize the poor as a homogenous group that is in some significant way different with regard to character than mainstream working society.

The debate about public welfare also reflected the genderized and racialized nature of welfare contributing to institutionalized gender bias and racism. Whether veiled or overt (such as Reagan's welfare queen), negative bias bestowed upon female public welfare recipients of color negates the disparity in social problems experienced by African American women, including increased incidences of poverty, violence, and untreated child sexual victimization, and their associated psychological and social problems (El-Bassel, Caldeira, Ruglass, & Gilbert, 2009; Martin, 2012; Siegel & Williams, 2003).

Although welfare reform was initiated by a Republican Congress, it was passed by the Democratic Clinton administration, in the form of the Personal Responsibility and Work Opportunity Act (PRWORA) of 1996, illustrating wide support not only for welfare reform but also for the underlying philosophical beliefs about the causes of poverty and effective poverty alleviation methods. PRWORA of 1996 reflected a marked shift away from its predecessor, the Aid to Families with Dependent Children (AFDC), an entitlement program created under the New Deal. Many social welfare advocates believe that the new program, Temporary Assistance for Needy Families (TANF), is very punitive in nature, with strict time limits for lifetime benefits (ranging between three and five years depending upon the state), stringent work requirements (often regardless of circumstances), and other punitive measures designed to control the behavior of recipients. Supporters of welfare reform and the passage of PRWORA relied on old arguments of Calvinism and social Darwinism, citing the need to control welfare fraud and welfare dependency, among a host of other behaviors exhibited by welfare recipients, such as sexual promiscuity and having children out of wedlock (Hudson & Coukos, 2005).

The Christian Right and Welfare Reform

A powerful voice within the Republican Party that was a big backer of welfare reform is often called the "Christian Right"—a group of individuals, often Evangelical Christians, who espouse conservative "family" values. Conservative Christian organizations, such as the Christian Coalition, the Eagle Forum, and Focus on the Family have wielded considerable influence within the Republican Party beginning in the 1980s, becoming a fringe core of the party in the 1990s (Green, Rozell, & Wilcox, 2003; Guth & Green, 1986; Knuckey, 2005). These groups were instrumental in the call for welfare reform, voicing significant concerns about moral decline in society and citing the need to defend and uphold traditional family values (Reese, 2007; Uluorta, 2008).

Uluorta (2008) points out that far too often "morality within the United States is a highly circumscribed concept that often confines itself to select individual behaviors such as those pertaining to sex and sexuality (e.g., abortion, abstinence), marriage (e.g., gay marriage) and social standing (e.g., welfare reform)" (pp. 253–254). Many within the Christian Right were fervent supporters of welfare reform, and specifically the PRWOA of 1996 because of its focus on behavioral reform, including the promotion of marriage and sexual abstinence (Reese, 2007). As an example of the Christian Right's focus on individualism, morality, and social responsibility, Uluorta (2008) references Evangelical pastor Rick Warren's book *The Purpose Driven Life: What on Earth Am I Here For?*, where Warren states that the only way to find true purpose in one's life is through "individual responsibility, discipline and being 'born-again'" (p. 254). While individual responsibility is certainly a trait worth achieving, it also can be a "code word" for philosophies that scapegoat the poor, and minimize long-standing social inequality.

The ability of the conservative Christian movement to mobilize its members into political action is notable. For instance, Uluorta (2008) points out the political lobbying success of Focus on the Family, a conservative Evangelical Christian organization that broadcasts its messages on over 1,600 radio stations and 16 television stations nationwide, has a frequently used website, and disseminates newsletters and political action alerts via email and physical mail to millions of members who are often asked to strongly advocate for the organization's policy positions reflecting its socially conservative values (focusing primarily on the support of traditionally "moral" behavior). This level and type of mobilization is of great concern to many within the human services fields and others who advocate for a more compassionate approach to helping the poor and disadvantaged, and who recognize the wide range of ways to frame social problems (and their causes), rather than focusing solely upon perceived behavioral patterns of those who are struggling. The rhetoric of the Christian Right and other socially conservative groups often frame their arguments in terms of tradition, yet their version of "American tradition," and patriotism often reflects the experiences of the majority population, the majority of whom have had the cumulative benefit of white privilege (Martin, 2012).

Box 2-1

White privilege is a social phenomenon where Caucasian members of society enjoy a distinct advantage over members of other ethnic groups. White privilege is defined as "unearned advantages of being white in a racially stratified society" and an expression of institutionalized power (Pinterits, Poteat, & Spanierman, 2009, p. 417). It is something that most Caucasians do not acknowledge leading many of those who benefit from this advantage to take personal credit for whatever they gain through white privilege (Neville, Worthington, & Spanierman, 2001). Unfortunately, this also means that many Caucasians also blame those from non-Caucasian groups for not being as successful. Yet, due to various forms of racial discrimination, it has typically been the white man who has benefited most from the best that life has to offer—gaining access into the best educational systems (or being the only ones to obtain an education at all), the best jobs, and the best neighborhoods. Even if white privilege were to end, the cumulative benefit of years of advantage would continue well into the future, just as the negative consequences of years of social exclusion will continue to negatively affect diverse groups who have not benefited from white privilege.

The Tea Party Movement

Another conservative social movement, which appears to overlap at least to some extent with the Christian Right, is the American Tea Party Movement, a social movement and a part of the Republican base that advocates for smaller government, lower taxes (the name of the group is a reference to the Boston Tea Party), state rights, and the literal interpretation of the U.S. Constitution. The Tea Party movement has quickly gained a reputation for advocating on behalf of very conservative policies, similar in many ways to the Christian Right. Michele Bachmann, a 2012 presidential candidate, was criticized for her alleged position on gay and lesbian rights, in relation to her and her husband's Christian counseling practice, which, according to a former gay client, claimed endorsed a "pray the gay away" approach to counseling homosexual clients (Bachmann denies this) (Ross, Schwartz, Mosk & Chuchman, 2011). There have also been allegations made against some members of the Tea Party movement for their stance on immigration and racial issues in general, which appear to be based upon negative racial stereotypes. The media has consistently highlighted the racially charged tone at some Tea Party political rallies, pointing out racial slurs on posters, many of which are directed at President Obama's ethnic background, although proponents of the Tea Party complain that the media is exaggerating racist elements at the protests and rallies by seeking out and over-focusing on the more extremist elements of the movement. Although "tea partiers" often deny racist or homophobic values, a recent study showed that about 60 percent of tea party opponents believed that the movement had strong racist and homophobic overtones (Gardner & Thompson, 2010). Currently the Tea Party is considered a part of the Republican base, but it appears to be creating some controversy within the party, particularly among the more moderate base. Whether the Tea Party remains a part of the Republican Party or braches off to its own party will depend upon many factors, including whether it can maintain its current momentum and increase the number of supporters.

> ### Human Services Delivery Systems
>
> Understanding and Mastery of Human Services Delivery Systems Political and ideological aspects of human services
>
> Critical Thinking Question: Issues such as abortion, same-sex marriage, and immigration are hotly contested in the United States today. What factors should human service professionals take into account in deciding whether, and how, to advocate for policies congruent with their political and/or religious beliefs?

A Time for Change: The Election of the First African American President

The 2008 presidential election was unprecedented in many respects. The United States had its first African American presidential candidate and its first female presidential candidate of a major party. Many people who have historically been relatively apathetic about politics were suddenly passionate about this election for a variety of reasons. Growing discontentment with the leadership of the preceding eight years coupled with a lengthy war in the Persian Gulf region and a struggling economy created a climate where significant social change could take root. Barack Obama's campaign slogans based upon hope and change (e.g., "Yes We Can!" and "Change We Can Believe In") seemed to capture this growing discontent. Many human service professionals and other advocates

> Some balance must be achieved between free market forces, which can stimulate the economy by creating a spirit of competition, ... [and] support for a strong nation-state that provides a safety net for all of its constituents.

hope that the election of President Obama has signaled a move toward a more compassionate treatment of social problems, including poverty.

The fledgling economy of 2007 evolved into an economic meltdown toward the end of the Bush presidency and extended into the Obama administration, evidenced by a plummeting stock market, the near-collapse of the banking industry, and the real estate bubble at a level not experienced since the Great Depression (Geithner, 2009). Whereas some economists have argued that this economic crisis is a result of the combination of many forces and trends, including globalization, technological innovation, and a changing workforce Kevin Doogan (2009), Jean Monnet Professor of European Policy Studies at the University of Bristol and author of *New Capitalism? The Transformation of Work,* discusses the effect of a primarily unregulated free market and unrestrained CEO compensation, resulting in federal financial bailouts of banks and a changing corporate structure. Doogan suggests that much of this most recent financial crisis is a result of manufactured insecurities and unrestrained exposure to market forces. Doogan's critique of neoliberal policies might seem "anticapitalist" to some political conservative advocates, but upon further analysis it seems clear that some balance must be achieved between free market forces, which can stimulate the economy by creating a spirit of competition, thus theoretically keeping prices low and quality high, with support for a strong nation-state that provides a safety net for all of its constituents.

President Obama and the 111th Congress responded to the economic crisis with several policy and legislative actions, including the passage of the American Recovery and Reinvestment Act of 2009 (often referred to as the Stimulus bill [Pub. L. No. 111-5]). This economic stimulus package, worth over $787 billion, includes a combination of federal tax cuts, various social welfare provisions, and increases in domestic spending, designed to stimulate the economy and assist Americans who were suffering economically. It will be some time before economists and the American public comes to a consensus on whether the stimulus package was successful in turning the economy around.

In the meantime, the lead-up to the 2012 presidential elections revealed the same debate about the causes of poverty and effective poverty alleviation strategies. After a brief display of compassion toward the poor at the height of the 2008 economic crisis, harsh sentiments reflecting historic stigmatization of the poor were strongly espoused, particularly among potential Republican primary candidates who continued their campaign against "big government," social welfare programs, and civil liberties in general. One Republican presidential hopeful, Newt Gingrich, even went so far as to challenge current child labor laws, calling them "stupid" (see Chapter 4 for a discussion on the history of child labor in the United States where African American and immigrant children faced life-threatening conditions working 12-plus hours a day in dangerous factories). In a campaign speech in Iowa in the fall of 2011, Gingrich scapegoated the poor, playing on negative racial stereotypes by characterizing poor ethnically diverse children living in poor neighborhoods as lazy and having no work ethic. In two different speeches (his initial speech and a subsequent speech where he was asked to clarify his earlier comments), Gingrich suggested that poor children in poor neighborhoods could start work early, perhaps as janitorial staff in their own schools. Characterizing most poor children in economically challenged neighborhoods Gingrich stated that these children have:

> No habits of working and nobody around them who works . . . they have no habit of showing up on Monday and staying all day or the concept of "I do this and you give me cash," unless it's illegal.

In his follow-up statements, he clarified his earlier comments by stating:

> You have a very poor neighborhood. You have students that are required to go to school. They have no money, no habit of work . . . What if you paid them in the afternoon to work in the clerical office or as the assistant librarian? And let me get into the janitor thing. What if they became assistant janitors, and their job was to mop the floor and clean the bathroom?

Framing his comments in religious terms, Gingrich concluded by stating:

> If we are all endowed by our creator with the right to pursue happiness, that has to apply to the poorest neighborhoods in the poorest counties, and I am prepared to find something that works, that breaks us out of the cycles we have now to find a way for poor children to work and earn honest money. (Dover, 2011, para 3–5)

Gingrich's comments clearly reflect the same rhetoric rooted in Calvinism and the Protestant Ethic theology, social Darwinism, and even Scientific Charity practiced through the centuries—the very philosophies that Jane Addams and other social reformers worked so hard to challenge. Such sentiments presume a level playing field in society, negating current and historic social forces, such as racial oppression and white privilege that have consistently given one group an unfair advantage for centuries.

Professional History

Understanding and Mastery of Professional History: Exposure to a spectrum of political ideologies

Critical Thinking Question: The relatively liberal ideology of President Obama and his supporters contrasts sharply with the conservatism of the Christian Right and the Tea Party movement. How have these competing views influenced the development of policies related to social welfare?

Concluding Thoughts on the History of Social welfare policy

As often happens in broad-based economic downturns, this most recent recession seems to have led to a softening of antipoverty rhetoric and a political discourse that recognizes the importance of an effective social welfare system for all Americans. Discussions of the need for universal healthcare, a federal living wage, and other policies designed to address the dire financial situation that so many in the United States found themselves facing, including employment lay-offs and home foreclosures at a rate not seen since the Great Depression, will be ongoing. The debate regarding how capitalism and a free market economy can be balanced with a social safety net for all members of U.S. society continues to rage among politicians and the public alike and will no doubt continue to continue for many years into the future. Unfortunately this sentiment appears to be short-lived, as anti-poor rhetoric and calls for cut backs in social welfare programs, including Medicare and Social Security, are increasing, alongside modest improvements in the U.S. economy. Only time will tell where U.S. society ultimately will fall in the philosophical spectrum of individual responsibility and social equity.

Social movements appear to be on the rise, with passionate supporters of both liberal causes, such as the "Marriage Equality," and more conservative social movements, such as "Pro-Life" and anti-immigration movements dominating political rhetoric, and leading to increased polarization within U.S. society. Human service professionals can positively engage in a variety of social movements by advocating for social equality in productive ways that does not necessarily contribute to existing polarization. Many of these advocacy techniques will be discussed in subsequent chapters focusing on specific social problems.

The following questions will test your knowledge of the content found within this chapter.

1. The feudal system was:
   a. England's primary manner of caring for the poor prior to the Middle Ages
   b. a system of care based upon feuds between rival communities where prevailing villages were compelled to provide care for those communities they conquered
   c. an elistist system where privileged and wealthy land-owners would parcel off small sections of their land, which would then be farmed by peasants or "serfs"
   d. Both A and C

2. The theory that competition over resources was necessary in life in order to weed out those who were ill-equipped to manage life's challenges and complexities is called
   a. predestination
   b. evolutionary Darwinism
   c. social Darwinism
   d. None of the above

3. The English Poor Laws of 1601 established three driving principles as the foundation for social legislation, including the belief that:
   a. the primary responsibility for provision lay with one's family
   b. poor relief should be handled at the local level
   c. no individual should be allowed to move to a new community if he or she was unable to provide for themselves financially
   d. All of the above

4. In what way was the settlement house movement different from the traditional charity organizations?

   a. Its goal was to no longer distinguish between the worthy and unworthy
   b. It provided only counseling rather than focusing on comprehensive care
   c. It focused on providing services to adults only rather than providing services to the entire family
   d. It worked diligently to prohibit immigrants from receiving the same benefits as U.S. citizens

5. Modjeska Simkins, Ida Wells, and Jane Hunter are examples of:
   a. former settlement house residents who went on to influence social policy by engaging in advocacy efforts on a policy level
   b. leaders in the American suffrage movement that gave the women the right to vote
   c. African American social workers who developed social programs for black communities since most social welfare programs often excluded African Americans
   d. Both A and B

6. Civil Works Administration (CWA), the Federal Emergency Relief Act (FERA), the Civilian Conservation Core (CCC), and the Social Security Act of 1935 are examples of:
   a. programs created by President Lyndon B. Johnson in response to the Great Depression
   b. programs created by President Roosevelt in response to the Great Depression
   c. programs created by President Hoover in response to the ongoing effects of World War II
   d. programs created in response to legislation passed by Jane Addams and Dorthea Dix

7. Describe the concept of the "Myth of the Welfare Queen," including the roots of this myth, and the short- and long-term effect of its existence.

8. Compare and contrast AFDC and TANF. What are the pros and cons of 1996 welfare reform?

Suggested Readings

Carlton-LaNey, I. B. (2001). *African American leadership: An empowerment tradition in social welfare history.* Washington, DC: NASW Press.

Katz, M. B. (1990). *Undeserving poor.* New York: Pantheon.

Katz, M. B. (1996). *In the shadow of the poorhouse: A social history of welfare in America.* New York: Basic Books.

Linn, J. W., & Scott, A. F. (2000). *Jane Addams: A biography.* Chicago: University of Illinois Press.

Martin, J. M., & Martin, E. P. (1985). *The helping tradition in the black family and community.* Washington, DC: NASW Press.

Reisch, M., & Andrews, J. (2002). *The road not taken: A history of radical social work in the U.S.* Washington, DC: Taylor & Francis.

Internet Resources

Jane Addams Hull House: http://www.hullhouse.org

Jane Addams Hull-House Museum: http://www.uic.edu/jaddams/hull/hull_house.html

The Social Work History Online Timeline: http://www.gnofn.org/~jill/swhistory

References

Beier, A. L. (1974). Vagrants and the social order in Elizabethan England. *The New England Quarterly, 43*(1), 59–78.

Birtles, S. (1999). Common land, poor relief and enclosure. *Past & Present [Great Britain],* (165), 74–106.

Chunn, D. E., & Gavigan, S. A. M. (2004). Welfare saw, welfare fraud, and the moral regulation of the 'Never Deserving' Poor. *Social & Legal Studies, 13*(2), 219–243.

Darwin, C. (2009). *The origin of species: By means of natural selection, or the preservation of favoured races in the struggle for life.* Boston: Cambridge University Press. (Original work published 1859.)

Davidson, J.W. (2008). They Say, Ida B. Wells and the Reconstruction of Race. New York: Oxford University Press.

Doogan, K. (2009). *New capitalism? The transformation of work.* Cambridge: Polity Press.

Dover, E. (2011, December 1). Gingrich says poor children have no work habits. ABC News. Retrieved December 23, 2011, from http://abcnews.go.com/blogs/politics/2011/12/gingrich-says-poor-children-have-no-work-ethic/

Duncan, C. M., & Moore, D. B. (2003). Catholic and Protestant social discourse and the American Welfare State. *Journal of Poverty, 7*(3), 57–83.

El-Bassel, N., Caldeira, N. A., Ruglass, L. M., & Gilbert, L. (2009). Addressing the unique needs of African American women in HIV prevention. *American Journal of Public Health, 99*(6), 996–1001.

Feagin, J. R. (1975). Subordinating the poor: Welfare and American beliefs. Englewood Cliffs, NJ: Prentice Hall.

Gardner, A. & Thompson, K. (2010). Tea Party group battles perceptions of racism. Washington Post-ABC News Poll. Retrieved June 12, 2012 from http://www.washingtonpost.com/wp-dyn/content/article/2010/05/04/AR2010050405168.html?hpid=moreheadlines

Geithner, T. F. (2009). Regulatory perspectives on the Obama administration's financial regulatory reform proposals-part two. House Financial Services Committee. Retrieved November 18, 2009, from http://www.house.gov/apps/list/hearing/financials-vcs_dem/geithner_-_treasury.pdf

Gettleman, M. E. (1963). Charity and social classes in the United States, 1874–1900. *American Journal of Economics & Sociology, 22*(2), 313–329.

GLAAD. (2010). GLAAD's Media Reference Guide: In Focus: Marriage. Retrieved November 12, 2011, from http://www.glaad.org/reference/marriage.

Gordon, L. (1991). Black and white visions of welfare: Women's welfare activism, 1890–1945. *Journal of American History, 78*(2), 559–590.

Green, J. C., Rozell, M. J., & Wilcox, C. (2003). The Christian right in American politics: Marching to the millennium. Washington, DC: Georgetown University Press.

Guth, J., & Green, J. C. (1986). Faith and politics: Religion and ideology among political contributors. *American Politics Quarterly, 14*(3), 186–199.

Hausman, D. M., & McPherson, M. S. (2006). *Economic analysis, moral philosophy, and public policy.* New York: Cambridge University Press.

Hofstadter, R. (1992). *Social Darwinism in American thought.* Boston: Beacon Press.

Hudson, K., & Coukos, A. (2005, March). The dark side of the Protestant Ethic: A comparative analysis of welfare reform. *Sociological Theory, 23*(1), 1–24.

Kim, H. C. (1977). The relationship of Protestant Ethic beliefs and values to achievement. *Journal for the Scientific Study of Religion, 16*(3), 252–262.

Kluegal, J. R. (1987). Macro-economic problems, beliefs about the poor and attitudes toward welfare spending. *Social Problems, 34*(1), 82–99.

Knuckey, J. (2005). A new front in the culture war? Moral traditionalism and voting behavior in U.S. House elections. *American Politics Research, 33,* 645–671.

Krugman, P. (2007). *Conscience of a Liberal.* New York: W.W. Norton & Co.

Kusmer, K. (1973). The functions of organized charities in the progressive era: Chicago as a case study. *Journal of American History, 60*(3), 657–678.

Lundblad, K. (1995, September). Jane Addams and social reform: A role model for the 1990s. *Social Work, 40*(5), 661–669.

Martin, M. (2012). Philosophical and religious influences on social welfare policy in the United States: The ongoing effect of Reformed theology and social Darwinism on attitudes toward the poor and social welfare policy and practice. *Journal of Social Work, 12*(1), 51–64. doi:10.1177/1468017310380088

Martin, M. E., (2012 January) Philosophical and religious influences on social welfare policy in the United States: The ongoing effect of Reformed theology and social Darwinism on attitudes toward the poor and social welfare policy and practice. *Journal of Social Work, 12,*1, 51–64.

Nelson, J. I. (1992). Social welfare and the market economy. *Social Science Quarterly, 73*(4), 815–828.

Neville, H., Worthington, R., Spanierman, L. (2001). Race, Power, and Multicultural Counseling Psychology: Understanding White Privilege and Color Blind Racial Attitudes. In Ponterotto, J., Casas, M, Suzuki, L, and Alexander, C.(Eds) *Handbook of Multicultural Counseling, Thousand Oaks,* CA: SAGE.

Pinterits, E.J., Poteat, V.P., & Spanierman, L.B. (2009). The White Privilege Attitude Scale: Development and initial validation. *Journal of Counseling Psychology,* 56, 3, 417–429.

Reese, E. (2007). The causes and consequences of U.S. welfare retrenchment. *Journal of Poverty,* 11(3), 47–63.

Ross, B., Schwartz, R., Most, M. Chuchman, M. (2011, May). *Michele Bachmann Clinic: Where you can pray away the gay? ABC News The Blotter.* Retrieved May, 2, 2012, from http://abcnews.go.com/Blotter/michele-bachmann-exclusive-pray-gay-candidates-clinic/story?id=14048691#.UIWryGnuW18

Schlabach, T. (1969). *Rationality & welfare: Public discussion of poverty and social insurance in the United States 1875–1935.* Social Security Commission, Research Notes and Special Studies. Retrieved September 18, 2005, from http://www.ssa.gov/history/reports/schlabachpreface.html

Schneiderhan, E. (2008, July). *Jane Addams and the rise and fall of pragmatist social provision at Hull-House, 1871–1896.* Paper presented at the annual meeting of the American Sociological Association, Sheraton Boston, and the Boston Marriott Copley Place, Boston.

Schram, S. F., Fordingy, R. C., & Sossz, J. (2008). Neo-liberal poverty governance: Race, place and the punitive turn in U.S. welfare policy. *Cambridge Journal of Regions, Economy and Society, 1,* 17–36.

Siegel, J., & Williams, L. (2003). The relationship between child sexual abuse and female delinquency and crime: A prospective study. *Journal of Research in Crime and Delinquency,* 40(1), 71–94.

Stacks, J. (1995, April 10). 100 days of attitude. *Time Magazine.* Retrieved online December 23, 2011, from: http://www.time.com/time/magazine/article/0,9171,982782,00.html.

Stephenson, C. (1943). Feudalism and its antecedents in England. *The American Historical Review, 48*(2), 245–265.

Trattner, W. (1998). *From poor law to welfare state* (6th ed.). New York: Free Press.

Tropman, J. E. (1986). The "Catholic ethic" versus the "Protestant ethic": Catholic social service and the welfare state. *Social Thought,* 12(1), 13–22.

Van Slyke, D. M. (2003). The mythology of privatisation in contracting for social services. *Public Administration Review, 63*(3), 296–315.

Uluorta, H. M. (2008). Welcome to the "All-American" fun house: Hailing the disciplinary neo-liberal non-subject. *Millennium: Journal of International Studies, 36*(2), 51–75.

Weaver, R. K., Shapiro, R. Y., & Jacobs, L. R. (1995). The polls–trends: Welfare. *Public Opinion Quarterly, 59,* 606–627.

Weber, M. (1958). *The Protestant ethic and the spirit of capitalism* (T. Parsons, Trans.). New York: Charles Scribner's Sons. (Original work published 1905.)

Weikart, R. (1998). Laissez-faire social Darwinism and individualist competition in Darwin and Huxley. *The European Legacy, 3*(1), 17–30.

Wright, T. (2000). Resisting homelessness: Global, national and local solutions. Contemporary Sociology, 29(10), 27–43.

Zucchino, D. (1999). *The myth of the welfare queen: a Pulitzer-prize winning journalist's portrait of women on the line.* New York: Touchstone.

CHAPTER 3

Professional Ethics and Values in Human Services

Tom Herzberg/Images.com

Ethics can be defined in many different ways, with most definitions including references to a set of guiding principles or moral values. In a professional context, *ethics* often refers to a set of standards that provide guidance to individuals within a particular discipline with the goal of assisting them in resolving ethical dilemmas they are likely to face. Regardless of how *ethics* per se is defined, ethical standards, within virtually all contexts, are by definition based on a foundational value system designed to tell us what good behavior is and what bad behavior is. Or, another more basic way of putting it is that ethical standards and principles tell us what we *ought* to do in any given situation.

Now you might be asking yourself—I'm a good person, so why do I need a detailed set of ethical values to tell me what to do? Don't good people behave "good" naturally? The answer may surprise you! Although it may be true that very few people wake up in the morning and say to themselves, "Hey! I think I'm going to lie, cheat, and steal today!," it is true that many people become hysterical or enraged, or are biased, selfish, naïve, or ignorant, and in the process of being so very human, they may very well behave quite unethically as they make decisions based on their urges, desires, passions, personal biases, negative stereotypes, or uninformed opinions.

Ethical values and principles are a very necessary part of life, both personally and professionally, and although some may argue that their personal ethical values are not necessarily tied to their professional ethics, a strong argument can be made that they are in fact very much a reflection of one another. Most of you probably remember former president Bill Clinton's impeachment hearings, which centered on his perjury in a sexual harassment claim filed against him, as well as his inappropriate relationship with a young White House intern. Many of his supporters argued fervently that what he did in his personal life had no bearing on his ability to be a good president. Yet others argued that poor character demonstrated in one's personal life will most definitely play out in one's professional life, and one cannot draw a line between these two domains.

Ethical standards and principles tell us what we ought to do in any given situation.

Moral, But by Whose Standards?

It would be very convenient if there were one long list of rules and all situations could be perceived in the same manner by everyone. But of course that is not how life works. Most people will argue that there are universal moral principles, particularly relating to issues such as murder, robbery, child abuse, and sexual assault. But even with these seemingly black-and-white moral issues, the gray seems to abound. Such is the case when someone kills in self-defense, or someone steals bread for a starving child. So, is morality absolute or relative? What I mean by this is, is there an absolute right and wrong in this world? Or, is the rightness and wrongness of a decision or action dependent on perspective, culture, or one's own truth? This is an age-old question and not one that I will answer definitively here. In fact, many moral theorists deal with this very issue, and although some argue for either extreme position, most will argue that both are true—there are ultimate moral principles that are universal (e.g., sexually abusing a child is always wrong) and there are many occasions when one must consider the appropriateness of a certain behavior within the context of one's culture (e.g., burping in public).

I want to address some of the issues that have the greatest potential of "muddying" the waters a bit when it comes to determining how we know whether an action is moral or immoral, which in turn will help us determine how we can ensure we're making moral decisions. We will then apply what we've learned to the professional arena, specifically the human services profession.

Ethical Values versus Emotional Desires: "I Know It Was Wrong, But We Were in Love!"

Other than the most rigid people, most people will find themselves caught in a tug-of-war between their ethical standards and their emotional desires, or feelings, with the latter often leading to breaking down of moral behavior at some point in their lives. I have a counseling practice, and I often tell my clients that feelings and emotions are like the interior design of a house—moving and poignant, even beautiful at times—but truly useful only if protected by the exterior and structure of the house—the walls and roof, which are the framework, like our ethical standards, values, and principles. Thus, although human beings are certainly emotional, individuals with high character are not driven to act solely on the basis of their desires and passions.

In fact, individuals who are motivated primarily by emotions are often emotionally unstable, not because their emotions are wrong, but because their values and principles are not well-enough defined and/or developed to contain or regulate their emotions, oftentimes leading to the inability to control their impulses. For instance, an employee might become angry with the boss and feel like striking the boss, but doesn't because the employee values nonviolence. A person's ethical values should then be the rudder of behavior, and although there are certainly times when people will be driven by passion, or will need to follow their hunches, emotions and desires serve people best when they aren't chief in the decision-making process.

Another reason why it is important to understand the relationship between our ethical values and our emotions is because we often use our emotions to justify our unethical behavior. Cheating on a test is wrong, unless the test is too hard and we hate our teacher; adultery is wrong, unless we're in a loveless marriage, are extremely lonely, and fall hopelessly in love with someone else; lying is wrong, unless we need the day off and will get paid only if we say we're sick, even though we're not; violence is wrong, unless we're provoked; and drinking too much alcohol is wrong, unless we've gone two weeks without and just had a very bad day. Thus, one of the primary functions of ethical values is to keep us on a good moral track, particularly when we find our ethical values at odds with our emotional desires and urges. Certainly there are times when emotions should lead, and we certainly do not want to become heartless in our application of rules. When someone is driven to act solely on the basis of their values or rules, they are often deemed rigid legalists. But when someone behaves in a manner that is solely driven by their feelings and desires, they are often deemed immature, volatile, and impulsive.

When Our Values Collide: "I Value Honesty, But What if Lives Are at Stake?"

Ethical behavior is not just made difficult because of competing emotions and desires, but oftentimes we find ourselves in situations where our values are competing with one another. We value family dinners with our kids, but what if that conflicts with our value of their extracurricular involvement? We value our friendships, but what if they are interfering in our marriage that we also value? Many times people act in a way that is later perceived to be unethical, when at the time they were committing the act they may have believed that they were acting in a very ethical manner, but were forced to choose among competing values. Employees who shred documents to protect their employers may very well believe they were acting ethically based on their ethical value of loyalty to their employer. Yet, they may later be charged with obstruction of justice because someone else perceived their behavior to be immoral. Perhaps in retrospect these employees will realize that their values were misguided, or they may forever believe as though they were behaving morally and the government was not.

In 1945 when Corrie ten Boom was hiding Jews in her attic, she chose to lie to the Nazi officers who came to her door questioning her, even though she believed lying to be wrong (ten Boom, Sherrill, & Sherrill, 1974). Corrie was put in a position where she had to choose the higher value. What did she value more? Complete honesty at all costs? Obedience to authority? Personal safety? Or interceding in matters of inhumane cruelty and injustice at all costs? In light of what we now know about Nazi Germany and the Holocaust, Corrie and her family are lauded as heroes, behaving in the highest moral fashion, refusing to stand by and do nothing as an evil government slaughtered millions of innocent people. Yet, does this mean that those who did not hide Jews acted immorally? What if you had the opportunity to interview a family who refused to hide a Jewish neighbor? What if this family told you that Nazis used the practice of dressing as Jews and going door-to-door asking for refuge and that the punishment for harboring a Jew was imprisonment in a concentration camp, and likely death? What if this family explained that they believed they behaved morally because their first responsibility was to protect their children? Would you still consider their behavior immoral? Or, what about the ruling authorities' perspective? Corrie ten Boom and her family broke the law. From the authorities' perspective, then, their behavior was immoral. What makes the ten Boom family's behavior moral now? Our belief that the Nazis were evil? So does this mean that if you or I believe that a particular law, or even our entire government, is evil— that we'd be justified in disobeying its laws? Many protective parents "kidnap" their children because they strongly believe that the family courts will not protect the children from the other abusive parent. If this is true, is their behavior justified? Many African American men believe that if they are pulled over by the police, it is because they are being racially profiled, and they may be arrested for no reason. Does this justify an attempt to flee? Would their behavior be any more or less moral than a slave who escaped prior to the Civil War?

I hope you are beginning to see that evaluating ethical behavior in retrospect, when we have the benefit of perspective and outcome, is a far easier task than determining what is truly ethical in the moment. And the lens that we use to evaluate the moral rightness of a behavior is often determined by the outcome—something that the person involved doesn't have the benefit of knowing or any control over, in many circumstances, when making decisions. This explains why some people who are initially perceived as highly immoral are later considered heroes, and some people who authentically believe they are behaving morally, end up in prison.

The Development of Moral Reasoning

Before developing a set of ethical values, it is important to understand the nature of moral development, and there is no shortage of theorists who have attempted to do just that. Obviously what people base their values on can vary dramatically. Value systems can be based on the values of one's family of origin, on one's culture, or on one's religious beliefs. Lawrence Kohlberg (Gibbs, 2003) believed that the capacity to reason morally developed along with cognitive development. Kohlberg conducted interviews with people of all ages and discovered that children (or immature adults) cited something as being immoral because they would get into trouble, thus relying on external references of right and wrong, whereas more mature adults could understand and grasp the various shades of gray

Human Systems

Understanding and Mastery of Human Systems: Emphasis on context and the role of diversity in determining and meeting human needs

Critical Thinking Question: We all face situations in which two or more of our values are at odds with each other. What factors should we take into account as we determine how to behave in such situations?

> **Evaluating ethical behavior in retrospect, when we have the benefit of perspective and outcome, is a far easier task than determining what is truly ethical in the moment.**

Human Systems

Understanding and Mastery of Human Systems: Theories of human development

Critical Thinking Question: According to Kohlberg's theory of moral development, the capacity for moral reasoning is connected with cognitive development and with the ability to think abstractly. How might this theory be used to inform the treatment of juveniles who break the law?

involved in a moral dilemma and cited the moral nature of a situation relying on internal references. Kohlberg theorized that the type of moral reasoning that adults use to evaluate the moral dilemmas in their lives is dependent on abstract reasoning ability, a cognitive function that children lack. And although the capacity for moral reasoning does not necessarily mean that someone will behave morally, it is important to consider someone's cognitive ability to apply moral reasoning before judging them.

Developing a Professional Code of Ethics

It is because of the difficult nature of determining what constitutes moral behavior—including the balancing of our ethical values and emotional urges, of knowing which competing values to choose in any given situation, of having the benefit of perspective when making moral decisions—that many professions have elected to develop foundational ethical standards and professional values to safeguard from emotion, bias, and misguided commitments being the primary motivators in ethical decision making. Many professions begin with some stated set of values or underlying guiding assumptions, oftentimes found reflected in their mission statement, and sometimes ethical standards are developed from some form of abuse. The ethical standard prohibiting human service professionals from dating a client was likely developed in response to some human service professionals' use of poor judgment in dating clients who later filed complaints because they felt exploited or abused.

Regardless of how standards are developed, virtually all professions rely on some form of ethical standards to maintain integrity and trust within their profession. Numerous professions espouse basic ethical principles, which serve as a foundation for their business practices and standards, but in addition to such values of choice, an increasing number of professions are bound by legally enforced ethical standards, which if violated can result in quite punitive consequences, ranging from professional or financial sanctions (such as license suspension or fines), to a wide range of criminal penalties (including incarceration).

Virtually every professional group operates under a professional organization or licensing entity that enforces ethical codes in some form. Attorneys operate within certain legal ethical standards administrated by the American Bar Association. Psychologists must abide by particular professional standards that are set forth by the American Psychological Association (APA). Even stockbrokers must not only abide by the ethical standards and values of their companies, which may include putting the client's needs first and not overcharging for services, but they must also abide by the legally binding ethical standards set forth by the Securities and Exchange Commission (SEC), which if violated can include both professional and financial sanctions, or in extreme cases, even a criminal indictment.

Resolving Ethical Dilemmas

It is very important that any professional code of ethics be considered an ever-growing and changing entity, never in final form, and always open for evaluation and debate.

West (2002) discussed the importance of "ethical mindfulness," citing several real-world examples of questionable ethical practices in the counseling and human services field, including issues related to informed consent (informing clients of their rights and making sure they know all that is involved in engaging in the counseling process), the use of real clients in therapist educational videotapes, and other ethical issues appropriate for discussion and evaluation.

But even if everyone agrees that having ethical standards is a good thing, and that constant evaluation is necessary, the next challenge is to determine how to respond when an ethical breach may have occurred. Kitchener's (1984) model of ethical decision making was designed to guide professionals in navigating the sometimes-murky waters of decision making in difficult situations. Kitchener's model is based on four assumptions that he maintains need to be at the heart of any ethical evaluation and can, in a sense, be used as a "litmus test" when attempting to determine whether a certain act was in fact unethical. These assumptions include: (1) autonomy, (2) beneficence, (3) nonmaleficence, and (4) justice.

In Kitchener's model, when a certain act is being evaluated to determine its ethical nature, the model would have the evaluators ask whether the professional acted with free will (autonomy); whether the professional's actions were intended to benefit the client (beneficence); whether the professional's actions involved evil, illegal, or harmful intentions (nonmaleficence); and whether these acts were carried out in a manner that respected the rights and dignity of all involved parties (justice).

Let us use Corrie ten Boom's actions as an example. The ruling government certainly considered her behavior unethical, and although we have the benefit of perspective and outcome in evaluating her behavior, as I mentioned before, rarely does one have this luxury when in the midst of a moral dilemma. If one were to use Kitchener's model in determining the ethical nature of ten Boom's behavior, it could be argued that she was not acting in a manner that was based on her free will (i.e., would she normally oppose government officials?), because although she was acting in autonomy, the Nazi regime forced her to hide her activities. Her behavior was beneficent in the sense that it involved acts of kindness toward her fellow human beings, she refused to do harm by standing by and allowing atrocities against her Jewish neighbors and friends, and she was motivated by her hatred for injustice. Thus ten Boom's behavior should be considered ethical regardless of the fact that history deems the Nazi Party an evil regime.

Cultural Influences on the Perception of Ethical Behavior

Cultural context is another very important variable to consider when evaluating the rightness or wrongness of behavior. Garcia, Cartwright, Winston, and Borzuchowska (2003) discussed a model of ethical decision making that stresses the importance of being culturally sensitive when evaluating any ethical decision-making process. Garcia et al. challenged the notion that all cultures value autonomy equally, arguing that many cultures operate on a very interdependent basis. They also cautioned that what one culture considers abnormal, another culture considers

Human Systems

Understanding and Mastery of Human Systems: Emphasis on context and the role of diversity in determining and meeting human needs

Critical Thinking Question: Ethical perspectives are subjective, varying across cultures, socioeconomic classes, and generations. How can an understanding of this fact help a human service professional to better serve her clients?

perfectly normal. But regardless of how one goes about determining what is ethical and how ethical decisions are made (or how unethical decisions are made), it is very important to remember to be sensitive to differences between *cultures, genders,* and *ages* (across the generations).

Again, it is also very important to remember that oftentimes what appears blatantly unethical in retrospect may have seemed quite ethical, or at the very least somewhat muddy, in the midst. Thus, taking the time to truly understand the behavior from the professional's perspective, keeping issues related to enculturation in mind, is absolutely imperative and undoubtedly very challenging.

Ethical Standards in Human Services

The ethical standards that govern the human services profession depend on many variables, including human service professionals' level of education, professional license, and even the state in which they practice. With the increasing popularity of the human services discipline, the National Organization for Human Services (NOHS) was founded in 1975. The NOHS website states that its purpose is to "unite educators, students, practitioners, and clients" within the field of human services. Although it has no enforcement powers, its members not only agree to abide by a code of ethics that is very similar to the one put forth by the NASW, but include a focus on the ethical standards as they apply to educators as well (NOHS, 1999). According to the NOHS its Ethical Standards of Human Service Professionals can be used as guidelines for human service professionals and educators in resolving ethical dilemmas they face both with clients and within the community-at-large.

The preamble of the NOHS ethical standards explain that the purpose of the standards is to provide human service professionals and educators with guidelines to help them manage ethical dilemmas effectively. Thus, while these standards are not legally binding they were established as guiding principles for ethical human service practice.

The guidelines are broken down into two sections, with section one focusing on standards for human service professionals, and section two focusing on standards for human service educators. In the section on human service professionals, the standards are broken down into categories pertaining to responsibilities to clients, responsibilities to community and society, responsibilities to colleagues, responsibilities to the profession, and responsibilities to employers.

The ethical standards for human service eductors include a reference to being accountable to other related professional disciplines, such as the "American Association of University Professors (AAUP), American Counseling Association (ACA), Academy of Criminal Justice (ACJS), American Psychological Association (APA), American Sociological Association (ASA), National Association of Social Workers

(NASW), National Board of Certified Counselors (NBCC), National Education Association (NEA), and the National Organization for Human Services (NOHS)" (NOHS, 2009).

Overall, the general theme of all of these ethical standards centers on respect for the dignity of others, doing no harm, honoring the integrity of others, and recognizing power differentials and avoiding exploitation of others, particularly clients and students. This is accomplished through maintaining self-awareness, engaging in all aspects of one's professional and personal life honestly and ethically, and by developing an awareness of past and current global dynamics, particularly those involving the marginalization and oppression of others. In fact, the NOHS human service ethical standards can be summed up in large part with Statement 28: "Human service professionals act with integrity, honesty, genuineness, and objectivity" (NOHS, 2009, para 31).

> The **NOHS** human service ethical standards can be summed up in large part with Statement 28: "Human service professionals act with integrity, honesty, genuineness, and objectivity" (NOHS, 2009, para 31).

Ethical principles are an integral part of everyday life, enabling us to conduct business, both personal and professional, in a respectful and safe manner, striving to respect the dignity of all persons, regardless of age, gender, race, and socioeconomic status (SES). Without ethical guidelines to help us navigate through various situations, we're all at risk for allowing emotions to rule, leaving each person open to the influence of personal biases. Ethical principles in the human services profession are foundational to the continued development of a helping profession that strives to objectively, professionally, and compassionately meet the complex needs of the most vulnerable members of our society, and without such guidelines, we are at risk of exposing clients to potential revictimization.

Concluding Thoughts on Professional Ethical Standards

I began this chapter by discussing how many professional fields have adopted ethical codes of conduct. Virtually all the helping professions have such ethical codes mandating how practitioners must conduct themselves professionally. There are significant similarities among the various counseling professional organizations, such as the NOHS, the APA, the ACA, and the NASW, but a review of each discipline's ethical standards reveals how the disciplines focusing on the human services (NOHS and NASW) tend to focus as much on macro responsibilities (communities and the broader society) as much as on the individual client. For instance, neither counselors, family therapists, clinical psychologists, nor licensed social workers can have a romantic relationship with clients, but one significant difference that sets the human services and social work fields apart from the other helping professions is the added responsibility to advocate for social justice—both on behalf of clients and on behavior of society as a whole, whereas the APA (2002), for instance, refers to justice in individual terms as it relates to every individual's right to benefit from the contributions of psychology.

> The focus on social justice in a broader context is important because it highlights the macro focus of the human services field, with the recognition that society and its social structures play a significant role in the relative mental and physical health of its members.

Human Services Delivery Systems

Understanding and Mastery of Human Services Delivery Systems: Political and ideological aspects of human services

Critical Thinking Question: The macro-level, social justice orientation of the human services profession sets it apart from other helping professions such as psychology. How might a human service professional's personal values support his work for social justice? In what ways might such personal values impede his work?

The focus on social justice in a broader context is important because it highlights the macro focus of the human services field, with the recognition that society and its social structures play a significant role in the relative mental and physical health of its members.

Human service professionals may have a greater likelihood of confronting complex ethical dilemmas than professionals working in other helping professions due to the broad range of human service practice settings and the broad range of clients with whom they work (many of whom may have quite complex individual and social problems). Thus it is imperative that anyone considering a career in human services become familiar with relevant laws and professional ethical standards of not only their specific discipline (pertaining to their academic degree and their licensing bodies) but of related fields as well.

The following questions will test your knowledge of the content found within this chapter.

1. Most people will find themselves caught in a tug-of-war between their ethical standards and their _____.
   a. religious beliefs
   b. emotional desires
   c. professional ethics
   d. personal experiences

2. Many times individuals act in ways that are later perceived to be unethical, when at the time they were committing the act they may have believed that they were acting in a very ethical manner, but were forced to choose among:
   a. competing values
   b. their emotions and their ethics
   c. being accepted or standing in isolation
   d. their friends and their job

3. According to Kohlberg, it is important to consider someone's _____ ability to apply moral reasoning to their behavior before judging them able to make moral decisions.
   a. cognitive
   b. cultural
   c. structural
   d. social

4. Numerous professions espouse basic ethical principles, which serve as a foundation for their business _____.
   a. negotiations
   b. practices and standards
   c. marketing strategies
   d. intervention and strategies

5. Violations of legally enforced ethical standards can result in
   a. professional sanctions, including license suspensions
   b. financial sanctions
   c. criminal penalties
   d. All of the above

6. Human service professionals face the increased likelihood of confronting ethical dilemmas of greater complexity because
   a. they work in settings lacking a formal set of professional standards
   b. education levels for human service professionals have declined in the past decade
   c. they work in a broad range of practice settings with a broad range of clients
   d. of emotional regulation

7. Describe Kitchener's model of ethical decision making and explore how it can help human service professionals successfully manage common ethical dilemmas.

8. Cite an NOHS ethical standard that you may have difficulty abiding by, and explain why.

Suggested Readings

Dolgoff, R., Lowenberg, F. M., & Harrington, D. (2004). *Ethical decisions for social work practice*. Belmont, CA: Wadsworth Publishing.

Kenyon, P. (1998). *What would you do? An ethical case workbook for human service professionals*. Belmont, CA: Wadsworth Publishing.

Nash, R. J. (1996). *"Real world" ethics: Frameworks for educators and human service professionals*. New York: Teacher's College Press.

Reamer, F. G. (1998). *Ethical standards in social work: A critical review of the NASW code of ethics*. Washington, DC: NASW Press.

Internet Resources

Josephson Institute of Ethics: http://www.josephsoninstitute.org

National Organization for Human Services Ethical Standards: http://www.nationalhumanservices.org/mc/page.do?sitePageId=89927&orgId=nohs

National Association of Social Workers Code of Ethics: http://www.naswdc.org/pubs/code/code.asp

References

American Psychological Association. (2002). *Ethical principles of psychologists and code of conduct.* Washington, DC: Author.

Garcia, J. G., Cartwright, B., Winston, S. M., & Borzuchowska, B. (2003). A transcultural integrative model for ethical decision making in counseling. *Journal of Counseling & Development, 81*(3), 268–277.

Gibbs, J. (2003). *Moral development and reality: Beyond the theories of Kohlberg and Hoffman.* London: Sage Publications Ltd.

Kitchener, K. S. (1984). Intuition, critical evaluation, and ethical principles: The foundation for ethical decisions in counseling psychology. *The Counseling Psychologist, 12,* 43–55.

National Association of Social Workers. (1999). *Code of ethics of the National Association of Social Workers.* Washington, DC: Author.

National Organization for Human Services. (1999). *Ethical standards of human service professionals.* Washington, DC: Author. Available online at: http://www.nationalhumanservices.org/ethical-standards-for-hs-professionals

ten Boom, C., Sherrill, J., & Sherrill, S. (1974). *The hiding place.* New York: Bantam Books.

West, W. (2002). Some ethical dilemmas in counseling and counseling research. *British Journal of Guidance & Counseling, 30*(3), 261–268.

CHAPTER 4

Skills and Intervention Strategies

Michael Newman/PhotoEdit

All professionals use tools to accomplish their job duties. A professional baseball player uses a bat, a ball, and a mitt. An accountant uses a calculator; an airline pilot uses an airplane. What is unique about the human services field is that the professional is the tool. Most people who enter this field do so because they possess some basic inclination toward counseling, advocacy, and caregiving. One might question, then, why someone who is a natural counselor needs a professional education to become a human service professional. Even the most naturally talented counselor needs training and refinement; needs to be taught useful techniques; and will be able to benefit from the results of broad-based research, the knowledge of others concerning professional issues, such as multicultural considerations, and collaboration with other professionals with years of practice experience.

Because the human service professional is the primary tool for intervention, it is very important for all human service professionals to gain insight into their own values and belief systems so that they can better understand how they influence their impressions of the clients they work with and the problems their clients face. Gaining personal insights into one's own life experiences, whether one was raised with privilege, or was a victim of oppression, for instance, will help the human service professional to consistently address and confront any personal biases toward or against certain groups of people, and social problems.

It is important to note though that human service professionals who do not have a license to counsel (such as licensed counselors, social workers, and psychologists) will not engage in counseling per se, and bachelor's level human service education does not train students to "counsel" clients in the traditional and legal sense of the word. Since human service professionals work on a variety of degree levels, I use the words *counsel* and *counseling* in this chapter in a general sense in reference to engaging

Learning Objectives

- Understand the nature and importance of informed consent and confidentiality, including the limits of confidentiality in the counseling relationship
- Recognize skills and competencies within oneself, such as empathy and active listening skills, which are important in human service generalist practice
- Develop an understanding of the concept of psychological boundaries and recognize situations when boundary setting is important
- Develop an understanding of the importance of client empowerment and self-determination germane to human services practice
- Understand the basic elements of a psychosocial assessment and common intervention strategies

in any type of direct practice with clients. This may involve the facilitation of support groups, providing general case management, or discussing a person's problems on the telephone as a crisis hotline worker. Direct practice may also include therapeutic counseling if the human service professional has a license to provide professional counseling services. Since the human services profession includes such a wide range of activities at so many levels (from paraprofessional "helpers" to professional licensed counselors), the term *counseling* within this chapter should be interpreted broadly.

Informed Consent and Confidentiality

Prior to any discussion on counseling, competencies and generalist skills, the important topics of informed consent, confidentiality, and the limits of confidentiality must first be discussed. Informed consent refers to disclosing to clients the nature and risks of the counseling relationship prior to their engaging in these services. According to the National Organization of Human Services (NOHS), human service professionals "negotiate with clients the purpose, goals, and nature of the helping relationship prior to its onset as well as inform clients of the limitations of the proposed relationship." According to the NOHS the client also has the right to terminate the counseling relationship at any time he sees fit (NOHS, 1996).

The NOHS also mandates the client's right to privacy and confidentiality (NOHS, 1996). Other professional organizations' ethical codes address confidentiality as well (National Association of Social Workers [NASW], 2002; American Counseling Association, 2005; American Psychological Association [APA], 2002). Confidentiality is an important aspect of the counselor–client relationship, where clients are assured that whatever they share with their counselor will not be shared with others. This commitment to keep whatever clients share within the counseling relationship private is not merely a clinical issue practiced by most in the mental health field—it is considered so vital to mental health treatment that confidentiality is a legal mandate in every state in the nation. Thus, any professional offering counseling services must by law maintain confidentiality or face losing their professional license or other sanction.

The importance of confidentiality is based on the belief that for trust to develop in the counseling relationship, clients must be assured that they have a safe place to discuss their most private thoughts, fears, and experiences. Without such a guarantee, clients might not be willing to discuss their fears that they are not good parents, their intermittent desire to abandon their families because they are so overwhelmed in life, or their histories of child sexual abuse. Knowing that they have a safe place to talk about their most private thoughts with someone who is not personally affected by their feelings, experiences, or choices makes this exploration possible for thousands of individuals, enabling them to become better parents, less overwhelmed in their lives, and learn how to turn childhood victimization into a survivor mentality.

> The importance of confidentiality is based on the belief that for trust to develop in the counseling relationship, clients must be assured that they have a safe place to discuss their most private thoughts, fears, and experiences.

The Limits of Confidentiality

There are *limits to confidentiality,* though, designed to ensure the safety of the client and the general public. Although there are no national standards for the limits of confidentiality in mental health services, each state has laws that establish exceptions of confidentiality related to both voluntary and involuntary disclosures. These laws determine how and when client information can be disclosed to other treatment providers, insurance companies, and caregivers, and typically require that the client sign an *authorization to release information,* a legal document that provides all relevant information about what information will be released and for what purpose. Statements 3 and 4 of the NOHS ethical standards stipulate the limits of confidentiality, as shown here:

STATEMENT 3: Human service professionals protect the client's right to privacy and confidentiality except when such confidentiality would cause harm to the client or others, when agency guidelines state otherwise, or under other stated conditions (e.g., local, state, or federal laws). Professionals inform clients of the limits of confidentiality prior to the onset of the helping relationship.

STATEMENT 4: If it is suspected that danger or harm may occur to the client or to others as a result of a client's behavior, the human service professional acts in an appropriate and professional manner to protect the safety of those individuals. This may involve seeking consultation, supervision, and/or breaking the confidentiality of the relationship. (NOHS, 1996)

In general, the limits of confidentiality involve the counselor's duty-to-warn and duty-to-protect and relate to situations where through direct disclosure clients share that they are a threat to themselves (suicidal) or others (homicidal). For instance, if a client shares with a human service professional that he plans on leaving the office and committing suicide, the practitioner has the legal obligation to disclose this information to the client's family or even the police to ensure the client's safety. If a minor child client discloses during the counseling session that she is being sexually abused by her uncle, the practitioner is legally obligated to report this information to child protective services to ensure the child's safety.

Disclosures are not always so clear-cut or direct, though, and there are many occasions where human service professionals find themselves needing to use their clinical skills to determine whether violating confidentiality is the appropriate course of action. For instance, consider the client who *may* be suicidal and who discloses a level of despair that *may* indicate suicidal ideation. Couple this with a disclosure that the client attempted suicide four months before and told no one; that he uses alcohol to "make the pain go away"; and that although he won't admit to a suicide plan, he doesn't always feel safe. A client who shares this type of disclosure—denying any outright plan to commit suicide, but appearing to manifest many signs of suicidal behavior—may very well be at real risk for committing suicide, but might be resistant to sharing this clearly either because of confusion about how he feels (wants to end life one moment and wants to live the next) or because he may have already planned to commit suicide and does not want anything or anyone to get in the way.

This scenario requires the practitioner to take a clinical risk—if the practitioner takes no action, the client may indeed commit suicide, but if the practitioner violates confidentiality and the client was not really at risk for suicide, then the counselor–client relationship might be seriously damaged. Because confidentiality does not bar professional discussions among practitioners within the same agency, clinical dilemmas are most appropriately explored in clinical supervision, where a team of counselors discusses the risks involved and as a group attempts to make the best decision possible.

Another challenging scenario involves a minor child client who discloses possible abuse—a spanking that seems to the practitioner to go beyond mere discipline, verbal abuse that might meet the criteria of child maltreatment, or some other indication that the child *may* be experiencing abuse at home. Determining when that line has been crossed is a clinical issue, best explored within clinical supervision, but it is important to note that legally, it is the practitioner who is responsible for complying with disclosure laws, and it is the practitioner's professional license that will be at risk if the appropriate actions are not taken. In some states a failure to report suspected child abuse can result in criminal misdemeanor charges; thus, although clinical supervision can be of significant assistance in making these types of clinical decisions, the practitioner must make the final decision on whether to violate confidentiality to protect the client's welfare.

Another limit to confidentiality involves a client who discloses during the counseling relationship that he or she has a plan to cause serious and immediate harm to another person. Laws in most states dictate specifically how, when, and to whom this information is to be disclosed. Duty-to-warn laws have been influenced greatly by a tragic incident that occurred on the University of California, Berkeley, campus when a student disclosed to a campus psychologist his intent to kill his girlfriend. Although the psychologist informed various individuals, including his supervisor and campus police, he did not inform the intended victim or her family. The girlfriend was later killed by the client. The family of the victim sued the university for the psychologist's failure to warn the victim. The case *Tarasoff v. The Regents of University of California* resulted in two decisions by the California Supreme Court in 1974 and 1976 (*Tarasoff I* and *II,* respectively). *Tarasoff I* found that a therapist has a duty to use reasonable care to give threatened persons a warning to prevent foreseeable danger. *Tarasoff II* was more specific in referencing the therapist's duty and obligation to warn intended victims if necessary to protect them from serious danger of violence. Virtually every state in the nation now uses the *Tarasoff* decisions as a foundation for the development of duty-to-warn laws (Fulero, 1988).

Although clients are told about the limits of confidentiality by the written informed consent, they may forget or be confused about what would warrant violation of the confidentiality privilege. Clients who share deeply personal information with their counselors may feel betrayed by the counselor who informs them that a disclosure is going to be shared with someone to protect the client or others. It is vital that this topic be fully discussed during the first counseling session so the client knows what disclosures do and do not limit confidentiality. For instance, disclosures of shoplifting, cheating on one's taxes, lying to an employer, having an affair, or *feeling* like attacking a coworker do not limit confidentiality, but admissions of plans to kill oneself or someone else, to set someone's house on fire, or admissions of child maltreatment (as defined in Chapter 5) do limit confidentiality requiring disclosure. Child and adolescent clients in particular may be taken by surprise when their confidentiality is violated; thus, it is often a good idea for the counselor to remind clients of these limits intermittently throughout the counseling relationship.

Skills and Competencies

Generalist practice has been defined as "a perspective focusing on the interface between systems with equal emphasis on the goals of social justice, humanizing systems, and improving the well-being of people" (Schatz, Jenkins, & Sheafor, 1990, p. 220). Generalist practice is also characterized as having a wide range of skills that are used with a diverse population. Therefore, the skills and intervention strategies referenced in this chapter will be general enough to be applied to a variety of situations and clients. More specific skills and intervention strategies will be discussed in successive chapters as they apply to clients seeking services in particular practice settings. Despite the generalist nature of the human services profession, and the fact that in most (if not all) states human services professionals working on a bachelor's level will not be permitted to work in the capacity of a professional licensed mental health provider, some direct practice with clients will occur in various contexts, as many who work in the helping fields will attest; thus, it is important for human service workers on all professional levels to become familiar with some basic theoretical modalities and counseling techniques.

Many of the skills included in this chapter could almost be considered personality characteristics. *Empathy* and *compassion* are powerful and necessary skills and often appear naturally engrained in someone's personality or character. Nevertheless, even if someone is naturally empathetic and a naturally good listener, it is imperative that these skills be sharpened and more fully developed to be truly useful in the human services field. Other skills are less natural and must be taught. For instance, although some people might be a good judge of character, they need to be taught various clinical assessment skills and techniques.

Sympathy and Empathy

Escalas and Stern (2003) discussed the traditional definition of both sympathy and empathy (commonly confused responses). They define sympathy as sorrow or concern for another's welfare, whereas empathy is defined as a person's absorption in the feelings of another. The difference between these two responses, although seemingly subtle, is significant when one considers that the response of empathy goes one step further, allowing oneself to actually feel what another feels.

In a counseling setting, empathy involves the willingness and ability to truly understand a client's beliefs, thoughts, feelings, and experiences from the client's own perspective. Sympathy is not a difficult emotional response to muster for the true victims of this world (Greenberg, Elliot, Watson, & Bohart, 2001). Imagine watching the news and hearing about the plight of a young couple whose five-year-old child was recently abducted. Your immediate response would likely be to express feelings of sorrow for them, and you might express concern for their welfare, wondering what will become of the little girl and her family as they search for her. You might stop short, though, of allowing yourself to feel the actual feelings of grief and fear that this couple is no doubt feeling. Allowing yourself to immerse so deeply in what you imagine their feelings to be might hit too close to home, particularly if you have children. You might feel compelled to distance yourself emotionally—to resist putting yourself in their place. You shiver as you watch your own five-year-old playing on the swing set in your backyard and will yourself not to give this situation another thought, lest you find it impossible to sleep tonight.

Effective practitioners cannot limit their emotional responses to sympathy alone, and to be effective counselors and advocates they must be willing to go on the emotional roller coaster ride with the client. This requires emotional maturity, the ability to be honest with oneself, the capacity for immersing oneself in another's emotional crisis without getting lost in the experience, and being able to keep the focus on the client, not on themselves. I have often referred to the empathetic response in counseling as having the emotional capacity to not only see the client's world through the client's eyes, but also be willing to walk alongside the client through a difficult time. This can be emotionally exhausting, but if I am working with a victim of rape, and if I want to be truly effective in helping my client navigate through this crisis, then I need to be willing to understand what it feels like to be sexually violated as best I can without having gone through this experience myself, what it feels like to be humiliated, and what it feels like to be filled with shame and embarrassment. Thus, although the concept of empathy might seem appealing, many practitioners resist truly empathizing with their clients because it requires them to search their own minds and hearts, to reflect on past hurts, and in this case, past times in their lives where they have been humiliated, shamed, and embarrassed—experiences many are not particularly inclined to revisit.

Another challenge in responding empathetically to clients is when one is working with clients who do not appear to deserve sympathy or empathy. How do counselors empathize with pedophiles, with parents who abuse their children, or with the drunk driver who drove into a family of five, killing a child? Unlike therapists in private practice who typically have complete control over their caseloads, human service professionals rarely have such control and are often given a caseload, depending on the practice setting, with clients who the general public might deem undeserving of anything other than a prison sentence.

Looking at the world through the eyes of a serial rapist, a domestic batterer, or a raging drunk might be the last thing any sane human being would want to do, but the willingness to do so is a requirement for human service professionals, who will likely find themselves working with *mandated clients,* individuals who are required by some governmental agency (e.g., the courts, department of probation, child welfare) to seek treatment.

So how does one accomplish this feat, when the behavior of such a client is often morally incomprehensible, or at the least abusive? The first step is in understanding that to empathize does not mean to condone. Consider the last motion picture that you watched. It is the director's job to help the viewer see the world through each of the characters' eyes. Considering the role of the director, although not a direct parallel, illustrates the concept of the human service professional essentially sitting alongside those in counseling and seeing the world through their eyes. You do not have to agree with their perspective, and you certainly do not have to agree with their actions, but to be a truly effective human service professional you must be willing and able to understand what it feels like to be them.

Although it might not make sense that a victim of abuse goes on to become the batterer, this dynamic is quite common. The boy who was sexually abused *may* grow up to be a pedophile, the child who was beaten *may* grow up to beat her own children, and the boy who witnessed his father beat his mother *may* grow up to beat his own wife. The nature of this dynamic will be discussed in later chapters, but understanding that most abusive behavior is borne out of pain might help to see mandated clients not as monsters, but as broken human beings who have suffered greatly themselves, yet rather than remaining vulnerable so healing could occur, their hearts were hardened and sometimes they become like those who hurt them.

Human Systems

Understanding and Mastery of Human Systems: An understanding of capacities, limitations, and resiliency of human systems

Critical Thinking Question: Most human service professionals will, at some point, work with clients whose values and actions they find difficult to accept, or even reprehensible. How might a professional increase her capacity to empathize with these individuals?

Boundary Setting

Any discussion of empathy and the need for emotional immersion in another's problems must be considered in the context of boundary setting. Although human services certainly is not the sort of career one can leave at the office, it would be imprudent to become so immersed in a client's problems that practitioners cannot distinguish the difference between their problems and the problems of their clients. It is probably easier to discuss good boundary setting by giving examples of poor boundary setting. The practitioner who counsels a victim of domestic violence and spends the majority of the session talking about her own abusive relationships has poor boundaries. The practitioner who becomes so upset about a mother abusing her child that he takes the child home with him is not setting good boundaries. The practitioner who becomes so upset at a client who projects anger in the counseling session that she cries and tells the client how she is having a horrible day and that the client just made it worse is not setting good boundaries. Finally, the practitioner who gets so immersed in his clients' problems that he becomes convinced his clients cannot survive without him is not setting good boundaries.

Personal boundaries are sometimes compared to physical boundaries such as the property line around one's house, porous enough that someone can enter the property, but solid enough that a neighbor knows not to set a shed up in another neighbor's yard (Cloud & Townsend, 1992). So too must human service professionals establish boundaries in their mental, physical, and emotional lives to determine what falls within their domain and responsibility and what does not.

Human service professionals [must] establish boundaries in their mental, physical, and emotional lives to determine what falls within their domain and responsibility and what does not.

In the human services field, some boundaries are determined by the ethical standards of the field. For instance, having a sexual relationship with a client violates an ethical boundary because this type of intimacy can exploit the practitioner–client relationship that grants the practitioner a significant measure of control—even authority—over the client. Violating the prohibition against having sexual relations with a client is so serious that it can result in suspension of one's professional license. Violating this ethical boundary might seem like an obviously bad idea to most people, but it occurs more often than many suspect. Counseling someone of the opposite sex creates a sense of intimacy that can sometimes foster romantic feelings, particularly on the part of the client.

Much like the child who develops a crush on a teacher, a client who is depressed and lonely may experience the practitioner's comfort, nurturing, and guidance as intimate love. But a sexual relationship when one party possesses power and control and the other is vulnerable and broken will always result in emotional and physical exploitation. A practitioner who respects this boundary will recognize the clinical nature of the client's feelings and will help the client see that experiencing intimacy can be a very positive experience, but developing a romantic relationship should only occur when it can be truly reciprocal. This is an example of a clearly marked boundary, and it is difficult to step over this boundary line without knowing one is in dangerous territory. However, other boundaries are not so clear and are frequently violated by human service professionals.

My first job in human services was as an adolescent counselor at a locked residential facility. I was 23 years old, fresh out of college, and excited to finally be making a difference in people's lives. I became too involved in my clients' lives, though, and quickly began to overidentify with the teens on my caseload. I was so flattered by my clients' expressed need for me that I was willing to work any hours necessary to make sure they knew how much I cared. If I worked a 3:00 PM to midnight shift, and one of the girls on my caseload told me that she needed me there in the morning, I would make sure I was there at 8:00 AM, even if it meant getting little sleep. If another counselor called me at home because a teen on my caseload was insisting that she would only talk to me, I dropped whatever I was doing and rushed down to the facility, feeling good that I was so needed.

This sort of behavior on my part indicated several problems. First, it led to a situation where I almost left the field of human services all together because after three years I was so burned out that I was no longer sure I wanted to be a human service professional in any respect. It also encouraged a sense of dependency among the girls on my caseload. Because it felt good to be needed, I neglected one of the fundamental values of human service professionals: empowering clients to be more self-sufficient. Setting boundaries would have encouraged my clients to develop relationships with other counselors and to rely on themselves and newly developed skills to cope with their struggles.

Since that point in my career I have developed some "rules for the road" for determining necessary boundaries and for making sure that I consistently enforce them. One rule is that I never work harder than my client. This does not mean that I do not advocate for clients, or that I do not assist clients in performing various tasks, but what it does mean is that I recognize that I am not truly helping clients who are not motivated to change because a counselor who overfunctions in a counseling relationship helps no one. Thus, when I begin to feel exhausted in my work with clients, I recognize this as a potential sign that I may be doing too much work, perhaps out of impatience and a need to see progress, and recognize that it is time to step back a bit and give my clients room to decide the best course of action for themselves.

I have also come to see my clients' lives as *their* journey, not mine. This conceptualization allows me to view myself as one of many individuals who will come alongside clients and help them at some point along their journeys, just as various people have helped and influenced me along my own life journey. This conceptual framework helps to remind me that my clients have free will to make whatever choices they deem fit. This self-determination means that they can accept my help and suggestions, or they can reject them.

A final conceptualization that can help establish and maintain healthy boundaries in a counseling relationship is to recognize that people grow and change at varying rates and in their own unique way. Thus, when I am working with someone and it appears as though nothing I am doing or saying is making a difference, I remember that I might be the *seed planter*. Seed planters do just as it sounds—they plant the seeds of future growth, but oftentimes they do not have the benefit of seeing these changes come to fruition. It is often this way when working with adolescents. I rarely witnessed the results of my work with my teen clients, but I had to trust that in five or ten years, something I said, some kindness I showed, some reframing I did would result in healthy personal growth.

The Hallmarks of Personal Growth

It is equally important to recognize the role of the *fertilizer* and the *harvester* in counseling relationships. These are the counselors who come into the lives of clients after the seeds have already been planted. The fertilizer is the practitioner who helps the client do productive work—this is no easy task, but the counselor has the benefit of seeing the results of the counseling and intervention strategies. The harvester is probably the most gratifying role a human service professional can have. This practitioner comes along when everything seems to align for the client. The client is ready to make the necessary changes for a healthy life, recognizes past negative patterns in relationships and choices, and has the necessary insight and motivation to effect true change.

I recently had a client who was at this point in her life. Fortunately, I was able to recognize that I could not take full credit for helping her to make the significant realizations and changes she was making in counseling. She'd had several prior counseling experiences, and my role was to help her to merge all that she had previously learned so that she could finally make the necessary, permanent changes in her attitude and approach to life, so that she could be a healthy and happy productive individual, recognizing her own right to self-determination and dignity and her responsibilities to herself, her family, and her community. If you are working productively with a client but see little to no progress, you may very well be the seed planter. If you are working productively with a client but it seems as though change is still a long way off, then you are probably the fertilizer, and if you are reaping changes left and right with a client, then you may very well be the harvester. Seeing yourself operating as a part of a team, even though you will likely never meet the practitioners who came before you or those who may come after, helps to ease the burden of feeling so responsible for a client's growth, as well as helping to resist the temptation to take full credit for the client's progress.

The Psychosocial Assessment

The process of assessing the psychosocial issues of a client utilizes a combination of numerous skills, such as *patience, active listening skills,* and *good observation skills,* as well as more tangible skills, such as being familiar with how to administer various psychological tests and assessments. The tools for conducting an effective assessment are numerous. The first session is often spent conducting an intake interview, which includes collecting basic demographic information about the client (e.g., age, marital status, number of children, and ethnicity). Other pertinent information includes the nature of the identified problem(s), employment status, housing situation, physical health status, medications taken, history of substance abuse, criminal history, history of trauma, any history of mental health problems (including depression, suicidal thoughts, or other mental illness), and any history of mental health services.

When I was in graduate school I recall being taught that the first five sessions with a client should be focused almost exclusively on assessment, but I quickly realized that if some intervention does not occur during those first few sessions, clients are not likely to return. Unlike many clients who come to see a clinical psychologist in private practice, many human service clients are in crisis, and they often need immediate intervention. Thus, human service professionals often find themselves jumping in with both feet, sizing up the client and the situation rather quickly so that some intervention strategies can be employed.

This by no means indicates that the assessment process should be shortchanged due to the frequent crisis nature of many human services agencies. Quite the opposite in fact—although it is true that the practitioner will be focusing more on assessment the first few sessions, the process of assessing the mental health functioning, as well as the client's situation, is ongoing and should continue throughout the counseling process. This is important for two reasons. First, before effective intervention strategies can be identified and used, the practitioner needs to know what the client's issues are. In addition, more often than not, new information will continue to emerge long after the formal assessment period is over, and if practitioners assume the assessment is complete, they might overlook important information about the client that emerges later in the counseling relationship.

Patience

Patience, therefore, is imperative in conducting an effective assessment. One reason why people enter the field of human services is because they love to figure other people out, but a seasoned professional will not allow this passion to result in a rush to judgment. It is important to hold at bay the intense desire to exclaim "Ah ha!" too quickly. I used to work with victims of domestic violence—a practice setting that I am passionate about because I am an advocate for those who are vulnerable. I recall one female client who shared stories of her controlling and abusive husband. Her stories seemed valid, and there was nothing in particular that would cause me to believe that she was not telling me the truth. In fact, what she shared about her husband's behavior met many of the hallmark signs of domestic violence relationships—her husband controlled the finances, and she had little or no access to the bank accounts; her husband appeared jealous and possessive, consistently demanding to know her constant whereabouts; and many of the arguments she reiterated to me reflected what appeared to be her husband's critical response to her in all respects, ranging from her housekeeping ability to the way she managed their children.

I quickly began to view her low self-esteem and depression as being the *result* of his abusive behavior and counseled her accordingly. Yet, several sessions into our counseling relationship she retold a story, which she apparently did not recall telling me before. This version, though, was considerably different. I knew she was reciting the same story, but this time the events illustrating her husband's abusive behavior were different. I was not sure whether she was simply merging stories accidentally or whether this was an indication that she was not being completely honest with me. I made a note to explore this area further and to be more diligent in determining the veracity of her stories.

After interviewing her husband and children and spending more time assessing my client, I discovered that she was actually the abusive member of the family! She feared her husband leaving her and seeking custody, and thus, she hoped to enter into a counseling relationship and manipulate the counselor, so that she was perceived as the victim, and the counselor would therefore support her version of events in court. If practitioners are not diligent in thoroughly assessing their clients, they will be far likelier to be manipulated by some of their clients, thus doing more harm than good.

Human service professionals should always approach clients with the understanding that the client's perspective is just that—the client's perspective. Moreover, I was once told that truth comes in three parts: what you said happened, what the other person said happened, and what really happened. Understanding this does not detract from the counselor's advocacy role of their clients, but rather supports the counselor's ability to help clients reframe various incidences and situations in their lives to help clients gain a healthier and more balanced perspective.

Thus, although a clinical assessment is a broad and ongoing process, it is also specific, where the human service professional is both assessing the mental health of clients and conducting a needs assessment to determine the quality and level of functioning in the various domains of their lives (interpersonal, work, family, social, spiritual, community). An affective clinical assessment depends on many skills, some of which have already been discussed earlier in this chapter. But two of the most important skills necessary for an effective clinical assessment relate to the practitioner's ability to listen well and be sufficiently observant.

Active Listening Skills

Active listening skills involve the ability to attend to the speaker fully, without distraction, without preconceived notions of what the speaker is saying, and without being distracted by thoughts of what one wants to say in response. Active listening in the counseling relationship also includes behaviors such as maintaining direct eye contact and observing the client's body language. It also involves considering virtually everything that the client says as relevant. It is often the subtle, offhand comment that yields the most information about the client's interpersonal dynamics.

I recall working with one female client for depression and parenting issues, who in response to my questions regarding her perception of the origin of her problems, spent a considerable amount of time discussing her troubled marriage and her difficulty making friends. In the midst of sharing a particularly painful story about her difficult college years, she made a casual comment about how one of her college roommates said something to her once that reminded her of something her mother always said. If I had not been actively listening, I could have missed the significance of that seemingly unimportant comment. It was stated as a joking aside, but I also noticed her brief pause and a quick, almost imperceptible, sadness in her eyes. The entire exchange lasted no more than a few seconds, but it completely turned the course of my assessment. I made a mental note to revisit the issue of her mother during a later session when we knew each other better and she knew she could trust me. Eventually it became clear that her core emotional issues resided in her tumultuous relationship with her mother, but she had previously been so protective of this relationship that it felt far too unsafe for her to recognize that her primary issues revolved around her relationship with a controlling, shaming mother. Over the course of the next several months I continued to revisit the issue, slowly at first and then more boldly once we were on solid ground in our own relationship, and she was finally able to recognize the hold her mother had on her all these years. Had I not been as attentive, responding instead to only what the client wanted to focus on, we would have spent our time together focusing on residual issues.

Observation Skills

Good *observation skills* are also an important part of the assessment process because individuals communicate as much through their bodies as they do through their words. Practitioners should observe their clients' eye contact, whether they are shifting uncomfortably in their seats when talking about certain subjects, crossing their arms self-protectively, or tapping their feet anxiously. All these behaviors can be clues or indicators of deeper dynamics. Employing good observation skills can also yield information about whether a client is being direct or evasive, genuine or masked, sincere or manipulative, open or guarded.

Family Genograms

A more comprehensive assessment tool involves constructing a *genogram* of the client's family. Murray Bowen (1978) developed Family Systems Theory, which is based on the premise that inter- and intrarelational patterns are transmitted from one generation to the next. Thus, one way to grasp the "big picture" of the client's life is to study this intergenerational transmission as it relates to issues such as communication style, emotional regulation, and various other "rules for living" (e.g., it is good to express emotions, it is bad to express emotions). Bowen believes that the goal for achieving positive well-being is to find the balance between achieving personal autonomy and individuation while maintaining appropriate closeness with one's family system. Those who are so close to their family system that they cannot make decisions without family approval without the fear of being considered betrayers of the family are considered *enmeshed*, and those who find it necessary to emotionally distance themselves to the point of estrangement to achieve independence are considered *cut off*.

Most people have some information about their parents, limited information about their grandparents, and oftentimes no information whatsoever about their great-grandparents. They may have grown up hearing one-sided (and unquestioned) versions of family feuds or odd distant relatives, but to gain accurate and valuable information about one's family system, information seeking must be intentional. This can be uncomfortable and may ruffle some feathers, because it is often the family members who have been cut off, or are considered the "black sheep," who hold the family secrets that will unlock the true underlying dynamics of a family system. Poking around the skeleton closet can often threaten families, particularly in closed family systems, but this information may also hold the key to truly unlocking hidden dynamics that have been in place sometimes for numerous generations.

Genograms use a variety of symbols designed to indicate gender, the type of relationship (married, divorced, etc.), *and* the nature of the relationships (cut off or enmeshed). Traumatic events, such as deaths, divorces, and miscarriages, are noted, as are the family's responses to these events (e.g., losses are openly talked about, never discussed, or denied). Typically, shameful events are also relevant, such as out-of-wedlock births (particularly relevant in earlier generations), abortions, extramarital affairs, domestic violence, alcohol abuse, sexual abuse and assault, and job losses. Such events are often kept secret but can affect family members for generations to come. The shame of an extramarital affair and an out-of-wedlock birth that was hushed up several generations back can have a profound effect on how emotions are handled and how feelings are communicated.

I once worked with a woman who struggled to understand why her mother never seemed to accept or approve of her. She had spent years in counseling attempting to understand her mother's and her own intense perfectionism and refusal to accept even the smallest of mistakes. My client was convinced that her mother was ashamed of her, and this belief affected every area of her functioning. A genogram revealed that my client's grandmother was raped, and my client's mother was the product of that rape. Both the grandmother and my client's mother lived their lives in constant shame, and their high expectations of my client were really a reflection of their desire to protect her from the shame they were forced to endure, not some statement of their disapproval of her. It was through the development of a genogram that my client was enabled to take a few emotional steps back and see her family system with more clarity.

A family genogram provides a structured way to obtain a comprehensive family history so that the practitioner and client can develop a more complete understanding of the family dynamics that are affecting the client in ways perhaps never before recognized or acknowledged. It also provides for an objective and nonshaming way

▶ **Human Systems**

Understanding and Mastery of Human Systems: Changing family structures and roles

Critical Thinking Question: Family structures and roles in the United States have evolved markedly over the past 50 years: Women have entered the workforce in large numbers, altering the balance of economic power in households; lifespans have increased, changing the roles of elders and their adult children; single-parent households have become more common; and gay and lesbian individuals and couples have come out of the closet. How can a deep exploration of family dynamics across several generations (e.g., the construction of a genogram) assist a client in developing insight into her current situation?

to gain a level of objective understanding of various issues within one's family system that can potentially pave the way for the client to view relationships and various events without personalizing hurtful experiences, including gaining an objective understanding of the nature of conflict-filled family relationships (Prest & Protinsky, 1993).

No longer is the client blaming himself for his father's seeming disapproval or feeling hurt because his mother seemed emotionally distant and rejecting. Instead clients can develop a greater understanding of the broader picture and can see their family members as individual people who are as much a victim of circumstances as the client. Thus, a family genogram is not merely an effective assessment tool, but also a very effective intervention tool that can be used to address long-standing issues that have potentially kept clients in emotional bondage for years.

Psychological Testing

Counselors have numerous other tools at their disposal as well, including various objective assessments tools, such as inventories designed to assess levels of depression, anxiety, social functioning, and personality style. Less objective measures, such as interpretive drawing exercises, free choice drawing, clay manipulation, and structured play therapy, can also be useful. These assessments work particularly well when working with clients who are either less verbal or dealing with particularly painful emotional issues.

When working with traumatized children, I would often ask them to draw a picture of their families. Although the results always need to be considered cautiously, and in the context of all other information gleaned during sessions, it is always interesting to see how children conceptualize themselves and their various family members. For instance, drawing a picture where the father is significantly larger than the rest of the family might indicate a perception that the father is overbearing. A child who draws himself floating away or standing separately from the rest of the family might indicate a feeling of being disconnected from the rest of the family members. Again, it is essential that practitioners use great caution when interpreting subjective techniques, and all assessment material should be considered as a whole, rather than giving too much weight to any one particular measure.

Clinical Diagnoses

Most licensed human service professionals use the *Diagnostic and Statistical Manual of Mental Disorders*, fourth edition, text revision (*DSM-IV-TR*), to diagnose the mental and emotional disorders of their clients. The *DSM-IV-TR* is a classification system developed by the APA (2000). It includes criteria for mental and emotional disorders, such as schizophrenia, depressive disorders, and anxiety disorders, and personality disorders such as narcissistic personality disorder and antisocial personality disorder (sociopathy). The *DSM-IV-TR* is a multiaxial diagnostic system, which means that individuals are diagnosed on five axes, or five different areas of functioning.

Clinical disorders requiring clinical attention, such as schizophrenia or depression, are diagnosed on Axis I. Personality disorders, such as borderline personality disorder and mental retardation, are diagnosed on Axis II. General medical conditions that might have an impact on one's mental health are diagnosed on Axis III. Psychosocial and environmental problems, such as problems with housing and employment, are diagnosed on Axis IV. Axis V is reserved for the client's global assessment of functioning (GAF). The GAF scale ranges from 0 to 100, with 0 indicating someone at a homicidal or suicidal level and 100 indicating a functioning level far higher than any of us will likely ever achieve. Although the assessment of one's GAF is somewhat subjective and arbitrary, the *DSM-IV-TR* contains a guide that assists practitioners in determining where their clients might fall in their overall functioning level. In general, individuals who are struggling in most areas of their lives and are in need of clinical intervention will be functioning somewhere in the range of 0 to 50.

Criticisms of the *DSM-IV-TR*

Although the diagnostic criteria of the *DSM-IV-TR* relies significantly on professional peer consensus and review and is backed by a large body of research, many professionals in the human services field have concerns about the *DSM-IV-TR* because it applies the medical model to emotional disorders. This paradigm in many respects pathologizes what might just be a broader range of human thoughts and behaviors, which in turn tends to create a stigma for those who are suffering from emotional problems. Consider someone who has recently been the victim of a violent crime. If he experiences mental flashbacks of the traumatic event, is he exhibiting behaviors that are adaptive and expected? Or, in the alternative, is he suffering from post-traumatic stress disorder? Is the angry adolescent whose parents were just divorced exhibiting a normal grief response to this loss? Or does he have oppositional defiant disorder? Even if human service professionals do not naturally view human behavior from a disease perspective, using the *DSM-IV-TR* can influence practitioners to view their clients from a pathological perspective (Duffy, Gillig, Tureen, & Ybarra, 2002).

Yet, even if one believes that the medical or disease model is appropriate to use when evaluating psychological disorders, an important distinction between the diagnostic system used to diagnose medical conditions and the system used to diagnose mental disorders is that the *DSM-IV-TR* uses criteria based on symptoms, whereas medical conditions are diagnosed based on the etiology (cause or origin) of the disorder. Thus rather than diagnosing a patient with a stomachache, which could potentially have many causes, the medical diagnosis would be a virus, an ulcer, or cancer. Yet, when considering mental disorders, one is not diagnosed with a neurotransmitter disorder, negative thinking, or an abusive childhood, but diagnosed with major depressive disorder based on the symptoms the client is experiencing, not on etiology.

Other criticisms of the *DSM-IV-TR* include questioning the process that determines what behaviors are deemed abnormal enough to be included in the *DSM-IV-TR* and which behaviors are not, and whether it is appropriate to categorize human behavior, pathologizing alternative understandings of human behavior (Duffy et al., 2002).

Many practitioners have also expressed concerns about health insurance companies' reliance on the *DSM-IV-TR* for the diagnoses of mental disorders required for reimbursement, which can put both practitioner and client in a precarious position—the practitioner might feel compelled to diagnose a client to get paid and the client may have difficulty obtaining insurance coverage in the future if diagnosed with a serious mental health disorder. Yet, despite the criticisms of the *DSM-IV-TR*, it remains the most well-researched, collaborative classification system for mental pathology currently in existence and does provide a means for organizing various emotional problems and mental disorders.

Many human service professionals use the *DSM-IV-TR* but in general rely on it less than other mental health disciplines, because the human services profession is based on empowerment theory, where clients are encouraged to recognize that they have more control over their lives than they may have previously thought. Self-determination is a related concept and refers to the rights of all individuals to make choices that they believe are in their own best interest. Self-determination can be empowering as clients realize that they have learned to have good judgment, which increases their sense of competency and self-reliance.

Human Services Delivery Systems

Understanding and Mastery of Human Services Delivery Systems: Major models used to conceptualize and integrate prevention, maintenance, intervention, rehabilitation, and healthy functioning

Critical Thinking Question: The *DSM-IV-TR* is commonly used by mental health professionals to diagnose emotional and mental disorders by examining five specific areas of functioning. In what ways does the five-axis system of the *DSM-IV-TR* parallel the human service professional's focus on the person within her environment? In what ways might it limit the professional's understanding of a client's situation?

Continuum of Mental Health

Another important consideration when evaluating someone's level of functioning and mental health status is to recognize that virtually all behaviors occur on a continuum. It is only when a particular behavior occurs frequently enough, and at an intensity level high enough to interfere with normal daily functioning for a significant amount of time, that it becomes the subject of clinical attention. All of us feel sad at times, but if we are so intensely sad that we stop eating and want to stay in bed all day, then we may be suffering from clinical depression. Similarly, many of us become concerned from time to time that our friends might be talking behind our backs or that one of our coworkers is trying to get us fired, but if we're convinced that everyone is out to get us, even people we've never met, then we may be suffering from some form of paranoia.

The *DSM-IV-TR* accounts for this continuum by including criteria relating to frequency and intensity of psychological experiences. For instance, to meet the criteria for major depressive disorder, an individual does not just have to be depressed, but must have a depressed mood nearly every day for at least a two-week period. An individual who meets the criteria for generalized anxiety disorder isn't someone who worries from time to time, but someone who worries *excessively,* more days than not, for at least six months.

In summary, the value of services provided depends on the effectiveness of the assessment. A good assessment defines the problem or problems the client is experiencing, develops a needs assessment to determine where the client's strengths and deficits lie, ascertains the client's social support system, and develops an appropriate treatment plan. It is also important to reassess the client at various points in the counseling process to monitor new or previously masked issues, and to make sure that treatment goals are consistent with the assessment.

> **The value of services provided depends on the effectiveness of the assessment.**

Case Management and Direct Practice

It is important to understand the qualitative differences between case management and direct counseling services. Although both encompass a broad range of activities, they are distinctly different. Direct practice with clients is focused more on an individual's psychological growth and the development of emotional insight and personal growth, whereas case management involves coordinating services with other systems impacting the life of the client. A case manager might coordinate services with a client's school social worker, the housing authority, the local rape crisis center, or even a court liaison, all in an attempt to meet the needs of the client who is interacting in some manner with each of these systems. The goal of the case manager is to assist the client in plugging in to necessary and supportive social services within the community and to learn how to improve the reciprocal relationship or transaction with each of these social systems. These efforts have many purposes and goals, but chief among them is the caseworker's proactive attempt to strengthen and broaden the client's social support network.

In the Human Services Board-Certified Practitioner Exam Handbook that is provided to human service professionals in preparation for the HSE-BCP examination, a description of case management includes the following tasks:

- Collaborate with professionals from other disciplines
- Identify community resources
- Utilizes a social services directory
- Coordinate delivery of services
- Participate as a member of a multidisciplinary team
- Determine local access to services
- Maintain a social services directory
- Participate in case conferences
- Serve as a liaison to other agencies
- Coordinate service plan with other service providers (Center for Credentialing and Education [CCE], 2011).

The NASW also provides a relatively comprehensive definition of case management:

> Social work case management is a method of providing services whereby a professional social worker assesses the needs of the client and the client's family, when appropriate, and arranges, coordinates, monitors, evaluates,

and advocates for a package of multiple services to meet the specific client's complex needs. A professional social worker is the primary provider of social work case management. Distinct from other forms of case management, social work case management addresses both the individual client's biopsychosocial status as well as the state of the social system in which case management operates. Social work case management is both micro and macro in nature: intervention occurs at both the client and system levels. It requires the social worker to develop and maintain a therapeutic relationship with the client, which may include linking the client with systems that provide him or her with needed services, resources, and opportunities. Services provided under the rubric of social work case management practice may be located in a single agency or may be spread across numerous agencies or organizations. (2002)

Direct Practice Techniques for Generalist Practice

Although the assessment process is ongoing, once the initial assessment is complete a treatment plan is developed that is designed to address the client's identified issues. I will cover direct practice and counseling techniques appropriate for clients served in particular practice settings in more detail in subsequent chapters, but there are basic techniques involved in generalist practice that apply in a broad way to most counseling and intervention situations.

Many individuals seeking services at a human services agency will need assistance with developing better coping skills. Regardless of whether the problems experienced by the client are pervasive or more limited, most clients can benefit from learning to manage high levels of stress, learning to prioritize the various problems in their lives, and learning how to manage the current crisis in a way that diminishes the possibility of a domino effect of crises. A crisis with one's child requiring a significant amount of time and attention can quickly result in a job loss, which can in turn result in the loss of housing. Confronting crises effectively, though, can have a positive impact on one's life, including an increase in self-esteem, the development of new and more effective coping skills, the gaining of wisdom and the development of new social skills, and the development of a better overall support system.

Most mental health experts recognize that one of the best opportunities for personal growth is a crisis, due to the possibility of shaking up long-standing and entrenched maladaptive patterns of behavior. Park and Fenster (2004) studied stress-related growth in a group of college students who experienced a stressful event and found that the struggle involved in a life crisis produced personal growth. This is true, though, only for those who expend the necessary energy to work through their struggles in a positive way. Those in the study who remain negative and avoided dealing with the problems borne out of the crisis did not take advantage of the growth-producing opportunities and thus did not experience any significant personal growth. Those who worked hard to manage the stress resulting from their crisis and were able to see the crisis as an opportunity for growth often developed better personal mastery skills and developed a changed and healthier perspective. Recognizing this potential for personal growth provides the practitioner with a framework for assisting clients in developing better coping skills that not only can better assist them in the management of concrete problems, but can also help them to shift their entire perspective of life struggles in general. For instance, clients who once saw themselves as powerless victims can begin to see themselves as empowered survivors.

Task-Centered Casework

When most individuals are confronted with a crisis, panic sets in, and it becomes difficult to address the problem in a healthy or meaningful way. Most of us can relate to feeling completely overwhelmed when facing a life crisis. We know there are things we need to do to manage the crisis, but all we see is a gigantic mountain looming before us. For some, this has a motivating effect, and they attack the mountain until every issue is resolved. But for some, particularly those with a long history of crises, those with poor coping skills, or those suffering from emotional or psychological problems with diminished personal management skills, the mountain can seem virtually insurmountable, and their response is to shrink away with a feeling of despair and defeat.

A counseling technique called the *Task-Centered Approach,* an intervention strategy developed by the School of Social Services at the University of Chicago (Reid, 1975), works well with clients who feel paralyzed in response to the challenges of various psychosocial problems. Treatment is typically short, lasting anywhere between two and four months, and is focused on problem solving. The client and counselor or caseworker define the problems together and develop mutually agreed-upon goals. Each problem is broken down into smaller and more easily manageable tasks. Goals can be as tangible as finding a new job or as intangible as more effectively managing frustration and anger. Rather than having one broad goal of obtaining a job, a client might have a week-one goal of doing nothing more than looking at the want ads in the local newspaper and a week-two goal of making one phone call to a prospective employer. Dividing large goals into smaller, specific, "stepping-stone" goals diminishes the possibility that clients will allow their anxiety to overwhelm them. By focusing on specific problems and breaking them into "bite-sized," manageable pieces, clients not only learn effective problem-solving skills, but also gain insight into the nature of their problems, develop increased self-esteem as they experience success rather than failure in response to meeting goal expectations, and learn to manage their emotions, such as anxiety and depression, without allowing such states to overtake and overwhelm them.

The counselor or caseworker assists clients in meeting goal expectations through a variety of intervention strategies specific to the actual problem, but can include planning for obstacles, role-playing (where the client can actually act out difficult situations in the safety of the counselor's office as a way of practicing communication, etc.), and mental rehearsal (similar to role-playing but involves the client thinking or fantasizing about some specific situation—such as an upcoming job interview or a difficult confrontation) (Reid, 1975). Revisiting original goals and evaluating client progress are also powerful tools in helping clients experience a sense of personal mastery and empowerment as they are helped to recognize their progress.

Consider the following case study.

CASE STUDY 4.1

Case Example of Task-Centered Approach

Mary is the 34-year-old single parent of a 5-year-old boy. She has been living with her mother since her own divorce three years ago. This is a negative situation because her mother is verbally abusive of Mary and her son, abuses alcohol, and smokes inside the home. In addition, their living space is small, and Mary and her son share a bedroom. Mary's original goal was to live with her mother for only six months, but whenever she considers moving out she becomes overwhelmed with the prospect of not only finding an appropriate apartment, but finding child care as well, because despite her mother's abusive behavior, Mary has been relying on her mother for before- and after-school child care while she works. Mary feels trapped but completely powerless to do anything about her situation. During Mary's intake interview she described her prior counseling experiences, sharing that she quit counseling because whenever she was faced with the prospect of finding an apartment, her fears would snowball into so many fears that she simply couldn't even bring herself to make the first phone call in search of housing. She ended up feeling embarrassed, as if she were letting the counselor down, and just decided she could not deal with any more failures, so she stopped going to counseling. Mary explained that through-out the past several years her mother has consistently reminded her that she would never make it on her own, that she would surely fail, and that she would end up destroying her life and the life of her son. Her mother also told Mary that if she moved out, and ran out of money, she would not bail Mary out again and would instead force Mary and her son to go to a shelter. Opening the newspaper to look for a rental advertisement resulted in a flood of worries and concerns—some specific and some she could not even put into words. She worried about everything from whether she would know what to say when calling on an apartment, to whether she would be able to support herself and her son. What if she was laid off from her job and could no longer afford her apartment and had to live in a shelter? What if she couldn't find a babysitter she could afford? What if she found an apartment and got a babysitter, but the babysitter ended up abusing her son worse than her mother did? She read about such things all the time in the newspaper, she reasoned. Or what if she found an apartment, but she had a financial emergency such as her car breaking down, and she started falling behind in her rent and was evicted? She couldn't fathom the thought of moving out and then having to move back in with her mother again, or worse what if her mother made good on her threat and refused to allow them to move back in with her? Once confronted with this slippery slope of catastrophizing, she would resist even taking the first step toward independence and could not bring herself to even look at rental ads. Mary's mood became increasingly melancholy over the years, and after years of verbal abuse from her mother, her ex-husband, and now her mother again, she had no confidence in her ability to financially support her own son or even to manage her own life without her mother's assistance. Mary's caseworker reassured her that there was absolutely no rush in finding an apartment. In fact, she reminded Mary that she was in charge of her own life and could make the choices she thought were best for her and her son. During the first two sessions, Mary and her caseworker developed realistic goals for her, including securing an apartment when Mary had the funds to ensure financial security. Mary and her caseworker developed a detailed budget and determined that she would need three months' salary put away in a savings account to ensure against any realistic financial emergencies. By identifying possible obstacles to Mary achieving independence, decisions were made based on facts and realistic risks, not on undefined and generalized fear. Once goals were developed and obstacles identified, Mary and her caseworker agreed on tasks to be accomplished by the following week. Mary's task for the first week was to look through the newspaper and circle rental advertisements within her price range. She was not to call any of them though, even if she found one that seemed ideal. Mary came in the second week with the newspaper filled with circled apartment ads. Mary and her caseworker spent the first portion of the session discussing how Mary felt while circling these ads. Mary explained that her initial excitement was quickly followed by intense anxiety, but that when she realized she could not call the apartments even if she had wanted to, she calmed down almost immediately. The next portion of the session was spent on determining tasks for the following week. The first task included circling all appropriate ads and calling on two apartments for informational purposes only. Because Mary had a significant amount of anxiety about calling and talking to a stranger, Mary and her caseworker wrote a script and rehearsed it by doing a role-play with her caseworker playing the part of the potential landlord. Mary's additional task for the week was to talk to her boss seeking reassurance that her employment was secure. Mary returned the following week excited. She called on two apartments and followed the script on the first one, but the second call went so well she did not even need the script. Her discussion with her boss also went well, and he reassured her that her job was secure. Mary shared excitedly that her boss was pleased that Mary showed assertiveness in approaching him and offered her an opportunity to attend some training courses so that she could be promoted. For the next three months Mary's counseling proceeded in a similar fashion with weekly tasks that inched her along slowly enough that she did not become overwhelmed by unreasonable fears, but quickly enough that she gained confidence and courage with each successive step. Mary rented an apartment during her fourth month of counseling with three months' income safely tucked away in a savings account, a promotion with a raise, and reputable and affordable day care.

Perceptual Reframing, Emotional Regulation, Networking, and Advocacy

Another general counseling method includes the reframing of a client's perception of a situation, emphasizing the importance of viewing various events, relationships, and occurrences from a variety of possible perspectives. For some reason it seems easier for human beings to assume the negative in many situations. Whether considering the intentions of a boyfriend or the prospects of getting a better job, most of us seem to gravitate toward negative assumptions. Many people in the midst of a physical or emotional crisis of any proportion will often resort to taking a somewhat polarized negative stance on an issue and would benefit from assistance in seeing situations and relationships from a different perspective. A client's perception that life is unfair and nothing good ever happens to her can be encouraged to see life struggles as normal and even good because they promote positive personal growth. Clients who feel shame because they were recently fired from a job they despised can be encouraged to see this incident as a disguised blessing opening the door to find a career for which they are far better suited.

Additional intervention goals include assisting clients with *emotional regulation,* teaching them how to sit with their emotions rather than immediately acting on them; developing a better *social support network* so that they can become emotionally independent and self-reliant; and *advocating for clients* who are being oppressed, either within their family systems or in society in general.

Cultural Competence and Diversity

Because human service professionals work with such a wide range of people, across various cultures, socioeconomic levels, coming from varying backgrounds, it is vital that human service education and training be presented in a context of cultural competence and cultural sensitivity. Cultural competence is reflective of a counselor's ability to work effectively with people of color and minority populations by being sensitive to their needs and recognizing their unique experiences and is a required component of working in the human services field. For instance, the NOHS ethical standards specify the requirements and competencies human service professionals are required to maintain. Specifically, standards 17 through 21 deal with issues related to cultural competence, focusing in particular on anti-discrimination, cultural awareness, self-awareness relating to personal cultural bias, and requirements for ongoing training in the field of cultural competence:

> Cultural competence is reflective of a counselor's ability to work effectively with people of color and minority populations by being sensitive to their needs and recognizing their unique experiences.

STATEMENT 17 Human service professionals provide services without discrimination or preference based on age, ethnicity, culture, race, disability, gender, religion, sexual orientation or socioeconomic status.

STATEMENT 18 Human service professionals are knowledgeable about the cultures and communities within which they practice. They are aware of multiculturalism in society and its impact on the community as well as individuals within the community. They respect individuals and groups, their cultures and beliefs.

STATEMENT 19 Human service professionals are aware of their own cultural backgrounds, beliefs, and values, recognizing the potential for impact on their relationships with others.

STATEMENT 20 Human service professionals are aware of sociopolitical issues that differentially affect clients from diverse backgrounds.

STATEMENT 21 Human service professionals seek the training, experience, education and supervision necessary to ensure their effectiveness in working with culturally diverse client. (NOHS, 1996)

The human services field is not the only discipline to require cultural training. Rather, most professional organizations require that their mental health professionals obtain cultural competency training based upon a foundation of respect for and sensitivity to cultural differences and diversity (Conner & Grote, 2008). Yet, cultural competency extends beyond that of ethnic differences. For instance, counselors who undergo cultural competency training will learn the importance of remaining sensitive to populations from different income levels, religions, physical and mental capacities, genders, and sexual orientations, as well as races, and as such, will learn the importance of avoiding what is commonly referred to as ethnocentrism—the tendency to perceive one's own background and associated values as being superior, or more "normal" than others. In recent years, the issue of cultural or multicultural competence has become so important that training protocols have been developed with recommendations that all those who work in the helping fields engage in some form of cultural competency training.

Cultural competence is somewhat of a general term though and is often used synonymously with other terms such as *cultural sensitivity*. Despite the relatively universal belief among human service and mental health experts that cultural competence is a vital aspect of practice, very little consensus exists as to what constitutes cultural competency on a practice level (Fortier & Shaw-Taylor, 2000). Although broad themes of respect and sensitivity tend to be universally accepted as foundational to cultural competent practice, the concept of cultural competency has tended to remain as an idea or general philosophy that has not yet been operationalized in a concrete way. For instance, Cunningham, Foster, and Henggeler (2002) surveyed counselors who considered themselves culturally competent and found that there was a vast difference in terms of which counseling methods they believed were most effective with culturally diverse clients. This last of consensus among experts on which specific counseling approaches and counselor responses constituted "cultural competence" makes it difficult, if not impossible, to determine what methods will have the greatest likelihood of having a positive outcome in counseling a particular ethnically diverse client group. Although recent research has attempted to develop what is called *evidence-based practice* with regard to cultural competence, to date there remains very little research on what constitutes *cultural competent practice*.

> ### Human Systems
>
> Understanding and Mastery of Human Systems: Emphasis on context and the role of diversity in determining and meeting human needs
>
> ***
>
> Critical Thinking Question: The text describes "ethnocentrism" as "the tendency to perceive one's own background and associated values as being superior, or more 'normal' than others." In what ways might ethnocentrism affect a human service professional's ability to effectively serve clients? How might he take steps to reduce his ethnocentrism?

Concluding Thoughts on Generalist Practice

Although human service professionals work with a very wide range of clients presenting with an equally diverse range of psychosocial problems, these skills and intervention techniques can be broadly applied in generalist practice. Understanding that people are not pathological by nature, but often are responding to real traumas, tragedies, and crises in a natural way (e.g., it is normal to become depressed after experiencing a loss) helps the human service professional look for a client's strengths, rather than solely assessing a client's perceived deficits.

The unique nature of the human services profession encourages practitioners to view the individual as a part of a greater whole; thus, a client's social world is assessed and evaluated, which enables human service professionals to help their clients better navigate their world. Essentially, it is the human service professional's commitment to working with displaced populations, assessing not only clients but the worlds in which they live, and then applying various culturally competent intervention techniques designed to encourage, empower, and integrate some of society's most broken and marginalized members helping them to become whole and functional, perhaps for the first time in their lives.

The following questions will test your knowledge of the content found within this chapter.

1. Disclosing the nature and risks of the counseling relationship to clients prior to their engaging in these services is called:
   a. the limits of confidentiality
   b. a duty-to-warn
   c. informed consent
   d. confidentiality

2. Keeping information shared by clients in the counseling relationship confidential is:
   a. mandated by law
   b. voluntary
   c. optional
   d. dependent upon the nature of the counseling relationship

3. Limits of confidentiality refers to:
   a. the counselor's legal right to share information disclosed by clients with colleagues for the purposes of clinical supervision
   b. the nature and purpose of counseling services
   c. the laws that determine how and when client information can be disclosed to other treatment providers, insurance companies, and governmental agencies
   d. Both A and B

4. A counselor's duty-to-warn and duty-to-protect relate to situations where through direct disclosure clients share:
   a. their intention to terminate counseling despite being ordered by the court to receive mental health services
   b. that they are a threat to themselves (suicidal) or others (homicidal)
   c. that they could potentially be a threat to themselves or others in the future, under certain theoretical conditions
   d. All of the above

5. Setting boundaries with clients encourages clients to
   a. develop relationships with other counselors
   b. rely on themselves and newly developed skills to cope with their struggles
   c. become self-destructive due to feelings of abandonment
   d. Both A and B

6. Patience, active listening, and observational skills are all aspects of:
   a. the psychological evaluation
   b. the psychosocial evaluation
   c. the clinical assessment
   d. emotional regulation

7. Describe the nature and purpose of creating a family genogram, including ways in which genograms aid clients in gaining a more objective perspective of family dynamics.

8. Compare and contrast direct practice and case management, including their respective techniques and goals.

Suggested Readings

Bowen, M. (1985). *Family therapy in clinical practice.* New York: Jason Aronson.

Epstein, L., & Brown, L. B. (2001). *Brief treatment and a new look at the task-centered approach.* Boston: Allyn & Bacon.

Fulero, S. M. (1988). Tarasoff: 10 years later. *Professional Psychology: Research and Practice, 19,* 184–190.

Nash, K. A., & Velazquez, J. (2003). *Cultural competence: A guide for human service agencies.* Atlanta, GA: CWLA Press.

Reamer, F. G. (2005). *Pocket guide to essential human services.* Washington, DC: NASW Press.

Russo, J. R. (2000). *Serving and surviving as a human-service worker.* Long Grove, IL: Waveland Press.

Tarasoff v. Regents of the University of California, 118 Cal. Rptr. 129, 529 P.2d.533 (Cal. 1974).

Tarasoff v. Regents of the University of California, 113 Cal. Rptr. 14, 551 P.2d.334 (Cal. 1976).

Wodarksi, J. S., Rapp-Paglicci, L. A., Dulmus, C. N., & Jongsma, A. E. (2001). *The social work and human services treatment planner.* Hoboken, NJ: John Wiley & Sons.

Internet Resources

American Counseling Association: http://www.counseling.org

Center for Credentialing & Education, Human Services Board Certified Practitioner: http://www.cce-global.org/ credentials-offered/hsbcp.

Genograms: http://www.genopro.com/genogram

National Organization for Human Services: http://www.national-humanservices.org

References

American Counseling Association. (2005). *ACA code of ethics.* Alexandria, VA: Author.

American Psychiatric Association. (2000). *Diagnostic and statistical manual of mental disorders* (4th ed., Text Revision). Washington, DC: Author.

American Psychological Association. (2002). *Ethical principles of psychologists and code of conduct.* Washington, DC: Author.

Bowen, M. (1978). *Family therapy in clinical practice.* New York: Jason Aronson.

Center for Credentialing & Education. (2011). Human Services-Board Certified Practitioner Exam Candidate Handbook. Retrieved January 1, 2011, from: http://www.cce-global.org/ Downloads/HS-BCPHandbook.pdf.

Cloud, H. C., & Townsend, J. (1992). *Boundaries.* Grand Rapids, MI: Zondervan.

Conner, K., & Grote, N. (2008, October). Enhancing the cultural relevance of empirically-supported mental health interventions. *Families in Society, 89*(4), 587–595. Retrieved September 14, 2009, from Academic Search Premier database.

Cunningham, P., Foster, S., & Henggeler, S. (2002, July). The elusive concept of cultural competence. *Children's Services: Social Policy, Research & Practice, 5*(3), 231–243. Retrieved September 14, 2009, from Academic Search Premier database.

Duffy, M., Gillig, S. E., Tureen, R. M., & Ybarra, M. A. (2002). A critical look at the DSM-IV-TR. *The Journal of Individual Psychology, 58*(4), 362–373.

Escalas, J. E., & Stern, B. B. (2003). Sympathy and empathy: Emotional responses to advertising dramas. *Journal of Consumer Research, 29*, 566–578.

Fortier, J. P., & Shaw-Taylor, Y. (2000). *Assuring cultural competence in healthcare: Recommendations for national standards and an outcomes-focused research agenda.* Resources for Cross-Cultural HealthCare and the Center for the Advancement of Health. Rockville, MD: U.S. Department of Health and Human Services, Office of Minority Health.

Fulero, S. M. (1988). Tarasoff: 10 years later. *Professional Psychology: Research and Practice, 19*, 184–190.

Greenberg, L. S., Elliot, R., Watson, J. C., & Bohart, A. C. (2001). Empathy. *Psychotherapy: Theory, Research, Practice, Training, 38*(4), 380–384.

National Association of Social Workers. (2000). Cultural competence in the social work profession. In *Social work speaks: NASW policy statements* (pp. 59–62). Washington, DC: NASW Press.

National Association of Social Workers. (2002). *NASW standards for social work case management.* Retrieved May 25, 2004, from http://www.naswdc.org/practice/standards/sw_case_mgmt. asp#intro

National Organization for Human Services. (1996). *Ethical standards of human service professionals.* Washington, DC: Author.

Park, C. L., & Fenster, J. R. (2004). Stress-related growth: Predictors of occurrence and correlates with psychological adjustment. *Journal of Social and Clinical Psychology, 23*(2), 195–215.

Prest, L. A., & Protinsky, H. (1993). Family systems theory: A unifying framework for codependency. *American Journal of Family Therapy, 21*(4), 352–360.

Reid, W. J. (1975). A test of a task-centered approach. *Social Work, 20*(1), 3–9.

Schatz, M., Jenkins, L., & Sheafor, B. (1990, Fall). Milford redefined: A model of initial and advanced generalist social work. *Journal of Social Work Education, 26*(3), 217–231. Retrieved June 24, 2009, from Professional Development Collection database.

CHAPTER 5

Child Welfare Services

. . . .

Overview and Purpose of Child and Family Services Agencies

Kyodo/Newscom

Learning Objectives

- Develop an understanding of the history of the child welfare system, recognizing the impact of historic policies and practices on the current child welfare system
- Develop an understanding of the demographic makeup of children currently in care of the child welfare system, understanding the reasons for overrepresentation of certain racial ethnic groups
- Understand how children enter the child welfare system, including having a basic understanding of the federal and state laws that govern child placement policies
- Develop an understanding of the nature of working with biological parents, children in placement, and foster parents, recognizing the complementary and conflictual roles of each
- Recognize the historic and current trends of bias and abuse of certain ethnic minority groups within the child welfare system, as well as understanding ways of avoiding such abuse through cultural competent practice

The field of child and family services generally involves the care and provision of children who cannot be appropriately cared for by their biological parents, as well as providing assistance for those who need support and assistance in the management and provision of their families. This practice setting is primarily concerned with children in foster care placement, but may also involve family preservation services and adoption services. A human service professional working in a child and family services setting may be involved in the following activities:

- Child abuse investigations
- Child abuse assessments
- Case management and counseling of the child in placement, foster families, and biological parents
- Case management and counseling of families in crisis
- Case management and counseling of potential adoptive parents, adult adoptees, and birth parents

The clinical issues involved in this field are quite broad but involve issues related to abandonment and loss, post-traumatic stress disorder (PTSD), cultural sensitivity, child development, parenting issues, substance abuse, anger management, and the ability to work with a broad range of life stressors and maladaptive responses that might lead to breakdowns within the family.

In addition to the wide range of activities in which a human service professional might engage within a child and family services agency, there is also a wide range of practice settings where the human service professional might work, the largest being a state's child protective services (CPS) agency. Human service professionals also work at not-for-profit agencies, some of which are contracted by the state to provide mandated services to children in substitute care and some of which provide voluntary services to any family in crisis. Within these agencies a human service professional may be involved in a number of activities, including counseling, case management, and writing grants for increased funding. Many human service professionals working in the field of child welfare may do so on a volunteer basis, and although these individuals are not paid professionals, the work they do is so vital that their role in the welfare of children must be mentioned. For instance, CASA (court-appointed special advocates) volunteers are court-appointed advocates for children who are placed into state care, working to protect the best interest of the children by being their voice in all court proceedings.

The History of the Foster Care System in the United States

The child welfare system in the United States has undergone significant changes in the last several hundred years due to numerous factors such as urbanization, industrialization, immigration, mass life-threatening illness, changes within the family system, changing social mores (including the reduction of shame associated with divorce, out-of-wedlock births, and single parenting by choice), and the eventual availability of government financial assistance for those in need. Thus, to truly understand the current child welfare system it is vital to understand its past.

In 2001, ABC's news show *Nightline* aired a documentary featuring the horrible plight of the street children of Romania (Belzberg, 2001). After the show, U.S. citizens flooded the network with telephone calls, expressing outrage and horror at the images that flashed across their television screens for almost two hours. The documentary revealed children as young as six years old living on the streets, with no food to eat, with only slightly older children and liquid glue to keep them warm at night. The reporter explained how political events in Romania created a situation where impoverished families could no longer care for their offspring, leading to the streets becoming flooded with marauding children, in desperate search of money and food. These children, who often resorted to pickpocketing and other petty crimes, were considered by most mainstream Romanians to be the scourge of society, pests to be avoided.

The U.S. response was one of literal horror, not only at the conditions in which the children were forced to live but also at the apparent apathy of most Romanians, particularly those in government, including the Romanian police force. The documentary showed numerous incidences of police mistreatment, including one young boy whose leg was broken in a scuffle with a police officer. This seeming indifference shocked viewers, who expressed outrage at the heartlessness necessary to not only accept orphans living on the street, but actually perceive these orphans as social pariahs. These concerned and outraged Americans are apparently unaware that our own recent past includes alarmingly similar conditions and attitudes toward orphans, with only 150 to 200 years separating the United States from Romania in this regard.

Historic Treatment of Children in Early America

There were many ways in which children were mistreated in Colonial America, but in this section I will be exploring primarily two areas of mistreatment of children in contemporary America and Great Britain, which in many respect served as the foundation for child welfare laws in the United States, including child labor laws. These two areas include the use of children in the labor market, otherwise known as "child labor," and the treatment of children who were, for whatever reason, without parents, most often referred to as "orphans" or "street children." Of course there are many other ways in which children were mistreated as well—without federal laws protecting children, there was rampant sexual abuse, physical abuse, and various other forms of maltreatment such as neglect (physical and emotional). By exploring primarily child labor and the treatment of orphans and street children, readers should not presume that other forms of maltreatment did not exist in America's history. The rationale for exploring these two areas of maltreatment (child labor and the treatment of orphans and street children) is based upon the fact that they represent a significant departure from how children are treated today, and also highlight key areas within child welfare, with regard to early child welfare advocacy and development of laws, policies, and programs intended to protect children.

Child Labor in Colonial America: Indentured Servitude and Apprenticeships

During Colonial America, all children were expected to work, whether bonded or not. In fact, children as young as 6 years old worked alongside their parents, and children as young as 12 years old were expected to work in adult-like capacities, often working in apprenticed positions outside of their homes, and away from their families. Children of poor families, particularly immigrants, were often forced to work alongside their parents either in indentured servitude or as slaves. During the many waves of early immigration, individuals, families, and minor children as young as 10 or 11 years old often paid for their passage to the United States through a process called indentured service. Indentured service contracts required that the servant—most often a poor individual, or families hoping for a better life in America—work off the cost of their travel by working for a master in some capacity once they arrived in America. If a family immigrated to the United States in this manner, then their children, regardless of age, were required to work as well.

The economic system of indentured servitude was extremely exploitative. Research indicates that it was the ship owners who would often recruit unsuspecting, yet desperate, individuals from other countries, with stories of abundant life in America. Many individuals and entire families accepted the call, believing that they could make a better life for themselves in Colonial America. They were told that the terms of their service would last for three years, and then they would be free—free to buy land and to make a life for themselves that was not possible in many European countries (Alderman, 1975). In reality, the cost of their passage would be paid off in only one year, and the remaining years of service were considered free work. Further, masters often treated their bonded labor quite poorly. Servants received no cash wages, but were supposed to be provided with basic necessities, which depending upon the nature and means of the master, might include anything from sufficient sustenance to meager sustenance and substandard shelter. Thus, while indentured servants were not considered slaves, the treatment of them was quite similar (Martin, in press).

Although most indentured servants were in their early 20s, Green (1995) notes that children who immigrated with their families on bonded contracts were expected to work as well, and were often treated no differently than their parents. Children were not allowed to enter into bonded labor contracts without the permission of their parents, but very poor and orphaned children, particularly in London, were often kidnapped and sold to ship captains, who then brought them to America and sold them as indentured servants, most often to masters who used them as house servants. Also, local governments that were responsible for the poor, would "bind out" poor and orphaned children in early America as a form of poor relief (Katz, 1996). Most local laws favored masters (since virtually all judges were in fact masters themselves), and stipulated that child bonded servants could often be kept until the age of 24, and if they ran away, their treatment became even more abusive and their time in bonded servitude was often doubled (Green, 1995).

Slavery and Child Labor

Indentured servitude eventually waned during the 17th century in favor of slavery, but the binding out of children who were poor and orphaned continued well into the 19th century. During the 300 years of the Atlantic slave trade over 15 million Africans were brought to the United States through the West Indies, or directly from Africa. Among these Africans were many children who were either forced or born into slavery along with their parents. In time masters realized that slaves who had once experienced freedom were far more difficult to control than those born into captivity; thus, a market developed for children who could work for a slave owner and essentially grow up as captive slaves, and be trained to be a submissive servant. According to Green (1995) children under the age of about seven were more often sold with their mothers, but once the children were between the ages of 7 and 10, they could and often were sold off and separated from their families, particularly to fill this growing need for young "negro" slave children born into captivity. Slavery was outlawed in 1865 with the passage of the Thirteenth Amendment to the U.S. Constitution, but the plight of African children did not improve significantly (and most human service professionals would argue that the legacy of slavery creates significant challenges for African American children to this day).

There were not as many African slave children born into captivity as one might expect, due in large part to extremely high rates of infant mortality of African slave children due to disease and poor nutrition. In fact, the infant mortality of African slave children under the age of four was double that of white children during the time when slavery was legal. Ironically, not only has this trend continued well into the 21st century, but it has gotten far worse with infant mortality among African American infants being about three times that of Caucasian infants (CDC, 2002).

Throughout early American history children worked within their own households and farms working alongside their families. The hope of parents was that their children would be able to afford to buy farms of their own when they grew to adulthood. Another form of work that children engaged in early America was apprenticeship. Apprenticeship involved the training of children in a craft. Some children went to live with the artisan who trained them and others did not. Essentially apprenticeship involved an artisan taking on an apprentice in early adolescence and teaching him a trade. The apprentice would serve as an assistant to the artisan (Schultz, 1985). Apprenticeships might involve learning to become a barber, making shoes, or woodworking. Children were not paid, and in fact parents often had to pay to have their children apprenticed. While most apprenticeships did not involve overt exploitation, the practice did reflect the focus on working children, rather than education. Apprenticeships eventually became less popular as Industrialization began in the late 18th century, as machines were developed replacing the need for many craftsmen.

Child Labor during the Industrial Era: Children and Factories

By the mid-19th century, virtually all apprenticeships and indentured contracts had disappeared, and most families could no longer support themselves through farming alone. The primary form of labor, particularly child labor, was factory work (Bender, 1975). Children were often recruited to work in factories, particularly orphans or those from poor families. By the early to mid-19th century, it is estimated that hundreds of thousands of children—some as young as six—were employed in the textile industry, including cotton mills. In fact, some scholars estimated that children were the bulk of the workforce in many factories throughout the 19th century, with some children working six days a week, 14 hours a day (Green, 1985). Excerpts of autobiographies written by individuals who worked in factories throughout their childhoods reference dismal conditions, with poor sanitation and air quality, repetitive work on machinery that left small hands bleeding, and very long days on their feet, which in many cases significantly shortened the life spans of these child workers (Green, 1985).

Garment industry sweatshops began to spring up throughout New York and other large cities in the mid- to late-19th century. Although sweatshops eventually occurred in factory-like settings, their origin involved what was called "outwork," where workers sewed garments and other textiles in their homes. Women and children were primarily hired for these tasks since they could be paid a lower wage. Since they were paid by the piece, they often worked 14 or more hours per day, seven days a week. Children worked alongside their mothers, because their small fingers enabled them to engage in detail work, such as sewing on buttons that was challenging for adults.

The U.S. Orphan Problem

The United States also experienced several waves of political, economic, and environmental tragedies that resulted in strikingly similar conditions as those experienced in Romania today. During the 1700s and 1800s in particular, attitudes toward children were harsh, and many orphaned or uncared for children roamed the streets, particularly in growing urban areas such as New York. These street children were often treated harshly and punitively. If children were on the streets because their parents were destitute, they were often sent to almshouses, regardless of their harsh conditions, to work alongside their homeless parents. Many homeless and orphaned children were forced into a form of indentured servitude called "apprenticeships," which taught them a trade and provided cheap labor during an era that saw many economic depressions and a shortage of available workers (Katz, 1996).

The plight of the orphan did not appear to tug at the heartstrings of the average U.S. citizen during that era, not only because of the vast amount of abandoned and orphaned children (which appears to have a desensitizing effect on the human psyche), but also because during the 17th through the mid-19th century, children were not perceived to be in need of special nurturing, because childhood was not considered a distinct stage of development until years later. The influence of Puritanical religious thought as well as the general mores of the times led to the common belief that children needed to be treated with harsh discipline or they would fall victim to sinful behaviors such as laziness and vice (Trattner, 1998).

A significant shift in child welfare policy occurred in the mid-1800s, though, when the Civil War left thousands of children orphaned, making tragedy a visitor in some respect to virtually every U.S. family. Coinciding with this increase in concern over the plight of disadvantaged children was a dramatic shift in the way children on the whole were viewed. The development of the field of psychology in the first quarter of the 20th century, as well as a transition in religious thought toward a more compassionate and loving God, led to the emerging belief that children were essentially good by nature and needed to be treated with kindness, love, and nurturing to enhance their development and ultimate potential as adults (Trattner, 1998).

> During the 17th through the mid-19th century, children were not perceived to be in need of special nurturing, because childhood was not considered a distinct stage of development until years later.

In addition to these changes, the Industrial Revolution reduced the need for apprenticeship, and at the same time, stories of abhorrent conditions and mass abuse in almshouses (particularly involving abuses against children) were being widely reported. Settlement house workers, Charity Organization Societies (COSs), and government officials alike were eager to address the problem of orphaned and abused children in the latter part of the 19th century, and the most commonly suggested solution was the creation of institutions designed solely for the care of orphaned and needy children.

The Orphan Asylum

Although some orphanages existed in the 1700s, they did not become the primary means for handling needy and orphaned children until the middle to late 1800s, and by the 1890s there were more than 600 orphanages in existence in the United States (Trattner, 1998). Orphanages, or orphan asylums as they were often called, did not house just children who lost both parents to death, but also became the solution for many of the economic and environmental conditions of the time. Even though mortality rates were down in both the United States and Europe during the Industrial age (Condran & Cheney, 1982), several factors existed that resulted in the increasing need for orphanages.

Poor safety conditions in factories resulted in a relatively high prevalence of work-related injuries and death among the poorest members of society, leaving many children orphaned or fatherless. Coupled with this was a significant influx of poor immigrants in the late 1800s and early 1900s, resulting in a vulnerable segment of society often not having an extended family on which to rely in cases of parental death or disability. This was often true of recently emigrated families, who left their extended families behind in their venture to the New World.

Families who were for whatever reason suddenly unable to support their children could leave them in the temporary care of an orphanage for a small fee, but if they missed some monthly payments, the children would become wards of the state, and the parents would lose all legal rights to them (Trattner, 1998). In addition, although infectious disease was nothing new to Colonial America, several infectious disease epidemics spread through urban United States between the mid-1800s and the early part of the 1900s, including smallpox, influenza, yellow fever, cholera, typhoid, and scarlet fever, leaving many children orphaned (Condran & Cheney, 1982).

Although the orphanage system was originally perceived as a significant improvement over placing children in almshouses or forcing them into indentured servitude, these institutions were not without their share of trouble, and in time, reports of harsh treatment and abuses were common in orphanages as well. Although some orphanages were government run, most were privately run with governmental funding, but had little if any oversight or accountability. Because the government paid on a per child basis, there was a financial incentive to run large operations, with some orphanages housing as many as 2,000 children under one roof. Obedience was highly valued in these institutions out of sheer necessity, whereas individuality, play, and creativity were discouraged through strict discipline and harsh punishment (Trattner, 1998).

The next wave of child welfare reform involved the gradual shift from institutionalized care to the substitute family foster care system, or the placing-out of children into private homes prompted by the development of compulsory public education, which meant that the education of an orphan was no longer linked with the provision of housing.

The Seeds of Foster Care: The Orphan Trains

Have you ever wondered where the expression "farming kids out" came from? The origin of this term is rooted in what is called the Orphan Train movement, a program developed by the first agency to utilize in-home placement rather than institutionalized care. The New York Children's Aid Society was founded by Rev. Charles Loring Brace, who recognized the serious problem of children growing up on the streets of New York due to several tragic events from the mid-19th century. Brace estimated that as many as 5,000 children were homeless and forced to roam the streets in search of money, food, and shelter. Brace was shocked at the cruel indifference of most New Yorkers, who called these children "Street Arabs" with "bad blood." He was also appalled at reports of children as young as five years old being arrested for vagrancy (Bellingham, 1984; Brace, 1967).

Many factors contributed to the serious orphan problem in New York. Historians estimate that approximately 1,000 immigrants flooded New York on a daily basis in the mid-1800s (Von Hartz, 1978). Mass urbanization remained the trend with poor rural families flocking to the cities looking for factory work. Industry safety standards were essentially nonexistent; thus, factory-related deaths were at an all-time high. An outbreak of typhoid fever also left many children orphaned or half-orphaned with new widows who had virtually no way to support their children because government aid was not yet available. These harsh social conditions, coupled with the absence of any organized governmental subsidy, left many children to fend for themselves on the streets of New York, resorting to any means for survival.

Brace feared that the temptations of street life would preclude any possibility that these children would grow up to be God-fearing, responsible adults, and he reasoned that children who had no parents, or whose parents could no longer care for them, would be far better off living in the clean open spaces of the farming communities out west, where fresh air and the need for workers were plentiful. Because the rail lines were rapidly opening up the West, Brace developed an innovative program where children would be loaded onto trains and taken west to good Christian farming families. Notices were sent in advance of train arrivals, and communities along the train line would come out and meet the train, so that families who had expressed an interest in taking one or more children could examine the children and take them right then, if they desired. Brace convened committees who would interview families to ensure that they met the standards for qualified adoptive or foster families.

Survivors of these Orphan Trains have talked about how they felt like cattle, being paraded across a stage. Interested foster parents would often feel the children's muscles and check their teeth before deciding what child they would take. Few parents would take more than one child, thus siblings were most often split up, sometimes without even a passing comment made by the child care agents or the new parents (Patrick, Sheets, & Trickel, 1990). It was almost as if the breaking of lifelong family bonds was considered trivial compared to the gift these children were receiving by being rescued from their hopeless existence on the streets.

Most children were not legally adopted, but were placed with a family under an indentured contract, which served two purposes. First, this type of contract allowed the placement agency to take the children back if something went wrong with the placement. Second, children placed under an indentured contract could not inherit property; thus, farming families could adopt boys to work on the farm or girls to assist with the housework, but didn't have to worry about them inheriting the family assets (Trattner, 1998; Warren, 1995).

The Orphan Trains ran from 1854 to 1929, delivering approximately 150,000 children to new homes across the west, from the midwestern states to Texas, and even as far west as California. Whether this social experiment was a glowing success or a miserable failure (or somewhere in between) depends on whom one asks. Some children were placed in wonderful, loving homes and grew up to be happy and responsible adults, who feel strongly that the Orphan Trains were a true blessing. But other survivors of the Orphan Trains shared stories of heartache and abuse. Some tell stories of lives no better than that of slaves, where they were taken in by families for no other reason than to provide hard labor for the cost of bed and board. Others tell stories of having siblings torn from their sides as families chose one child, leaving brothers and sisters on the train. And still others tell stories of failed adoptions, where farming families exercised their one-year return option, sending the children back to the orphanage or allowing the children to drift from farm to farm to earn their keep (Holt, 1992).

Typical wanted advertisement posted throughout the Midwest by the Children's Home and Aid Society between 1854 and 1929. Nemaha County Herald/nebraska State Historical Society

Children on the Orphan Train.
Riis, Jacob A. (Jacob August), 1849–1914/Library of Congress Prints and Photographs Division [LC-USZ62-17233]

Eventually new child welfare practices caught up with new child development theories, leading to a general focus shifting from one of work virtue to one of valuing childhood play. By the early 20th century the practice of "farming out" children received increasing criticism, and the last trainload of children was delivered to its many destinations in 1929. Despite the controversy surrounding the Orphan Train movement and the many similar outplacement programs that followed across the country, even its harshest critics agreed that it was a far better alternative than allowing children to fend for themselves on the streets of New York. Also, despite the program's many shortcomings, including poor oversight and insufficient screening of the families, it is considered the forerunner of the current foster care system in the United States, where children are placed in available private homes, rather than in institutions (Trattner, 1998).

Jane Addams and the Fight for Child Labor Laws

At around the same time that Charles Loring Brace was sending New York orphans out west, Jane Addams and her friend Ellen Gates Starr were busy founding Hull-House of Chicago, the first U.S. settlement house providing residential and what we would now call "wrap around" services, as well as advocacy to marginalized populations working in sweatshop conditions in Chicago. Addams was appalled by the conditions of those living in poverty in urban communities, particularly the plight of recently arrived immigrants, who were forced to live in substandard tenement housing and work long hours in factories, often in very dangerous working conditions.

Hull-House offered several services for children and their widowed mothers, including after-school care for those children whose mothers worked long hours in factories. Providing comprehensive services to those in need, and living among them in their own community were some of the ways in which Addams became aware of the plight of children forced to work in the factories. In her autobiography *Twenty Years at Hull-House*, Addams wrote of her first encounter with child labor:

> Our very first Christmas at Hull-House, when we as yet knew nothing of child labor, a number of little girls refused the candy which was offered them as part of the Christmas good cheer, saying simply that they "worked in a candy factory and could not bear the sight of it." We discovered that for six weeks they had from seven in the morning until nine at night, and they were exhausted as well as satiated … during the same winter from a Hull-House club were injured at one machine in a neighboring factory for a lack of a guard which would have cost but a few dollars. When the injury of one of these boys resulted in his death, we felt quite sure that the owners of the factory would share our horror and remorse, and that they would do everything possible to prevent the recurrence of such a tragedy. To our surprise they did nothing whatever, and I made my first acquaintance then with those pathetic documents signed by the parents of working children, that they will make no claim for damages resulting from "carelessness." (Addams, 1911, pp. 198–199)

Addams and her colleagues began an advocacy campaign against sweatshop conditions in Chicago factories early in the Hull-House's existence, advocating in particular for the women and children who were most often hired to work in them. Their activism seemed to pay off quickly when the Illinois legislature passed a law limiting the word *day* to just eight hours (from the typical 12- to 14-hour day). Their excitement though was soon tempered when the law was quickly overturned by the Illinois Supreme Court as unconstitutional. In her autobiography Addams discussed how the greatest opposition to child labor laws came from business sector—business men from large corporations (such as Chicago glass companies), who considered such legislation as "radicalism," arguing that their companies would not be able to survive without the labor of children (Addams, 2011; Martin, in press).

Addams and the Hull-House networked quite extensively joining efforts with trade unions and even the Democratic Party, which in 1892 adopted into its platform union recommendations to prohibit children under the age of 15 years old from working in factories. Addams and her Hull-House colleagues increased the focus of their activism to the federal level with their support for the *Sulzer Bill*, which when passed allowed for the creation of the Department of Labor. In 1904 the National Child Labor Committee was formed, and Addams served as chairman for one term. In 1912, one of Addams's Hull-House colleagues, Julia Lathrop was appointed chief of a new federal agency by President William Taft, focusing on child welfare, including child labor. As chief of the Children's Bureau Lathrop was responsible for investigating and reporting on all relevant issues pertaining to the welfare of children from all classes, and spent a considerable amount of time extensively researching the dangers of child labor (Martin, in press).

After several failed attempts federal legislation barring child labor was finally passed in 1938, and signed into law by President Franklin D. Roosevelt, three years after Addams's death. The Fair Labor Standards Act is a comprehensive bill regulating various aspects of labor in the United States, including child labor. The act defined "oppressive child labor" and set minimum ages of employment and the number of hours children were allowed to work. This act is still in existence today and has been amended several times to address such issues as equal pay (Equal Pay Act of 1963), age discrimination (Age Discrimination in Employment Act of 1967), and low wages (federal minimum wage increases) (Martin, in press).

Professional History

Understanding and Mastery of Professional History: Historical roots of human services

Critical Thinking Question: It is clear that the treatment of children throughout the history of the United States (as in all cultures across time) is shaped by religious and philosophical beliefs, societal structures, and economic systems. How is current child welfare policy and practice shaped by these same factors? In 50 or 100 years, what will historians find laudable about our current policies? What will they find short-sighted or harmful to children?

Overview of the Current U.S. Child Welfare System

Children living in contemporary western societies face very different challenges than children living 100 years ago. Child labor laws preclude child exploitation in the workforce, and federal and state social welfare programs now exist, which have helped not only to alleviate poverty but also have helped protect families from the effect of various catastrophes, such as natural disasters and pandemics. Also, vulnerable groups of children are far better protected from disparity in treatment through the passage of such federal legislation as the Civil Rights Act of 1964 and the Americans with Disabilities Act; yet, there remains disparity in treatment of children from certain ethnic groups, such as African Americans, Latinos, and Native Americans.

There also remain serious issues with how some children are treated within U.S. society. For instance, few truly effective systems are in place to assist runaway and homeless youth. Far too often adolescents who experienced physical and sexual abuse in their homes are typically not served well by child protective custody services, and often choose to live on the streets rather than remain in their homes, or trust the "system" to provide for their care. Far too many children are charged as adults for crimes they committed as children, and most of these are children of color—primarily African American boys. African American girls also experience disparity in treatment by organizations charged with the responsibility for their protection. For instance, there is a growing recognition that African American girls are far more likely to be victims of domestic sex trafficking; yet, if they are apprehended, rather than being treated as victims, they are far more likely to be charged as prostitutes and sent back to the streets (Martin, in press).

With regard to child protection and the care of orphaned and abused children, care has slowly transitioned from institutionalized care, to primarily substitute family care or foster care over the past 100 years. By 1980, virtually no children remained in institutionalized care in the United States, excluding group homes, treatment centers, and homes for developmentally disabled children (Shughart & Chappell, 1999). Government public assistance programs, which developed in the 1960s, reduced the necessity for the removal of children from their homes due to poverty, because single mothers now had someplace to go for financial help in raising their children (Trattner, 1998).

The demographic makeup of children currently in the foster care system differs considerably from the children institutionalized in orphanages in the 1800s, as well as the children of the Orphan Train era. Thus, gone are the days where the majority of children being placed into substitute care were orphaned due to industrial accidents, war, or illness. Instead, the majority of children currently in child protective custody have been removed from their homes due to serious maltreatment. Also, unlike earlier eras when orphanage placements were most often permanent, almost half of all children currently in foster care have the goal of reunifying with their biological parents (U.S. Department of Health and Human Services, 2008).

As of September 1, 2010 (the most recent statistics available), there were approximately 408,325 children in the U.S. foster care system. This represents a decrease of almost 105,000 children since 2006, and it also represents a continued pattern of reduction of children in out-of-home placement since 1998 (U.S. Department of Health and Human Services, 2010). Approximately 41 percent of all children in foster care are Caucasian, followed by 29 percent African American children, and 21 percent Hispanic children. These demographics indicate an overrepresentation of African American children in the foster care system in particular because African Americans constitute only

15 percent of the general population, whereas Caucasians constitute 61 percent of the general population.

The average age of children in care is about nine years old, with the greatest number of children in foster care placement between the ages of 11 and 15 years, followed by children ages one through five years. About half of all children in placement are in nonrelative foster care placement, followed by about a quarter of all children placed in relative care. The median length of stay in foster care is about 18 months, but it appears that if children aren't placed in the first 18 months of placement, chances increase that children will remain in placement for several years. The greatest number of children who left the child welfare system in 2010 were infants under 3 years of age, and those exiting the system ages 17 and above (U.S. Department of Health and Human Services, 2010).

A primary goal of the foster care system is to reunite foster care children with their biological parents whenever possible.

The U.S. child welfare system exists to provide a safety net for children and families in crisis. A primary goal of the foster care system is to reunite foster care children with their biological parents whenever possible (Sanchirico & Jablonka, 2000). Federal and state laws have established three basic goals for children in the U.S. child welfare system:

- *Safety* from abuse and neglect
- *Permanency* in a stable, loving home (preferably with the biological parents)
- *Well-being* of the child with regard to their physical health, mental health, and developmental and educational needs

How these goals are met depends on the specific issues involved in each case, but before these various alternatives are considered, it is important to understand how a child enters the child welfare system in the first place.

Getting into the System

So, how does a child end up in foster care? Made-for-television movies might have the public thinking that child welfare workers have the power to remove children from homes with minimal evidence of abuse. Yet, in reality, several criteria must be met to place a child into protective custody, and a child cannot be removed from a family home without a judge's approval. The U.S. Constitution guarantees certain liberties to parents by giving them the right to parent their child in the manner they see fit. But such liberties are balanced by the parents' duty to protect their child's safety and ensure their well-being. If parents cannot or will not protect their children from *significant* harm, the state has the legal obligation to intervene (Goldman & Salus, 2003).

The U.S. Congress has passed several pieces of legislation that support the state's obligation to protect its youngest residents, including the Child Abuse Prevention and Treatment Act (CAPTA) of 1974, which was established to ensure that children of maltreatment are reported to the appropriate authorities. This act (which was most recently amended in 2010) also provides minimum standards for definitions of the different types of child maltreatment. The Adoption Assistance and Child Welfare Act of 1980 requires that states develop supportive programs and procedures enabling maltreated children to remain in their own homes and to assist family reunification following out-of-home placements.

Other legislation is aimed at (1) improving court efficiency so that child abuse cases will not languish in the court system for years, (2) providing assistance to foster care children approaching their eighteenth birthday, and (3) bolstering family preservation programs designed as an early intervention program in the hopes of circumventing out-of-home placement (Goldman & Salus, 2003).

In 1997 the president signed into law the Adoption and Safe Families Act, which amended and made improvements to the Adoption Assistance and Child Welfare Act of 1980. Among the amendments the act provides are incentives for families adopting children in the foster care system and mandates that states provide evidence of adoption efforts. Amendments also set a new accelerated time line for terminating the rights of parents whose children are in foster care placement. As we will see in subsequent sections of this chapter, there are both positive and negative aspects of this legislation. Certainly no one wants abused and neglected children to languish in temporary placement, but expediting the finding of permanent homes should not be at the expense of biological parents' rights to have an appropriate amount of time to meet the state's criteria for regaining the custody of their children. Balancing the rights of the biological parents with the best interest of their child is challenging, particularly in light of the complexity involved in many foster care cases.

In 2006, the Safe and Timely Interstate Placement of Foster Children Act (Pub. L. No. 109-239) was passed, which made it easier to place children in another state, if necessary. This legislation holds states accountable for the orderly, safe and timely placement of children across state lines by requiring that home studies be completed in less than 60 days, and that the children be accepted within 14 days of completion. The legislation also provides grants for interstate placement and requires caseworkers to make interstate visits, when necessary.

Quite likely, the most significant federal legislation passed recently is the Fostering Connections to Success and Increasing Adoptions Act of 2008 (Pub. L. No. 110-351), which former president Bush signed into law in October 2008. This law amends the Social Security Act by enhancing incentives, particularly in regard to kinship care, including providing kinship guardian financial assistance as well as providing "family connection" grants designed to facilitate and support kinship care. This legislation also includes provisions related to education and healthcare particularly for children in kinship care, many of whom were not eligible for special assistance programs unless they were in nonrelative care.

Child Abuse Investigations

Mandated Reporters

There are several ways that a child abuse investigation may be initiated, but all have their origin in a concern that a child is being mistreated in some manner. Many professionals, such as counselors, teachers, physicians, and even Sunday school teachers, are required by law to call their state's child abuse hotline immediately if they suspect that a child is being abused or neglected. *Mandated reporters* typically fall into one of several categories and include professionals who work with children as a part of their normal work duties. Mandated reporters include personnel in the following fields: medical, schools, social service, mental health, law enforcement, child care, and members of the clergy.

Most states have strict laws that define the parameters of child abuse reporting, including delineating what constitutes a reportable concern, the time frame in which a mandated reporter must report the suspected abuse, and the consequences of failing to report suspected abuse, such as the suspension of one's professional license. In fact, in most states, the failure to comply with mandated reporting requirements is a crime (a misdemeanor or even a felony for repeated failures). In many states, the majority of calls made to the child abuse hotline are from mandated reporters, but this does not preclude anyone from calling the child abuse hotline if they suspect that a child is being abused or neglected by a parent or caregiver. Thus, it is not uncommon for neighbors, friends, or even relatives to report suspected child abuse, and those who are not mandated reporters are allowed to call anonymously.

Sequence of Events in the Reporting and Investigation of Child Abuse

A child abuse investigation is initiated when someone, either a concerned individual or a mandated reporter, places a call to the state child abuse hotline. Due to the intrusive nature of an abuse investigation, federal and state laws exist to protect the privacy of family life. Thus, hotline workers must adhere to strict guidelines regarding what reports can and cannot be accepted. If the report of alleged abuse meets the stated criteria, then the report will be accepted and investigated in a timely manner.

For state CPS agencies to receive federal funding, the federal law mandates that all child abuse reports be screened immediately and investigated in a timely manner (CAPTA, 2010). Although federal law does not specify a particular time frame, most states have compliance laws stipulating specific guidelines mandating that reports of abuse be investigated anywhere from immediately after receiving a report for cases involving imminent risk, to 10 days in some states for reports with moderate to minimal risk to the child (Kopel, Charlton, & Well, 2003).

Once a hotline worker makes the decision to accept a child abuse report, the case is sent to the appropriate regional agency and assigned to an abuse investigator, who is a licensed social worker or other licensed human service professional. The actual investigation will vary depending on the specific circumstances of the allegations, but most investigations will involve interviewing the child, the nonoffending parent(s), and the alleged perpetrator. Although the sequence of the interviews might alter depending on the specific circumstances of the case, most investigators prefer to interview the child before the parents or caregivers to avoid the potential for influencing or intimidating the child.

Types of Child Maltreatment

Child maltreatment is a crime regardless of who the perpetrator is and should always be reported to authorities, but a state's CPS agency becomes involved when the abuse is perpetrated by someone who is acting in a caregiving role to the child. This includes a parent, a relative, a parent's boyfriend or girlfriend, a teacher, or even a babysitter.

Although each state is charged with the responsibility for defining child abuse and neglect according to state statute, the federal government has developed a definition of what constitutes the minimum standard for child abuse and neglect and has created four general categories of child maltreatment, including neglect, physical abuse, sexual abuse, and emotional abuse. The following is the U.S. Health and Human Services' definition of each type of abuse, but again it is important to remember that each state, although bound to this minimum standard, will likely have additional criteria and scenarios that qualify as abuse (National Clearinghouse on Child Abuse and Neglect, 2005).

Neglect is failure to provide for a child's basic needs. Neglect may be

- Physical (e.g., failure to provide necessary food or shelter or lack of appropriate supervision)
- Medical (e.g., failure to provide necessary medical or mental health treatment)
- Educational (e.g., failure to educate a child or attend to the child's special education needs)
- Emotional (e.g., inattention to a child's emotional needs, failure to provide psychological care, or permitting the child to use alcohol or other drugs)

Because cultural values, standards of care in the community, and poverty may be contributing factors related to caregiving challenges, the existence of some of these problems does not necessarily indicate that the legal abuse of a child is occurring. Rather, the manifestation of certain problems within a family system, such as not sending a child to school, may indicate an overwhelmed family's need for information and general assistance. Yet, if a family fails to utilize the information, assistance, and resources provided and the child's health or safety is determined to be at risk, then CPS intervention may be required.

Physical abuse includes physical injury (ranging from minor bruises to severe fractures or death) as a result of punching, beating, kicking, biting, shaking, throwing, stabbing, choking, hitting (with a hand, stick, strap, or other object), burning, or otherwise harming a child. An injury is considered abuse regardless of whether the caretaker intended to hurt the child.

Sexual abuse includes activities by a parent or caretaker that include fondling a child's genitals, penetration, incest, rape, sodomy, indecent exposure, and exploitation through prostitution or the production of pornographic materials.

Emotional abuse involves a pattern of behavior that impairs a child's emotional development or sense of self-worth. This may include constant criticism, threats, or rejection, as well as withholding love, support, or guidance. Emotional abuse is often difficult to prove, and therefore, CPS may not be able to intervene without evidence of significant harm to the child. Emotional abuse is almost always present when other forms of abuse are identified.

The Forensic Interview

In the past 25 years, allegations of child abuse, particularly child sexual abuse, have skyrocketed. Reasons for this include increased public awareness, mandatory reporting requirements, and a significant change in attitudes regarding child abuse, with an increasing sentiment that abuse is no longer a private family matter. Yet, as the pendulum swung, the 1970s witnessed a sort of frenzy in child sexual abuse reporting, and a popular contention among mental health experts was that children were incapable of making false allegations. This belief fostered a sense of overeagerness on the part of some therapists, who sometimes used inappropriate interviewing techniques, with leading questions: "Did he touch you on your privates?", forced choice: "Did he touch you under your clothing, or over your clothing?", option posing: "I heard that your uncle has been bothering you", or suggestive questions: "Many kids at your school have said that your teacher has touched them, did he touch you too?".

Eventually this method of questioning was met with overwhelming criticism, particularly by members of the legal community, who were charged with defending those individuals falsely accused of sexually abusing children in their charge. These types of questions significantly increased the likelihood of erroneous disclosures, particularly with preschool-aged children (Hewitt, 1999; Peterson & Biggs, 1997; Poole & Lindsay, 1998).

In response to such criticism, CPS agencies across the country developed pilot programs that combined the resources from several investigative branches, including CPS agencies, police departments, and district attorneys' offices. This coordinated approach not only prevents the trauma of duplicative interviews by separate enforcement agencies, but also allows for the highly specialized training of investigators on forensic interviewing techniques that avoid any type of suggestive or leading questions.

Although there is a general understanding among investigators of what constitutes a forensic interview, there was still concern that many interviewers used types of questions that were somewhat leading in nature, including an interviewer's inadvertent reaction to a child's response that either encouraged or discouraged an honest disclosure. For instance, an investigator who strongly believes that a child has been abused may inadvertently respond with frustration if a child denies the abuse, which may influence the child, who wants to please the investigator, to give a false disclosure of abuse. Even an expression of sympathy on the part of the interviewer, in response to disclosures of abuse, can inadvertently encourage a child to embellish somewhat to receive more of the interviewer's compassion.

The National Institute of Child Health and Human Development (NICHD) developed a forensic interviewing protocol that teaches interviewers how to ask open-ended questions, using retrieval cues that rely on free recall. "Tell me everything you can remember" is an example of an open-ended question. "Tell me more about the room you were in" is an example of a retrieval cue (Bourg, Broderick, & Flagor, 1999; Sternberg, Lamb, & Orbach, 2001).

To Intervene, or Not Intervene: Models for Decision Making

Many variables influence the outcome of an investigation, including the criteria with which a CPS agency uses to determine (1) whether abuse is occurring and (2) whether the abuse rises to the level of warranting intervention. In other words, it is possible for some abuse reports to be determined as *unfounded,* even though the investigator may strongly suspect that an unhealthy home environment does exist. But another reason for not substantiating an incident of child abuse relates more to poor or inconsistent decision-making policies within a CPS agency due to human errors in decision making. DePanfilis and Scannapieco (1994) discussed the vital importance of CPS agencies developing and adhering to a consistent and realistic decision-making model when determining whether family intervention is warranted in order to avoid the inherent problems in making bias-free and fact-based decisions. Child abuse investigators are responsible for:

1. assessing the safety of children who are at risk of maltreatment,
2. deciding what types and levels of services may be immediately needed to keep children safe, and
3. determining under what conditions children must be placed in out-of-home care for their protection. (p. 229)

According to the Child Welfare League of American (CWLA) there are several approaches to making risk assessments of child maltreatment in child protection. The approaches are either statistically based or based upon consensus of experts in the field, as well as research on the area of child maltreatment. Actuarial models of risk assessment and decision making assess families based upon factors and characteristics that are statistically associated with the recurrence of maltreatment. Because the inventory is based upon a statistical calculation, the validity of the inventory may be considered higher than the consensus-based model risk assessments; yet, many within the child welfare fields express concern that actuarial models do not allow enough for clinical assessment. An example of an actuarial model for risk assessment and decision making includes the CRC Actuarial Models for Risk Assessment (Austin, D'Andrade, Lemon, Benton, Chow & Reyes, 2005).

Consensus-based approaches include the theoretically-empirically guided approach that ranks a series of factors that have empirical support for their association with child maltreatment, and Family Assessment Scales (CWLA, 2005). Some examples of consensus-based models for risk assessment and decision making include the Washington Risk Assessment Matrix (WRAM), the California Family Assessment and Factor Analysis (CFAFA, or the "Fresno Model"), and the Child Emergency Response Assessment Protocol (CERAP) (Austin, D'Andrade, Lemon, Benton, Chow, & Reyes, 2005).

The *Child at Risk Field System* (CARF) is an example of a consensus-based risk-assessment model that has been tested in the field. The CARF provides the following guidelines for abuse investigators making a determination about abuse:

Where children were determined to be maltreated and unsafe, the offending parents

1. were out of control,
2. were frequently violent,
3. showed no remorse,
4. may actually request placement,
5. did not respond to previous attempts to intervene, and/or
6. location was unknown.

And the caseworker believed that

1. the parents were a flight risk,
2. the child had special needs the parents could not meet,
3. the conditions in the home are life-threatening, and/or
4. the nonoffending parent could not protect the children.

Where children were determined to be maltreated and *safe,* the parents

1. possessed a sufficient amount of impulse control,
2. accepted responsibility for the situation in their home,
3. had appropriate understanding of the child, showed concern for the child and remorse for the maltreatment,
4. had a history of accessing help and services, and
5. exhibited knowledge of good parenting skills.

Thus, although definitions of child maltreatment are statutorily defined, there is a tremendous amount of latitude that an investigator has in determining whether child maltreatment is occurring and whether the extent of the abuse warrants intervention. Primarily, it is through the use of an effective and well-tested decision-making model that an abuse investigator will have the greatest likelihood of making an appropriate determination in a child abuse investigation.

Working with Children in Placement

Permanency Plans

When an abuse investigator determines that a child must be placed into protective custody, the child is removed from the home and placed in one of many environments, including relative foster care, nonrelative foster care, or an emergency shelter pending more permanent placement. The case is then transferred to a family caseworker, who evaluates all the relevant dynamics of the case (i.e., reason for placement, nature of abuse, attitude of the parents), as well as assesses the strengths and weaknesses of the biological parents and the family structure. A permanency goal for the child must then be determined and can include:

1. Reunification with the biological parents
2. Living with relatives
3. Guardianship with close friends
4. Short-term or long-term foster care
5. Emancipation (with older adolescents)
6. Adoption with termination of parental rights

Although reunification with the biological parents remains the most common permanency plan, recent changes in many state and federal laws have shifted the focus from protecting the biological family unit to considering the "best interest of the child."

Human Services Delivery Systems

Understanding and Mastery of Human Services Delivery Systems: Major models used to conceptualize and integrate prevention, maintenance, intervention, rehabilitation, and healthy functioning

Critical Thinking Question: A number of tools exist for assessing the occurrence of child abuse or maltreatment and for gauging the likelihood of a reoccurrence of the abuse. Yet, these tools are not perfect, and critics argue that they should not be used as a substitute for the professional experience and expertise of human service workers. How might human service professionals balance the use of these tools with their own practice wisdom?

The reason for this shift can be traced to several high-profile cases in the mid-1990s where children were either seriously abused or killed after being reunified with their biological parents. Well-meaning child advocates launched campaigns in Washington, DC, appealing to Congress to do something about the horrible plight of children who were returned to their biological families only to face further abuse and sometimes their deaths in a failed effort to save troubled families.

Although there was no documented increase of child maltreatment during this time period, newspaper and magazine articles highlighting tragic (but rare) cases of continued abuse or deaths when children were reunited with their families were passed around Congress, and articles such as "The Little Boy Who Didn't Have to Die" were utilized in an effort to make an emotional appeal to legislators to shift priorities from family reunification to parental termination and subsequent adoption (Spake, 1994). The result of this campaign was the passage of the American Adoption and Safe Family Act of 1997, which marked a clear departure away from abuse prevention and family preservation and toward paving the way for termination of biological parents' rights, clearing the way for adoption of children in foster care placement.

The *best interest of the child* standard may sound great on the surface, but it has been the subject of significant scrutiny, with critics questioning just how this standard is being applied. In other words, best interest of the child according to whom? According to the foster parents? The courts? The caseworker? It doesn't take much analysis to see how easily this standard can be abused. For instance, what if the caseworker determines that it is in the best interest of the child to be placed permanently with a two-parent financially secure home rather than to be returned to the child's poor single mother, regardless of how diligently this parent works to regain custody? The potential to make permanency plans that discriminate against biological parents who are marginalized members of society, such as parents who are poor, single, of a minority race, homosexual, and perhaps even undocumented immigrants, is significant.

Dorothy Roberts, author of *Shattered Bonds: The Color of Child Welfare* (2002), cautions that the new federal law creates many problems, including a conflict created when caseworkers are required to pursue two permanency plans at the same time to comply with the new permanency plan time frames—reunification with the family and possible adoption. What many caseworkers do to accomplish this task is to place foster children in *preadoptive* homes while at the same time planning for reunification with the biological parents. This creates a situation where the biological parents' rights are often in conflict with the children's rights, and where foster care families, who are by definition charged with the responsibility of fostering a relationship between the children and their biological parents, are now competing for the children.

Another possible conflict according to Roberts includes the act's adoption incentive program, where states are given financial incentives of $4,000 for each child placed for adoption (above a baseline) and $6,000 for a special needs adoption. The potential for agency abuse is evident as states scramble to replace lost revenue due to the poor economy. Roberts warns that this new legislation was not directed at effecting faster termination of parental rights in cases with severe abuse because these cases were always relatively "open-and-shut." Rather, it is the cases involving poverty-related maltreatment, most often in African American and Native American homes, that have been most affected by this new federal law, which Roberts fears has led to increased social injustice in many CPS agencies.

For this reason as well as many others, the caseworker must be careful in determining what criteria to use in making permanency determination recommendations. For instance, some experts have suggested using attachment ties as a guide in deciding a permanent placement plan (Gauthier, Fortin, & Jéliu, 2004). These researchers suggest that a child should remain with the family who they appear to have the greatest attachment with to avoid further emotional ruptures. Yet, the potential for foster parent bias is great, particularly in light of the fact that the foster parents will have a greater advantage over the biological parents because children will, of course, have a greater likelihood of developing a stronger attachment to the family they are living with, particularly if biological parents are restricted from participating regularly in their children's lives through regular visitation. U.S. history is filled with reports of abuses of this sort, where parents considered unworthy have experienced unfair treatment by CPS agencies (see discussion on Native Americans), and this legislation risks escorting in a new dawn of similar abuses.

Working with Biological Families

A caseworker works with the biological parents most closely when it is determined that the most appropriate permanency plan is parent reunification. Once a child has been placed into foster care, the caseworker must prepare a detailed service plan, typically within 30 days, outlining goals that the biological parents must accomplish before regaining custody of their child. The specific goals must be related to the identified parenting deficits, but can include goals such as:

1. Counseling
2. Parenting classes
3. Treatment for substance abuse
4. Anger management
5. Securing employment
6. Securing housing
7. Maintaining regular contact with children

It is then the responsibility of the caseworker to facilitate the biological parents achieving these goals. This might involve giving referrals to the parents or securing services for them, as well as monitoring their ongoing progress.

It is also important for caseworkers to be aware that biological parents who have had their children removed may be enduring emotional trauma in response to this loss, which may result in them behaving in ways that could be uncharacteristic for them. The strain of having to be accountable to external forces exerting control over their lives makes many biological parents vulnerable to feeling overwhelming shame, which may manifest in defensiveness that could be misinterpreted as indifference or a lack of remorse. An effective caseworker will understand this possible dynamic and will create an environment where biological parents will be able to overcome the barrier of defensiveness and shame and work on the issues identified in their service plan.

The intergenerational nature of child abuse has been well documented in research (Bentovim, 2002, 2004; Ehrensaft, Cohen, & Brown, 2003; Newcomb, Locke, & Thomas, 2001; Pears & Capaldi, 2001), and although the majority of individuals who have been abused in childhood do not go on to abuse their own children, parents who are abusive to their children have likely been abused in their own childhoods. Homes marked by violence, drug abuse, neglect, and sexual abuse create patterns that can be passed down to the next generation. Although it might not initially make sense that someone who endured the pain of abuse would inflict this same abuse on their own child, the complex nature of child abuse oftentimes renders abuse patterns beyond the control of the batterer without some form of intervention. For instance, consider Case Study 5.1 about Rick.

> Although the majority of individuals who have been abused in childhood do not go on to abuse their own children, parents who are abusive to their children have likely been abused in their own childhoods.

CASE STUDY 5.1

Case Example of the Intergenerational Cycle of Child Abuse

Rick grew up in a home marked with domestic violence, which oftentimes extended to the children. Rick's mother was chronically depressed and often resorted to using alcohol to avoid dealing with her feelings. Rick recalls days and sometimes weeks where she refused to get out of bed, and he was responsible for caring for his younger siblings. His father also had an alcohol problem and would fly into nightly rages where he would physically abuse his mother. When Rick got older, he attempted to intervene and protect his mother, which only resulted in his father physically abusing him. In addition to physical abuse, Rick was also the victim of emotional abuse and neglect. Rick's father would often call him derogatory names and humiliate him by telling him that he would amount to nothing in life. It seemed as though Rick could do nothing right, and when he was about 12 years old, he promised himself that he would never allow anyone to hurt or humiliate him again. Rick married when he was 21 and was hopeful that his life of being victimized was over. He loved his wife very much and was determined to be the best husband and father he could possibly be. He vowed not to repeat the mistakes of his parents. But deep inside he was plagued with fears that he wasn't good enough for his wife and that she would eventually leave him. He became increasingly jealous and accused his wife of wanting to leave him. If she tried to convince him otherwise, he accused her of lying. When she became pregnant he was thrilled, but after the baby was born he became upset because his wife seemed to want to spend all her time with the baby, leaving him to fend for himself. One day Rick's boss called him into his office and pointed out a mistake that Rick made. All Rick could think of was the promise he had made to himself years ago to never allow anyone to hurt or ridicule him again. Even though his boss's comments would have seemed reasonable to most people, to Rick it was a recreation of the abuse he endured as a child. He lost control of his temper, slammed his fist into the wall, and quit his job. When he got home he told his wife and fully expected her to sympathize with him and support his decision to not tolerate such abuse, but instead she complained that his act was selfish, particularly in light of his responsibilities as a father. Rick completely lost his temper and in a blinding rage accused his wife of betraying him. In the blur that followed, Rick accused her of cheating on him, of caring about the baby more than him, and of even getting pregnant by another man. In the midst of his angry outburst he shoved his wife against the wall. All he could think of was how this woman who he thought was his savior was really his enemy, and at that moment he hated her for allowing him to lower his guard and trust her. All the pain of his childhood, with all the hurt and humiliation, came rushing back, and he began to choke her. When his baby interrupted his rage, he screamed at his son to shut up. When his baby's crying got louder, he picked him up and shook him violently.

The case study about Rick illustrates some of the dynamics at play with the intergenerational transmission of abuse, and why it is so important for caseworkers to understand what may occur in the mind of someone who has endured physical, emotional, and sexual abuse at the hands of parents and other caregivers. Individuals who have suffered significant childhood abuse often suffer from low frustration tolerance, displaced anger, inability to delay gratification, impulse control problems, problems with emotional regulation, difficulty attaching to others, and an unstable self-identity (Bentovim, 2002, 2004). Issues such as poor parental modeling, lack of understanding about normal child development, and an individual's level of residual anger and frustration tolerance affect a person's ability to positively parent their children.

BIOLOGICAL PARENTS AND THEIR CHILDREN: MAINTAINING THE CONNECTION   A part of any good reunification plan will involve a visitation schedule that supports and encourages the child's relationship with the biological parents and provides them with applying new parenting techniques that they've learned in parenting classes and counseling (Sanchirico & Jablonka, 2000).

An effective caseworker will give consistent feedback to the biological parents about their progress toward meeting service plan goals, will balance constructive feedback with encouragement, will protect the parent–child relationship, and will do whatever possible to remove barriers to complying with their service plan, such as finding alternate mental health providers when waiting lists would cause unreasonable delays and resolving conflicts between goals, such as not scheduling visitation during the parents' working hours when maintaining stable employment is a service plan goal.

Working with Foster Children: Common Clinical Issues

Foster children obviously come in all "shapes and sizes," so it is difficult to summarize the issues and experiences of the majority of children in foster care in a page or two. But certain generalizations can be made, particularly with regard to the types of experiences that bring a child into substitute care, as well as the range of short-term and long-term emotional and psychological manifestations many children in foster care may experience. The clinical issues that a caseworker may deal with will vary depending on variables such as the age of the child, the length of time in placement, the reasons for placement, and the plan for permanency (i.e., adoption or family reunification).

Younger children are typically easier to place and may display less oppositional behavior than adolescents, who are often placed in group homes.

Children who have been sexually abused often manifest emotional problems that require sophisticated handling on the part of the caseworkers, therapists, and foster parents. Sexually abused children may act out sexually with their foster parents as well as other children, which can create an uncomfortable situation, particularly for those who are unfamiliar with such acting out behavior. In addition, most children who have been mistreated in some manner may behave well during the honeymoon period of placement, but then act out once they begin to feel more secure. This phenomenon can lead to disrupted placements if the foster parents are unaware of the dynamics behind this shift in behavior.

A recent national survey of approximately 4,000 foster care children, aged 2 through 14, who had been removed from their homes due to maltreatment, revealed that nearly half of these children had clinically significant psychological and/or behavioral problems. Alarmingly though, only about half of these children had received any counseling in the past year. The children who were the most likely to receive mental health services were younger children who had been victims of sexual abuse. African American children were the least likely to receive mental health services, as were children who remained living in their biological homes (Burns et al., 2004). Siu and Hogan (1989) identified five clinical themes experienced by most children in foster care and made recommendations for how child welfare caseworkers should respond. These include issues related to (1) separation; (2) loss, grief, and mourning; (3) identity issues; (4) continuity of family ties; and (5) crisis.

Separation

Children involved in the child welfare system are contending with either issues related to separation from their biological family members or the threat of separation. Siu and Hogan (1989) recommended that caseworkers be familiar with the psychological dynamics involved in such separations as they relate to each developmental stage. It is important for caseworkers to acknowledge that these children are not just being separated from their biological parents, but are experiencing multiple separations, such as separation from their extended family, perhaps their siblings and their familiar surroundings, including their bedroom, house, neighborhood, and even their family pets. Caseworkers need to confront these separation issues head on with the children, resisting the temptation to avoid them in response to their own separation anxiety.

Children often go through different stages when confronted with significant separation, beginning with the *preprotest* stage, where children accept removal from their home with little protest. But this stage is ultimately followed by the *protest* stage, where children can respond with outright combative and oppositional behavior or with a more subtle uncooperative attitude. The third stage is marked by *despair,* where the child often submits to the placement with a sense of brokenness and hopelessness. The final stage involves *adjustment* to the placement, but involves a sense of detachment to that which the child had been attached—namely, their biological families (Rutter, 1978).

Caseworkers can respond to children dealing with separation issues by being honest with them (in an age-appropriate manner) regarding what is happening with their families and by helping to prepare them for the upcoming changes in order to reduce the anxiety associated with anticipating the unknown. Younger children are far more likely to be operating in the "here and now"; thus, it is important for the caseworker to reassure the child that the separation is only temporary (if the goal is family reunification) and that the feelings of sadness and discomfort experienced after being separated will not last forever.

Children who have been removed from their homes also need to be reassured that they are not the cause of the family disruption. It is quite common for foster care children to feel responsible for their parents "getting into trouble," and they may even be tempted to recant their disclosures of abuse in the hope that they can return home. Such children often reason that enduring the abuse is better than having their family torn apart and their parents in trouble. In fact, many abused children have been told for years that if they ever did disclose the abuse that the parents would go to jail and the children would be taken away. Thus, it is important that the caseworker anticipate the possibility of such prior conversations between children and parents and address this by encouraging the children and reassuring them that the current course of action will actually benefit and strengthen the entire family.

Loss, Grief, and Mourning

Coming alongside children who have experienced a loss and permitting them to grieve involves having a high tolerance for a wide range of emotions. Lee and Whiting (2007) discuss the concept of ambiguous loss with regard to children in foster care. Ambiguous loss is defined as loss that is unclear, undefined, and in many instances, unresolvable. Ambiguous loss in foster care situations can involve losses that are confusing for the child, such as the loss of an abusive parent. Children who are removed from an abusive home and placed in a foster home with caring, nonabusive parents may feel conflicted about the loss of the parent and entry into the child welfare system. Feelings may include confusion, ambivalence, and guilt, for instance.

Earlier research studies have found that people who endure ambivalent loss tend to experience similar feelings, such as:

- "Frozen" (unresolved) grief, including outrage and inability to "move on"
- Confusion, distress, and ambivalence
- Uncertainly leading to immobilization
- Blocked coping processes
- Experience of helplessness, and therefore, depression, anxiety, and relationship conflicts
- Response with absolutes, namely, denial of change or loss, denial of facts
- Rigidity of family roles (maintaining that the lost person will return as before) and outrage at the lost person being excluded
- Confusion in boundaries and roles (e.g., who the parent figures are)
- Guilt, if hope has been given up
- Refusal to talk about the individuals and the situation (Boss, 2004 as cited in Lee & Whiting, 2007, p. 419).

With these feelings in mind Lee and Whiting (2007) interviewed 182 foster children, ages two through 10. Children were asked about each of the feelings identified in Boss' study as typical responses to ambiguous loss. The study showed that virtually all of the children interviewed exhibited these typical feelings, particularly feelings associated with confusion, ambiguity, and outrage about their situation. Several children noted confusion about their future—not knowing when they would see their parent(s) again, or how long they would be in foster care. The children also expressed feelings of uncertainty, guilt, and immobility.

Lee and Whiting (2007) recommend using the model of ambiguous loss when working with children in foster care, cautioning against pathologizing their feelings (and the consequential behaviors). In describing the application of this model of loss, Lee and Whiting state:

> Therapists, case managers, officers of the court, and foster family members need not see these externalizing and internalizing behaviors as pathology, but as active coping strategies appropriate to the children's circumstances. Attempts to squelch these behaviors in the interest of tranquil foster placements are unrealistic and may exacerbate underlying psychosocial conditions. (p. 426)

In referencing therapy goals they continue:

> The immediate goal is to make understandable those things that are disruptive to the foster placement. The diverse stakeholders, including the children, need to appreciate how unresolved grief leads to ambivalence about and fears of interdependency, relationship testing, and self-fulfilling prophecies of non-lovableness. In short, all invested members must move from deficit detecting to appreciating that many of these otherwise disturbing behaviors are signs of ego strength. (p. 426)

Siu and Hogan (1989) also cite the importance of caseworkers understanding the nature of grieving and thereby assisting foster care children to grieve the loss of their families. It is vital for caseworkers to be familiar with the possible expressions of depression among grieving children, which often manifest as irritability and can easily be mistaken for misbehavior. It is also quite common for children to express heartfelt grief for parents who have horribly abused them. Even children who have been sexually abused often express missing their abusive parent. Caseworkers must be careful to allow these children to grieve their parents, despite the fact that the parents have hurt them.

Identity Issues

Identity is a multifaceted concept referring primarily to one's self-knowledge, self-appraisal, and self-assessment. Developmental theorist Erik Erikson (1963, 1968, 1975) believed that identity formation involves the integration of numerous and sometimes conflicting childhood identities. Erikson believed that this convergence of identities took place during the adolescent stage of development, when the adolescent developed an internal continuity and consistency that integrated all different aspects of the self, allowing one's real identity to emerge. Our individual identities are based on several factors, some involving internal traits and some involving external traits. As individuals mature, their basis for identity becomes more internally based. But children, particularly younger children, will typically base their identity more on external, rather than internal attributes. For instance, if someone were to ask you to describe yourself, you might begin by saying that you are a college student (external). You might then share that you are a soccer player (external) and on student counsel (external). But, you might then describe yourself as an extrovert (internal), who is courageous (internal), loyal (internal), and kind (internal). The more internally based one's identity is, the more resilient a person will be in times of crisis and transition.

Children tend to be far more external in their self-identity, and their self-appraisal can be quite fragile, varying dramatically if their external structure is removed. Siu and Hogan (1989) suggested that caseworkers become familiar with the process of identity development and how the removal of children from their family of origin can significantly affect their sense of personal identity. The nature of this impact will depend, of course, on the age of the children and their stage of development, but can also be affected by several other variables. Some of the factors involved in identity formation include one's gender, ethnic and cultural identity, extracurricular activities, talents, socioeconomic status, and relationships with others. Children are often unaware of how they are affected by things such as their socioeconomic status, but it affects them nonetheless.

One's positive identity is dependent on an affirming reciprocal exchange between the various aspects of identity and one's environment. Consider this reciprocity as a mirror reflecting back either a positive or negative image of how one is perceived and valued by others. Essentially, the positive or negative nature of one's identity is based at least in part on how these various aspects of one's self are valued by others. Individuals who are extremely talented musically may only perceive this talent as a positive part of their identity if their family and community perceive musical talent as valuable. Children who are intelligent but are raised in families that value athletic prowess may not perceive their intellectual ability as a positive and valuable trait. Children who are removed from their home for maltreatment and are placed in a new environment will struggle with identity issues because despite being in a more positive environment, they are no longer the youngest sibling, no longer the owner of a small dog, no longer the funniest student in the class, and no longer the best bike rider in the neighborhood. Now they are foster children, different and set apart, perhaps living in a home much nicer than their own, leaving their feelings somewhat deficient and "less than"; they are no longer funny because they know no one in class, and they are not the youngest kids because they are only foster children in new homes.

Because so much of children's identities reside outside the self and are dependent on external validation and encouragement, an effective caseworker must understand the various dynamics of identity development, understanding how removing children from their homes, even abusive homes, can undermine children's identity development. Any acting out behavior on the part of the child should

Human Systems

Understanding and Mastery of Human Systems: Theories of human development

Critical Thinking Question: Removing a child from his biological parents can contribute to difficulties in the child's identity development. On the other hand, abuse, neglect, and maltreatment can also erode a child's development of a strong, internally focused sense of identity. How might these effects be exacerbated by placement with a foster or adoptive family, which is significantly different (in terms of ethnicity, religious beliefs, or socioeconomic status) from the child's family of origin?

be viewed through this lens of identity disruption, and the caseworker can then respond by providing comfort and encouragement to the child during this transition. Children who have only received praise for their ability to play good basketball are going to struggle immensely with their identity if placed in homes that value academic performance or musical ability. A caseworker can assist these children in recognizing that their worth is internal and should not be based solely on the approval and affirmation of others.

Continuity of Family Ties

Picture yourself in a boat moored to a dock on the shore of a large lake. Being anchored here provides you with a connection to the mainland and a sense of security, without fearing becoming adrift at sea. But what if you need to get to the other side of the lake? You would have to pull up your anchor and drift across the water, and it wouldn't be until you reached the other side and safely anchored yourself against that shore that you would feel secure and stable again. Many significant life transitions are like this time adrift at sea—caught between two shores, where continuity and stability are temporarily lost. Children who have been removed from their biological homes will undoubtedly lose their sense of continuity with their biological families and will feel adrift at sea during the time period when they have not yet established new bonds with their foster family.

Siu and Hogan (1989) strongly recommend that caseworkers consider the importance of continuity and stability when considering where to place a child. Ready access to the biological family and even close friends should always be a priority in placement decisions, and although this can become challenging, particularly in low-income areas where there may be a limited number of available foster families, consideration should still be given to a placement that will facilitate ongoing parental involvement.

At times siblings must be placed in separate foster homes, and consideration to continuity issues needs to be extended to this situation as well. Far too often, siblings in foster families do not visit with each other regularly because of the geographic constraints placed on foster families, who are often responsible for providing transportation.

Caseworkers may find themselves in double-bind situations, though, where they must make difficult choices regarding keeping siblings together by placing them in a foster home that is a significant distance away from a parent who does not have transportation, or placing the children in different foster homes that are closer to their biological parents, but precludes family visitation due to the difficulty in coordinating visits among various foster families. Caseworkers must rely on their clinical skills in deciding on the right course of action and should then recognize and acknowledge how this interruption of family continuity and stability will affect the children, particularly early in the placement.

Far too often the foster care system, with all its complications, does not do an effective job of *fostering* a relationship between children in placement and their biological families, because if children do not have ready access to their biological families, they will most likely search for continuity and connectedness with their foster families, which, although necessary and important, can pose a risk to the continuing bond with their biological parents.

Research has clearly shown that children who visit their biological parents more frequently have a stronger bond with them and have fewer behavioral problems, are less apt to take psychiatric medication, such as antidepressant medication, and are less likely to be developmentally delayed, which underscores the importance of strengthening the attachment between foster children and their biological parents through regular and consistent visitation (McWey & Mullis, 2004). Restricting visitation for any reason other than the safety of the child will have a negative effect on this attachment and might even be subsequently used against the biological parents when it comes time to make reunification plans.

Crisis

Removing children from their biological homes and placing them into foster care constitutes a crisis. Siu and Hogan (1989) referred to this crisis as a critical transition, which throws an already fragile family into complete disequilibrium. In fact, most child welfare experts put foster care placement in the category of a *catastrophic crisis*. Crises are not always bad though, and a popular contention among mental health experts discussed in Chapter 4 is that a crisis provides the best opportunity for personal growth and authentic change.

Ordinary coping skills are typically not going to be enough to help a child deal with the trauma associated with being placed in foster care. But an effective and seasoned caseworker can help a child develop more effective coping skills that can help them respond to the multiple crises of being removed from their home and placed with strangers.

Working with Foster Parents

Foster care can refer to many placement settings, including kinship care, an emergency shelter, a residential treatment center, a group home, or even an independent living situation (with older adolescents), but most frequently foster care involves placing a child with a licensed foster family (two-parent or single-parent family). Every state has certain guidelines and standards that prospective foster parents must meet to qualify to become licensed (Barth, 2001). Licensure typically requires that families participate in up to 10 training sessions focusing on topics such as the developmental needs of at-risk children, issues related to child sexual abuse, appropriate disciplining techniques for at-risk children, ways that foster parents can support the relationship between the foster children and their biological parents, and ways to manage the stress of adding new members to their family. In addition, individuals who will be foster parenting children of a different ethnicity will undergo training focusing on transcultural parenting issues.

Foster parents provide an invaluable service by accepting troubled children into their homes and providing love, nurturing, and security, even though they know the children may be in their homes for only a short time. In addition to good training, foster parents benefit from caseworkers who are consistently supportive and available to them, particularly during high stress times when foster children are acting out. Foster placement will be far less likely to fail if the foster parents feel sufficiently well prepared and supported by their caseworker.

Because the majority of foster children return to their biological parents, foster parents must be supported in their role in the reunification process. The success of a reunification plan depends largely on the cooperation of the foster parents. A foster parent who eagerly facilitates visitation and the sharing of vital information with the biological parents will help protect and maintain the continuity between the foster children and their biological parents. The caseworker plays a pivotal role in providing support and assistance to foster parents. A foster parent who feels unsupported will be far more likely to either purposely or inadvertently undermine the relationship between the foster child and the biological parents. Much of the time this action comes in the form of advocacy for the child. Unfortunately, though, this advocacy, as well meaning as it may be, has the potential of disrupting the necessary process of reunification. Thus, although it is certainly understandable that the process of emotional bonding with the foster child makes foster parents vulnerable to advocating for the best interest of their foster children, foster parents who take it upon themselves to protect their foster child by discouraging the relationship with the biological parents in any way are violating their designated roles, and their effectiveness as foster parents will most likely be seriously compromised.

The Public Broadcasting Service (PBS) documentary entitled *Failure to Protect: The Taking of Logan Marr* documents the removal of five-year-old Logan and her baby sister, Baily, from their young biological mother, Christie Marr. The documentary reveals how Maine's child welfare system, the Department of Human Services (DHS), removed Logan from her mother's care on the presumption that the child might be abused at some *future* time based on some dynamics in the home. After years of jumping through hoops and getting Logan back, Christie had another child, but ultimately lost both of her girls after marrying someone whom DHS did not approve. Regardless of Christie's compliance with her parenting plan, the caseworker placed her girls with another DHS worker who was also a licensed foster parent. The foster mother wanted to adopt the Marr girls and actively hindered the relationship between the girls and their mother. In this situation, as well as many others, the foster mother was responsible for providing transportation for visitation, as well as for keeping Christie comprised of major events in the girls' lives. Thus, she had tremendous power to limit visitation if she so desired or to be begrudging with vital information about the girls.

Logan ultimately died in this foster mother's care, and her death led to an uproar over the treatment of Christie, the apparent "cozy" relationship between the foster mother and the DHS caseworker, as well as the caseworker's refusal to investigate Logan's earlier complaints that her foster mother had abused her. This tragic case illustrates how vital it is for foster parents to be well trained and sufficiently supported by their caseworker. An effective caseworker will be able to sense when a foster parent is either burning out or overstepping appropriate boundaries and will respond with support and limit setting as necessary.

Reunification

The decision of whether or when to reunify foster children with their biological parents is based on many factors, including the biological parents' success in meeting their service plan goals. Even if these goals are sufficiently met, the timing of reunification may depend on minimizing disruptions in the child's life, such as switching schools in the middle of the school year. If reunification is the plan from the beginning of placement, then the caseworker should be planning for this event from the initial stages of the case. Problems arise when issues such as court postponements, additional service plan goals, changes in caseworker assignments, and other factors lead to delays in reunification. A judge may deem it perfectly reasonable to postpone a reunification hearing so that a child can complete the final four months of school without disruption, but such a decision can be devastating for the biological parents who have worked diligently to reach all service plan goals and go to court expecting to leave with their biological child, only to be told they must wait an additional four months to avoid their child changing schools in the middle of the school year. The potential for a biological parent to give up attempting to regain custody and to relapse into unhealthy behaviors out of discouragement and frustration is great, and caseworkers must be sensitive to the possibility of such frustrations leading to despair or relapse.

Therefore, even though reunification with biological parents is associated with several changes in the child's life, many of which may be negative in nature (Lau, Litrownik, Newton, & Landsverk, 2003), an effective caseworker will begin preparing the child for these transitions from the beginning of placement in foster care. Simply verbalizing what is going to happen, telling the child what to expect in the future, and giving such children a voice in expressing their fears and frustrations, even if they do not have decision-making power, will go a long way in minimizing the negative effect of reunification, particularly for children who have been in placement for a significant amount of time.

Reunification is not just stressful for the child, it is stressful for the biological parents as well, and many biological parents are the most vulnerable to stress-related relapse in the weeks following reunification. The combination of increased stress and the acting out of the child due to yet another transition can create a potentially volatile situation where negative behavior patterns resurface. Any good reunification plan involves ongoing monitoring and provision of in-home services to prevent any such problems during the reunification transition. These services can be provided by the county child welfare office directly or by a contracted agency-based practice that specializes in providing services such as in-home case management and support. With good support services, many reunifications go quite smoothly, and in time the children and parents settle in to a regular routine where healthier communication patterns and positive parenting styles will lead to a positive response from the children.

Family Preservation

Because the number of children placed in substitute care rose consistently since the 1980s, particularly in most urban communities, there has been an increasing focus on early intervention and prevention programs since the early 1990s. Family preservation programs are designed to reduce the need for out-of-home placement by intervening in a family process before the dynamics deteriorate to the point of requiring the removal of the children. These programs are comprised of a variety of short-term, intensive services designed to immediately reduce stress and teach important skills that will reduce the likelihood of out-of-home placement. Services can include family counseling, parenting training, assistance with household budgeting, stress management, child development, respite care for caregivers, and in some cases, cash assistance (Child Welfare League of America, n.d.).

Although there has been some controversy surrounding the success of these programs in reducing foster care placements, the federal government remains committed to early intervention programs, and many counties report that approximately 80 percent of families who have participated in family preservation programs remained intact in the year following the suspension of services (Child Welfare League of America, n.d.).

Relevant to any discussion on family preservation is the importance of human rights as they relate to children, particularly those who are living in environments that are fragile, thus increasing the already vulnerable nature of dependence. The United Nations Convention on the Rights of the Child (UNCRC), adopted in 1989 and enacted in 1990, is considered by most in child welfare to be one of the most significant international treaties establishing and enforcing human rights for all children. Every country in the world has signed and ratified the UNCRC except the United States and Somalia, both of which have signed but not ratified the treaty. The UNCRC consists of 41 articles setting forth basic rights of children (as well as the means for ensuring the enforcement of these rights) based upon the "best interest of the child" principle, which places the needs of children, particularly in decisions relating to their care, as a primary concern above all other interests. The ultimate goal of the UNCRC is to protect the survival, health, education, and development of children securing their well-being (UNCRC, 1989).

> **The ultimate goal of the United Nations Convention on the Rights of the Child is to protect the survival, health, education, and development of children and to secure their well-being.**

The UNCRC guarantees children the most basic rights, including the right to live, to develop in a healthy manner (including the right to play and enjoy a wide range of child-appropriate activities), to have a legal name and identity that is registered with the government (such as a birth certificate), to reside with parents (as long as this is in the child's best interest), to have access to appropriate healthcare, to have an education, and to have an adequate standard of living free from profound poverty. Several articles also guarantee a child's freedom of expression including having a voice in choices that affect them (as is deemed developmentally appropriate), appropriate freedom of expression, privacy, and access to information, with indigenous children even having the right to practice their own cultural traditions. Children are guaranteed the right to protection, including protection from violence, child labor, exposure to the drug trade, drug abuse, sexual exploitation, abduction, trafficking, excessive detention, and punishment. Relevant to the discussion on family preservation, several articles of the UNCRC set forth the rights of children who for whatever reason cannot reside with their families, including the right to be cared for in a manner that respects their religion, ethnic group, and cultural traditions, and the right to have all aspects of the UNCRC applied to them regardless of their residential or family status (UNCRC, 1989).

Clearly, the international community recognizes the value of the biological family unit and supports all governmental efforts designed to support families maintain their bonds, particularly with their children. Such support can be in the form of "family-friendly" policies, financial and case management support for kinship care (increased since the passage of the Fostering Connections to Success and Increasing Adoptions Act of 2008), as well as other measures that focus on prevention and preservation rather than solely intervention.

Minority Populations and Multicultural Considerations

Children of color are overrepresented in the foster care system, comprising nearly 60 percent of all placements in the year 2004. This is nearly twice their representation in the general population. Of all children requiring child welfare intervention, the majority of African American children requiring care are placed in foster care, whereas the majority of Caucasian children receive in-home services (Child Welfare League of America, 2002). In addition, African American children remain in foster care far longer and are reunited with their families far less often. This overrepresentation of children of color in the foster care system, particularly African American children, is fueled by other long-standing factors such as social oppression, negative social conditions, racial discrimination, and economic injustice. For instance, African American children were initially excluded from the child welfare system, but are now the most overrepresented of all racial groups (Smith & Devore, 2004).

Some reasons for this overrepresentation relate to complex social issues such as institutionalized racism, intergenerational poverty, and culturally based drug abuse. But other possible causes include racism within the child welfare system.

Types of racial discrimination include:

1. *Racial bias in referring families for family preservation programs versus out-of-home placement.* Certain special populations, including African American families, are not consistently targeted for family preservation programs. Reasons for this include caseworker bias based on the belief that the needs of the African American community may be too great to be appropriately handled by this program (Denby & Curtis, 2003).

2. *Racial partiality in assessing parent–child attachment leading to delays in returning children to their biological parents.* A 2003 study of approximately 250 black and white children in foster care placement found that racial partiality existed in assessing the parent–child attachment when the caseworker was of a different race than the biological parent. Although this result was reciprocal (i.e., black caseworkers showed partiality to black families and white caseworkers show partiality to white families), the effect of this trend has particular relevance to the African American community because the majority of caseworkers are Caucasian, and African American children are disproportionately represented among children in foster care. The results of this study revealed that Caucasian caseworkers might have erred when they concluded that African American mothers were poorly attached to their children because of the caseworker's lack of understanding of cultural differences between Caucasian and African American customs (Surbeck, 2003).

3. *Caseworkers who are poorly trained in cultural competencies.* For a caseworker to accurately assess many of the factors necessary in determining whether out-of-home

placement is warranted, such as the level of violence in the home, the ability of parents to protect their children, or the level of parental remorse, a caseworker must be aware of commonly held negative stereotypes of various racial groups. It is unacceptable for a member of the majority culture to claim not to hold any negative stereotypes, and it is only through the honest admission of overt and subtle negative biases toward other cultures that a caseworker can begin to work effectively with a variety of ethnic groups.

Placing Children of Color in Caucasian Homes

Considerable controversy exists surrounding the placement of children of color in Caucasian homes. Many advocacy organizations do not support this practice, whereas others claim that it is not in the best interest of children to experience placement delays simply because there are no foster families available that are the same race as the child. From a "micro" perspective, this latter argument makes sense. If an African American child is in desperate need of a long-term foster home, how much sense would it make to have a policy in place that prevents placement in a suitable home only because the foster family is Caucasian? After all, all children deserve loving homes, and the color of their skin should not keep them from being placed in one. Right?

Yet, from a "macro" perspective, a different viewpoint is revealed. Consider the equity of a majority culture systematically destroying an entire race, as the United States has done to the African American population during the slavery and post–Civil War era or to the Native American population during colonial times and the era of early occupation of the United States. How do you think these racial groups would perceive this same majority culture then rushing in to "rescue" the children who were maltreated in great part because of this cultural genocide and the resultant social breakdown?

Advocates of placing children of color in homes of the same race cite such cultural genocide in their arguments. Alternatives to transracial placement include the development of kinship care programs, where members of a child's extended family act as foster parents, often made possible through financial assistance. The National Association of Black Social Workers (NABSW) cites the long-standing tradition of informal kinship care within the African American community extending back to the Middle Ages and solidified during the slavery era, when many African Americans acted in the informal capacity of parents for children whose biological parents were sold and sent away. Such cultural traditions can serve as a precursor for federally funded programs that promote kinship care foster programs, which respect cultural identity and tradition (NABSW, 2003).

Recent studies support the concerns expressed by the NABSW and others about the difficulties faced by even the most well-meaning white adoptive parents to appropriately and accurately teach their black adopted children lessons about race in a culturally appropriate manner. A recent study by Smith, Juarez, and Jacobson (2011) found that the majority of adoptive families of black adoptees were white, middle to upper-class families from primarily white communities, and despite their attempts to teach their children about matters of race and instill in them a sense of cultural pride, most of the black adoptees were often left to struggle with racial discrimination and racial enculturation on their own. The primary reason for this dynamic was that their white adoptive families more often than not experienced race quite differently than their black adopted children, viewing racial dynamics through a white Eurocentric lens (Smith, Juarez & Jacobson, 2011).

In their study on the attempts of white parents to teach their black adopted children about race and racism in America, Smith, Juarez, and Jacobson (2011) state:

> As members of U.S. society's dominant mainstream, White adoptive parents are positioned to transmit collective understandings, interpretations, knowledge, and memories about Whiteness, not Blackness. They are well positioned to teach lessons about race that reflect and give privilege to the interests, values, experiences, and perspectives of Whites. (p. 1198)

Their study revealed that while a majority of white transracial adoptive parents cited the importance of their children developing a sense of pride in their cultural heritage, they framed "cultural pride" as an individual process, not a collective one. Since the majority of transracial families interviewed in the study lived in primarily white communities, their black children did not participate or engage in communities of color; thus, any development of cultural pride was done in collective isolation.

Most of the white parents in this study taught their children about African American culture, including the nature of race relations in America, through books, films, and cultural events, such as attending black camps. For instance, several white adoptive parents shared that they taught their black adoptive children about overcoming racism through the telling of stories of famous black individuals who became successful despite racial barriers through personal fortitude and a lot of hard work. Yet Smith, Juarez, and Jacobson (2011) point out how this type of racial framing illustrates Western notions of individualism, rather than community efforts more reflective of African American culture and history, and did not teach black adoptees about racial inequality involved in "structural relations within society that enable the hard work of some to pay off more than that of (racialized) others" (p. 1214).

This study revealed just how committed the white adoptive parents who were interviewed were in their attempts to appropriately validate their black adoptive children's racial heritage and culture pride, but they did so in ways that were distinctly white. For instance, the white adoptive parents taught their black children to:

- Affirm and feel positively about racial differences,
- Subvert personal needs and responses to racial discrimination to help Whites learn about race and racism, and
- Develop a thick skin to deflect the consequences of race-based discrimination in a way that avoids conflict and does not disrupt harmony with Whites. (pp. 1221–1222)

Framing racial and cultural dynamics in such a White Eurocentric individualist way contradicted sharply with how most African American parents handled matters of race with their children. Although the white parents in this study clearly loved their black adopted children and appeared very committed to addressing matters of race, with regard to cultural pride and dealing with racial prejudice, by presuming that racism was the result of white ignorance that could be overcome only through education and hard work, the white parents were inadvertently drawing from historic white cultural narratives of racial inequality, not black ones, which are far more likely to emphasize the purposeful agenda of racial oppression and inequality within American society, and the collective struggle of African Americans to fight against it.

Although Smith, Juarez, and Jacobson (2011) do not specifically advocate against transracial adoption, they do caution white parents to be very careful about the ways in which they choose to teach lessons about race to their adopted children, in order to avoid even the inadvertent inculcation of white racist framing of the black experience in America. They suggest doing this through the reframing of race and racial issues through the experiences of the black community, and not through the lens of White America. Whether this is possible, is difficult to say, but further research on ways in which race lessons can be taught to black adoptees will inform this growing area of research, particularly if informed by black adoptees themselves.

Native Americans and the U.S. Child Welfare System

The British colonization of North America involved an organized and methodical campaign to decimate the Native American population through invasion, trickery (such as trading land for alcohol), and ultimately the forced relocation of all Native Americans onto government-designated reservations, where the assimilation into the majority culture became a primary goal of the U.S. government (Brown, 2001). The few Native Americans who survived this genocide were broken physically, emotionally, and spiritually, suffering from alcoholism, rampant unemployment, and debilitating depression.

In the early part of the 19th century the U.S. government assumed full responsibility for educating Native American children. It is estimated that from the early 1800s through the early part of the 20th century, virtually all Native American children were forcibly removed from their homes on the reservations and placed in Indian boarding schools, where they were not allowed to speak in their native tongues, practice their cultural religion, or wear their traditional dress. During school breaks many of these children were placed as servants in Caucasian homes rather than being allowed to return home for visits. The result of this forced assimilation amounted to cultural genocide where an entire generation of Native Americans was institutionalized, deprived of a relationship with their biological families, and robbed of their cultural heritage.

This ongoing campaign to assimilate the Native Americans into European American

Student body assembled on the Carlisle Indian School Grounds.
Buyenlarge/Archive Photos/Getty Images

culture became even more aggressive between 1950 and 1970, when social workers with governmental backing removed thousands of Native American children from their homes on the reservations for alleged maltreatment, placing them in adoptive Caucasian homes. In reality, many of the problems on the reservations were the product of years of governmental oppression resulting in extreme poverty and other commonly associated social ills, and the U.S. government response to this was to tear Native American families apart rather than intervene with mental health services.

Old Sun Residential School.
Library of Congress

Between 1941 and 1978, approximately 70 percent of all Native American children were removed from their homes and placed either in orphanages or with Caucasian families, many of whom later adopted them (Marr, C. 2002). In truth, few of these children were removed from their homes due to maltreatment as it is currently defined. Rather, approximately 99 percent of these children were removed because social workers believed that the children were victims of social deprivation due to the extreme poverty common on most Indian reservations (U.S. Senate, 1974). The result of this government action has been nothing short of devastating. Native Americans have one of the highest suicide rates in the nation, with Native American youth, particularly those who have spent time in U.S. boarding schools, having on average five to six times the rate of suicide compared to the non-Native American population. When these children graduated from high school, they were adults without a culture—no longer feeling comfortable on the reservation after years of being negatively indoctrinated against their cultural heritage, yet not being accepted by the white population either. The response of many of these individuals was to turn to alcohol in an attempt to drown out the pain.

In 1978, the Indian Child Welfare Act (Pub. L. No. 95-608) was passed, which prevented the unjustified removal of Native American children from their homes. The act specifies that if removal is necessary, then the children must be placed in a home that reflects their culture and preserves tribal tradition. Tribal approval must be obtained prior to placement, even when the placement is a result of a voluntary adoption proceeding (Kreisher, 2002). This act has for the most part successfully stemmed the tide of mass removal of thousands of Native American children from their homes on the reservations, but unfortunately many caseworkers still do not understand the reason why such a bill was passed in the first place, or why it is necessary, and mistakenly believe that this act hampers placing needy children in loving homes.

Gaining a fuller understanding of the history between people of color and the U.S. child welfare system will make it easier to understand why some minority groups may not trust human service professionals in issues regarding allegations of abuse. The social worker might not be aware of the long-standing negative history between government child welfare agencies and a particular racial group, but members of that particular group are most likely aware of this history. It is vital that human service professionals develop cultural competencies, regardless of whether they are actively working with ethnic minority populations. It is only through a comprehensive understanding of the history of child welfare policies and abuses of power that the U.S. child welfare system will truly achieve its goal of respecting the autonomy and dignity of all people, regardless of race, gender, age, nationality, and sexual orientation.

Concluding Thoughts on Child Protective Services

Human service professionals who work with troubled families have the opportunity to effect change that positively affects not only the present families, but all future generations within that family system as well. CPS caseworkers often experience high caseloads and can feel overwhelmed and burned out in the face of such immensely complicated dynamics commonly involved in child welfare cases.

An increased focus on family preservation programs and other early intervention programs offer the best opportunity for reducing out-of-home placements, but these programs must be offered to all potentially appropriate families without bias. This can occur through sufficient federal and state funding of child welfare programs and the effective recruitment and training of human service professionals willing to work with a variety of families, from various cultures dealing with a wide range of life challenges.

The following questions will test your knowledge of the content found within this chapter.

1. Prior to the Civil War, the common belief about children was that they needed
   a. dedicated play time in order to develop psychosocially
   b. to be treated with harsh discipline or they would fall victim to laziness and vice
   c. to be in school at least six hours a day
   d. to be treated with tenderness and understanding

2. Prior to the Industrial Revolution, orphans were often
   a. forced to live on the streets
   b. sold into apprenticeships that were sometimes no better than slavery
   c. sent to almshouses to work alongside adults
   d. All of the above

3. One significant difference between child welfare programs of 100 years ago and those of today is that
   a. alcohol was the chief cause of child removal 100 years ago and today it is drug abuse
   b. the majority of children in substitute care today are not orphans but are victims of child maltreatment
   c. caring for orphaned and abused children has slowly transitioned from institutionalized care to primarily substitute family care, or foster care over the past 100 years
   d. Both B and C

4. The Child Abuse Prevention and Treatment Act (CAPTA) of 1974 was established to ensure that
   a. children of maltreatment are reported to the appropriate authorities
   b. parental rights are terminated on a timely basis
   c. Both A and C
   d. None of the above

5. Children of color are not just disproportionately represented in the foster care system in the United States, but
   a. far fewer children of color are reunited with their families
   b. far more children of color are placed into institutionalized care
   c. far more children of color are emancipated prior to their 17th birthday
   d. far fewer of these children receive regular visitation with their biological parent(s)

6. In 1978 the Indian Child Welfare Act (PL 95-608) was passed, which
   a. prevented the unjustified removal of Native American children from their homes
   b. required that Native American children be placed in Native American homes
   c. required tribal approval prior to adoptive placement, even when the placement was a result of a voluntary adoption proceeding
   d. All of the above

7. Explore some of the psychological dynamics experienced by many biological children removed from their biological homes and placed into nonrelative foster care, and some ways human service professionals can assist foster care children with this transition.

8. Explore the advantages of family preservation programs, including ways in which human service professionals can ensure that all families can benefit from this program equitably.

References

Addams, J. (1912). Twenty Years at Hull House; with autobiographical notes. Kindle Edition.

Alderman, C.L. (1975). Colonists for Sale: The Story of Indentured Servants in America, New York: Macmillan, 1975.

Austin, M. J., D'Andrade, A., Lemon, K., Benton, A., Chow, B., & Reyes, C. (2005). *Risk and safety assessment in child welfare: Instrument comparisons.* University of California, Berkeley, School of Social Welfare (BASSC), 2, 1–16. Retrieved January 1, 2012, from http://cssr.berkeley.edu/bassc/public/risk_summ.pdf

Barth, R. P. (2001). Policy implications of foster family characteristics. *Family Relations, 50*(1), 16–19.

Bellingham, B. (1984). *Little wanderers: A socio-historical study of the nineteenth century origins of child fostering and adoption reform, based on early records of the New York Children's Aid Society.* Unpublished doctoral dissertation, University of Pennsylvania. (Available from University Microfilm Incorporated (UMI), Ann Arbor, MI.)

Belzberg, E. (Director). (2001). *The forgotten children underground* [Docudrama]. Childhope International.

Bender, T. (1975). *Toward an urban vision: Ideas and institutions in Nineteenth century America.* Baltimore, MD: The John Hopkins University Press, 1975.

Bentovim, A. (2002). Preventing sexually abused young people from becoming batterers, and treating the victimization experiences of young people who offend sexually. *Child Abuse & Neglect, 26*(6–7), 661–678.

Bentovim, A. (2004). Working with abusing families: General issues and a systemic perspective. *Journal of Family Psychotherapy, 15*(1–2), 119–135.

Bourg, W., Broderick, R., & Flagor, R. (1999). *A child interviewer's guidebook.* Thousand Oaks, CA: Sage Publications.

Brace, C. L. (1967). *The dangerous classes of New York and twenty years work among them.* Montclair, NJ: Patterson Smith.

Brown, D. (2001). *Bury my heart at wounded knee: An Indian history of the American west.* New York: Henry Holt and Company.

Burns, B. J., Phillips, S. D., Wagner, H. R., Barth, R. P., Kolko, D. J., Campbell, Y., et al. (2004). Mental health need and access to mental health services by youths involved with child welfare: A national survey. *Journal of the American Academy of Child & Adolescent Psychiatry, 43*(8), 960–970.

CAPTA Reauthorization Act of 2010 (P.L. 111-320) Retrieved September 9, 2012 from http://www.govtrack.us/congress/bills/111/s3817/text

Centers for Disease Control and Prevention. Infant mortality and low birth weight among black and white infants—United States, 1980–2000. MMWR, July 2002; 51:589–592.

Child Welfare League of America. (n.d.). Family preservation and permanency planning: About this area of focus. Retrieved March 2, 2004, from http://www.cwla.org/programs/familypractice/fampractabout.htm

Child Welfare League of America. (2002). Minorities as majority: Disproportionality in child welfare and juvenile justice. *Children's Voice.* Retrieved March 4, 2005, from http://www.cwla.org/articles/cv0211minorities.htm

Child Welfare League of America. (2005). *A comparison of approaches to risk assessment in child protection and brief summary of issues identified from research on assessment in related fields.* Retrieved online January 1, 2012, at http://www.childwelfare.gov/responding/iia/safety_risk/

Condran, G. A., & Cheney, R. A. (1982). Mortality trends in Philadelphia: Age and cause-specific death rates 1870–1930. *Demography, 19*(1), 97–123.

Convention on the Rights of the Child, Resolution adopted by the U.N. General Assembly, 44th Session, November 20, A/RESZ (1989).

Denby, R. W., & Curtis, C. M. (2003). Why special populations are not the target of family preservation services: A case for program reform. *Journal of Sociology & Social Welfare, 30*(2), 149–173.

DePanfilis, D., & Scannapieco, M. (1994). Assessing the safety of children at risk of maltreatment: Decision-making models. *Child Welfare, 73*(3), 229–246.

Ehrensaft, M. K., Cohen, P., & Brown, J. (2003). Intergenerational transmission of partner violence: A 20-year prospective study. *Journal of Consulting & Clinical Psychology, 71*(4), 741–753.

Erikson, E. H. (1963). *Childhood and society* (2nd ed.). New York: W. W. Norton & Co.

Erikson, E. H. (1968). *Identity: Youth and crisis.* London: Faber & Faber.

Erikson, E. H. (1975). *Life history and the historical moment.* New York: Norton.

First Nations Orphan Association. (n.d.). Retrieved March 28, 2006, from http://www.angelfire.com/falcon/fnoa

Flango, V., & Flango, C. (1994). *The flow of adoption information from the States.* Williamsburg, VA: National Center for State Courts.

Fostering Connections to Success and Increasing Adoptions Act of 2008, Pub. L. No. 110-351, 122 Stat. 3949 (2008).

Gauthier, Y., Fortin, G., & Jéliu, G. (2004). Clinical application of attachment theory in permanency planning for children in foster care: The importance of continuity of care. *Infant Mental Health Journal, 25*(4), 379–396.

Goldman, J., & Salus, M. (2003). *A coordinated response to child abuse and neglect: The foundation for practice.* Washington, DC: U.S. Department of Health and Human Services, National Center on Child Abuse and Neglect.

Hewitt, S. K. (1999). *Assessing allegations of sexual abuse in preschool children: Understanding small voices.* Thousand Oaks, CA: Sage Publications.

Holt, M. (1992). *The Orphan Trains: Placing out in America.* Lincoln: University of Nebraska Press.

Katz, M. B. (1996). *In the shadow of the poorhouse: A social history of welfare in America.* New York: Basic Books.

Kopel, S., Charlton, T., & Well, S. J. (2003). Investigation laws and practices in child protective services. *Child Welfare, 82*(6), 661–684.

Kreisher, K. (2002, March). Coming home: The lingering effects of the Indian adoption project. *Children's Voice.* Child Welfare League of America. Retrieved July 10, 2004, from http://www.cwla.org/articles/cv0203indianadopt.htm

Lau, A. S., Litrownik, A. J., Newton, R. R., & Landsverk, J. (2003). Going home: The complex effects of reunification on internalizing problems among children in foster care. *Journal of Abnormal Child Psychology, 31*(4), 345–358.

Lee, R. E., & Whiting, J. B. (2007). Foster children's expressions of ambiguous loss. *American Journal of Family Therapy, 35*:417–428.

Marr, C. (2002). Assimilation through education: Indian boarding schools in the Pacific Northwest. Seattle, WA: University of Washington Libraries Digital Collections Retrieved February 22, 2002, from http://content.lib.washington.edu/aipnw/marr/marr.html

McWey, L., & Mullis, A. K. (2004). Improving the lives of children in foster care: The impact of supervised visitation. *Family Relations: Interdisciplinary Journal of Applied Family Studies, 53*(3), 293–300.

National Association of Black Social Workers. (2003). *Kinship care.* Retrieved October 23, 2005, from http://www.nabsw.org/mserver/KimshipCare.aspx?menuContext=760

National Clearinghouse on Child Abuse and Neglect. (2005). *Definition of child abuse and neglect state statutes.* Series 2005. U.S. Department of Health and Human Services, Administration for Children & Families. Retrieved May 30, 2004, from http://www.childwelfare.gov/systemwide/laws_policies/statutes/define.cfm

Newcomb, M., Locke, D., & Thomas F. (2001). Intergenerational cycle of maltreatment: A popular concept obscured by methodological limitations. *Child Abuse & Neglect, 25*(9), 1219–1240.

Patrick, M., Sheets, E., & Trickel, E. (1990). *We are a part of history: The Orphan Trains.* Virginia Beach, VA: The Donning Co.

Pears, K. C., & Capaldi, D. M. (2001). Intergenerational transmission of abuse: A two-generational prospective study of an at-risk sample. *Child Abuse & Neglect, 25*(11), 1439–1461.

Peterson, C., & Biggs, M. (1997). Interviewing children about trauma: Problems with "specific" questions. *Journal of Traumatic Stress, 10*(2), 279–290.

Poole, D. A., & Lindsay, D. S. (1998). Assessing the accuracy of young children's reports: Lessons from the investigation of child sexual abuse. *Applied & Preventive Psychology, 7*(1), 1–26.

Roberts, D. (2002). *Shattered bonds: The color of child welfare.* New York: Basic Civitas Books.

Rutter, B. (1978). *The parents' guide to foster family care.* New York: Child Family League of America.

Sanchirico, A., & Jablonka, K. (2000). Keeping foster children connected to their biological parents: The impact of foster parent training and support. *Child and Adolescent Social Work Journal, 17*(3), 185–203.

Schultz, C.B. (1985). Children and childhood in Eighteenth Century. In Joseph M. Hawes and N. Ray Hiner (eds) *American Childhood: A Research Guide and Historical Handbook.* Westport: CT: Greenwood Press, 70, 79–80.

Shughart, W. F., & Chappell, W. F. (1999). Fostering the demand for adoptions: An empirical analysis of the impact of orphanages and foster care on adoptions in the U.S. In R. D. McKenzie (Ed.), *Rethinking orphanages for the 21st century* (pp. 151–171). Thousand Oaks, CA: Sage Publishers.

Siu, S., & Hogan, P. T. (1989). Common clinical themes in child welfare. *Social Work, 34*(4), 229–345.

Smith, C. J., & Devore, W. (2004). African American children in the child welfare and kinship system: From exclusion to over inclusion. *Children & Youth Services Review, 26*(5), 427–446.

Smith, D. T., Juarez, B. G., & Jacobson, C. K. (2011). White on Black: Can White parents teach Black adoptive children how to understand and cope with racism? *Journal of Black Studies, 42*(8), 1195–1230. doi:10.1177/0021934711404237

Spake, A. (1994, November). The little boy who didn't have to die. *McCall's,* p. 142.

Sternberg, K. J., Lamb, M. E., & Orbach, Y. (2001). Use of a structured investigative protocol enhances young children's responses to free-recall prompts in the course of forensic interviews. *Journal of Applied Psychology, 86*(5), 997–1005.

Surbeck, B. C. (2003). An investigation of racial partiality in child welfare assessments of attachment. *American Journal of Orthopsychiatry, 73*(1), 13–23.

Trattner, W. (1998). *From poor law to welfare state* (6th ed.). New York: Free Press.

United Nations Children's Fund [UNICEF] (2007, November 27). UNICEF's position on inter-country adoption [Press Release]. Available online at: http://www.unicef.org/media/media_41918.html.

U.S. Department of Health and Human Services, Administration for Children & Families, Administration on Children, Youth and Families, Children's Bureau. (2008). *The AFCARS report.* Retrieved June 22, 2009, from http://www.acf.hhs.gov/programs/cb/stats_research/afcars/tar/report14.htm

Intercountry Adoption: Country Information. (2011). U.S. Department of State: Bureau of Consular Affairs. Retrieved from http://adoption.state.gov/country_information.php.

U.S. Senate. (1974). *Hearings before the Subcommittee on Indian Affairs of the Committee on Interior and Insular Affairs,* 99th Cong., 2nd Session (testimony of William Byler). Washington, DC: U.S. Government Printing Office.

Von Hartz, J. (1978). *New York street kids.* New York: Dover.

Warren, A. (1995). *Orphan Train rider: One boy's true story.* Boston: Houghton Mifflin Co.

CHAPTER 6

Adolescent Services

ZUMA Wire Service/Alamy Limited

Learning Objectives

- Understand the stage of adolescent development from both a historical and contemporary perspective, recognizing how structural events have affected the course of adolescent development
- Compare and contrast concrete and abstract reasoning, recognizing how abstract reasoning in adolescence affects both thinking and behavior
- Identify major psychosocial dynamics experienced within the adolescent population and ways in which human services professionals can intervene
- Describe key social problems experienced by adolescents in mainstream United States, and explain the role and function of human service professionals working a variety of practice settings
- Describe ways in which culture affects the experience of adolescence, identifying the nature and dynamics associated with at-risk groups

One second she's curled up in my lap asking me to stroke her hair as she cries about a fight she had with one of her girlfriends, and the next second she's screaming at me, telling me she doesn't need a mother, and that her father and I are ruining her life. She is so dramatic and her moods shift from moment to moment. She's driving us crazy, and I'm wondering where my sweet little girl went. So complains one of my neighbors about her 15-year-old daughter. The stage of adolescence is as confusing for adults as it is for the adolescents. This stage of development serves as the bridge from childhood to adulthood, and crossing this bridge often involves several circuitous routes that sometimes appear to parents as though no progress toward maturity is being made.

Adolescence is an interesting stage of development for many reasons. The concept of this stage is rather new as there was little acknowledgment or understanding of adolescence as a separate stage of development until the latter part of the 19th century. But even now, when adolescence is accepted as a distinct stage of development, there are significant differences in how the stage of adolescence is perceived among various cultures, both within the United States and internationally. In addition, many societal changes have occurred in the last 150 years that have had a dramatic impact on adolescents themselves, creating new dynamics and issues reflected in developmental theories.

Adolescence: A New Stage of Development?

It has been widely reported among psychologists, sociologists, and historians that the stage of adolescence is relatively new, not having been formally acknowledged until psychologist G. Stanley Hall began his study of adolescence in 1882, culminating in his groundbreaking book on adolescence published in 1904. Yet, it would be misleading to assume that because society did not formally acknowledge the stage of adolescence that it did not exist. There was little acknowledgment of childhood being a distinct stage of development prior to the late 1800s, but that does not mean that children did not throw tantrums, play, and essentially act and feel like children. Hall's earliest writing on the study of adolescence sounds strikingly similar to contemporary descriptions of adolescent behavior. Hall described adolescents as possessing a "lack of emotional steadiness, violent impulses, unreasonable conduct, lack of enthusiasm and sympathy" (as cited in Demos & Demos, 1969, p. 635).

But even if adolescents have always behaved as adolescents, there have been significant shifts in child and adolescent developmental theories, influenced by the societal changes that have occurred over the past few hundred years. These changes have influenced not only how the stages of childhood and adolescence are perceived, but also the course of development itself. Lifestyles were quite different 200 years ago when the United States was a new country. The U.S. economy was different, livelihoods were different, neighborhoods were different, and families were different. An important question to consider is what kind of impact these changes have had on adolescent development and whether adolescent behavior has changed or whether society's expectations and perception of adolescents have changed.

There is no question that the mass urbanization of the past 200 years has had an impact on individual and family lives, including the lives of adolescents, who at one point in history worked alongside family members on the family farm, but who in contemporary times have far less vocational responsibility, as an increasing amount of focus is placed on the academic education of adolescents. Even the way in which many adolescents are educated has changed, likely influencing adolescent development, as teens spend significantly more time with their peers in large school environments, with increasing exposure to violence (Larsen, 2003; National Center for Education Statistics, 1999; Raywid, 1996).

Thus, although adolescents of the past acted in ways that are strikingly similar to the ways in which they act today, the many profound changes within U.S. society, including changes in family structure, the public educational system, and expectations of adolescents within these systems, have influenced the ways in which many contemporary adolescents both develop and behave.

Developmental Perspectives

To understand the behavior of adolescents, it is important to understand the developmental stages that children and adolescents progress through on their way to adulthood. Development occurs within various domains, including the intellectual, emotional, psychosocial, moral, and even spiritual spheres. Many theories of development propose that individuals progress through distinct stages of growth with earlier stages acting as foundations for successive stages. Because the course of development is influenced by many factors, both on an individual and on a broader societal level, it is important to consider both developmental theories and the course of developmental growth and maturity of children and adolescents within various contexts. For instance, in the previous section we discussed changes that have occurred in families in the United States since the mid-19th century. It is likely that what was considered "normal" behavior for adolescents in 1900 would not necessarily be considered "normal" in contemporary society.

In other words, it is important to consider the normative aspects of adolescent development within a *historical context*. What is expected of an adolescent, and what is considered adaptive and healthy behavior, depends on what is occurring in the world during the time in which the adolescent lives. A world war with a mandatory draft forces adolescents to grow up quickly, just as the Great Depression shortened childhoods across the country as adolescents were looked upon to help support their families. Yet, in contemporary society childhoods are often considered lengthened by a good economy, which reduces the need for adolescent employment, an increase in educational requirements required for professional employment, and the cessation of a mandatory draft, all of which have led to many believing that contemporary society has lower expectations of adolescents than in past eras. Adolescents who did not work during the Great Depression would likely have been considered irresponsible for not being willing to assist in the support of their families, but adolescents who do not work in contemporary society are likely presumed to be focused solely on their academic studies in preparation for college.

It is also important to consider developmental issues within a *cultural context*. What is considered normative and emotionally healthy within one culture may be considered maladaptive in another, and what is considered respectful and honorable behavior in one culture may be a sign of an emotional disorder in another. For instance, in many cultures, remaining in the family home until marriage is considered the norm. It is common in collectivist cultures, such as Asian, Latino, and even some European cultures, for single adult children as old as 30 years to live at home with their parents. In many of these cultures it would be considered a sign of disrespect for a single adult to move from the family home to gain independence prior to getting married. The United States is, for the most part, an individualistic society that values independence and autonomy; thus, many within the U.S. culture may perceive the 30-year-old male still living with his parents as a sign of unhealthy emotional enmeshment, where the boundaries between parents and adult child are blurred.

Finally, it is important to consider development within a *regional context*. Although urbanization over the last 200 years within the United States has resulted in the majority of people living in urban or suburban communities, rural life still exists in the United States and some research suggests that there are significant differences between adolescent life in rural communities and adolescent life in urban communities. Although there is not a wide body of research comparing urban and rural adolescents, a study conducted in 2001 found that rural adolescents felt less pressure to become involved in gang activities, were confronted with less violence both on and off campus, and felt less academic pressure, from both their school and their parents, compared with adolescents residing in urban areas (Gandara, Gutierrez, & O'Hara, 2001).

Understanding the natural course of development will assist the human service professional to correctly evaluate an adolescent's behavior, framing it as either adaptive or maladaptive, depending on the context within which the behavior is exhibited. For instance, understanding that it is normal for an adolescent to act in a self-centered and dramatic manner will aid the human service professional in framing behavior that, in an adult, would be indicative of a personality disorder.

Keeping historical, cultural, and regional contexts in mind will assist the human service practitioner in not mischaracterizing certain behaviors because their origin is either misunderstood or not valued by the majority culture. Adolescents in contemporary culture may act in a different manner than adolescents in past generations; yet, this does not necessarily mean that adolescents today are any less respectful than those of the past. It is also important for those in human services to understand that adolescents who immigrated to the United States from a Latin American country might act in a different manner than adolescents who have lived in the United States their entire lives, or that adolescents who recently moved from a farming community to a large city school might act differently than adolescents who grew up in an urban community.

> Keeping historical, cultural, and regional contexts in mind will assist the human service practitioner in not mischaracterizing certain behaviors because their origin is either misunderstood or not valued by the majority culture.

Having a competent grasp of normative development can be a guide for human service professionals who work with adolescents and must evaluate and assess their behavior before determining the appropriate level of intervention or whether intervention is warranted at all. Most of the developmental theorists agree that adolescence is a time of searching for one's own identity and developing a sense of autonomy. Trying on different "selves" is a common mental and behavioral activity of adolescents who are in the process of developing an internally anchored sense of who they are, rather than defining themselves by what others think or expect of them (including their parents) (Erikson, 1968; Kerpelman & Pittman, 2001). Many normal and healthy adolescents can be quite dramatic and egocentric in their behavior, and although this might give many parents cause for concern, most adolescents grow out of this stage to become giving and compassionate adults.

Human Systems

Understanding and Mastery of Human Systems: Theories of human development

Critical Thinking Question: Chapter 5 noted that, in terms of identity development, younger children tend to focus on their relationship to others (external identity), while mature adults tend to identify themselves by internal characteristics. How might the egocentrism, drama, and changeability that characterize adolescence fit into this framework of identity development?

Common Psychosocial Issues and the Role of the Human Service Professional

The common stereotype of adolescents being generally rebellious and out of control is both true and untrue. Many adolescents are quite responsible and do not have mental health problems. But adolescence is a time of stress; of trying on different "selves"; and of exploring undiscovered issues, attitudes, and behaviors. There are many reasons for these dynamics. Most developmental theorists consider this time in one's life to be transitional, and typically all transitions can be stormy. But there are other relevant issues that make adolescence unique among the various developmental stages of life, which has an impact on providing counseling services to those adolescents who are troubled.

Abstract Reasoning: A Dangerous Weapon in the Hands of an Adolescent

Jean Piaget (1950), a Swiss-born biologist turned psychologist, developed a theory of cognitive development that is still the dominant theory of intellectual development today. Among Piaget's many findings is his discovery that children, adolescents, and adults each think differently. Most notably, Piaget discovered that younger children think concretely, meaning that they lack the ability to understand many adult concepts such as parables and analogies, as well as other abstract concepts. If a group of adults were asked what it meant to "let the chips fall where they may," they will most likely explain that this is an idiom meaning to let things happen naturally. But if a group of children were asked what this statement meant, they will most likely reply that it means that if chips fall on the ground, one should not pick them up.

Piaget (1950) believed that as children approached adolescence, they began to develop the ability for logical reasoning involved in abstract thought. Abstract thought or reasoning enables us to have empathy by "putting ourselves in someone else's shoes." It allows us to think metaphorically, to understand sarcasm, to deduce, to analyze, to synthesize, and to rationalize. It also allows us to understand, and thus internalize, moral standards: to not just know that something is wrong, but to understand *why* it is wrong. If children of the age of five are asked why it is wrong to hit another child on the playground, they might state that it is wrong because they will get in trouble. But most adults would be able to explain that this act is wrong because it violates another person's personal rights, that violence does not resolve conflict, and that they would not want to be hit, even if someone else was angry with them. This type of reasoning requires empathy, the ability to see situations from multiple perspectives, the ability to draw on other experiences, and the ability to connect the immediacy of hitting someone to the generalized concept of violence—all of which require abstract reasoning ability.

It is through the development of abstract reasoning ability that adolescents discover that their parents might not always be right, that lying can be rationalized, that breaking the rules can sometimes be fun, and that authority can be questioned. When a child asks, "Why?" the question usually relates to why the sky is blue and the grass is green. But when adolescents ask, "Why?" it often relates to asking why sex before marriage is wrong, why education must occur in a 20′ × 20′ classroom, why drinking alcohol is bad, and perhaps even existential questions such as whether God is real or why they were put on this earth.

Abstract reasoning is a useful and powerful intellectual tool and can be a lethal weapon in the hands of an unstable and angry adolescent. Existential questions about the meaning of life can quickly spiral into questioning why one should exist at all, and questions about the concept of authority can quickly evolve into abandoning the concept of obeying authority altogether. The necessary skill of logical or abstract reasoning often enables a troubled adolescent to rationalize away reasons not to rebel.

Adolescent Rebellion

As long as there have been adolescents, one can be assured that there has been adolescent rebellion. Casually defined, adolescent rebellion can include any behavior on the part of an adolescent that is in marked opposition to standard rules, either within the family or within society in general. Determining what specifically constitutes rebellious behavior, though, can be a bit more challenging and often depends on current social mores, as well as one's own personal value system. Behaviors that involve outright destruction and the breaking of laws are easily characterized as rebellious. But whether the more subtle challenging of rules is considered rebellious is certainly in the eye of the beholder, where one person's rebellion is another person's sign of autonomy and individuation. For instance, most would agree that behaviors such as taking illegal drugs, habitual lying, and engaging in chronic truancy are rebellious, but what about the occasional drinking of alcohol or the intermittent breaking of a curfew? Many mental health experts and even some parents might normalize this behavior as being typical of the majority of adolescents who are striving for increased independence and testing limits along the way.

In general, though, any behavior in adolescents should be considered maladaptive if it is interfering with normal functioning and causing problems in the adolescent's everyday life. For instance, adolescents who skip one day of school in an entire year would not be considered "rebellious," but adolescents who are truant several times per week, thereby affecting their ability to pass their classes, would likely be characterized as rebellious.

EXTERNALIZING BEHAVIORS   Conduct disorder and oppositional defiant disorder are disorders included in the *Diagnostic and Statistical Manual of Mental Disorders,* fourth edition, text revision (*DSM-IV-TR*) that are diagnosed during adolescence to describe behavioral problems in children and adolescents. Conduct disorder, the more serious of these two disorders, involves a consistent pattern of behaviors in which social mores and rules are habitually broken and the rights of others are consistently violated without regard for the other person's feelings. To avoid a child being diagnosed with conduct disorder in response to uncharacteristic or minor rebellion, children cannot receive this diagnosis unless they meet at least three of the following four criteria in the preceding 12-month period:

1. Exhibiting aggression to people and animals, such as bullying, threatening or intimidating others, initiating fights, using weapons, exhibiting physical cruelty toward people or animals, stealing from a victim (e.g., armed robbery), or forced sexual activity.
2. Destroying property, such as destructively setting a fire, or deliberately destroying another person's property, such as fire setting with the intention of causing serious damage.
3. Deceitfulness or theft, such as breaking into someone's home or car; lying to obtain something desired; or nonviolent stealing such as shoplifting.
4. Serious violations of rules, such as frequently staying out at night despite parental curfew, running away from home, and frequent truancies from school (American Psychiatric Association, 2000).

Again, what most often determines the difference between the adolescents who are harmlessly spreading their wings and adolescents with conduct disorder is the *frequency, persistence,* and *seriousness* of the maladaptive behaviors. A 12-year-old who "runs away" to the next-door neighbor's house or a 16-year-old who breaks curfew by 30 minutes on just a few occasions would certainly not be diagnosed with this disorder. But a 12-year-old who runs away for weeks at a time or a 16-year-old who comes home whenever he pleases certainly might.

Oppositional defiance disorder is another emotional disorder commonly diagnosed in adolescents and is characterized by a milder set of behavioral problems, including negative, hostile, and defiant behavior such as losing one's temper, arguing with adults, and consistently refusing to obey rules. Other criteria include blaming others for personal mistakes, being easily annoyed, frequent feelings of anger and resentment, spite, and vindictiveness.

Because human service professionals always evaluate the mental health of individuals *within the context of their environment,* it is vital to examine any potential environmental causes or influences of an adolescent's maladaptive behavior. For instance, socioeconomic status, gender, parenting styles, environment, genetic influences, cognitive deficits, and temperament have all been associated with juvenile delinquency (Lahey, Moffitt, & Caspi, 2003). It is important to note that although such research indicates some type of a relationship between conduct disorders and these various influences, they do not specify whether any of these variables actually *cause* conduct disorders in adolescents. Thus, it would be incorrect to assume that because a child is from a lower socioeconomic background that she will engage in juvenile delinquency. More likely, families that are chaotic, perhaps even abusive, are likely to be from a lower socioeconomic level because such behaviors are often not amenable to the skill sets required to be a high wage earner.

> It is vital to examine any potential environmental causes or influences of an adolescent's maladaptive behavior.

I have worked with adolescents for years—first in a residential setting, then in a school setting, and now in my private practice, and I have found that adolescents typically act out for specific reasons. Clinically evaluating the entire picture is extremely important as many children and adolescents who meet the criteria for conduct disorder or oppositional defiance disorder come from homes where maladaptive behavior abounds (Frick, 2004). Such behaviors are often a manifestation of earlier abuse, neglect, and general chaos in the home environment. In general, if children and adolescents cannot talk out their feelings, they will likely act them out, often in a negative manner. Thus, if adolescents have neither the opportunity nor the maturity to connect behaviors with feelings, they will be at greater risk of expressing negative feelings in a destructive way.

INTERNALIZING BEHAVIORS   Adolescents, like children and adults, do not always manifest their emotional problems in outward ways. In fact, some of the most emotionally disturbed adolescents turn their anxiety, anger, and sadness inward with behaviors that reflect forms of depression. These adolescents are often overlooked, particularly within a school system, because they are not disruptive, often sitting in the back of the class quietly, disturbing no one. Yet, emotional disturbances turned inward can often be the most serious of all, putting these adolescents at higher risk of depression, self-abuse, and suicide.

DEPRESSION AND ANXIETY. Everyone experiences depression from time to time, but when feelings of sadness become so pronounced and long-standing that these emotions become barriers to normal functioning, the individual may be suffering from clinical depression, also referred to as major depressive disorder. The *DSM-IV-TR* lists several criteria for major depressive disorder, including abnormally depressed mood; loss of interest and pleasure; inappropriate guilt; disturbances in sleep, appetite, energy level, memory, and concentration; and, in serious cases, frequent thoughts of suicide. In children and adolescents the melancholy can often appear as irritability, which can lead to confusion in diagnosing the appropriate disorder because an irritable teenager can look far more oppositional than a sad or melancholy one.

The term used to indicate the existence of two emotional disorders simultaneously is *comorbidity,* and the comorbidity of depression and anxiety is quite high, with approximately 80 percent of those with depression also suffering from anxiety of some type (Gorwood, 2004). Although anxiety has a completely different set of diagnostic criteria (see the *DSM-IV-TR*), if one examines the possible origin of mood disorders, then it makes sense that the emotional issues that can make someone feel depressed could likely lead to feelings of anxiety as well.

There are many treatments for depression and anxiety, ranging from counseling to drug therapy, including antidepressants and antianxiety medication. However, working with adolescents is a special challenge because adolescents can be impulsive, dramatic, and narcissistic as a normal part of development, but a depressed adolescent who is impulsive, dramatic, and narcissistic can be dangerous.

I discussed earlier how some adolescents express their negative, uncomfortable emotions by acting *out* in aggressive and destructive ways toward others, but another way that adolescents deal with their problems is by turning all their emotions inward. Adolescents suffering from depression and anxiety often manifest many self-destructive behaviors, the most serious being suicide. But there are many other self-abusive behaviors that emotionally disturbed adolescents may engage in that, although certainly not as serious as suicide, still warrant serious clinical intervention.

DELIBERATE SELF-HARM. Deliberate self-harm (DSH), also sometimes called *self-injury, self-abuse,* or *self-mutilation,* is defined in various ways in research studies. One definition of self-injury includes any deliberate, repetitive attempt to harm one's own bodily tissue without a conscious desire to commit suicide (Nock & Prinstein, 2005). Hicks and Hinck (2009) use a more narrow definition for DSH, which they define as "the intentional act of tissue destruction with the purpose of shifting overwhelming emotional pain to a more acceptable physical pain" (p. 409). They describe the purpose of DSH stating that the "[t]issue damage is a visual demonstration of extreme emotional distress, and the physical act of mutilation seems to reconcile this emotion" (p. 409). DSH most often includes cutting the arms and legs with a razor blade or any sharp object (such as a paper clip), but can also include burning, picking at wounds, and even head-banging. People who self-mutilate using a sharp object are commonly called *cutters.*

Although self-injury occurs in the adult population (occurring in about 4 percent of the general population), adolescents are at increased risk for self-injury, with 39 percent of the adolescent population admitting to having self-injured at some point in their lives and 61 percent of adolescents in a psychiatric in-patient setting having self-injured (Nock & Prinstein, 2005). Approximately 40 percent of college students have admitted to engaging in self-injury (Whitlock, Purington, Gershkovich, 2009).

DSH can be a difficult issue to treat because so little is known about its causes. In addition, this type of behavior tends to be resistant to treatment. What is known is that females tend to engage in DSH far more than males, with some studies indicating that of all those who self-abuse, 97 percent are women (Nock & Prinstein, 2005). One reason for this may be due at least in part to how females are socialized to internalize their negative feelings, whereas males are socialized to externalize their negative feelings.

The precise reasons why adolescents engage in DSH behaviors is unknown, but DSH has been associated with a host of emotional and psychological problems, including suicidal thoughts, eating disorders, chronic feelings of hopelessness and despair, depression and anxiety, sexual abuse, physical abuse, severe emotional abuse, perfectionism, and a pervasive sense of loneliness (Nock & Prinstein, 2005). The National Institute of Mental Health estimates that approximately 50 to 60 percent of cutters were sexually abused as children (Crowe & Bunclark, 2000). Many adolescents who engage in DSH cite many reasons for physically harming themselves, including the belief that the cutting or burning allows them to feel something in the midst of emotional numbness. In fact, in order for self-mutilation to be considered DSH the pain and/or the sight of blood caused by the self-mutilation must result in some relief of emotional pain, and psychological reintegration—in other words, not in pleasure as is the case with masochism. Additionally, the self-mutilation of tissue must not reflect a suicide attempt or a desire to adorn oneself, such as the case of tattooing or piercing (Clarke & Whittaker, 1998; Favazza, 1996).

Other reasons for self-injury relate to the internal expression of rage and relieving intolerable tension resulting from deep feelings of anger, frustration, despair, and loneliness. Adolescents who are survivors of sexual molestation often claim that they cut in response to the shame.

A human service professional will likely encounter adolescent clients who engage in DSH in a variety of practice settings, including adolescent residential facilities, group homes, foster homes, schools, and any other settings where adolescents are served. It is important that clinicians always be on the lookout for common warning signs of self-injury, even if the adolescent or the parents deny the behavior. Adolescents who self-mutilate for attention will often flaunt their "work" by showing off what frequently amounts to superficial cuts on the forearm or thighs. But as mentioned earlier, serious self-mutilators will often hide their wounds; thus, a human service professional would be wise to note suspicious behaviors, such as consistently wearing long sleeves and pants, even on warm days. More obvious signs of self-injury may include parallel scars on the forearm or thighs, burn marks in these same places or even on the fingertips, or any unexplained or suspicious wound, particularly wounds that tend not to heal (due to chronic reinjury).

The most successful treatment programs include a combination of individual, group, and family therapy with the goal of increasing the adolescent's personal insight and awareness of the dynamics underlying the compulsion to self-injure. Issues such as *impulse control* and *emotional regulation* are paramount in any successful treatment plan, as is assisting the adolescent client in learning how to understand and effectively manage intense or uncomfortable emotions in a direct manner. This approach will allow self-abusive adolescents to own their emotions, rather than deny or suppress them.

SUICIDE.   The ultimate internalizing behavior is, of course, the killing of one's self, and although people have been committing suicide for centuries, understanding the dynamics of suicidal behavior, or suicidal ideation, remains a relatively new area of study. Of particular interest to social scientists and mental health practitioners is discovering how to most effectively prevent suicide attempts. As with self-injury, adolescents are at particularly high risk of suicide and suicidal ideation for several reasons, including their propensity for impulsivity, as well as their frequent feelings of omnipotence.

Between 1999 and 2006 (the most recent data available), 11 percent of all deaths of adolescents between the ages of 12 and 19 were caused by suicide, making suicide the third leading cause of death, behind unintentional accidents and homicide (Miniño, 2010). Adolescent suicidal behavior can include suicidal gestures, suicide attempts, and serious suicide attempts and suicide completions. Each of these behaviors can result in a completed suicide, even if that is not the intention of the adolescent, but it is important to distinguish between each of these types of suicidal behavior for the purposes of intervention, as well as developing an understanding of what goes on in the mind of an adolescent who engages in any type of suicidal behavior.

SUICIDAL GESTURES.   A suicidal gesture typically involves behavior on the part of an adolescent that is unlikely to result in a completed suicide, but is more often a cry for help or attention. Even if a practitioner does not believe that her adolescent clients truly wish to kill themselves, these gestures should not be taken lightly, because it is always possible that adolescents will kill themselves even if death wasn't the intended outcome.

SUICIDE ATTEMPTS AND COMPLETE SUICIDE.   Certainly the most serious of all suicidal behavior involves actions that are intended to end one's life. As with the adult population, it is not necessarily the adolescents who scream their suicidal intentions from the rooftop who clinicians need to be the most concerned about, but the sad, hopeless, and depressed adolescents who quietly slink away, without drawing any attention, determined to kill themselves in a manner that precludes intervention. Fortunately, not all serious attempts are successful. Some adolescents experience a last-minute change of heart and call a family member or friend, reach out to a suicide hotline, or call 9-1-1.

The types of adolescents who attempt suicide are different than those who complete suicide. For instance, research indicates that about 85 percent of "attempters" are female (Andrus et al., 1991), whereas about 80 percent of suicide completers are typically male (Arias, Anderson, Kung, Murphy, & Kochanek, 2003). Reasons for this might be related to the social acceptance of males completing suicide rather than making an attempt (Moskos, Achilles, & Gray, 2004). Other reasons may relate to gender-related methods for committed suicide, such as the male tendency to elect for far more lethal methods such as the use of firearms, whereas women tend to use less lethal methods, such as drug overdoses (Vörös, Osváth, Fekete, 2004). Among adolescent populations, those who admitted having attempted suicide were up to 30 percent more likely to be addicted to drugs and alcohol (Vörös, Fekete, Hewitt, & Osváth, 2005).

ASSESSMENT, INTERVENTION, AND TREATMENT OF SUICIDAL BEHAVIOR.   Recognizing whether an adolescent is at real risk of attempting suicide is an important clinical skill that develops with education and experience. One of the most intimidating issues facing any human service professional is knowing how to predict suicidal behavior. The answer to that question is that it is virtually impossible to definitively predict when anyone will make an attempt to end her life, but there are indicators and precursors that practitioners can look for, such as the psychosocial risk factors discussed in the previous section.

> It is virtually impossible to definitively predict when anyone will make an attempt to end her life, but there are indicators and precursors that practitioners can look for.

Although any human service professional should have a "safety first" approach to treatment, there are valid concerns for not calling 9-1-1 each time an adolescent client sounds hopeless or immersed in despair, including not wanting to destroy the counseling relationship by overreacting. When adolescent clients share that they sometimes wonder what it might feel like to die, and an anxious practitioner responds by having the adolescent involuntarily hospitalized, trust can certainly be destroyed. But in light of the alarming increase in adolescent suicides since the mid-1990s, particularly within the adolescent male population, safety is of paramount importance. Thus, some sort of balance must be struck between honoring the privacy and safety of the counseling relationship and making sure that the adolescent remains safe.

Before any successful intervention strategy can be developed, the questions of why so many teenagers are killing themselves and who is most at risk must be addressed. Suicide rates of African American males is increasing dramatically, particularly among those in higher socioeconomic status, and suicide rates in the adolescent Native American population are exceedingly high (Moskos et al., 2004).

Rutter and Behrendt (2004) conducted a study of 100 at-risk adolescents, focusing on psychosocial risk factors. Their research revealed that those adolescents who were plagued by feelings of *hopelessness,* had little to no *social support,* had feelings of *hostility,* and had a *negative self-concept* were at the greatest risk for committing suicide. This research is consistent with the research on self-injury, which revealed that self-mutilation was often the manifestation of rage and hostility turned inward, and as previously mentioned, suicide is the most injurious of all self-abusive behaviors.

Other risk factors for suicide include having a friend commit suicide (Hazell & Lewin, 1999), and for males having a gun available was a significant risk factor and for girls low self-esteem. Research also showed that deep involvement in school activities markedly decreased the potential for suicidal behavior (Bearman & Moody, 2004). Treatment will then emanate directly from any deficits found in these areas of functioning and will include the development of emotional insight and better coping skills to deal with all these emotions and insights.

If an adolescent is assessed to be a suicide risk, a safety plan must be developed with the parents or primary caregivers, because the desire to commit suicide can only come to fruition if there is opportunity. Thus, it is important for the adolescent's environment to be as free of risk as possible. For instance, a good home safety plan will include the removal of all pharmaceutical drugs, guns, kitchen knives, and loose razor blades. A depressed and socially isolated adolescent who is not actively suicidal but who thinks about dying from time to time may not need to be hospitalized, but should be monitored at all times so that any escalation in depressive symptoms can be addressed immediately. At any time that adolescent clients acknowledge suicidal intent, admit to feeling frightened of their desire to harm themselves, or disclose having a suicide plan, the human service professional may decide hospitalization is warranted, and in that case, the family will be directed to either call 9-1-1 or take their teen to their local emergency room.

Spirito and his colleagues found that the single most powerful predictors of continued suicidal behavior are the existence of depression and family dysfunction. Therefore, any treatment plan designed to address suicidal behavior must seriously address what is most likely the interplay between negative family relations and the adolescent's feelings of depression (Spirito, Valeri, Boergers, & Donaldson, 2003).

Current treatment intervention focuses on school-based suicide prevention education programs, crisis centers including teen suicide hotlines, screening programs aimed at identifying high-risk adolescents within their community, peer support programs, and public awareness campaigns, including pleas to remove guns from homes with at-risk adolescents. Suggestions for future programs include recommendations that the juvenile justice system coordinate efforts with the school-based programs and other youth outreach agencies, because over 60 percent of adolescents who committed suicide also had a history of involvement with the justice system (Moskos et al., 2004).

Human service professionals must be prepared to deal with the growing trend of suicidal behaviors in the adolescent population. Through education, prevention, and intervention strategies, including a multidisciplinary approach that addresses depression from an emotional and social, as well as a medical, perspective, mental health experts are optimistic that adolescent suicide can be successfully addressed.

> ## Human Services Delivery Systems
>
> Understanding and Mastery of Human Services Delivery Systems: Range of populations served and needs addressed by human services
>
> **Critical Thinking Question:** The text cites a range of clinical issues that disproportionately affect adolescents; it also notes that adolescence is often a time of turbulence, drama, and trying on new roles and behaviors. What are some ways in which a human service professional can walk the tightrope between protecting an adolescent client from harm and allowing the client to navigate normal adolescent changes?

Eating Disorders in the Adolescent Population

Another set of disorders common to adolescents is eating disorders, including anorexia nervosa and bulimia nervosa. Although individuals of all ages suffer from eating disorders, the primary onset of eating disorders occurs during adolescence (Ray, 2004). Females tend to suffer from eating disorders far more often than males, comprising approximately 85 to 90 percent of all documented cases of eating disorders, but the incidence of eating disorders in males is increasing, particularly among male athletes (Walcott, Pratt, & Patel, 2003). Additionally, men who have eating disorders tend to overeat, whereas women tend to under-eat (Striegel-Moore, Rosselli, Perrin, DeBar, Wilson, May, & Kraemer, 2009).

Anorexia involves the intentional starving of oneself and the refusal to maintain expected body weight. The *DSM-IV-TR* criteria for anorexia includes a body weight of less than 85 percent of normal body weight, an intense fear of gaining weight, distortion of how one's body is perceived, and the absence of a menstrual cycle for at least three months (American Psychiatric Association, 2000).

Among the various theories of the causes of anorexia, the most popular tend to focus on maladaptive family patterns where the adolescent's anorexia is presumed to help protect unhealthy family dynamics. These maladaptive patterns can include conflict avoidance, rigidity, and family enmeshment (Lock & le Grange, 2005). It is for this reason that family counseling is the most common recommended treatment for adolescents suffering from anorexia, in addition to in-patient treatment for adolescents who are at risk of serious health complications (Fairburn, 2005).

Bulimia involves a pattern of binge eating, indicating a lack of control followed by purging in the form of self-induced vomiting, use of laxatives, or excessive exercise in an attempt to rid oneself of the abundance of food (American Psychiatric Association, 2000). Bulimia is far more prevalent than anorexia in the adolescent population (van Hoeken, Seidell, & Hoek, 2003).

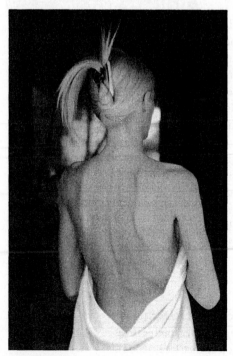

Common risk factors of adolescents suffering from bulimia include perfectionism, body dissatisfaction, and low self-esteem (Vohs et al., 2001). Adolescents who engage in binging behavior often experience significant shame once the binging phase is over. These feelings of shame are often dealt with by purging to rid the body of the excess food. This binging-purging cycle often becomes a compulsion, robbing the adolescent of the ability to stop the behavior.

Treatment for bulimia often includes insight therapy, family therapy, and cognitive behavioral therapy (CBT), which focuses on the negative self-statements the adolescent thinks in response to life events, as well as negative self-appraisals (Gowers & Bryant-Waugh, 2004). Depression and anxiety are often associated with both anorexia and bulimia; thus, a course of antidepressant or antianxiety medication is often considered appropriate.

Other Clinical Issues Affecting the Adolescent Population

Other issues that are commonly diagnosed in adolescents include substance abuse, discussed in detail in Chapters 11 and 12. Attention deficit issues, such as attention deficit disorder (ADD) and attention deficit/hyperactivity disorder (ADHD), are also a growing concern in the adolescent population, particularly in school settings, and thus are discussed in detail in Chapter 12.

A sufferer of anorexia nervosa who is clearly below her ideal body weight.

Prisma/SuperStock

Adolescence is a time of sexual discovery and experimentation and thus is an issue that must be acknowledged and addressed in a counseling setting. Issues related to sexual behavior and teen pregnancy are explored in Chapter 12. Adolescents are also at increased risk for homelessness and for academic failure and sexual exploitation once homeless. The problem of homelessness among the adolescent population is explored in Chapter 9.

Practice Settings Specific to Adolescent Treatment

There are many practice settings where adolescents receive clinical services, as well as many ways in which these services are provided. Some adolescents may receive individual counseling from therapists who are in private practice. These counseling services can be provided by anyone who has a license to provide independent counseling services such as *psychiatrists, psychologists, marriage and family therapists* (MFTs), *licensed clinical professional counselors* (LCPCs), and *licensed clinical social workers* (LCSWs). Counseling typically occurs in the counselor's office as often as the practitioner and parents deem necessary, but once a week is the most common schedule.

Counseling also occurs in many other settings, such as in schools by school social workers, counselors, and psychologists (see Chapter 12); human service agencies that specialize in adolescent issues; outreach organizations such as after-school programs; religious organizations such as Jewish Community Centers (JCC) (see Chapter 13); organizations that provide therapeutic foster care; and the juvenile justice system.

Residential care is a practice setting often utilized for adolescents who are severely behaviorally disordered and at high risk of self-harm and destructive behaviors. Although institutionalized care has steadily decreased for most segments of the population, this institutionalized care for the adolescent population has literally skyrocketed since the 1980s (Wells, 1991). These institutions can be locked or open, private or governmental, short or long term, therapeutic or more punitive in nature, but all provide some level of mental health services in relatively large, dormitory-like settings, where the adolescent residents sleep and attend school.

Residential treatment programs vary widely in type and nature, with some residential programs offering services making them sound more like a boarding school than a treatment facility, boasting equine programs, river rafting, and "therapeutic" skiing programs, whereas others are far more sterile offering few extracurricular activities. One reason for this difference can be directly related to the range of populations served. For instance, behavior on the part of an adolescent that results in court intervention and juvenile detention in a residential facility would not necessarily be conducive to a therapeutic ski trip to Vail, Colorado.

Placement times can also vary, with some adolescents being placed in a residential facility for a few months to some who require several years, again depending on the severity of their problems. One popular short-term residential program is Outward Bound, a wilderness therapy program that uses physical challenges to help adolescents deal more effectively with their emotional problems. These programs are offered in various locations within the United States and range from 21 to 28 days in length.

Group homes (or therapeutic foster homes) offer less-structured residential care, where various community services are often accessed and where adolescents attend the local public high school and are not isolated from the general community.

More structured residential treatment programs are a bit more sterile in nature, offering services to adolescents whose conduct problems or self-destructive behaviors require a more long-term, in-depth, and controlled environment. Adolescents in these programs are isolated from the general population and even attend school within the facility where they are housed. Treatment modalities in these facilities often include a combination of behavior modification where desirable behaviors are rewarded and undesirable behaviors are punished, individual therapy, group therapy, and family therapy. Referrals to such programs can be made by parents, public schools, or the juvenile court system.

The most structured and most serious of all residential treatment programs include correctional institutions for adolescents, most commonly referred to as juvenile hall or juvenile detention centers. These facilities are reserved for adolescents who have been convicted of breaking some law, and although there is far more of an emphasis on rehabilitation than in adult correctional facilities, there is a far greater emphasis on corrections and punishment than in a therapeutic treatment center.

A creative version of the juvenile correctional institution that has received mixed reviews is "boot camp" programs, which offer rehabilitation (as well as restraint) in the form of a military-like, highly structured environment. The philosophy behind these boot camps is that adolescents or young adults who suffer from poor impulse control, low self-esteem, and high rates of acting out behavior can benefit from a military-like structured setting that pushes them to their limit (both physical and emotional). The high emphasis on structure and self-discipline, coupled with the push to achieve, is believed to have a positive impact on both self-esteem and self-respect, which is hoped to generalize into more respectful behavior in society. Many parents and participants commonly claim dramatic changes in the behavior of participants after a boot camp experience, but research appears to indicate that boot camps do not necessarily reduce recidivism rates in young offenders (Peters, Thomas, & Zamberlan, 1997).

Another type of treatment facility for adolescents experiencing mental health problems is in-patient psychiatric hospitals. These programs tend to be acute (short term), focusing on stabilizing the adolescent's high-risk behaviors, such as suicidal behavior, self-abuse, substance abuse, and eating disorders. Some in-patient programs specialize in one or more of these disorders or are more general in nature, offering short-term acute services to any adolescent who cannot be maintained safely outside a hospital setting. Many of the same type of therapies are available in an in-patient setting as in a residential treatment center, with the exception that drug therapies may be more prevalent in a psychiatric hospital. In-patient hospitals also rely heavily on discharge planning, a task that typically falls to a hospital social worker or other human service

Human Services Delivery Systems

Understanding and Mastery of Human Services Delivery Systems: Range and characteristics of human services delivery systems and organizations

Critical Thinking Question: A wide variety of therapeutic interventions exist for adolescents, ranging from weekly outpatient sessions with a counselor, through programs such as Outward Bound, to inpatient facilities with a range of programs and supports. Ideally, adolescent clients would receive the level of treatment most appropriate to their situation; in reality, what other factors play a role in the type and level of treatment which adolescents receive?

professional who works with the family and community resources to ensure that the adolescent will transition back to home and school with enough outpatient support to minimize the need for rehospitalization.

Multicultural Considerations

It would be naïve to assume that race and ethnicity did not have a significant effect on adolescent development, including the types of problems adolescents of various races experience as well as the various responses to those problems, both within the family and within the community. Human service professionals must be aware of the way in which race and ethnicity affect adolescent development and behavior, as well as any negative stereotypes that might affect the types of diagnoses adolescents receive.

A 2001 study found that African American adolescents were more commonly diagnosed with conduct disorders, whereas Caucasian adolescents more often received a diagnosis of depression (DelBello, Lopez-Larson, & Soutullo, 2001). But is this because more African American adolescents actually have conduct disorders? Or is it because the negative stereotype that African American males are typically more violent influenced the practitioner rendering the diagnosis? DelBello et al. (2001) doubted that the difference in diagnosing reflected any real variation in disorders among adolescents of different races, but was more likely attributable to variables such as misdiagnosing based on cultural differences and misperceptions.

Other research indicates that Latino adolescents, specifically Mexican Americans, are at higher risk for delinquency, depression, and suicide than Caucasians (Roberts, 2000). African American youth tend to show the greatest need for mental health services, yet were severely underserved, and although most mentally ill African American adolescents had a long history of diagnosable mental health problems, often their first exposure to treatment was within the juvenile justice system. One reason for this might be that there is a negative stigma associated with mental health disorders in certain ethnic minority groups. But another equally significant reason is likely the lack of affordable mental health services in ethnically diverse neighborhoods, as well as issues such as poor or no insurance coverage for mental health services in ethnically diverse populations. In fact, a recent study showed that very little has changed in this trend in the last few decades, despite considerable research in this area and policy recommendations. For instance, a 2011 study revealed that African American, Latino, and Asian adolescents with major depression were significantly less likely to receive mental health treatment, including prescription medication, than non-Hispanic white adolescents, regardless of income levels and health insurance (Cummings & Druss, 2011). It is interesting to note that Latino adolescents were rated as the most underserved of all racial groups, despite the fact that they had significant needs, and Caucasians were reported to have the highest rate of mental health utilization, although they have less serious mental health diagnoses compared to other racially diverse groups (Rawal, Romansky, & Jenuwine, 2004).

> A 2011 study revealed that African American, Latino, and Asian adolescents with major depression were significantly less likely to receive mental health treatment, including prescription medication, than non-Hispanic white adolescents, regardless of income levels and health insurance.

Certainly not all differences in adolescent diagnoses can be attributed to cultural misperceptions, misdiagnoses, and underutilization of services. Social conditions, such as poverty, high crime neighborhoods, and unemployment likely contribute to a significant proportion of mental health problems in racially ethnic youth. Rawal et al. (2004) noted that African American adolescents are far more likely to be raised in single-parent households, be placed in foster care, and experience significantly higher rates of familial abuse and neglect, all of which can be expected to have a negative impact on their mental health. Latino adolescents also exhibited higher incidences of acting out and antisocial behaviors, such as juvenile delinquency, compared to Caucasians; yet, they also had greater familial support, with their caregivers exhibiting greater understanding and involvement in their mental health issues, which might act as an intervention negating the necessity of more serious intervention.

Regardless of the reasons for the differences in mental health issues among adolescents of different ethnic groups, it is imperative that human service professionals be trained to deliver culturally competent counseling. Education that addresses all these issues, including institutionalized racism, both within the community and within the juvenile justice system; culturally based stigmas associated with mental health issues; social conditions affecting adolescents of all races; and the relevant histories of various racially ethic minority groups within the United States (e.g., the history of slavery among African Americans or the history of forced institutionalized care among Native American youth) will assist the human service professional render a bias-free mental health evaluation and provide the most appropriate treatment for the adolescent client.

> ### Human Systems
>
> Understanding and Mastery of Human Systems: Changing family structures and roles
>
> Critical Thinking Question: In some cases, and particularly among certain ethnic and cultural groups, the family can serve as a strong source of support for troubled adolescents; on the other hand, families may sometimes stand in the way of a teen's accessing professional help. How might a human service professional build on the strengths that clients' families have to offer, and break down barriers to cooperation?

Concluding Thoughts on Adolescents

Clearly, our society will continue to change and evolve, affecting all its members, including adolescents. As our society becomes more technologically based, it will become more complex as well, which will no doubt mandate increasing levels of education—a trend that the United States has seen steadily increase in the last 50 years at least. This does not mean that juvenile violence will continue to rise. Most mental health experts refuse to adopt such a fatalistic attitude. History reveals that adolescence has always been a difficult stage to navigate, long before it was even recognized as an official stage of development. The greatest hope one can offer parents and educators alike is that adolescents who often seem destined for a lifetime of narcissistic obsession most often evolve into loving, caring, and responsible adults. Human service professionals can help families ensure that this is the path for as many adolescents as possible through effective program development and supportive services on all levels.

The following questions will test your knowledge of the content found within this chapter.

1. One of the first theorists to study the stage of adolescence and who in 1904 described adolescents as possessing a "lack of emotional steadiness, violent impulses, unreasonable conduct, lack of enthusiasm and sympathy" was
   a. Erik Erikson
   b. Sigmund Freud
   c. G. Stanley Hall
   d. Jean Piaget

2. When considering the normative nature of adolescent behavior, what contexts must one keep in mind?
   a. Historical, cultural, and regional
   b. Historical, cultural, and contemporary
   c. Cultural, regional, and socioeconomic
   d. Cultural, socioeconomic, and contemporary

3. Oppositional defiance disorder is an emotional disorder commonly diagnosed in adolescence and is characterized by
   a. negative, hostile, and defiant behavior
   b. angry, rebellious, and rage-filled behavior
   c. depressed, anxious, and socially phobic behavior
   d. melancholy, stoicism, and apathetic behavior

4. Human service professionals working with the adolescent population should look for signs of self-mutilation, which may include
   a. wearing long sleeves and pants on warm days
   b. parallel scars on the forearm or thighs
   c. wounds that tend not to heal
   d. All of the above

5. A majority of adolescents who committed suicide also had a history of involvement with
   a. drugs and alcohol
   b. a negative peer group
   c. the mental health system
   d. the juvenile justice system

6. A 2001 study found that African American adolescents were more commonly diagnosed with _____ whereas Caucasian adolescents were more often diagnosed with _____.
   a. depression/conduct disorder
   b. conduct disorder/depression
   c. conduct disorder/anxiety
   d. oppositional defiance disorder/conduct disorder

7. Describe some of the reasons why many adolescents within ethnic minority populations do not receive necessary mental healthcare services compared to their Caucasian counterparts. Explore some of the vulnerabilities experienced by ethnic minority youth and ways that human services professionals can address these unmet needs.

8. What are some common ways that rebelliousness manifests in the adolescent population? Make sure to frame your response within appropriate context. As a human service professional what are some evidence-based ways that you might consider responding to an adolescent client experiencing psychosocial problems?

Suggested Readings

Cloud, H., & Townsend, J. (2001). *Boundaries with kids.* Grand Rapids, MI: Zondervan.

Mattaini, M. A. (2001). *Peace power for adolescents strategies for a culture of nonviolence.* Washington, DC: NASW Press.

Roles, P. (2005). *Facing teenage pregnancy: A handbook for the pregnant teen.* Atlanta, GA: CWLA Press.

Ungar, M. (2002). *Playing at being bad: The hidden resilience of troubled teens.* Washington, DC: NASW Press.

Ungar, M. (2003). *Nurturing hidden resilience in troubled youth.* Washington, DC: NASW Press.

Internet Resources

Adolescence and Peer Pressure: http://ianrpubs.unl.edu/family/nf211.htm

Child and Adolescent Mental Health: http://www.nimh.nih.gov/health/topics/child-and-adolescent-mental-health/index.shtml

Mental Health Risk Factors for Adolescents: http://education.indiana.edu/cas/adol/mental.html

Outward Bound: http://www.outwardbound.org

WHO Adolescent Health: http://www.who.int/child-adolescent-health/OVERVIEW/AHD/adh_over.htm

References

American Psychiatric Association. (2000). *Diagnostic and statistical manual of mental disorders* (4th ed., Text Revision). Washington, DC: Author.

Andrus, J. K., Fleming, D. W., Heumann, M. A., Wassell, J. T., Hopkins, D. D. Y., & Gordan, S. (1991). Surveillance of attempted suicide among adolescents in Oregon, 1988. *American Journal of Public Health, 81*, 1067–1069.

Arias, E., Anderson, R. N., Kung, H. C., Murphy, S., & Kochanek, K. D. (2003). Deaths: Final data for 2001. *National Vital Statistics Reports, 52*(3). Hyattsville, MD: National Center for Health Statistics.

Bearman, P. S., & Moody, J. (2004). Suicide and friends among American adolescents. *American Journal of Public Health, 94*(1), 89–95.

Clarke, L., & Whittaker, M. (1998). Self-mutilation: Culture, contexts and nursing responses. *Journal of Clinical Nursing, 7,* 129–137.

Crowe, M., & Bunclark, J. (2000). Repeated self-injury and its management. *International Review of Psychiatry, 12*(1), 48–53.

Cummings, J. R., & Druss, B. G. (2011). Racial/ethnic differences in mental health service use among adolescents with major depression. *Journal of the American Academy of Child & Adolescent Psychiatry, 50*(2), 160–170.

DelBello, M., Lopez-Larson, M. P., & Soutullo, C. A. (2001). Effects of race on psychiatric diagnosis of hospitalized adolescents: A retrospective chart review. *Journal of Child and Adolescent Psychopharmacology, 11*(1), 95–103.

Demos, J., & Demos, V. (1969). Adolescence in a historical perspective. *Journal of Marriage and the Family, 31*(4), 632–638.

Erikson, E. H. (1968). *Identity: Youth and crisis.* New York: Norton.

Fairburn, C. G. (2005). Evidence-based treatment of anorexia nervosa. *International Journal of Eating Disorders, 37*(Suppl.), S26–S30.

Favazza, A. R. (1996). *Bodies under siege: Self-mutilation and body modification in culture and psychiatry* (2nd ed.). Baltimore: Johns Hopkins University Press.

Frick, P. (2004). Developmental pathways to conduct disorder: Implications for serving youth who show severe aggressive and antisocial behavior. *Psychology in the Schools, 41*(8), 823–834.

Gandara, P., Gutierrez, D., & O'Hara, S. (2001). Planning for the future in rural and urban high schools. *Journal of Education for Students Placed at Risk, 6*(1–2), 73–93. (ERIC Document Reproduction Service No. UD522844)

Gorwood, P. (2004). Generalized anxiety disorder and major depressive disorder comorbidity: An example of genetic pleiotrophy? *European Psychiatry, 19*(1), 27–33.

Gowers, S., & Bryant-Waugh, R. (2004). Management of child and adolescent eating disorders: The current evidence base and future directions. *Journal of Child Psychology & Psychiatry, 45*(1), 63–83.

Hazell, P., & Lewin, T. (1999). Friends of adolescent suicide attempters and completers. *Journal of American Academy of Child & Adolescent Psychiatry, 32*(11), 76–81.

Hicks, K., & Hinck, S. M. (2009). Best-practice intervention for care of clients who self-mutilate. *Journal of the American Academy of Nurse Practitioners, 21*(8), 430–436. doi:10.1111/j.1745-7599.2009.00426.x

Kerpelman, J. L., & Pittman, J. F. (2001). The instability of possible selves: Identity processes within late adolescents' close peer relationships. *Journal of Adolescence, 24*(4), 491–512.

Lahey, B. B., Moffitt, T. E., & Caspi, A. (Eds.). (2003). *Causes of conduct disorder and juvenile delinquency.* New York: Guilford Press.

Larsen, M. (2003). *Violence in U.S. public schools: A summary of findings.* New York: ERIC Digest. (ERIC Document Reproduction Service No. ED482921)

Lock, J., & le Grange, D. (2005). Family-based treatment of eating disorders. *International Journal of Eating Disorders, 37*(Suppl.), S64–S67.

Miniño A. M. (2010). Mortality among teenagers aged 12–19 years: United States, 1999–2006. NCHS data brief, no 37. Hyattsville, MD: National Center for Health Statistics.

Moskos, M. A., Achilles, J., & Gray, D. (2004). Adolescent suicide myths in the U.S. *Journal of Crisis Intervention & Suicide Prevention, 25*(4), 176–182.

National Center for Education Statistics. (1999). *Digest of education statistics.* Washington, DC: National Research Council Panel on High Risk Youth, National Academy of Sciences.

National Institute of Mental Health. (2005). *Schizophrenia.* Bethesda, MD: National Institutes of Health. Retrieved November 15, 2005, from http://www.nimh.nih.gov/publicat/schizoph.cfm#definition

Nock, M. K., & Prinstein, M. J. (2005). Contextual features and behavioral functions of self-mutilation among adolescents. *Journal of Abnormal Psychology, 114*(1), 140–146.

Peters, M., Thomas, D., & Zamberlan, C. (1997). *Boot camps for juvenile offenders.* Office of Juvenile Justice and Delinquency Prevention, U.S. Department of Justice. Washington, DC: U.S. Government Printing Office.

Piaget, J. (1950). *The psychology of intelligence.* London: Routledge & Kegan Paul.

Rawal, P., Romansky, J., & Jenuwine, M. (2004). Racial differences in the mental health needs and service utilization of youth in the juvenile justice system. *Journal of Behavioral Health Services & Research, 31*(3), 242–254.

Ray, S. L. (2004). Eating disorders in adolescent males. *Professional School Counseling, 8*(1), 98–102.

Raywid, M. (1996). *Downsizing schools in big cities.* New York: ERIC Clearinghouse on Urban Education. (ERIC Document Reproduction Service No. ED393958)

Roberts, R. E. (2000). Depression and suicidal behaviors among adolescents: The role of ethnicity. In I. Cuéllar & F. A. Paniagua (Eds.), *Handbook of multicultural mental health* (pp. 360–389). San Diego, CA: Academic Press.

Rutter, P. A., & Behrendt, A. E. (2004). Adolescent suicide risk: Four psychosocial factors. *Adolescence, 39*(154), 295–302.

Spirito, A., Valeri, S., Boergers, J., & Donaldson, D. (2003). Predictors of continued suicidal behavior in adolescents following a suicide attempt. *Journal of Clinical Child and Adolescent Psychology, 32*(2), 284–289.

Striegel-Moore, R. H., Rosselli, F., Perrin, N., DeBar, L., Wilson, G., May, A., & Kraemer, H. C. (2009). Gender difference in the prevalence of eating disorder symptoms. *International Journal of Eating Disorders, 42*(5), 471–474. doi:10.1002/eat.20625.

van Hoeken, D., Seidell, J., & Hoek, H. (2003). Epidemiology. In J. Treasure, U. Schmidt, & E. van Furth (Eds.), *Handbook of eating disorders* (2d ed., pp. 11–34). Chichester, UK: Wiley.

Vohs, K. D., Voelz, Z. R., Pettit, J. W., Bardone, A. M., Katz, J., Abramson, L. Y., et al. (2001). Perfectionism, body dissatisfaction, and self-esteem: An interactive model of bulimic symptom development. *Journal of Social & Clinical Psychology, 20,* 476–497.

Vörös, V., Fekete, S., Hewitt, A., Osváth, P. (2005). Suicidal behavior in adolescents—psychopathology and addictive comorbidity. *Neuropsychopharmacol Hung, 7*(2):66–71. Hungarian. PMID: 16167457 [PubMed—indexed for MEDLINE]

Vörös, V., Osváth, P., Fekete, S. (2004). Gender differences in suicidal behavior. *Neuropsychopharmacol Hung, 6*:65–71.

Walcott, D. D., Pratt, H. D., & Patel, D. R. (2003). Adolescents and eating disorders: Gender, racial, ethnic, sociocultural, and socioeconomic issues. *Journal of Adolescent Research, 18,* 223–243.

Wells, K. (1991). Long-term residential treatment for children: Introduction. *American Journal of Orthopsychiatry, 61,* 324–326.

Whitlock, J.L., Purington, A., Gershkovich, M. (2009). Influence of the media on self injurious behavior. In Understanding non-suicidal self-injury: Current science and practice, edited by M. Nock. *American Psychological Association Press.* 139–156.

CHAPTER 7

Aging and Services for the Older Adult

· ·

Ariel Skelley/Corbis

Learning Objectives

- Understand the changing demographics of the U.S. population, often referred to as the "Aging of America"
- Identify the impact that the aging baby boomers are having and will likely continue to have on various aspects of U.S. society and culture
- Recognize various elements of successful aging and be able to describe lifestyles that lead to a successful aging lifestyle
- Become familiar with the nature of ageism in a variety of contexts and describe ways in which human service professionals can work to combat discrimination based upon age
- Identify trends associated with grandparents parenting, including developing a basic awareness of current demographic patterns, common causes of these patterns, and the various issues facing custodial grandparents

Carrie looked at the sea of faces before her. They looked empty—almost as if they had no souls. The only sounds in the camp were the incessant, never ceasing buzzing of hungry flies. Even the children were quiet. Carrie reasoned that the calm was due to hunger—people were often subdued when they hadn't eaten well in days, but she knew this calm was related to something far removed from hunger.

Just three days ago the people in this camp were victims of an Arab militia known as the Janjaweed. These bands of marauding fighters combed the countryside, indiscriminately killing black Africans. As the villagers looked on in horror, Janjaweed militia began to systematically slaughter the innocent villagers one by one. Not even infants were spared; some militia tossed babies and toddlers into the air, calling them future enemies, as they shot them with machine guns. The few villagers who managed to escape joined other escaping villagers running through the desert and were eventually picked up by the American Red Cross.

Carrie is a missionary with an organization that specializes in sending retirees abroad. When Carrie became a widow at the age of 71, she thought her life was over. However, the pastor at her church approached her, and after months of talking, he finally convinced her that her years of nursing experience need not go to waste. Carrie was initially skeptical when her pastor shared the stories of other retirees, many of whom were widows, who served in clinics and refugee camps overseas in countries like Guatemala, Burma, and Sudan, but it wasn't until she met some older adult missionaries at home on sabbatical that she finally realized that this was something she could do.

Of course, Carrie's adult children thought she'd lost her mind, they even questioned whether her decision to become a missionary was a sign of early Alzheimer's, but they eventually grew to understand her decision and even respect it, although she was certain that they never felt truly comfortable with the thought of their old mother living in a refugee camp in the middle of a war-torn country. Carrie's contemplative thoughts were interrupted with the announcement of the most recent influx of shell-shocked and injured refugees, and she ventured out of the makeshift hospital to meet the new arrivals.

Dan was shocked as he walked down La Salle Avenue, in the heart of the business district in Chicago. He was used to seeing homeless people, either standing or sitting along the side of the road with signs asking for money, but he had never seen an old couple begging for money before. What was unique about this couple was that they looked as though they could be his own mother and father.

He began to walk by them, avoiding their stare as he usually did when people begged for money, but this time was different, and he could not resist approaching this couple. "Hi, my name is Dan, and I'd love to give you some money." The couple looked at him sheepishly, and he noticed the shame in their eyes. "Thank you," the woman said quietly, diverting her glance downward. Dan handed them a $10 bill and started to walk away, but curiosity got the better of him. He turned around and asked them if he could talk to them about their situation. The husband and wife looked at each other, and Dan did not know if it was with suspicion or simple caution, but they eventually agreed once Dan offered to buy them lunch.

Over their meals of hot soup and sandwiches, Rosemary and Donald shared about their all-American lives. They raised two children in a suburb of Chicago, owned a home, and even had a family dog. They were like anyone else in the neighborhood or their church, until Donald was laid off two years before his scheduled retirement when the company he had worked for for 40 years downsized. Donald was unable to find a job due to his age, and eventually they had to let their health insurance lapse because they could no longer financially handle the extremely high monthly premiums.

Unfortunately, Rosemary became ill the following month with a bout of influenza that ultimately developed into pneumonia. The hospital bill for her two-week stay was almost $10,000. With no retirement and only Social Security benefits to count on, and with their two adult children serving overseas in the military, Donald and Rosemary began a downhill financial descent that didn't stop until they depleted their life savings and ultimately lost their house. Thus, although most couples like Donald and Rosemary spend their golden years playing golf in Florida, Donald and Rosemary spend their days sitting outside the train station, begging for money.

These two vignettes highlight the vast range of experiences Americans living in the United States can have in the last decades of their lives—what is normally called *old age*. And although there are some similarities between the older adults of today and the older adults of 100 years ago, there are also significant differences brought on by many of the societal changes referenced in earlier chapters, particularly in relation to social welfare policy explored in Chapter 2, including ongoing urbanization, changes in the family structure, and the dawning of the technological era. However, there have also been transitions in culture and society that have affected the older adult community in a unique way. These include issues such as increasing longevity, the global community and economy, other economic shifts, the advent of long-distance travel enabling family members to move further and further away from one another, the healthcare "crisis," significant demographic shifts, as well as the increasing complexity of society in general. Each of these issues and their impact on the older adult population affects the human services field as it attempts to meet the complex needs of this growing population.

The Aging of America: Changing Demographics

The opening vignette illustrates the vast range of experiences of those considered "old" in the United States. Today's older adults in the United States experience a broader range of lifestyles than ever before, but they experience a greater range of challenges as well. There are several reasons for this vast array of lifestyle choices and options, including the increase in the human life span, changes in the perception of old age in general, changes in the economy, and finally changes in the nature and definition of the American family in the United States, including a dramatic increase in divorce and two-parent working families.

> **Today's older adults in the United States experience a broader range of lifestyles than ever before, but they experience a greater range of challenges as well.**

Read just about any scholarly article relating to the older adult population, and you will likely read about the *Graying of America*. This term relates to the increase in the older adult population in the United States (as well as in most parts of the world). This dramatic increase, as well as the projected increase in the U.S. older adult population between now and 2050, is directly related to the aging of a cohort of individuals referred to as the *baby boomers*. The baby boomers are popularly defined as those having been born between 1946 and 1964. The name refers to the *boom* of births after World War II, which caused an unusual spike in the U.S. population. Approximately 76 million individuals (roughly 29 percent of the U.S. population) fall into the cohort of baby boomers. It is obvious why this cohort has been the focus of particular interest to social scientists, the media, politicians, and others. For one thing, despite the somewhat broad range of ages within this cohort, similarities between members are numerous, including their socioeconomic status, which tends to be higher than earlier cohorts, consumer habits, and political concerns. As the boomers age, their tastes and concerns transition, and in recent years their collective focus has included discussions regarding the consequences of this large cohort heading into their retirement years. The *graying of America*, then, refers to the projected increase in the older adult population because of the aging boomers.

The aging of the baby boomers is not the only variable leading to the increase in the older adult population. Other factors include the 50 percent increase in the human life span the United States has experienced during the 20th century. In 1900 the average human life span in the United States was about 47 years. But by 1999 it had increased to about 77 years, which is where it stands today, although it is expected to increase at least another 15 years by the year 2100 (Arias, 2004). This life expectancy increase is due to many variables, including improved medical technology, medical discoveries such as antibiotics and immunizations for various life-threatening diseases, and generally safer lifestyles.

Currently there are approximately 40 million people over the age of 65 living in the United States (NHSTA, 2009), but that number is projected to double by the year 2050, growing to more than 88 million (Passel & Cohn, 2008). Additionally, the U.S. Census Bureau projects that the population of those aged 85 and older is expected grow from 5.8 million in 2009 to approximately 20 million by the year 2050 (Department of Health and Human Services, 2010). When one considers that from 1900 to 2050 the over-65 population in the United States will grow from about 3 million to almost 90 million, it is not difficult to understand why the field of gerontology has received so much attention in recent years!

So far this all sounds pretty good—we're living longer, and in the next 10 or 20 years a third of the population will be classified as older adults, which will no doubt increase the attention paid to social and political issues important to those in their retirement years. However, the landscape for older adults in the United States is not completely rosy; quite the opposite, in fact. Some will no doubt enjoy their longer life span, but for many, their extra years on this earth may be spent in a long-term care facility with chronic health problems far too complex to make remaining in their home a possibility. Increases in rates of dementia, depression, and alcohol abuse are valid concerns for older adults and their family members, as they face a multitude of challenges in a rapidly changing world.

The most recent economic crisis starting in 2008, also called the Global Financial Crisis, resulted in the forced retirement and unanticipated layoff of many aging individuals within the workforce. In addition, changes in the U.S. and global economies risk leaving many individuals approaching retirement in economically vulnerable positions as companies shift away from offering employees lifelong careers with permanent and secure retirement plans. Sharp increases in the cost of medical care and possible changes in Social Security benefits are also putting some older adults at risk of financial vulnerability. Thus, an increasingly older population will no doubt have an impact on the financial, housing, medical, mental health, and even transportation needs of the older adult population. Add to that, changes in the U.S. family structure, such as the significant increase in divorce rates, have put some older adults in the position of having to provide day care for their grandchildren and, in some cases, even parenting their grandchildren. Thus, although some older adults will be able to take advantage of the many medical advances, healthier lifestyles, and increased opportunities for enjoying life, many others will not.

This chapter will explore the wide range of issues confronting the older adult population in the United States, as well as exploring some issues projected to be relevant in the future. The role of the human service professional will be explored as well, with a special focus on how the field of gerontology has changed in recent years, expanding the role of the human service professional in various practice settings.

Old and Old-Old: A Developmental Perspective

Before beginning any real discussion about clinical issues affecting older adults or the role of the human service professional, it is important to understand the various aspects of physical, social, and emotional development common to individuals in the last quarter or so of their lives. Although there is no specific age limit marking the end of middle age and the beginning of older adult years, most contemporary developmental theorists consider old age to begin at around the age of retirement.

Many theorists have argued that adults do not go through systematic and uniform developmental stages in the same way that children do; thus, earlier developmental theories typically stop at early adulthood or lump all adult development into one category stretching from post-adolescence and beyond. One reason for this approach is that if development consists of the combined impact of physical, cognitive, and emotional maturity, then certainly one can see that children who are spurred on to extend their social boundaries will be motivated to push themselves from a crawl to a walk in their quest to explore their social worlds. Yet, once one has reached physical and cognitive maturity, this interplay between physical ability and emotional desire (where one dynamic acts as the incentive for the other) subsides, and the motivation to pursue a particular life course becomes based more on personal choice and internal motivation, making adult maturity anything but systematic or universal.

Nevertheless, should we assume that adults do not continue to develop in any sort of consistent or predictable way? Would it be correct to assume that once individuals have reached all physical developmental milestones (somewhere after puberty) their emotional development occurs in a completely unique and individualistic manner? Most of us have heard about the infamous "midlife crisis" marking the entry into middle age, or "empty nesting," the universal life crisis some women experience in response to their adult children leaving home, and regardless of the validity of the universality of such life events, it does seem reasonable to assume that individuals within a particular society will respond and adapt to both internal and external demands and expectations placed on them by cultural mores and norms and that there would be some interplay between their physical development (or physical decline), their emotional and cognitive development, influenced by their social worlds, which give meaning to their experiences.

Cultural expectations in the United States, such as marriage, child rearing, employment, and home ownership, certainly have an impact on those in early and middle adulthood, just as retirement, increased physical problems, and widowhood will have an impact on those in later adulthood. Yet, because the options and choices available to adults are so broad, any developmental theory must be considered in somewhat broad and descriptive terms, rather than the narrower and more prescriptive terms often used to evaluate and consider child developmental theories.

Erik Erikson (1959, 1966), a psychodynamic theorist who studied under Sigmund Freud (the father of psychoanalysis), developed a theory of psychosocial development, beginning with birth and ending with death. According to Erikson, each stage of development presented a unique challenge or crisis brought about by the combining forces of both physiological changes and psychosocial need. Successfully resolving the developmental crisis resulted in being better prepared for the next stage. The eighth stage of Erikson's model is *integrity versus despair* and spans from age 65 to death. Erikson believed that individuals in this age range needed to reflect back on their lives, taking stock of their choices and the value of their various achievements. If this reflection resulted in a sense of contentment with one's choices and life experiences, then the individual will be able to accept death with a sense of integrity, but if he does not like the choices made, the relationships developed, and the wisdom gained, then he will face death with a sense of despair.

Because the successful navigation of each stage is dependent on the successful navigation of the preceding stages, Erikson believed that individuals who did not develop a sense of basic trust in others or in the world (Stage 1), struggled developing a sense of personal autonomy (Stage 2), had difficulty developing any personal initiative (Stage 3) or a sense of accomplishment (Stage 4), faced challenges in adolescence when attempting to discover a personal identity (Stage 5), making it difficult to develop truly intimate relationships with others (Stage 6), leaving them incapable of offering true guidance and generativity to the younger generations (Stage 7), which would likely mean that they would not reflect back on their life with any sense of contentment and satisfaction, and would then likely face impending death with a true and deep sense of despair.

Daniel Levinson (1978, 1996) is probably one of the most well-known adult developmental theorists, having developed a life span theory extending from birth through death. Levinson wrote two books explaining his theory, *The Seasons of a Man's Life* (1978) and *The Seasons of a Woman's Life* (1996), where he focused on middle adulthood, but what was revolutionary about his theory was his argument that adults do continue to grow and develop on an age-related timetable. Levinson noticed that adults in the latter half of their lives are more reflective, and as they approached a point in their lives where they had more time behind them than ahead of them, this reflection intensified. Levinson also believed that individuals progress through periods of stability that are followed by shorter stages of transition. The themes in his theory most relevant to human service professionals include this notion of life reflection—the taking stock of one's life choices and accomplishments, the need to be able to give back to society, which encompasses an acknowledgment that at some point the goal in life is not solely to focus on one's own driving needs, but to give back to others and community through the sharing of gained wisdom and mentoring.

Finally, Levinson's belief that as people age they need to become more intrinsically focused rather than externally based is equally relevant. Consider the man who in his 30s gains self-esteem and a sense of identity through working 80 hours per week and running marathons. How will this same man define himself when he is 70 and no longer has the physical stamina or agility to perform these activities? Levinson believed that a developmental task for aging adults was to become more internally anchored, more intrinsic in their self-identify, lest they develop a sense of despair and depression later in life when they are no longer able to live up to their own youthful expectations.

Another theory that purports to describe the changes individuals experience from middle to older adulthood on emotional, cognitive, and physical levels is called *gerotranscendence* (Tornstam, 1994). This theory explores how an individual moves from a strong connection to the material world to transcending above the material aspects of the world into a more existential approach to the world. In a similar way as Levinson, Tornstam describes how individuals progressing from midlife onward transition from an externalizing perspective, where they are focusing outward toward the world, to a more internally focused approach in life.

Tornstam (2003) describes three dimensions of transcendence, including the "cosmic level" where individuals change their notions of time and space, such as reorienting themselves in regard to how they view life and death, ultimately accepting death with a sense of peace. The second realm relates to "the self" where individuals increasingly move away from self-centeredness, transcendending above a focus on the physical and move toward more altruism. The third level of transcendence is a realm involving social and individual relationships, where the relationships are viewed in a new light with new meaning, including developing new insights into the differences between "the self" (who they really are), and the roles they play in life (mother/father, son/daughter, friend, etc.), and the ability to rise above black-and-white thinking, embracing the gray in life (Degges-White, 2005).

Degges-White (2005) discusses implications of Tornstam's theory of gerotranscendence for counselors working with the older adult population, highlighting key issues involved in the process of personal transcendence across the three dimensions (cosmic, self, and relationships with others). For instance, Degges-White cites the importance of counselors becoming comfortable with the concept of death within themselves, so that they can help their aging clients accept the inevitability of death without fear and anxiety. With regard to transcendence in the "self" domain, Degges-White describes how counselors can help their older clients conduct a "life review" where they seek to better understand and accept their life choices, thus finding a level of peace and self-acceptance about their choices and experiences, particularly the challenging and painful ones. The ultimate focus of counseling older adults using a gerotranscence model is to assist older adults move toward increased self- and other-acceptance and wisdom in various dimensions and domains in life, and in a sense, giving them permission to drawn intrinsically inward as they let go of the more transitory dimensions of life, and toward a more existential framework.

Human Services Delivery Systems

Understanding and Mastery of Human Services Delivery Systems: Theories of human development

Critical Thinking Question: Theorists such as Erikson, Levinson, and Tornstam produced models describing stages of development among older adults. How might human service professionals utilize these models in their work with the aging population?

Successful Aging

A relatively recent concept that has become popular in relation to the study of geriatrics is the concept of *successful aging*, which is used to describe the process of getting the most out of one's life in later years. Successful aging literally means to add years to one's life and to get the most out of living (Havighurst, 1961). Researchers have examined individuals who age better than others to determine what differences might account for their "success," and some of the variables at play include maintaining a moderately high physical and social activity level, including keeping active with hobbies, social events, and regular exercise (Warr, Butcher, & Robertson, 2004). A study in 2007 found that when older adults participated in some type of social activity, such as paid or unpaid work, religious activities, and political involvement, mortality and cognitive function impairment were reduced, yet disparity in opportunities for meaningful social activities left some older adult groups more vulnerable to physical and cognitive decline (Hsu, 2007).

The natural aging process, though, seems to discourage high activity levels in virtually all domains. Employment provides most people with one of the greatest opportunities for social interaction, and when individuals retire, a significant portion of their social life is lost along with their career. Physical limitations also encourage disengagement. Few older adults play on intramural softball teams, and even something like poor night vision can keep an older adult from being able to hop in the car and visit family. Thus, many older people naturally begin withdrawing from the world, both physically and socially, in response to diminished capability and opportunity, and with such disengagement comes an increase in physical and emotional problems, such as depression and even alcohol abuse to combat loneliness.

> **Many older people naturally begin withdrawing from the world, both physically and socially, in response to diminished capability and opportunity, and with such disengagement comes an increase in physical and emotional problems, such as depression and even alcohol abuse to combat loneliness.**

A very recent study seems to indicate that good psychological health is the most important factor of all in ensuring good quality of life in later years (Bowling, & Iliffe, 2011). For instance, the ability of older adults to rely on their psychological resources, such as a good self-efficacy (one's perception of personal competency) and resilience were more strongly linked to successful aging than were biological and social factors. This does not mean that good physical health and an active social life aren't important, but as Tornstam (2005) and Degges-White (2005) suggest, older adults who can mentally and emotionally transcend beyond the physical and social limitations inherent in the aging process seem to age more "successfully" and have a better quality of life compared to older adults who lack these psychological resources.

Consider an individual's level of psychological resilience, which encompasses one's coping strategies that can be relied upon during challenging times. For instance, many older adults must not only face increased health problems and physical limitations, but also deal with multiple losses as they begin to lose friends, siblings, and even their spouse to death. Many older adults must move from their longtime home into residential care or the home of a family member, and even the loss of independence can create a situation where their mental health is determined by the veracity of their coping mechanisms. Psychological resilience enables older adults to manage these multiple losses in a healthier manner, even perhaps finding some existential meaning in facing these losses with a sense of wisdom and acceptance, despite the deep pain and sense of powerlessness many older adults may feel.

Current Issues Affecting Older Adults and the Role of the Human Service Professional

In anticipation of the increase in the older adult population as well as an increase in the needs and complex nature of the issues facing many older adults, the Older Americans Act was signed into federal law in 1965. This act led to the creation of the Administration on Aging, and it funded grants to the states for various community and human service programs and provided money for age-related research and the development of human service agencies called Area Agencies on Aging (AAA) operating on the local level. The Administration on Aging also acts as a clearinghouse, disseminating information about a number of issues affecting the older adult population in the United States.

Numerous issues affect today's older adult population, including elder abuse, age-based discrimination, housing needs, biopsychological problems (such as depression, anxiety, and alcoholism), adjustment to retirement, and grandparenting. Those in the human services field are often included in the group of professionals most likely to come into contact with the older adult population, either through direct service or through providing counseling services to a family member of an older adult, and therefore they must be familiar with these key issues, knowing how they affect older adults and their family.

Ageism

Ask some typical young Americans what they think it is like to be a man of 70, and they may well tell you that an average 70-year-old man is in poor health, drifts off to sleep at a moment's notice, talks of nothing but the distant past, and unproductively sits in a rocker, rocking back and forth all day long. They might even throw in a comment or two about his general grouchy disposition. Ask if older adults still have the desire for sexual intimacy, and you might get a good hearty laugh in response. However, this description of older adults is a myth based on deeply entrenched negative stereotypes and can serve as a foundation of a form of prejudice and discrimination of older adults called *ageism*.

The term *ageism* was first coined by Robert Butler (1969), chairman of a congressional committee on aging in 1968. He defined ageism as "a systematic stereotyping of and discrimination against people simply because they are old, just as racism and sexism accomplish this with skin color and gender." Butler theorized that the basis of this negative stereotype is a fear of growing old. This fear and the resultant negative stereotyping can often result in the discrimination of the older adult population in all areas of life and is the basis of many forms of elder abuse.

Ageism typically involves any attitude or behavior that negatively categorizes older adults based either on partial truth (often taken out of context) or on outright myths of the aging process. Such myths often describe old age as involving (1) poor health, illness, and disability; (2) lack of mental sharpness and acuity, senility, and dementia; (3) sadness, depression, and loneliness; (4) an irritable demeanor; (5) a sexless life; (6) routine boredom; (7) a lack of vitality and continual decline; (8) an inability to learn new things; and (9) loss of productivity (Thornton, 2002).

Gerontologists caution that the promotion of such negative stereotypes of old age and older adults not only trivializes older individuals, but also risks displacing the older adult population as communities undervalue them based on the perception that older adults are nothing more than a drain on society. A further risk of ageism is that older adults may internalize this negative stereotype, creating a self-fulfilling prophecy of sorts (Thornton, 2002). This is similar to what happens with other vulnerable populations, such as minority groups, who internalize the negative perceptions of them held by the majority population (Snyder, 2001).

Old age has not always been something those in the United States have viewed negatively. In fact, earlier in the 20th century, societal attitudes reflected a relatively positive view of older adults and of the aging experience. Older adults were respected for their wisdom and valued for their experience. They were not typically perceived as being a drain on society or as a burden to the community. Yet, sometime around the mid-1900s, as life expectancy began to grow and medical technology improved dramatically, professionals such as physicians, psychologists, and gerontologists began discussing older adults in terms of the *problems* they posed (Hirshbein, 2001).

Many social psychologists and gerontologists cite the media as a major source of negative stereotypes of older adults. These critics claim that the consistent negative portrayal of older adults in both television shows and commercials, for example, portraying them as dimwitted, foolish individuals living in the past, has a dehumanizing effect on the entire older adult population and has a negative effect on the self-concept of older adults. Yet, the results of a study conducted in 2004, which reviewed television commercials from the 1950s to the 1990s, did not support this critical view of the media (Miller, Leyell, & Mazacheck, 2004). In fact, Miller and his colleagues found that the media depiction of older adults has been relatively positive, particularly in the latter two decades.

It is vital that human service professionals make certain that they do not hold any of these misconceptions of old age. For instance, assuming that someone over the age of 70 is incapable of being productive and of learning something new, of gaining a new insight, whether in the counseling office or in life in general, would undoubtedly affect the dynamic between the counselor and the older adult client. In fact, research shows that negative stereotypes about aging are often internalized by older adults and can actually increase feelings of loneliness and dependency (Coudin & Alexopoulos, 2010). Practitioners then must address any misconceptions they have of old age and of the older adult population in general. Practices such as talking down to older adult clients and not directly addressing difficult issues for fear that they lack the capacity to understand will undoubtedly affect the level of investment the client makes in the counseling relationship. This type of behavior on the part of the practitioner can also encourage a self-fulfilling prophecy within older adult clients, where they begin to act the part of the incapable, unproductive, and cognitively dull individual. Making positive assumptions about older adult clients will increase the possibility of bringing out the most authentic and dynamic aspects of older adult clients.

Housing

Contrary to the common belief of many in the United States, most older adults remain in their homes until death and are cared for by family members (Bergeron & Gray, 2003). But as medical technology allows people to live longer albeit not necessarily healthier lives, coupled with the fact that more women than ever are in the workforce and therefore unavailable to care for their older and chronically ill relatives, many older adults find themselves needing to move out of their homes once they reach a certain level of physical and/or cognitive decline. They might move into the home of a family member, which was far more prevalent when the United States was an agricultural society, and both men and women were home based in their work, or they might move into a *retirement community,* where they can still enjoy their independence while enjoying many facility-offered services to meet their needs, such as shuttle service, handicapped-accessible facilities, and child-free living.

Government-subsidized older adult housing can make housing costs more affordable for the older adult population, whether in the form of a subsidy provided directly to older adults in the form of tax credits, loans, or rental vouchers or subsidies provided to the housing community, which then passes on this discount to the renter. One problem with many of these programs, though, is that they require older adults to find their own housing in the community, much of which is older and not appropriate for older adult residents who often need special age-related accommodations. Another concern relates to government-subsidized communities that are designed for older adult populations but tend to be wrought with problems related to safety, including problems with poor physical upkeep of the property.

A 2003 longitudinal study that followed 1,200 older adults in their transition from independent living to age-restricted housing in 1995 found that those older adults who transitioned to more expensive communities fared the best with regard to physical health and overall life satisfaction and those who transitioned to government-subsidized housing programs fared the worse. Although the study investigators acknowledged that levels of life satisfaction might be related to a cumulative effect of a lifetime of poverty, they concluded that overall quality of housing has a direct relationship to life satisfaction (Krout, 2003).

Older adults needing more consistent care with their activities of daily living (ADL) sometimes enter *assisted-living facilities*. These facilities offer apartment-like living in a more structured environment. In many respects assisted-living facilities act as a bridge between independent living and nursing home care. Assisted-living facilities offer assistance with eating, bathing, dressing, housekeeping, and medication, and some even have fully functioning medical centers. Many assisted-living apartments have alarm systems in every unit, offer a restaurant-style cafeteria, a club for social activities, a hairdresser, a medical staff, home healthcare, and a relatively full array of human services. The services are far more intensive than in a retirement community, as residents in assisted-living facilities are there because they cannot manage their ADL without daily assistance. Human service professionals provide many of the same services as provided in retirement communities, but at a more comprehensive level.

Human Services Delivery Systems

Understanding and Mastery of Human Services Delivery Systems: Range of populations served and needs addressed by human services

Critical Thinking Question: Many older adults need to change their living accommodations at some point, for a variety of reasons including affordability, physical challenges, proximity to other family members, or a need for specialized care. What additional factors should a human service professional take into account while working with older adult clients who are making such a transition?

Homelessness and the Older Adult Population

One of the opening vignettes of this chapter highlighted the issue of homelessness in the older adult community. Although older adults are at a lower risk for homelessness than other age groups, homelessness in the older adult population is a growing concern because the percentage is expected to grow as the baby boomer generation ages (Gonyea, Mills-Dick, & Bachman, 2010). Additionally, for years the problem of homelessness among the older population has been essentially ignored by policy makers and legislators, rendering this population relatively invisible (Gonyea, Mills-Dick, & Bachman, 2010).

The common causes of homelessness in the general population apply to older adult subgroups as well, such as a lack of affordable housing, too few jobs for unskilled workers, and a reduction in human services support (Hecht & Coyle, 2001; Kutza & Keigher, 1991), but the older adult population in general has additional risk factors such as being too old to sufficiently recover from a job loss, enter a new career, or reenter the workforce, as well as experiencing chronic illnesses that either are costly or bar older adults from being self-supporting (Kutza & Keigher, 1991).

An aged homeless woman with all of her belongings in two garbage bags.
Joseph Sohm/Visions of America/Corbis

For statistical purposes, individuals above the age of 50 to 55 are usually considered in the older adult category, but generally the lower threshold for what is considered "elderly" is increasing. Homeless older adults are a particularly vulnerable subgroup because of age-related physical vulnerability, which is often exacerbated by poor nutrition and difficult living conditions either on the streets or in a homeless shelter. They are also at a much higher risk of becoming a victim of crime while living on the streets (Hecht & Coyle, 2001).

A research study based in Los Angeles found that unlike the homeless in the general population, 85 percent of the older adult population was white (versus 61 percent in the younger homeless population) and 59 percent were veterans (versus 27 percent in the younger homeless population). Older homeless adults were far more likely to be socially isolated and suffer from a physical illness, but less likely to suffer from substance abuse, mental illness, or domestic violence (Linn & Mayer-Oakes, 1990). Older homeless adults between the ages of 50 and 65 are often the most vulnerable group because they are frequently the target of ageism when attempting to reenter the workforce, but too young to qualify for Medicare and Social Security benefits (Hecht & Coyle, 2001).

The differences between younger homeless and older homeless populations become important when considering programs designed to assist the older adult homeless population. Many human services' homeless assistance programs focus on root causes of homelessness more common in younger populations, such as providing assistance with substance abuse and domestic violence. Any human services programs designed to assist the older adult subgroups with housing issues need to focus more on issues related to insufficient income, health concerns, and low-income housing, offering supportive services to the older adult population with declining health.

Adjustment to Retirement

The concept of retirement is so common to the 21st century that it rarely needs explanation. When an individual comments on his or her upcoming retirement, others seem to instinctually understand that what is being discussed is the practice of leaving one's employment to permanently enter a phase of chosen nonemployment, and even though some might choose to dabble in part-time employment from time to time, the most common conceptualization of retirement involves an employee permanently surrendering his or her position, at approximately age 65, and drawing on a pension or retirement account that has likely been accruing for years. Of course, there are numerous variations on this theme—some people don't ever formally retire, and some people work in fields that have mandatory retirement ages, such as the airline industry, which requires that all pilots retire at the age of 60, and for some, retirement is a luxury they cannot afford. Also, it would be incorrect to assume that everyone in the workforce has accrued a pension sizeable enough to permit them to live on for years. But despite the range of retirement experiences, certain generalizations can be made about the retirement experience for the majority of those living in the United States during the 21st century.

Robert Atchley (1976) was one of the first researchers who attempted to describe the retirement experience for men and women. He identified five distinct, yet overlapping, stages that most retirees progress through on formal retirement. These stages are as follows:

1. The Honeymoon Phase: Retirees embrace retirement and all their newfound freedom in an optimistic but unrealistic manner.
2. Disenchantment: Retirees become disillusioned with what they thought retirement was going to be like and get discouraged with what often feels as though is too much time on their hands.
3. Reorientation: Retirees develop a more realistic view of retirement, with regard to both increased opportunities and increased constraints.
4. Stability: Retirees adjust to retirement.
5. Termination: Retirees eventually lose independence due to physical and cognitive decline.

There has been some controversy about whether retirees actually progress through such distinct phases or whether there is just too much of a range of experiences among retirees in the United States to categorize experiences in a stage theory. A study by Reitzes and Mutran (2004) appears to support Atchley's stage theory, finding that retirees experience a temporary lift right after retiring (for about 6 months), but then develop an increasingly negative attitude after about the 12-month mark, with some retirees starting to experience increased optimism after about two years. The study also found that an individual's level of self-esteem preretirement seemed to have an effect on their overall mental health after retirement, with those who had higher levels of self-esteem faring better. A more recent study on postretirement dynamics seems to support some of Atchley, and Reitzes and Mutran's findings, while refuting others. The study, which was funded by the National Institute on Aging, found that men and women who continued to work for a period of time after retirement, on a part-time or temporary basis (called "bridge employment") had much better physical and psychological quality during their elder years, indicating that sudden and complete retirement, without any transition, may have negative side effects for an older adult's physical and mental health. Interestingly, the positive effects gained from bridge employment existed regardless of the retiree's preretirement mental and physical health (Zhan, Wang, Liu, & Shultz, 2009).

Because nearly 50 percent of the U.S. population is now over the age of 50, the implications of retirement preparation and adjustment to retirement for the human services field obviously cannot be ignored. Human service professionals will likely come into contact with retired or retiring adults in many different settings, thus it is important to realize that impending retirement can become an issue for someone even in middle adulthood.

> **Because nearly 50 percent of the U.S. population is now over the age of 50, the implications of retirement preparation and adjustment on the human services field obviously cannot be ignored.**

Finally, race and gender have a significant effect on retirement experiences. Research has shown that women and minority workers often have different attitudes and experiences surrounding retirement issues due to disparity in income and education levels (McNamara & Williamson, 2004). Thus, the human service professional must understand that most factors affecting a client's retirement experience are going to be influenced by the client's gender and racial background.

Grandparents Parenting

The practice of grandparents raising grandchildren has increased dramatically over the past several years, signaling many problems within U.S. society that have emerged since the 1970s. The U.S. Congress became interested in this issue in the mid-1990s, and in 1996, it passed legislation that required the 2000 U.S. Census to include questions regarding whether grandparents were residing with grandchildren, whether they had primary responsibility for them, and what length of time they had acted in a parental role (i.e., revealing whether the situation was temporary or permanent).

Current (as of 2009, the most recent statistics available) figures estimate that approximately 6.6 million U.S. households (about 5 percent of the population) are comprised of grandparents coresiding with grandchildren under the age of 18; 64 percent of these are female grandparent-headed households. Approximately 2.7 million of these families involved grandparents who were primarily responsible for their grandchildren (U.S. Census Bureau, 2012). This represents a significant increase over past years, and means that 28 percent of grandparents in the United States are responsible for raising their grandchildren. About two-thirds of these grandparents are between the ages of 50 and 59, and about a third are over 60. Some of these households included at least one of the parents, but many of them included one or both grandparents acting in the role of surrogate parent(s).

Although the demographics of grandparent-headed households vary considerably, such households are far likelier to be an ethnic minority, suffer from poverty, and have low education levels. Households led by only a grandmother are far more likely to face economic hardship. Grandparent caregivers in the Southeast and in urban areas have the highest levels of poverty and the lowest levels of education (Simmons & Dye, 2003; Whitely & Kelley, 2007). Even though older African American grandmothers are disproportionately represented in custodial grandparent arrangements, a recent research study indicates that older grandparents may experience less emotional strain related to their primary parenting role than do younger grandparents, likely related to their increased ability to manage stressful life situations (Conway, Jones, & Speakes-Lewis, 2011).

TABLE 7.1 Grandparents Living with Grandchildren, by Race

Characteristic	Total	Race							Hispanic origin		
		White Alone	Black or African American Alone	American Indian and Alaska Native Alone	Asian Alone	Native Hawaiian and Other Pacific Islander Alone	Some Other Race Alone	Two or More Races	Hispanic or Latino (of any race)	Non-Hispanic or Latino	
										Total	White Alone, Non-Hispanic or Latino
Population 30 years old and over	158,881,037	125,715,472	16,484,644	1,127,455	5,631,301	169,331	5,890,748	2,862,086	14,618,891	144,262,146	119,063,492
Grandparents living with grandchildren	5,771,671	3,219,409	1,358,699	90,524	359,709	17,014	567,486	158,830	1,221,661	4,550,010	2,654,788
Percent of population 30 years old and over	3.6	2.5	8.2	8.0	6.4	10.0	9.6	5.5	8.4	8.2	2.2
Responsible for grandchildren	2,426,730	1,340,809	702,595	50,765	71,791	6,587	191,107	63,076	424,304	2,002,426	1,142,006
Percent of coresident grandparents	42.0	41.6	51.7	56.1	20.0	38.7	33.7	39.7	34.7	44.0	43.0
By duration of care (percent)a Total	100.0	100.0	100.0	100.0	100.0	100.0	100.0	100.0	100.0	100.0	100.0
Less than 6 months	12.1	12.6	9.8	13.0	13.6	12.7	15.6	13.5	14.6	11.5	12.4
6 to 11 months	10.8	11.6	9.3	10.5	11.0	8.4	11.4	11.2	11.2	10.7	11.6
1 to 2 years	23.2	23.8	21.2	22.5	25.2	23.8	26.1	23.4	25.1	22.8	23.6
2 to 4 years	15.4	15.8	14.6	13.9	17.6	11.7	15.7	16.0	15.8	15.3	15.7
5 years or more	38.5	36.3	45.2	40.0	32.7	43.3	31.1	35.9	33.3	39.6	36.6

Data based on sample. For information on confidentiality protection, sampling error, nonsampling error, and definitions, see www.census.gov/prod/cen2000/doc/st3.pdf.

apercent duration based on grandparents responsible for grandchildren. Percent distribution may not sum to 100 percent because of rounding.

Source: U.S. Census Bureau, Census 2000, Summary File 4.

Source: U.S. Census Bureau, American Community Survey 2009, Subject Table S1002, "Grandparents," <http://factfinder.census.gov/>, accessed February 2011

Ethnic minority children are far more likely to be raised by a grandparent than Caucasian children, with African American children in the Southeast states having a significantly higher rate of living with custodial grandparents than children in other regions in the United States. African American grandparents are far likelier to experience poverty, despite the fact that the majority are in the labor force. They are also far likelier to not have health insurance, and experience greater physical and emotional stressors (Whitely & Kelley, 2007).

There are several reasons why grandparents become surrogate parents, but the chief reasons include the following:

1. The high divorce rate, leaving many women facing potential poverty, resulting in them returning home to live with parents
2. The sharp rise in teen pregnancies, resulting in the mother residing with her parents for economic (and oftentimes emotional) reasons
3. The increase in relative foster care in response to a sharp increase in child welfare intervention due to child abuse
4. The increase in parents serving time in prison, primarily for drug and drug-related offenses punishable by high prison sentences due to the U.S. government's War on Drugs
5. The sharp increase of drug use, particularly among women of color whose use of crack cocaine has literally exploded over the past 10 years
6. The AIDS crisis, which has devastated many communities, leaving children orphaned and in need of permanent homes. These cases are complicated when the children have contracted HIV, particularly when one considers their complex medical needs (de Toledo & Brown, 1995).

The issues facing grandparents raising grandchildren are complex involving emotional as well as financial, legal, and physical challenges. Many grandparent caregivers are often forced to live in a type of limbo not knowing how long they will remain responsible for their grandchildren, particularly when the biological parents are either in jail or suffering from drug addiction that prevents them from resuming their primary parenting role.

The choice to act as a surrogate parent is in many instances made in a time of crisis; thus, older adults who may have been planning their retirement for years often find themselves in a position where they either take on this parenting role in the face of the situation that rendered the biological parents unable to continue parenting or allow their grandchildren to enter the county foster care system. Parenting younger children has its unique challenges, but often comes with some level of social support, at least within the elementary school system, but this is often not the case with older children, particularly adolescents.

Parenting adolescents can often present significant challenges for grandparents, particularly those who are very old. Parenting adolescents can be an exhausting endeavor for the young or middle-aged parent, but imagine the demands placed on someone who is an older adult, has limited physical capacity, and even more limited financial means. Adolescents who have endured significant loss through death or abandonment,

> ## Human Systems
>
> Understanding and Mastery of Human Systems: Changing family structures and roles
>
> ---
>
> Critical Thinking Question: The phenomenon of grandparents parenting occurs disproportionately among families of lower socioeconomic status and places a variety of additional strains on the aging grandparent. How might a human services professional work at the micro (individual/family), mezzo (community), and macro (larger policy) levels to build support for these grandparents and the children they are raising?
>
> . . . .

have been raised in abusive homes, or have been raised by parents who abuse drugs or are serving time in prison are likely to act out emotionally and even physically, putting even greater stress on an already vulnerable family system.

Human service professionals may enter a grandparent-led family system in numerous ways—they could be the school social worker working with the children, they might be the child welfare caseworker assigned to assist the grandparents who are serving as relative foster care parents, or they might work for a human services agency offering outreach services to grandparent caregivers.

Depression

Another significant concern affecting the older adult population is the increased incidence of depression. In fact, the National Institute of Mental Health (NIMH) estimates that approximately 2 million individuals over the age of 65 suffer from some form of depression, and as many as 5 million more suffer from some form of depressive symptoms, although they may not meet all the criteria for clinical depression. Although prevalence rates can vary rather widely within the population, due in part to how depression is defined, these statistics indicate that at any given time anywhere from 5 to 30 percent of the older adult population may suffer from some form of depression, compared to a 1 percent prevalence rate in the general population (Birrer & Vemuri, 2004).

Depression rates in nursing homes are even higher, with some studies finding up to 50 percent of the residents meeting the criteria for clinical depression. Older adults are also disproportionately at risk for suicide. Although individuals aged 65 years and older make up about 12 percent of the U.S. population, they account for nearly 16 percent of all those who committed suicide in the year 2004, which is the highest rates of all age groups. Surprisingly, older adults at the highest risk for suicide are white males over the age of 85, many of whom are widowed (Birrer & Vemuri, 2004; Kraaij & de Wilde, 2001; McIntosh, 2004; NIMH, 2007).

Many believe that depression is just a normal part of the aging process caused by the natural course of cognitive and physical decline and the multiple losses associated with growing old. But depression is not a natural part of growing older and can be avoided. Unfortunately, many in the medical and mental health fields, even older adults themselves, believe that it is, and thus many in the older adult population who are suffering from depression remain undiagnosed and untreated. Misdiagnosis is also relatively common, with depression often being mistaken for dementia or some other form of cognitive impairment (Birrer & Vemuri, 2004).

Human service professionals working with the older adult community must be observant of the signs of depression. They must also be aware of the many risk factors for depression, including anxiety; chronic medical conditions such as heart disease, stroke, and diabetes; dementia; being unmarried; alcohol abuse; stressful life events; and minimal social support (Birrer & Vemuri, 2004; Lynch, Compton, Mendelson, Robins, & Krishnan, 2000; Waite, Bebbington, Skelton-Robinson, & Orrell, 2004).

Dementia

The American Psychiatric Association defines dementia as progressive, degenerative illnesses experienced during old age that impair brain function and cognitive ability. Dementia is an umbrella term encompassing most likely numerous disorders. Two of the most common forms of dementia are Alzheimer's disease and multi-infarct dementia (small strokes in the brain).

The general symptoms of dementia include a comprehensive shutting down of all bodily systems indicative by progressive memory loss, increased difficulty concentrating, a steady decrease in problem-solving skills and judgment capability, confusion, hallucinations and delusions, altered sensations or perceptions, impaired recognition of everyday objects and familiar people, altered sleep patterns, motor system impairment, inability to maintain ADL (such as dressing oneself), agitation, anxiety, and depression. Ultimately, the dementia sufferer enters a complete vegetative state prior to death.

According to the NIMH, multi-infarct dementia accounts for nearly 20 percent of all dementias, affecting about 4 in 10,000 people. Even more individuals suffer from some form of mild cognitive impairment, but do not yet meet the criteria for full-blown dementia (Palmer, Fratiglioni, & Winblad, 2003). Alzheimer's disease affects approximately 4.5 million Americans, or about 5 percent of the population between the ages of 65 and 74 years, and the incident rate increases to 50 percent for those over 85 years of age. Diagnosis is based on symptoms, and it is only through an autopsy that a definitive diagnosis is made. The United States has experienced a dramatic increase in the incidence of dementia in the latter part of the 20th century, primarily due to the increased human life span. It is theorized that dementia did not have an opportunity to develop prior to the 1900s, when the average life span was about 47 years. There is no known cure for dementia, thus treatment is focused on delaying and relieving symptoms.

Human service professionals may work directly with the sufferer of dementia or with the caregiver (typically a spouse or adult child) if they work in a practice setting that serves the older adult community. However, dealing with dementia as a clinical issue can occur in any practice setting because any client may have a relative suffering from one of these disorders and will therefore need counsel and perhaps even case management. Consider the practitioner who assists clients in managing an ailing parent, questioning whether their parent is suffering from cognitive impairment, grieving the slow loss of the parent they love, and needing support in making difficult decisions such as determining when their parent can no longer live alone. Or, consider the school social worker who is counseling a student whose grandfather was recently diagnosed with Alzheimer's disease. The pressure on the entire family system will affect the student in numerous ways—academically, emotionally, perhaps even physically—and will frequently magnify any existing issues with which the student is currently struggling.

Elder Abuse

Older adults are a vulnerable population due to factors such as their physical frailty, dependence, social isolation, and the existence of cognitive impairment, and as such are at risk of various forms of abuse and exploitation. The National Center on Elder Abuse (NCEA) defines elder abuse as any "knowing, intentional, or negligent act by a caregiver or any other person that causes harm or a serious risk of harm to a vulnerable adult." The specific definition of elder abuse varies from state to state, but in general can include physical, emotional, or sexual abuse; neglect and abandonment; or financial exploitation.

Although elder abuse is presumed to have always occurred, just as other forms of abuse such as child abuse and spousal abuse, it was not legally defined until addressed within a 1987 amendment of the Older Americans Act. Reports of elder abuse have increased significantly over the last several years not only due to an increase in reporting requirements, but also due to societal changes that are putting more older adults at risk. In 1986 there were 117,000 reports of elder abuse nationwide, and by 1996 the number of abuse reports increased to 293,000 (Tatara, 1997). By the year 2000 (the most recent reported data) the number of elder abuse reports had risen to an alarming 472,813 among all 50 states, Guam, and Washington, DC. One reason for the rise in abuse reports is that the newest figures include not only abuse in domestic settings, but abuse in institutional settings as well, but despite the more comprehensive data collection methods, there is no escaping the fact that elder abuse is increasing within the United States (Teaster, 2000). Elder abuse is projected to continue to rise in the coming years due to the increased life span and the resultant increase in chronic illnesses, changing family patterns, and the complexity involved with contemporary caregiving.

Sixty percent of all reported abuse victims are women, 65 percent of all abuse victims are white, more than 60 percent of abuse incidences occurred in domestic settings, and about 8 percent of abuse incidences occurred in institutionalized settings. Family members were the most commonly cited perpetrators, including both spouses and adult children (Teaster, 2000). After years of failed attempts, in March 2010, the Elder Justice Act of 2009 (EJA) was passed and signed into law by President Obama as part of the Patient Protection and Affordable Care Act (PPACA). The act sets forth numerous provisions for addressing the abuse, neglect, and exploitation of older adults, both in the form of preventative and responsive measures. For instance, the legislation provides grants for a number of training programs focusing on prevention of abuse and exploitation of older adults; provides measures for expanding long-term care services, including a long-term care ombudsman program; establishes mandatory reporting requirements for abuse against older adults occurring in long-term care facilities; and includes provisions for creating national advisory councils (National Health Policy Forum, 2010).

Despite the fact that there remain relatively limited mechanisms on a national level regulating how elder abuse is to be handled (primarily due to a lack of funding), every state in the United States has an adult protective services (APS) agency. There is significant variation between states, particularly related to reporting laws and investigation methods and policies. Some states have separate agencies handling elder abuse, and some combine the protection of older adults with the protection of disabled adults of all ages. One significant difference between state policies involves who is considered a mandated reporter. Sixteen states require anyone who is aware of elder abuse to report it. About half the states require medical personnel, the clergy, and mental health personnel, including all human service professionals, to report elder abuse. Some states specify that only medical personnel are mandated reporters. Yet five states—Colorado, Delaware, New York, South Dakota, and Wisconsin—do not mandate that anyone report elder abuse (Teaster, 2000).

Elder abuse tends to be grossly underreported for several reasons, but many cite the lack of uniform reporting requirements as a primary reason. Because of the wide range of elder abuse reporting requirements, as well as differences in adult protective services' investigation policies and enforcement powers, it is essential that those working in the human services field be aware of the elder abuse reporting laws and requirements in their state. Many human service professionals may be in a position to protect an older adult client but may not be aware that their state has an elder abuse hotline.

Caregiver burnout is one of the primary risk factors of elder abuse. The most common scenario involves a loving family member who becomes intensely frustrated by the seemingly impossible task of caring for a spouse or parent with a chronic illness such as dementia. Providing the continuous care of someone with Alzheimer's disease, for example, can be frustrating, provoking an abusive response from someone with no history of abusive behavior.

One of the most effective intervention strategies is *caregiver support groups*. These groups are typically facilitated by a social worker or other human services practitioner and focus on providing caregivers, many of whom are older adults themselves, a safe place to express their frustrations, sadness, and other feelings related to caring for their dependent older adult loved one.

Practice Settings Serving Older Adults

Unfortunately working with older adults remains a rather unpopular career pursuit for human service professionals across the globe (Weiss, 2005). Yet, a commitment on the part of several educational accreditation agencies is to increase infusion of gerontology issues in educational curriculum, as well as financial incentives such as scholarships and stipends for students in gerontological field placements, is believed to be making a difference in the number of students who select gerontology as their career focus.

For human service professionals wishing to provide direct service to the older adult population, a wide array of choices in practice settings awaits them. Virtually all practice settings delivering services to older adults have certain treatment and intervention goals, including the promotion of the health and well-being of older adults, special attention to the needs of special populations such as women and ethnic minority groups, providing effective services at an affordable price, identifying the common needs of all elders, and removing existing social barriers so that elders can be empowered to seek assistance in meeting those needs.

AAAs, discussed earlier in this chapter, often serve as human service agencies offering direct service to the older adult community on a local level. Generally these agencies offer a multitude of services for older adults, such as nutrition programs, services for homebound older adults, low-income minority older adults, and other programs focusing on the needs of older adults within the local community. Many AAAs also act as a referral source for other services in the area. For instance, the Mid-Florida AAAs offer programs for those suffering from Alzheimer's (including caregiver respite), a toll-free elder hotline that links older adults in the area with resources, an emergency home energy assistance program, paralegal services, home care for older adults, Medicaid waivers, and practitioners who work with older adults in helping them make informed decisions. Most AAAs offer both in-house services, many of which are facilitated by human service professionals, and off-site programs. Human service professionals working at an AAAs-funded center might facilitate caregiver respite programs, or they might provide case management services for an agency that provides employment services for clients over 60 years old. Even at centers where services are primarily medical in nature, human service professionals often provide adjunct counseling and case management as a support service.

Other practice settings include adult day cares, geriatric assessment units, nursing home facilities, veterans' services, elder abuse programs, adult protective services, bereavement services, "senior" centers, and hospices. A human service professional will likely perform similar types of direct service, consultation, and educational services focused on assisting older clients maintain or improve their quality of life, independence, and level of self-determination. Tasks are typically performed using a multidisciplinary team approach and can include conducting psychosocial assessments, providing case management, developing treatment plans, providing referrals for appropriate services, and providing counseling to older adult clients and their families. Services are also provided to family caregivers offering support and respite care.

> ## Human Services Delivery Systems
>
> Understanding and Mastery of Human Services Delivery Systems: Range and characteristics of human services delivery systems and organizations
>
> Critical Thinking Question: What are some of the major ways in which the needs of the older adult population are different from those of other groups of human service clients? In what ways are older adults' needs similar to those of other groups?

Special Populations

As the frail older adult population has increased in numbers, the government has shifted its priorities and began developing programs aimed at long-term healthcare needs, with a particular focus on vulnerable populations such as women, ethnic minorities, and older adults living in rural communities. It is difficult to define who is "special" or particularly vulnerable within the older adult population because in many senses all older adults could conceivably be considered special in that they are vulnerable to social, economic, physical, and psychological harm or exploitation simply by virtue of their advancing age and corresponding dependency needs. But many gerontologists classify various subpopulations as more vulnerable for various reasons. For instance, successful aging has been linked to good economic status, good healthcare, relatively low stress levels, and high levels of social connections. A 2004 study also showed a link between good health and financial stability, finding that Caucasians tend to have greater economic wealth and better health than African American and Latino populations (Lum, 2004).

> Successful aging has been linked to good economic status, good healthcare, relatively low stress levels, and high levels of social connections.

Women are often considered a special population because as a group they are more prone to depression and typically have a worse response to antidepressant medication (Kessler, 2003). Women often experience greater financial vulnerability, particularly if divorced or widowed, and are often in lower-wage jobs, undereducated, and underinsured. Widowhood is a common occurrence for women because they live an average of seven years longer than men, and although the majority of women in the United States marry, 75 percent of women are unmarried by the age of 65. Widowhood puts women at increased risk for lower morale and other mental health problems, even though these symptoms abate with time and intervention (Bennett, 1997).

Research has also shown a link between stress and racism that affects quality of life. A study conducted in 2002 found that racism, and particularly institutionalized racism (such as government-sanctioned racism through discrimination in housing, employment, and healthcare), had a detrimental effect on older African Americans, particularly men, who tend to experience worse racial discrimination than women (Utsey, Payne, Jackson, & Jones, 2002). Other research has shown how institutionalized racism can lead to feelings of invisibility, stress, depression, and ultimately despair as the person experiences a sense of futility in combating a lifetime of discrimination and White privilege (Franklin, Boyd-Franklin, & Kelly, 2006).

Other special populations could conceivably include any subgroup that is vulnerable at any point across the life span because of physical or mental disability, veterans' status, and those individuals living in isolated rural areas. Identifying special populations within the older adult population will allow the human service professional to explore issues that can potentially render older adult clients at increased risk and vulnerability during old age. For example, research has shown that veterans are at special risk for depression, post-traumatic stress disorder (PTSD), and alcohol abuse. Thus, older adult veterans will be at particular risk for these conditions. An older adult client who is developmentally disabled will also face increased vulnerability compared to those in the older adult population who have intelligence in the normal range. A human service professional who is well versed on typical risk factors for older adults in the United States, as well as for the increased risk factors facing special populations, will be far more effective in protecting and advocating for their older adult clients.

Concluding Thoughts on Services for Older Adults

The older adult population is increasing at a dramatic rate in the United States, rendering this one of the fastest growing target populations of human service agencies. As the baby boomers continue to age and as life continues to become more complex, many within the older adult population will rely on human service professionals to meet many of their basic needs. Many human service educational programs are adding the field of older adult care, or social gerontology, as an area of specialization in response to the growing need for practitioners committed to work with this population in a variety of capacities.

Future considerations include the continued effort to identify vulnerable populations, as well as addressing ongoing concerns such as the shortage of available affordable housing, the availability of long-term care and healthcare services directed to the older adult population, and the increased role of parenting responsibilities placed on the older adult population. Human service professionals can make a significant positive impact on the lives of older adults and their family members by addressing both ongoing and anticipated needs of this population.

The following questions will test your knowledge of the content found within this chapter.

1. The term that relates to the increase in the elderly population in the United States is
   a. the Aging Increase
   b. the Baby Boomers Blast
   c. the Graying of America
   d. the Elderly Explosion

2. The *baby boomers* is a cohort of people born
   a. before World War II
   b. after World War II
   c. between 1946 and 1964
   d. Both b and c

3. In 1900 the average human life span in the United States was about _____ years but by 1999 it had increased to about _____ years.
   a. 47/77
   b. 47/86
   c. 62/77
   d. 62/86

4. The eighth stage of Erikson's model spanning from age 65 to death is called
   a. integrity versus despair
   b. intimacy versus isolation

   c. generativity versus stagnation
   d. industry versus inferiority

5. Researchers have examined individuals who age better than others to determine what differences might account for their "success," and some of the variables at play include
   a. having hobbies and attending social events
   b. maintaining a moderately high physical and social activity level
   c. exercising regularly
   d. All of the above

6. Robert Atchley's study of the stages of retirement found that directly after the honeymoon phase many retirees experienced
   a. reorientation, where retirees developed a more realistic view of retirement, both with regard to increased opportunities, but also with regard to increased constraints
   b. disenchantment, where retirees became disillusioned with what they thought retirement was going to be like
   c. stability, where retires adjusted to retirement
   d. All of the above

7. Describe the recent trend in grandparents parenting, including reasons for the increase in custodial grandparents, demographics, and effect on grandparent(s) and children.

8. Describe the dynamic of elder abuse, its characteristics, associated dynamics, common demographics, and causes and consequences.

Suggested Readings

Bergling, T. (2004). *Reeling in the years: Gay men's perspectives on age and ageism.* Binghamton, NY: Southern Tier Editions.

Davis, R. (1989). *My journey into Alzheimer's disease.* Wheaton, IL: Tyndale House.

de Toledo, S., & Brown, D. E. (1995). *Grandparents as parents: A survival guide for raising a second family.* New York: The Guildford Press.

Kaye, L. W. (2005). *Perspectives on productive aging: Social work with the new aged.* Washington, DC: NASW Press.

McGowin, D. F. (1994). *Living in the labyrinth: A personal journey through the maze of Alzheimer's.* New York: Delta Books.

Osborne, H. (2002). *Ticklebelly hill: Grandparents raising grandchildren.* Bloomington, IN: Authorhouse.

Rosenthal, E. R. (1990). *Women, aging and ageism.* Binghamton, NY: Huntington Park Press.

Internet Resources

Alzheimer's Disease Education & Referral Center: http://www.nia.nih.gov/alzheimers

AARP: http://aarp.org

Arthritis Foundation: http://www.arthritis.org

Elder Hostel: http://www.elderhostel.org

The Grandparent Foundation: http://www.grandparenting.org

National Indian Council on Aging: http://www.nicoa.org

References

Bergeron, R.L., & Gray, B. (2003). Ethical dilemmas of reporting suspected elder abuse. *Social Work*, 48(1), 96–106.

Conway, F., Jones, S., & Speakes-Lewis, A. (2011). Emotional strain in caregiving among African American grandmothers raising their grandchildren. *Journal of Women & Aging*, 23(2), 113–128. doi:10.1080/08952841.2011.561142

Coudin, G., & Alexopoulos, T. (2010). "Help me! I'm old!" How negative aging stereotypes create dependency among older adults. *Aging & Mental Health*, 14(5), 516–523. doi:10.1080/13607861003713182

Arias, B. (2004). United States life tables, 2002. *National Vital Statistics Reports*, 53(6). Hyattsville, MD: National Center for Health Statistics.

Atchley, R. C. (1976). *The sociology of retirement*. New York: John Wiley.

Bennett, K. M. (1997). Widowhood in elderly women: The medium- and long-term effects on mental and physical health. *Mortality*, 2, 137–148.

Birrer, R. B., & Vemuri, S. P. (2004). Depression in later life: A diagnostic and therapeutic challenge. *American Family Physician*, 69(10), 2375–2382.

Bowling, A., & Iliffe, S. (2011). Psychological approach to successful ageing predicts future quality of life in older adults. *Health & Quality of Life Outcomes*, 9(1), 13–22. doi:10.1186/1477-7525-9-13

Butler, R. N. (1969). Ageism: Another form of bigotry. *Gerontologist*, 9, 243–246.

de Toledo, S., & Brown, D. E. (1995). *Grandparents as parents: A survival guide for raising a second family*. New York: Guilford Press.

Degges-White, S. (2005). Understanding gerotranscendence in older adults: A new perspective for counselors. *Adultspan: Theory Research & Practice*, 4(1), 36–48.

Department of Health and Human Services: Administration on Aging (2010). *Projected future grown of the older adult population*. Retrieved online January 20, 2012 http://www.aoa.gov/AoARoot/Aging_Statistics/future_growth/future_growth.aspx#age.

Erikson, E. H. (1959). Identity and the life cycle. *Psychological Issues*, 1, 1–171.

Erikson, E. H. (1966). Eight ages of man. *International Journal of Psychiatry*, 2, 281–300.

Franklin, A., Boyd-Franklin, N., & Kelly, S. (2006, June). Racism and invisibility: Race-related stress, emotional abuse and psychological trauma for people of color. *Journal of Emotional Abuse*, 6(2/3), 9–30. Retrieved September 14, 2009, from doi:10.1300/J135v06n02-02

Gonyea, J. G., Mills-Dick, K., & Bachman, S. S. (2010). The complexities of elder homelessness, a shifting political landscape and emerging community responses. *Journal of Gerontological Social Work*, 53(7), 575–590. doi:10.1080/01634372.2010.510169

Havighurst, R. J. (1961). Successful aging. *Gerontologist*, 1(1), 8–13.

Hecht, L., & Coyle, B. (2001). Elderly homeless: A comparison of older and younger adult emergency shelter seekers in Bakersfield, California. *American Behavioral Scientist*, 45(1), 66–79.

Hirshbein, L. D. (2001). Popular views of old age in America, 1900–1950. *Journal of American Geriatrics Society*, 49, 1555–1560.

Hsu, H. (2007, November). Does social participation by the elderly reduce mortality and cognitive impairment? *Aging & Mental Health*, 11(6), 699–707. Retrieved September 14, 2009, from doi:10.1080/13607860701366335

Kessler, R. C. (2003). Epidemiology of women and depression. *Journal of Affective Disorders*, 74(1), 5–13.

Kraaij, V., & de Wilde, J. (2001). Negative life events and depressive symptoms in the elderly life: A life span perspective. *Aging & Mental Health*, 5(1), 84–91.

Krout, J. A. (2003). *Residential choices and experiences of older adults: Pathways for life quality*. New York: Spring Publishing Company.

Kutza, E. A., & Keigher, S. M. (1991). The elderly "new homeless": An emergency population at risk. *Social Work*, 36(4), 283–293.

Levinson, D. (1978). *The seasons of a man's life*. New York: Knopf.

Levinson, D. (1996). *The seasons of a woman's life*. New York: Knopf.

Linn, G. L., & Mayer-Oakes, S. A. (1990). Differences in health status between older and younger homeless adults. *Journal of American Geriatric Society*, 38(11), 1220–1229.

Lum, Y. (2004). Health-wealth association among older Americans: Racial and ethnic differences. *Social Work Research*, 28(2), 106–116.

Lynch, T. R., Compton, J. S., Mendelson, T., Robins, C. J., & Krishnan, K. R. R. (2000). Anxious depression among the elderly: Clinical and phenomenological correlates. *Aging and Mental Health*, 4(3), 268–274.

McIntosh, J. L. (2004). *U.S.A. suicide: 2004 official final data*. Retrieved December 21, 2009, from http://www.ct.gov/dmhas/lib/dmhas/prevention/cyspi/AAS2004data.pdf

McNamara, T. K., & Williamson, J. B. (2004). Race, gender, and the retirement decisions of people ages 60 to 80: Prospects for age integration in employment. *International Journal of Aging and Human Development*, 59(3), 255–286.

Miller, D., Leyell, T., & Mazacheck, J. (2004). Stereotypes of the elderly in U.S. television commercials from the 1950s to the 1990s. *International Journal of Aging and Human Development*, 58(4), 315–340.

National Health Forum Policy. (2010, November). *The Elder Abuse Act: Addressing elder abuse, neglect and exploitation*. Retrieved online January 20, 2012 http://www.nhpf.org/library/the-basics/Basics_ElderJustice_11-30-10.pdf

National Institute of Mental Health. (2007). *Older adults: Depression and suicide facts* (NIH Publication No. QF 11-7697). Bethesda, MD: National Institutes of Health.

National Highway Safety and Traffic Administration [NHSTA]. (2009). *Traffic Safety Facts 2009 Data: Older Population*. Retrieved online January 20, 2012 http://www-nrd.nhtsa.dot.gov/Pubs/811391.pdf.

Palmer K, Winblad B, Fratiglioni L. (2003). Detection of Alzheimer's disease and dementia in the preclinical phase: population based cohort study. *BMJ: British Medical Journal* 326(7383).

Passel, J.S. & Cohn, D. U.S. Population Projections: 2005–2050. *Pew Research Center: Social & Demographic Trends*. Retrieved June, 11, 2010 from: www.pewhispanic.org/files/reports/85.pdf

Reitzes, D. C., & Mutran, E. J. (2004). The transition to retirement: Stages and factors that influence retirement adjustment. *International Journal of Aging and Human Development, 59*(1), 63–84.

Simmons, T., & Dye, J. L. (2003, October). *Grandparents living with grandchildren: 2000*. Washington DC: U.S. Bureau of the Census.

Snyder, M. (2001). *Self and society*. Malden, MA: Blackwell Publishers.

Tatara, T. (1997). *Summaries of the statistical data on elder abuse in domestic settings*. Washington, DC: National Center on Elder Abuse.

Teaster, P. B. (2000). *A response to the abuse of vulnerable adults: A 2000 survey of state adult protective services*. Washington, DC: National Center on Elder Abuse.

Thornton, J. E. (2002). Myths of aging or ageist stereotypes. *Educational Gerontology, 28*, 301–312.

Tornstam, L., 1994, Gerotranscendence—A Theoretical and Empirical Exploration, in Thomas, L.E., Eisenhandler, S.A., eds., Aging and the Religious Dimension, *Westport: Greenwood Publishing Group*

Tornstam, L. (2003). *Gerotranscendence from young old age to old old age*. Retrieved January 20, 2012, from www.soc.uu.se/Download.aspx?id=SpeY85XbP%2Bg%3DShare

Tornstam, L. (2005). *Gerotranscendence: A Developmental Theory of Positive Aging*. New York: Springer Publishing.

Utsey, S. O., Payne, Y. A., Jackson, E. S., & Jones, A. M. (2002). Race-related stress, quality of life indicators, and life satisfaction among elderly African Americans. *Cultural Diversity and Ethnic Minority Psychology, 48*(3), 224–233.

Waite, A., Bebbington, P., Skelton-Robinson, M., & Orrell, M. (2004). Life events, depression and social support in dementia. *British Journal of Clinical Psychology, 43*, 313–324.

Warr, P., Butcher, V., & Robertson, I. (2004). Activity and psychological well-being in older people. *Aging & Mental Health, 8*(2), 172–183.

Weiss, I. (2005). Interest in working with the elderly: A cross-national study of graduating social work students. *Journal of Social Work Education, 41*(3), 379. Retrieved September 14, 2009, from MasterFILE Premier database.

Whitley, D. M. & Kelley, S. J. (2007, January). Grandparents raising grandchildren: A call to action. (Prepared for the Administration for Children and Families, Region IV.) Retrieved January 20, 2012, from http://www.acf.hhs.gov/opa/doc/grandparents.pdf

Zhan, Y., Wang, M., Liu, S., & Shultz, K. S. (2009). Bridge employment and retirees' health: A longitudinal investigation. *Journal of Occupational Health Psychology, 14*, 374–389.

CHAPTER 8

Mental Health and Mental Illness

cloki/2010/Used under license from Shutterstock.com

Every society has its mentally ill—those members whose behavior is considered outside what is normal and appropriate. Each society has also developed ways in which to handle or manage such individuals so that healthy societal function is not disrupted. But because the criteria for what is considered *normal behavior* changes from era to era, as well as from culture to culture, it is important to keep cultural mores and generational issues in mind when characterizing someone's behavior as abnormal or unhealthy.

It would be difficult to imagine human service professionals who do not at some point in their career come into contact with clients suffering from some form of mental illness. *Mental illness* is a term that, in its broadest sense, refers to a wide range of mental and emotional disorders, such as depression and anxiety disorders and, in its most narrow sense, refers to those individuals who suffer from severe and chronic mental illness, requiring at least intermittent custodial care. Because of the broadness of this term, it can be challenging to reach a consensus on just how many people suffer from mental illness at any one time within the United States.

A recently published report found that close to 27 percent of the U.S. adult population suffer from some diagnosable mental disorder, about 40 percent suffered from a mental illness of a moderate severity, and about 25 percent suffered from mental illness that was considered severe (SAMHSA, 2010). The term *severely mentally ill* typically refers to those individuals who suffer from schizophrenia, bipolar disorder, severe and recurrent depression, and other mental disorders that prevent normal functioning such as maintaining employment or performing activities of daily living. Individuals suffering from severe mental illness are often unable to consistently provide self-care, think clearly, reason, relate to others, and cope with the demands of daily life. Research has also shown

Learning Objectives

- Understand the reasons for deinstitutionalization and its impact on the mentally ill population
- Become familiar with the community mental health model currently in place in the United States, and identify its key strengths and deficits
- Develop a basic understanding of the basic criteria of serious mental illness that a human service professional may encounter, becoming familiar with practice settings where human service professionals will most likely encounter individuals suffering from mental illness
- Become familiar with special populations suffering from mental illness such as the homeless, prisoners, and ethnic minority populations
- Understand the current state of mental health legislation, and become familiar with how such legislation impact the funding and treatment of current mental health programs

that there is a correlation between poverty and mental illness (as cited in SAMHSA, 2010), which is important to remember when considering the complexity of mental illness, and how the mentally ill are treated within the United States and other places around the globe.

The History of Mental Illness: Perceptions and Treatment

To understand the current climate with regard to perceptions of mental illness as well as treatment paradigms commonly used in the United States, it is important to have some understanding of the historic treatment of the severely mentally ill. It has been said that the measure of a truly civil, ethical, and compassionate society is reflected in how it treats its most vulnerable members. The mentally ill, particularly the severely and chronically mentally ill, certainly fall into this category, and if this statement is in fact true, then the U.S. society has undoubtedly gone through some periods that were uncivilized, unethical, and compassionless.

Early in human history, mental illness, or madness as it was often called, was commonly believed to be caused by demonic possession. Skulls dating back to at least 5000 BCE were found with small holes drilled throughout, presumably to allow the indwelling demons to escape. Demonic possession and witchcraft were still thought to be the cause of insanity and "lunacy" throughout the Middle Ages and well into the 17th and 18th centuries. A common "cure" for madness in the Middle Ages involved tying up those suspected being witches or demon-possessed with a rope and lowering them into freezing cold water. If they floated, they were believed to be witches and were then killed in some horrible way. If they sunk, they were not witches, but the cold water was believed to be a cure for madness, so either way the problem of insanity was resolved (Porter, 2002).

During colonial times the problem of the insane and "feebleminded" was considered a family matter, but as populations in the cities grew, those suffering from some form of mental illness increasingly became a problem for the community. Almshouses, typically used as poorhouses or workhouses for those unable or unwilling to find work on their own, were often used to house the insane as well. By the mid-1700s many towns in Colonial America were following the trend in Europe of building separate almshouses and even specialized hospitals for the insane (Torrey & Miller, 2002). Yet, reports of mistreatment were common. In fact, the trend of abuse noted in the Middle Ages continued throughout the 19th century, where members of society whose behavior was not in line with social mores and the general expectations of society were subjected to public beatings, incarceration, and sometimes death, particularly if their strange behavior was perceived as threatening. Typical "treatment" in asylums, in almshouses, and even in the new state hospital system included, among other things, beatings with chains and rods. Chains were also used to contain patients in insane asylums—some for most of their lives (Torrey & Miller, 2002).

By early 18th century, mental health reform had begun, led in part by Philippe Pinel of France, who when appointed chief physician at a hospital for the incurably mentally insane was appalled at the barbaric conditions of the hospital. He found patients chained to walls, some for up to 40 years, and a system where community residents could pay an admission fee to see the insane patients as if they were animals in a zoo. In 1792 Pinel was memorialized for his decision to unchain up to 5,000 patients. This event marked the beginning of the era of "Moral Treatment" of the mentally ill. Pinel later became chief of another hospital in Paris, where he consistently pushed for reform for more compassionate care.

Dorothea Dix, a U.S. social activist, was a leader in advocating for more compassionate treatment of the mentally ill in insane asylums. Her plea to the Massachusetts state legislature in 1843 poignantly described the deplorable conditions those with mental illness were forced to endure, including being held in cages by chains, often naked, beaten with rods, and whipped to ensure obedience. Dix pleaded for the legislators to intercede on behalf of society's most vulnerable members. Dix's efforts resulted in an improvement in the conditions of both hospitals and asylums (Torrey & Miller, 2002).

By the beginning of the 20th century most of the almshouses and insane asylums had closed, and state mental institutions became the primary facilities housing the mentally ill. Yet although institutionalized care was considered revolutionary, compassionate, and far better than the plight of the mentally ill in former generations, rampant abuses involving cruel treatment, neglect, and physical and emotional abuse were increasingly reported throughout the early 1900s.

The Deinstitutionalization of the Mentally Ill

Although horrible abuses in state and private mental hospitals were well documented through the mid-1900s, institutionalized care remained the primary method of treatment for the seriously mentally ill for another 50 years. The U.S. government's first legislative involvement in the care of the mentally ill occurred in 1946, when former president Harry Truman signed the National Mental Health Act. The signing of this act allowed for the creation of National Institute of Mental Health (NIMH) (one of the first four institutes under the National Institutes of Health) in 1949.

In 1955 the Mental Health Study Act was passed, which directed the convening of the Joint Commission on Mental Health and Illness (under the auspices of the NIMH), charged with the responsibility of analyzing and assessing the needs of the country's mentally ill, as well as making recommendations for a more effective and comprehensive national approach to their treatment. The committee was comprised of professionals in the mental health field, such as psychiatrists, psychologists, therapists, educators, and representatives from various professional agencies, including the American Academy of Neurology, American Academy of Pediatrics, American Psychological Association, National Association of Social Workers (NASW), and National Association for Mental Health.

The mentally ill were often housed in inhumane conditions, sometimes restrained for extended periods of time.
Peter Turnley/Corbis

In general, in addition to making recommendations for increasing funding for both research and training of professionals, the committee recommended transitioning from an institutionalized treatment model to an outpatient community mental health model, where patients were treated in the *least restricted environment* within the community. This report led to the creation of the Community Mental Health Centers (CMHC) Act of 1963, which was passed under the Kennedy administration. This act enabled funding of a new national mental healthcare system focusing on prevention and community-based care, rather than on institutionalized custodial care (Feldman, 2003). The passage of the CMHC Act set the deinstitutionalization movement into motion, prompted by an overall dissatisfaction with public mental hospitals in general, the development of new psychotropic medications, and a new focus on the brain–behavior connection that fostered a sense of hope and optimism among those in the mental health field (Mowbray & Holter, 2002).

> Mental illness remains a pervasive problem in today's society regardless of significant efforts to curb its devastating impact on individuals, families, and society.

Several decades after President Kennedy described the CMHC program as a "bold new approach" to dealing with mental illness, many in the mental health field cite frustration and discouragement with what many perceive as numerous failures of the program. The replacement of hope with discouragement is in part due to the reality that mental illness has been a far more worthy opponent than early advocates for change suspected. Early proponents of deinstitutionalization had hoped that through early detection, increased research, psychotropic medication, and better intervention strategies, mental illness could be greatly reduced and perhaps even eliminated. Yet, mental illness remains a pervasive problem in today's society regardless of significant efforts to curb its devastating impact on individuals, families, and society. The most serious criticisms are leveled at the federal government, which many claim fell short of funding commitments, resulting in far fewer community mental health centers being opened across the United States, which in turn resulted in the burden of care for the country's mentally ill shifting from the public mental hospital system to nursing homes, the streets, and the prison system (Sullivan, 1992).

Professional History

Understanding and Mastery of Professional History: Historical roots of human services

Critical Thinking Question: We have seen that societal attitudes about vulnerable populations have a marked impact on policies and practices. This is clearly illustrated by the historical treatment of the mentally ill. How do current stereotypes and attitudes about individuals with mental illnesses affect the treatment options available to them?

Common Mental Illnesses and Clinical Issues

Human service professionals may encounter mental illness directly when clients seek therapy for previously diagnosed disorders, or they may encounter mental illness indirectly when clients seek services from a human services agency for reasons unrelated to their mental health and symptoms of mental illness begin to surface in the midst of the counseling relationship. Whether clients present with prior diagnoses or have no previously identified mental health issues, practitioners must be able to recognize the common signs and symptoms of mental illness in their clients.

In the United States individuals are diagnosed using the *Diagnostic and Statistical Manual of Mental Disorders*, fourth edition, text revision (*DSM-IV-TR*), which categorizes mental disorders in a manner similar to physical disorders. Certain criteria must be met to diagnose someone with a particular mental or emotional disorder. It is important to remember, though, that mental and emotional disorders are diagnosed based on symptoms, not causes or etiology, as with medical illness. There is some controversy surrounding the possibility that the *DSM-IV-TR* contributes to pathologizing people rather than focusing on their strengths, but it is to date the most effective system available for assessing individuals in a systematic, organized, and universal manner.

The first two axes of the *DSM-IV-TR* are the ones most commonly used by clinicians in diagnosing clients. Clinical or mental disorders are diagnosed on Axis I, in 14 different categories. Disorders such as anxiety disorders, eating disorders, mood disorders (depression and bipolar disorder), and substance-related disorders are all diagnosed on Axis I. Many clinical disorders are amenable to treatment through psychotherapy and psychotropic medication, but they are diagnosed on the first axis because they are serious enough to warrant clinical attention.

Axis II is reserved for personality disorders and mental retardation. Personality disorders differ from clinical disorders in many respects, but most notably, many clinicians believe that personality disorders can be resistant to treatment because most of the problems that individuals with personality disorders experience are by definition ingrained in their personalities, thus authentic change is challenging because changing one's personality requires pervasive transformation. Examples of personality disorders include antisocial personality disorder (sociopathy), borderline personality disorder, and narcissistic personality disorder.

Serious Mental Disorders Diagnosed on Axis I

The following section includes some of the more serious mental illnesses that human service professionals often encounter in clients who are functionally impaired and in need of extensive case management and counseling services. Depending on the severity of their illness, these individuals might be in and out of inpatient psychiatric facilities, referred through the court system, or even living on the streets. It is important, then, that even entry-level human service professionals be generally familiar with these disorders so that their cases can be as effectively managed as possible and referred for appropriate services.

PSYCHOTIC DISORDERS   Psychotic disorders include a number of illnesses where contact with reality is severely impaired. Common symptoms of a psychotic disorder include hallucinations, delusions, and generally bizarre and eccentric behavior.

The most common psychotic disorder is *schizophrenia,* which is actually an umbrella term referencing what is theorized to be a number of disorders with similar symptoms, but with many different causes, such as genetic anomalies, brain chemistry disturbances, and brain damage. Recent research has even suggested that some forms of schizophrenia may be caused by exposure to the Borna virus (Terayama et al., 2003). Schizophrenia usually manifests during the teen and early adulthood years. Schizophrenia is *not* a split personality, and a diagnosis of schizophrenia does not automatically mean that someone will become violent (despite sensationalized media reports).

Schizophrenia is not caused by a bad childhood, although stress and trauma can trigger a psychotic episode.

According to the *DSM-IV-TR* (APA, 2000), symptoms of schizophrenia can include the following:

1. *Delusions:* False beliefs or misperceptions, many of which could not possibly be true, such as believing that the government is monitoring one's activities through the television set or that one has special powers, such as speaking to others through mental telepathy.
2. *Hallucinations:* Sensations that are experienced but do not exist, such as hearing voices, seeing things that are not there, smelling smells that do not exist, or feeling sensations when nothing is present.
3. *Disorganized thinking and speech:* The frequent trailing off into incoherent talk often referred to as *word salad*. Speech often reflects thinking that makes no sense. Some people with schizophrenia may even make up their own words, and some will stop speaking all together (as is the case with catatonia).
4. *Negative symptoms:* The absence of normal behavior such as the *lack* of emotion (often referred to as affective flattening), alogia (complete *lack* of any speech), and extreme apathy (complete *lack* of interest or drive).

Treatment used to consist solely of custodial care and heavy tranquilizers to minimize symptoms, particularly destructive ones. Antipsychotic medication has been available since the mid-1950s, but negative side effects of the medication, such as sexual impotence, tardive dyskinesia (involuntary jerking spasms of the muscles), and tranquilizing effects, kept many individuals with schizophrenia from taking their medication consistently. Yet new "atypical" antipsychotic drugs, such as risperidone, have shown great promise in significantly reducing schizophrenic symptoms such as hallucinations and delusion without nearly the number of side effects.

AFFECTIVE DISORDERS   Affective disorders include disorders of one's mood and emotions and include depression and bipolar disorder. People who suffer from clinical depression, referred to in the *DSM-IV-TR* as major depressive disorder, often feel sad, anxious, empty, hopeless, irritable, guilty, worthless, helpless, and feeling tired all the time. Major depression also often involves sleep and eating disturbances, difficulty concentrating and remembering things, various somatic symptoms such as head and body aches, and may also include thoughts of suicide. Someone with clinical depression may experience all of these feelings or a combination of them (e.g., they may be feeling sad and guilty but not anxious, or they may be eating relatively normally but they can't sleep at night). In order for a diagnosis of clinical depression to be given, the *DSM-IV-TR* stipulates that the individual experience five or more of these symptoms for at least a two-week period (APA, 2000). Depression is quickly becoming one of the most profound disorders affecting the population, and it tends to co-occur with many other disorders and conditions, including poverty (SAMHSA, 2010). In fact, the World Health Organization has projected that depression will continue to be widespread globally, becoming a leading cause of disability by 2020 (Michaud, Murray, & Bloom, 2001).

References to depression date back to the beginning of recorded time. Hippocrates wrote about melancholy in the 4th century, citing an imbalance in the body's "humors" or liquids (blood, bile, phlegm, and black bile) as the cause of melancholy. Everyone feels sad at times, and grieving over a loss is perfectly normal and in fact healthy, despite the pain and discomfort involved. A productive depression can motivate people to change both themselves and their circumstances, where complacency might otherwise keep someone in an unhealthy situation. But debilitating depression is rarely productive and can leave people feeling ashamed, particularly in a productivity-oriented society such as the United States. Such shame and guilt just serves to add an increased burden to the depressed person, exacerbating depressive symptoms and often leading to a downward emotional spiral.

A popular theory of depression is the cognitive-behavioral theory, which is somewhat of a hybrid model (incorporating aspects of Aaron Beck's cognitive theory of depression and the theory of behaviorism). This theory hypothesizes that depression is related to negative or irrational thinking. Thoughts such as "I'm a horrible person" or "Nothing good will ever happen to me, I will always fail" if thought consistently enough can ultimately lead to feelings of sadness, despair, and hopelessness (Beck, 1964).

Another popular theory, particularly with human service professionals and social workers, is a social-contextual model of depression in which environmental conditions such as negative life events, racial discrimination, and poverty impacting an individual is believed to contribute to depression, particularly if the depressed individual does not have the coping skills to deal with them in a positive manner (Swindle, Cronkite, & Moos, 1989).

In the last several decades a biological model of depression has emerged in which a predisposition to depression is believed to be genetically related, and depressive symptoms are believed to be caused by neurohormonal irregularities, such as problems with neurotransmitter functioning. Most human service professionals embrace a biopsychosocial model of depression that recognizes the biological basis of many depressions, the emotional nature of depression, and the impact that one's environment, including factors such as an abusive childhood, and even social oppression can have on depression.

Because depression often *co-occurs* with other disorders, such as anxiety, eating disorders, substance abuse disorders, and even psychotic disorders, it is essential that all human service professionals involved in direct service be able to screen for depression, even if a client is not seeking services for this purpose.

Serious Mental Disorders Diagnosed on Axis II

Personality disorders include generally rigid and inflexible patterns of inner experience and outward behavior. Personality disorders often involve unhealthy and maladaptive patterns of perceiving things, difficulty controlling or regulating emotions, and difficulty controlling emotional impulses. Someone with a personality disorder will often perceive things differently than others and often misperceive another's behavior and intentions. Up to 30 percent of all individuals seeking mental healthcare services have at least one personality disorder (Dingfelder, 2004).

But just because someone has personality traits that are irritating or somewhat eccentric, it does not mean that they have a personality disorder. One of my best friends can be defensive if someone is criticizing her children. She often misperceives innocent comments as slights or criticism of her parenting. But does this mean that she has a personality disorder? Of course not. But what if her defensiveness was so intense that she started arguments constantly with friends and family members? What if she could not enjoy going out socially because all she could think about was protecting her children? What if she perceived insults everywhere and could not get along with anyone, including her children's teachers? This behavior might then push her in the direction of a personality disorder—a collection of maladaptive and rigid personality traits that are exhibited across different contexts and interfere with one's ability to function effectively in life, including interfering with one's ability to enjoy reasonably healthy relationships with others.

For instance, it might be perfectly normal for a woman to feel emotionally attacked whenever she gets into an argument with her husband if he has an attacking way of expressing his needs and frustrations. But it is not necessarily healthy or normative for a woman to feel emotionally attacked whenever she receives constructive feedback that she perceives as criticism from her husband, friends, family, coworkers, supervisor, teachers, and children. It also might be perfectly healthy for a man to consistently focus on himself in certain situations where perhaps he feels somewhat insecure, such as in large social environments. But it might not be considered healthy if this person excessively focused on himself in virtually all areas of his life—at home, at work, with family, in social situations large and small, often at the expense of others.

The relative level of health or adaptive aspects of one's personality traits are judged on a continuum, like so many other mental and emotional conditions. If someone is a bit on the rigid side with certain issues, it wouldn't necessarily be appropriate to diagnose this person with a personality disorder. Yet, if someone gets far enough out on the continuum with regard to rigidity, for example, so that it interferes with an ability to function at work, with family, or with social situations, then this person might have what is considered a disordered personality. The key difference according to the *DSM-IV-TR* (American Psychiatric Association, 2000) is that a personality disorder must cause distress and impairment of functioning in several important areas of functioning.

The *DSM-IV-TR* categorizes personality disorders into three groups or clusters with three to four personality disorders in each cluster. Although all personalities share some factors in common, such as misperception, rigidity, pervasive problems in interpersonal relationships, and emotional regulation, each cluster of personality disorders varies considerably with both symptoms and cause. For instance, many of the Cluster A personality disorders, such as schizoid personality disorder, are strongly believed to be precursors of psychotic disorders. Obsessive compulsive personality disorder is also theorized to be obsessive compulsive disorder in the early stages. Yet the Cluster B and C personality disorders such as borderline personality disorder and dependent personality disorder are theorized to have some biological influences, but are believed to be related to abuse in childhood, particularly sexual and physical abuse (Bandelow et al., 2005).

Counseling individuals with personality disorders is often frustrating for practitioners because progress is slow, and clients are often resistant to change. Yet many new counseling techniques are being developed, some with significant success, but progress is always slow because authentic change requires that clients actually change the entire way they perceive the world, and themselves within it. They must also learn how to "sit with their emotions" rather than act on them and to control their impulses rather than indulging them. Thus, teaching self-discipline is a significant component of counseling most individuals with personality disorders. Antidepressant and antianxiety medication can help with the co-occurring depression and anxiety common with many personality disorders.

Mental Health Practice Settings and Counseling Interventions

Human service professionals involved in the practice of caring for the mentally ill was formally marked by involvement in the *aftercare movement* of the late 1800s and early 1900s. Aftercare, a social reform issue of the time, involved the short-term care of the formerly "insane" and "lunatics" (Vourlekis, Edinburg, & Knee, 1998). Aftercare was typically managed by private charitable societies who offered temporary assistance and housing for those coming out of the state asylum system. Social workers were on the forefront of this helping model, which was really before its time because this type of "continuum of care" was not a part of the psychological mainstream during that era. It wasn't long before aftercare programs were considered the sole domain of social workers, who were paid by the state, and ultimately by public or private hospitals. This program set the foundation for the contemporary role of those in the human services field who provide both advocacy and direct service to those who suffer from mental illness.

Intervention Strategies

A chief complaint of many in the human services field is the mental health community's general tendency to approach mental illness from a pathological perspective. This inclination to see human behavior in what often amounts to polarized terms of good and bad, acceptable and unacceptable, desirable and undesirable has only served to promote the social stigma of mental illness. Viewing mental illness through the lens of biology can also contribute to the tendency to pathologize the mentally ill where individuals are seen as "sick" and "broken." So although the discovery that many forms of mental illness have biological roots can relieve the mentally ill and their family of unnecessary guilt, it also suggests limited potential on the part of the mentally ill, increasing both social stigma and social rejection (Sullivan, 1992).

> **A chief complaint of many in the human services field is the mental health community's general tendency to approach mental illness from a pathological perspective.**

An alternative approach to viewing mental illness is to use a *strengths perspective,* a model commonly used in the human services field. This theoretical perspective encourages the practitioner to recognize and promote a client's strengths, rather than focusing on deficits. A strengths perspective also presumes clients' ability to solve their own problems through the development of self-sufficiency and self-determination. Although there are several contributors to strengths-perspective research in the human services field, Saleebey (1996) has developed several principles for practitioners to follow that can help clients experience a sense of empowerment in their lives. Saleebey encourages practitioners to recognize that all clients:

1. have resources available to them, both within themselves and their communities;
2. are members of the community and as such are entitled to respect and dignity;
3. are resilient by nature and have the potential to grow and heal in the face of crisis and adversity;
4. need to be in relationship with others in order to self-actualize; and
5. have the right to their own perception of their problems, even if this perception isn't held by the practitioner.

Sullivan (1992) was one of the first theorists to apply the strengths perspective to the area of chronic mental illness where clients suffering from mental illness are encouraged to recognize and develop their own personal strengths and abilities. Sullivan compared this approach to one often used when working with the physically challenged, where focusing on physical disabilities is replaced with focusing on and developing one's physical abilities. Sullivan claimed that by redefining the problem (rather than continuing to search for new solutions), by fully integrating the mentally challenged into society, and by focusing on strengths and abilities rather than solely on deficits, an environment can then be created that is truly consistent with the early goals of mental health reforms who sought to remove treatment barriers promoting respectful, compassionate, and comprehensive care of the mentally ill. Operating from a strengths perspective is important regardless of what intervention strategies a human service professional uses in direct practice.

Human service professionals utilize many tools and interventions when working with mentally ill clients. Some of these intervention strategies include *insight counseling*, where clients develop self-awareness skills intended to help them cope more effectively with their various mental health–related challenges. *Group counseling* assists mentally ill individuals gain strength and support from others in similar situations—some a few steps ahead of them and some a few steps behind. *Psychotropic medication* based on recent brain research offers many clients hope of controlling the often debilitating symptoms common to many serious mental illnesses.

Common Practice Settings

Human service professionals working with the mentally ill population do so in a variety of practice settings, including outpatient mental health clinics, not-for-profit agencies, outreach programs, job training agencies, housing assistance programs, prisoner assistance programs, government agencies, such as departments of mental health and human services, and probation programs. Human service professionals might be case managers responsible for conducting needs assessments and coordinating the mental healthcare of clients, they might be providing psychotherapy services on an individual and/or group basis, or they may provide more concrete services such as job training. In truth, a human service professional will likely encounter clients with serious mental illnesses in just about any practice setting, but in this section I will focus on those settings where the seriously and persistently mentally ill is the target population.

Community mental health centers provide direct services to the seriously and chronically mentally ill population. They are typically licensed by the state and designated to serve a certain catchment area within the community. Services offered often include outpatient services for adults and children, 24-hour crisis intervention, case management services, community support, psychiatric services, alcohol and drug treatment, psychological evaluations, and various educational workshops. They might also offer partial hospitalization and day treatment programs. Although most community mental health centers operate on a sliding scale, they cannot turn away clients who have no ability to pay, thus they are highly reliant on public funding.

Another practice setting that often encounters the seriously mentally ill is the *full-service human service agency*. It is difficult to define *human service agencies* because they come in all shapes, sizes, and colors, but essentially a full-service human or social service agency is a not-for-profit organization, meaning that any financial profits must be reinvested in the agency. This distinction also means that the agency is exempt from paying state and federal taxes, allowing more money to be directed back into the agency. Human services agencies typically offer an array of services aimed at various target populations, including the seriously mentally ill. The agency might provide general counseling services or might target more specific services, such as providing job skills training, housing assistance, or substance abuse counseling.

A human service professional might work in a number of capacities within a human services agency, depending in large part on what types of programs the agency offers. For instance, a human services worker might offer general case management services coordinating all the care the client is receiving and act as the point person for the psychologist, psychiatrist, and any other service providers involved. They might provide direct counseling services or run support groups focusing on a number of psychosocial and daily life issues. If the agency provides outreach services, the human services worker might be out in the community providing emergency crisis intervention services for the local police department or other emergency personnel. Obviously the list of program services is almost endless, particularly because a part of the role of the human services worker and agency is to identify needs within a community and fulfill those needs if not otherwise met.

An alternative to inpatient hospitalization is *partial hospitalization* or *day treatment programs*. These programs, often operated within a hospital setting, are intensive and offer services for individuals who are having difficulty coping in their daily lives, but are not at a point where inpatient hospitalization is a necessity. Clients attend the program five days a week, for approximately seven hours a day, and typically work with a multidisciplinary team of professionals, including a psychiatrist, psychologist, and social worker. Family involvement is highly encouraged. Certain partial hospitalization programs narrowly focus on specific issues, such as eating disorders, self-abuse, or substance abuse, whereas others focus on a wider range of clinical issues such as severe depression, anger management, and past abuse issues. The nature of the program will also vary depending on whether the target population is adults, adolescents, or children.

**Human Services
Delivery Systems**

Understanding and Mastery of Human
Services Delivery Systems: Range of
populations served and needs addressed
by human services

Critical Thinking Question: Mental
illnesses cover a broad spectrum of
symptoms, and their severity can range
from minimal to completely debilitating.
What steps can a human service
professional take to ensure that clients
with mental illnesses receive the level
and nature of treatment they need?

These structured programs can either serve as an alternative intervention to inpatient hospitalization or they can be utilized in the transition from inpatient hospitalization.

Although deinstitutionalization of the mentally ill has resulted in a dramatic reduction in long-term hospitalization of the severely mentally ill, some individuals who are acutely disturbed or suicidal are hospitalized on a short-term basis for diagnostic assessment and stabilizing in inpatient or *acute psychiatric hospitals.* Psychiatric units are typically locked for the safety of the patients who are often either actively psychotic or a danger to themselves or others. Again, services are focused on assessment and stabilizing with focus on discharge planning. Human service professionals and paraprofessionals, such as licensed social workers, counselors, and psychologists often provide case management and discharge planning services in inpatient settings providing adult services and will likely provide more intensive counseling services such as facilitating individual and group counseling, as well as behavioral management if the program is focused on children and adolescents.

Mental Illness and Special Populations

Mental Illness and the Homeless Population

One unanticipated consequence of deinstitutionalization was the shifting of literally thousands of mentally ill patients from institutions to the streets. In fact, a 2005 study found that nearly one in six mentally ill individuals are homeless (Folsom, Hawthorne, & Lindamer, 2005). Such individuals would have previously been hospitalized, but with the closing of the majority of public mental hospitals and the transitioning of most psychiatric units to a focus on short-term stays, the severely mentally ill who do not have a network of supportive and able family members are often left with no place to live. Even individuals who do have supportive families will often live on the streets due to the nature of psychosis, which clouds judgment and impairs the ability to think without distortion, leading some individuals to disappear for literally years at a time.

This link between homelessness and mental illness is not solely related to deinstitutionalization. Certainly warehousing the mentally ill kept them off the streets, but the nature of this link is far more complex and likely reciprocal in nature, meaning that severe mental illness leaves many incapable of providing for their basic needs, and the stressful nature of living on the streets, not knowing where one will lay their head at night, dealing with exposure to violence as well as inclement weather, and not knowing where their next meal will come from would put the healthiest of individuals at risk for developing some mental illness.

Government sources estimate that approximately 26 percent of the homeless population is severely mentally ill (U.S. Conference of Mayors, 2011), and if mental illness is broadened to include clinical depression and substance abuse disorders (often used to self-medicate), that percentage jumps to an astounding 50 to 80 percent, and this number is continuing to rise (North, Eyrich, Pollio, & Spitznagel, 2004; Shern et al., 2000). The mentally ill homeless population is a somewhat diverse group, but African American single men and veterans are most likely to be homeless and suffering from mental illness (Folsom et al., 2005; Koerber, 2005; Shern et al., 2000). Although deinstitutionalization is credited for being the primary cause of this increase in the homeless population, the increase in homelessness did not occur until the 1980s, thus other issues are at play as well, including a shortage in affordable housing and again a lack of funding of housing assistance programs targeted to middle-aged men and veterans.

One of the biggest challenges in getting individuals with severe mental illness off the streets is engaging them in treatment. One of the problems noted after deinstitutionalization was the common difficulty of mentally ill individuals exercising their newly won right to refuse treatment. But a deeper look into this issue reveals that it may not be as simple as individuals in need not wanting help, but rather may be far more related to the difficulty and complexity of *accessing* needed services (Shern et al., 2000).

Barriers to accessing services often include difficulties in applying for government assistance such as Medicaid and Medicare to pay for both treatment and medication. Another barrier involves the actual service delivery model most popular in counseling and mental health centers, where the client comes to the psychologist. History clearly reveals that this model simply does not work with seriously mentally ill individuals, particularly those living on the streets. Such individuals are often confused, disoriented, and frequently distrusting of others, particularly if they are suffering from some sort of paranoid disorder. To expect a person who is homeless and suffering from some mental illness to remember a weekly appointment and somehow figure out how to navigate transportation is clearly unrealistic.

Another barrier to seeking treatment involves the many stipulations and requirements common in standard treatment models used by many community mental health centers. Most standard mental health programs have strict participation requirements, particularly related to behavioral issues such as maintaining sobriety to remain in a housing assistance program, or program requirements such as requiring clients to participate in weekly counseling support groups to receive other services. In fact, most standard programs are directive with seriously mentally ill clients, often determining treatment goals and interventions for the client, rather than empowering clients to assist in determining their own treatment goals and interventions (Shern et al., 2000).

The problem of homelessness among the mentally ill population will not be resolved until sufficient long-term housing assistance can be provided. Housing assistance programs typically have long waiting lists and often allow only women with children accelerated access to the program. Because African American men and veterans are overrepresented in the mentally ill homeless population, more programs need to be developed that target these populations most at risk for homelessness. Such programs must also be designed to address issues related to alcohol and substance abuse problems as well because many within the mentally ill homeless population have co-occurring substance abuse problems.

Mental Illness and the Prison Population: The Criminalization of the Mentally Ill

Another unintended by-product of deinstitutionalization is what has effectively amounted to the inadvertent shifting of chronically mentally ill patients from public hospitals to jails and prisons. In fact, many mental health advocates have argued that prisons have now become one of the primary institutions warehousing the United States' most severely mentally ill individuals (Palermo, Smith, & Liska, 1991; Torrey, 1995). Thus, although this was never the intention of policy changers and proponents of deinstitutionalization, it appears that the United States has in many respects returned to the era where the mentally ill were locked away in almshouses.

> Prisons have now become one of the primary institutions warehousing the United States' most severely mentally ill individuals.

Human Rights Watch reports that the number of mentally ill has quadrupled in the last six years, from 283,000 prisoners in 1998 to over 1.25 million in 2006. In fact, it has recently been reported that there are more mentally ill individuals in prisons and jails than in hospitals. Further, approximately 40 percent of the mentally ill population will come into contact with the criminal justice system at some point in their lives (Torrey, Kennard, Eslinger, Lamb, & Pavle, 2010).

Women are particularly overrepresented in the prison population with approximately 31 percent of women in state prisons suffering from some form of serious mental illness (compared to about 14 percent for men) (Torrey, Kennard, Eslinger, Lamb, & Pavle, 2010). Most mentally ill prisoners are poor and were either undiagnosed prior to their incarceration or untreated in the months prior to entering the prison system. Mentally ill inmates were twice as likely to have a history of physical abuse and four times as likely to have been the victim of sexual abuse. In fact, almost 65 percent of mentally ill female inmates reported having been physically and/or sexually abused prior to going to prison (Ditton, 1999).

But what does this really mean? Could it simply mean that some mentally ill individuals break the law more than mentally healthy individuals? Couldn't it be argued that one must certainly be mentally ill to kill a string of women or one's entire family? After all, what sane person sexually abuses children? Depending on how mental illness is defined, it could be argued that those who commit heinous crimes are by definition mentally ill, and their mental illness does not and should not negate the appropriateness of sending them to prison for their crimes. But even in situations where offenders clearly should be incarcerated, a retrospective look at their mental health histories might reveal a history of poor service utilization, treatment refusal, or an outright inability to access much needed mental health treatment.

While prisons have always held mentally ill prisoners, the number of incarcerated mentally ill has increased sharply, in large part because of a decrease in treatment options available for the mentally ill population in the general population. The reasons for this decrease include a reduction in funding of community mental health centers, barriers to access treatment for certain segments of the population, and increasing difficulty in involuntary hospitalization of the severely mentally ill (The Sentencing Project, 2008).

The majority of those mentally ill who are incarcerated have been convicted of nonviolent petty crimes, related to the mentally ill. In fact mentally ill individuals in prisons and jail are targets for violence, such as assault, robbery, and sexual assault

(Marley & Buila, 2001). Far too often mentally ill prisoners, particularly those in the general prison population, are consistent targets of victimization, particularly sex-related crimes, many of which go unreported.

The incarceration of the mentally ill is not a simple problem, thus it has no simple answers. Mental health and prison advocates cite barriers to accessing mental health services and problems with early intervention as direct causes of seriously mentally ill individuals ending up in the penal system, rather than in psychiatric facilities. Once again the controversial issue of an individual's right to refuse treatment is relevant in this matter as well evidenced by the many family members of the mentally ill who consistently complain that the courts have refused to order involuntary treatment, only to have their mentally ill family member commit a violent crime some time later. What is so unfortunate in these incidences is that the majority of mentally ill defendants are amenable to treatment, but many were not receiving any treatment at the time of their incarceration (Marley & Buila, 2001).

MENTAL HEALTH COURTS   Many steps are currently being taken by those in the criminal justice system and mental health and human services fields to address the issue of the incarceration of the mentally ill as the many factors that create this complex cycle of re-incarceration. The development of mental health courts program is an example of a significant step in the right direction. The Mental Health Courts Program was developed pursuant to the America's Law Enforcement and Mental Health Project (Pub. L. No. 106-515 passed in November 2000) and is administered under the Bureau of Justice Assistance (BJA), a component of the U.S. Department of Justice, in cooperation with the Substance Abuse & Mental Health Services Administration (SAMHSA).

The goal of the BJA is to encourage, lead, and fund the development of comprehensive programs run by criminal justice systems across the country that offer alternatives to incarceration as well as helping to avoid future court involvement.

Mental health court program goals include the following:

- Increased public safety for communities—by reducing criminal activity and lowering the high recidivism rates for people with mental illnesses who become involved in the criminal justice system
- Increased treatment engagement by participants—by brokering comprehensive services and supports, rewarding adherence to treatment plans, and sanctioning nonadherence
- Improved quality of life for participants—by ensuring that program participants are connected to needed community-based treatments, housing, and other services that encourage recovery
- More effective use of resources for sponsoring jurisdictions—by reducing repeated contacts between people with mental illnesses and the criminal justice system and by providing treatment in the community when appropriate, where it is more effective and less costly than in correctional institutions

Even though the goal is for all court jurisdictions to have a mental health court, as of 2007, only 175 mental health courts were in existence across the United States (Council of State Governments, 2008). But their numbers are growing, from only a handful in the 1990s, and preliminary research indicates that they are successfully diverting the mentally ill from jail to programs offering much needed services. One study researching one of the first mental health courts (located in Broward County, Florida) found that participants spent 75 percent less time in jail, received needed mental health services on a more frequent basis, and were no more likely to commit a new crime, compared to mentally ill defendants who proceeded through the traditional court process (Christy, Poythress, Boothroyd, Petrila, & Mehra, 2005).

Multicultural Considerations

Early studies have shown that ethnic minority populations are often poorly served in mental health centers because of a lack of culturally competent counselors and bilingual counselors (Sue, 1977). Other early studies showed that whereas those in the Latino and Asian populations were underrepresented in community mental health center settings, those within the African American and Native American populations were overrepresented (Sue & McKinney, 1975; Diala, Muntaner, Walrath, Nickerson, LaVeist, & Leaf, 2001). This pattern may be partly due to cultural acceptance or rejection of psychotherapy within different cultural groups, and it might also be related to the relative complexity of issues facing the populations served, particularly Native Americans, who traditionally have high rates of substance abuse and depression and often reside in remote areas.

A 1997 study found that African American caregivers of mentally ill individuals face a number of barriers, making it difficult for them to be involved in their family member's treatment, including a failure on the part of practitioners to recognize them as an integral part of the treatment team. Mental health clinicians need to partner with the family members of mentally ill clients and keep an open line of communication so that family caregivers do not feel marginalized in the treatment process. The authors of the study suggested that by working hard to engage family caregivers in treatment, common negative assumptions of family members of African American clients can be countered and overcome (Biegel, Johnsen, & Shafran, 1997).

Mental health providers and social justice advocates have increasingly expressed concerns about the impact of anti-immigration policies on the Latino community, particularly with regard to the additional stress such legislation and the associated *xenophobia* (an irrational fear of immigrants or those presumed to be foreigners) can cause the immigrant community (Ayon, Marsiglia & Bermudez-Parsai, 2010). A recent study of Latino youth and their families in the Southwest, which has the greatest number of anti-immigrant policies and attitudes, showed that the majority of the Latino youth and parents surveyed had experienced significant discrimination related to their ethnic backgrounds, and immigrant status, even if they were born in the United States. Yet, their strong ties to family (immediate and extended) and their communities (referred to as familismo) seemed to counter the effects of discrimination (Ayon, Marsiglia & Bermudez-Parsai, 2010). Mental health providers, including human service professionals who embrace Euro-American values of individualism, could potentially view the cultural tradition of familismo as something negative rather than as a cultural strength that can serve as protection from the negative effects of discrimination and xenophobia.

It is important that human service professionals be aware of their negative biases, whether they are toward people of color, sexual orientation, or socioeconomic status. Most people, particularly within the majority culture, deny having negative or stereotypical biases toward cultures different than their own, because few want to be characterized as racist, homophobic, or elitist, but all individuals possess some negative biases, and if not directly confronted both through personal awareness and in clinical supervision within their agency, even subtle biases will unfold within the counseling relationship. Racial bias can influence many factors associated with mental healthcare and counseling. For instance, a relatively recent study found that African Americans were far more likely to be diagnosed with disruptive behavioral disorders in mental health counseling, compared to Caucasians who were far more likely to be diagnosed with less serious clinical disorders, such as adjustment disorder (Feisthamel & Schwartz, 2009). Racially disproportionate clinical diagnostic assessment may be due to personal racial biases on the part of human service professionals, but may also be due to counselors not taking into consideration the disproportionate challenges facing many ethnic minorities, such as increased levels of poverty, racial oppression, and higher rates of unemployment compared to Caucasians in America (Feisthamel & Schwartz, 2009).

Other types of bias can enter the counseling relationship as well. Consider the bias that many in the United States (particularly Euro-Americans) have about time. The U.S. culture tends to highly value time and promptness. When someone is timely, they are often considered to be respectful, considerate of others, and organized. Conversely, those who are consistently late are often presumed to be disrespectful, inconsiderate of others, disorganized, and perhaps even lazy. Yet not all cultures value time in the same manner, and a stereotyped bias is that individuals from these cultures (e.g., Latino and East Indian cultures) are lazy, disrespectful, and disorganized. Human service professionals who have been enculturated in U.S. values might not even realize that they hold this stereotype and might unconsciously attribute negative traits to clients who consistently show up late to their appointments. Thus, although it might be worth exploring whether this pattern is related to lacking motivation, it may be racist to make negative assumptions about a client's character based solely on the fact that the client is from an ethnic culture that does not value time in the same manner as those embracing U.S. values.

Hence, although rarely is someone eager to admit holding negative stereotypes about certain races, cultures, or lifestyles, it is imperative, particularly when working with the seriously mentally ill population, that these negative stereotypes are explored, challenged, and discarded. Otherwise they will remain powerful forces in how human service professionals subtly or overtly evaluate and assess client actions and motivations, strengths, and deficits, including assessing accountability and causation for their life circumstances.

Current Legislation Affecting Access to Mental Health Services

Mental Health Parity

Some mental illnesses take a lifetime to develop. Others seem to hit out of nowhere, such as schizophrenia. Mental illness cuts across all socioeconomic, racial, and gender lines; in fact, one could say that mental illness is an "equal opportunity" affliction. I have worked with both the lower-income and undereducated population and the upper-income and highly educated population, and my only observation about the difference regarding these two groups is that oftentimes those on the upper end of the income/education continuums do a better job at hiding their mental illnesses and emotional disorders, at least for a time. For this reason, as well as the increasing evidence of the biological basis of many mental illnesses formerly believed to be solely psychological in nature, most mental health advocates argued the importance of requiring health insurance companies to cover mental health conditions in the same manner as they cover general medical conditions. Yet in the 1980s, when managed care became the norm in health insurance coverage, many advocates complained that managing costs became synonymous with limiting much-needed benefits, particularly in the area of mental health coverage.

> **Mental illness cuts across all socioeconomic, racial, and gender lines; in fact, one could say that mental illness is an "equal opportunity" affliction.**

Through bipartisan efforts, the Mental Health Parity Act was passed in 1996. It bars employee-sponsored group health insurance plans from limiting coverage for mental health benefits on a greater basis than for general medical or surgical benefits. This initial bill removed annual and lifetime dollar limits commonly used by insurance companies to limit mental health benefits. Unfortunately, the majority of health insurance companies have found loopholes, allowing them to avoid complying with this legislation.

Human Services Delivery Systems

Understanding and Mastery of Human Services Delivery Systems: Range and characteristics of human services delivery systems and organizations

Critical Thinking Question: There is a disturbing correlation between mental illness and incarceration in contemporary U.S. society. Mental health courts appear to be one promising alternative to incarcerating individuals with mental illnesses who have committed certain crimes. What strategies might human service professionals use in reaching out to incarcerated and recently released individuals who have mental illnesses?

The Mental Health Parity and Addiction Equity Act of 2008 (sponsored by President Obama when he was a senator), which was attached to the 2008 federal bail-out legislation, promises significant reform of mental health parity in the United States. The act went into effect January 1, 2010, and requires group health plans (covering 50 or more employees) that already provide medical and mental health coverage to provide mental health and substance abuse benefits at the same level as provided medical benefits (i.e., it does not require employers to provide mental health and substance abuse coverage). Thus, if an employer-sponsored insurance plan offers mental health benefits, the benefits must be consistent with what is offered in the medical plan with regard to deductibles, co-pays, number of visits allowable per year, and so on. Although some exemptions do exist in this act, it goes a long way in securing parity of mental health and substance abuse benefit coverage.

Other Federal Legislation

Former president George W. Bush announced the establishment of the *New Freedom Initiative* designed to identify and remove barriers to community living for all individuals with mental disabilities and long-term mental illness. This initiative led to the formation of the Commission on Mental Health on April 29, 2002. The commission was charged with the responsibility of studying the mental health delivery system and making recommendations on ways for adults and children with serious mental illnesses to integrate into their communities as fully and as effectively as possible. Referring to the current system as offering a "piecemeal" approach to mental healthcare, the commission made recommendations for change based on the contention that people can recover from mental illness and are not destined to accept a life of long-term disability. The commission promised to transform mental healthcare in America by promoting access to educational and employment opportunities to individuals with mental disabilities, as well as promoting full access to community life (President's New Freedom Commission on Mental Health, 2003).

One of the chief complaints of the commission's report was that the current system did not offer much hope of recovery to those suffering from a mental illness. In addition, the commission noted that it took sometimes years for new treatment strategies discovered at research institutes to be used in clinical settings. Thus, although thousands of government dollars were being spent identifying new treatment modalities, those suffering from mental illness often did not benefit from these discoveries, due to many factors, including poor communication between research facilities and clinical settings. The commission then made several recommendations for removing barriers to treatment and lifting the stigma often associated with mental illness.

Ironically, the final report for the committee studying the CMHC program in 1963 had goals that were very similar in nature and just as admirable as the commission's goals, yet clearly the implementation of those goals did not unfold as anyone had hoped or planned. Former president Bush's administration did move forward on some of the commission's goals, including providing some assistance to states in developing a more effective mental health delivery system, as well as increasing the number of screening programs designed to increase early detection and treatment of serious mental illness. Yet, in the midst of these ambitious goals and promises of sweeping reforms, federal funding for mental health programs under former president Bush was significantly cut by billions of dollars over several years.

Perhaps one of the most significant federal laws to be passed in years is the Patient Protection and Affordable Care Act of 2010 (PPACA) signed into law by President Obama in March of 2010 after a fierce public relations war waged by Republicans and health insurance companies designed to prevent its passage. The PPACA, taking effect incrementally from 2010 to 2014, is a comprehensive healthcare reform bill. It will have an impact on behavioral and mental healthcare coverage as well, thus while I will be exploring this legislation in more detail in Chapter 10, I will touch on its relevance to mental healthcare here. Overall this legislation is designed to make it easier for individuals and families to obtain quality health insurance, despite pre-existing conditions, and will make it more difficult for health insurance companies to deny coverage. It also expands Medicare in a variety of ways, including bolstering community and home-based services, and provides incentives for preventative, holistic, and wellness care. With respect to behavioral and mental healthcare, the PPACA provides increased incentives for coordinated care, school-based care including mental healthcare and substance abuse treatment, and it includes provisions that will require the inclusion of mental health and substance abuse coverage in benefits packages, including prescription drug coverage, and wellness and prevention services.

One of the most powerful ways that the federal government can influence policy and program development is through sufficient funding and a chief complaint about the many failures of community-based mental healthcare programs is the lack of sufficient federal funding. Thus, the success or failure of any new federal legislation or program focusing on mental healthcare reform is dependent on broad-based government financial commitment. Some of the funding that was cut under the George W. Bush administration was reinstated under President Obama, and the success of the PPACA remains to be seen, but advocates are hopeful that it will address many of the challenges faced by those struggling with mental illness, as they face significant challenges in attempting to get their holistic needs met.

Some human service professionals might question the importance of understanding federal trends in funding programs designed to meet the needs of the mentally ill, yet virtually all human service professionals will be affected one way or another if state and federal budget cuts continue to be implemented. Attempting to facilitate much-needed mental health–related programs without proper funding can involve everything from understaffing and high caseloads to inadequate office space and the general inability to meet the comprehensive needs of the chronically mentally ill. Thus, although human service professionals involved in direct practice may prefer to steer clear from administrative and policy concerns, such involvement particularly on an advocacy level is important because the effective facilitation of vital mental health programs is dependent on effective legislation and appropriate funding.

> **Professional History**

Understanding and Mastery of Professional History: Historical and current legislation affecting services delivery

Critical Thinking Question: The Mental Health Parity Act of 1996 and the Mental Health Parity and Addiction Equity Act of 2008 expanded access to mental health services for individuals and families who had employer-sponsored health insurance. In general, which populations benefited most from these pieces of legislation? Which populations did not benefit?

Ethical Considerations

Human service professionals working with the chronic and seriously mentally ill must manage several ethical considerations. Although most human service workers will not be formally diagnosing clients using the *DSM-IV-TR*, unless they are licensed to engage in professional counseling, it is still important to be aware of the ethical challenges facing those counselors who do, since often human service professionals will be working within multidisciplinary team, as well as with clients who have received one or more *DSM-IV-TR* diagnoses. Challenges in diagnosing clients abound, but many relate to conducting a thorough assessment of clients and diagnosing them accurately without "upcoding" (rendering a more serious diagnosis for insurance reimbursement purposes) (Kress, Hoffman & Eriksen, 2010), diagnosing based upon the "trendy" disorders (e.g., not allowing pharmaceutical companies to drive diagnoses), or diagnosing based upon a client's ability to pay. For instance, a recent research study found that counselors were more likely to render a *DSM-IV-TR* diagnosis to managed care clients than those who pay out of pocket (Lowe, Pomerantz, & Pettibone, 2007). Other ethical considerations include avoiding reductionist approaches to clients based upon a client's diagnoses (e.g., "my borderline client," "my OCD client"), and avoiding assessing, diagnosing, and treating clients in ways that emanate from personal bias related to gender, income level, education level, ethnicity, or immigration status.

Concluding Thoughts on Mental Health and Mental Illness

The field of mental health is a dynamic practice area for the human service professional for many reasons. Human service professionals have the ability to truly make an impact while working with some of society's most vulnerable members. Because *mental illness* is such a broad term, encompassing such a wide array of psychological, emotional, and behavioral issues, the human service professional works as a true generalist whether in a direct service capacity or whether providing advocacy within the community. The United States has experienced dramatic shifts in its mental health delivery system during the past 50 years and will no doubt continue to experience future changes, some intended and some unintended. Human services workers are on the front lines of these intended changes lobbying for increased funding, developing new programs to meet the complex needs of the severely and chronically mentally ill population.

The following questions will test your knowledge of the content found within this chapter.

1. Approximately 26 percent of the U.S. adult population suffers from some diagnosable mental disorder, most of which
   a. are relatively serious, requiring formal psychological intervention
   b. are relatively minor, not requiring formal psychological intervention
   c. are serious enough to require at least short-term hospitalization
   d. affect primarily Caucasian women

2. Early in human history mental illness was commonly believed to be caused by
   a. alcoholism
   b. remaining unmarried
   c. demonic possession
   d. rebellion

3. The passage of the Community Mental Health Centers (CMHC) Act
   a. increased the amount of time patients remained in psychiatric hospitals
   b. started what is often called the era of "deinstitutionalization"
   c. started the HMO phenomenon
   d. None of the above

4. Believing that the government is monitoring one's activities through the television set is an example of
   a. delusion
   b. splitting
   c. dissociation
   d. hallucination

5. Saleebey encourages human service professionals to recognize that the chronically mentally ill are
   a. dangerous and should be treated with caution
   b. members of the community and are entitled to respect and dignity
   c. resilient by nature and have the potential to grow and heal in the face of crisis adversity
   d. Both B and C

6. The group most likely to be homeless and to suffer from some mental illness is
   a. African American single men
   b. veterans
   c. older adults in residential care
   d. Both A and B

7. Describe the process of the deinstitutionalization of the mental ill, citing the reasons, goals, and short- and long-term effects of the transition from an institutional model to a community mental health model.

8. Discuss the reciprocal relationship between mental illness, homelessness, and incarceration.

Suggested Readings

Kreisman, J. J., & Straus, H. (1991). *I hate you, don't leave me.* New York: Avon.

Lachenmeyer, N. (2001). *The outsider: A journey into my father's struggle with madness.* New York: Broadway.

Mason, P. T., & Kreger, R. (1998). *Stop walking on eggshells: Taking your life back when someone you care about has borderline personality disorder.* Oakland, CA: New Harbinger Publishing.

Porter, R. (2002). *Madness: A brief history.* New York: University Press.

Torrey, E. F., & Miller, J. (2001). *The invisible plague: The rise of mental illness from 1750 to the present.* New Brunswick, NJ: Rutgers University Press.

Internet Resources

Affective disorders: http://www.pendulum.org

Anxiety Disorders Association of America: http://www.adaa.org

Borderline personality disorder: http://www.bpdcentral.com

Children and adults with attention deficit/hyperactivity disorder: http://www.chadd.org

Depression central: http://www.psycom.net/depression.central.html

Eating disorders: http://www.something-fishy.org

Internet mental health: http://www.mentalhealth.com

National Alliance on Mental Illness: http://nami.org

Personality disorders: http://personalitydisorders.mentalhelp.net

PsyWeb mental health site: www.psyweb.com

Schizophrenia: http://www.schizophrenia.com

References

American Psychiatric Association. (2000). *Diagnostic and statistical manual of mental disorders* (4th ed., Text Revision). Washington, DC: Author.

Ayon, C., Marsiglia, F. F., & Bermudez-Parsai, M. (2010). Latino family mental health: Exploring the role of discrimination and familismo. *Journal of Community Psychology, 38*(6), 742–756.

Bandelow, B., Krause, J., Wedekind, D., Broocks, A., Hajak, G., & Rüther, E. (2005). Early traumatic life events, parental attitudes, family history, and birth risk factors in patients with borderline personality disorder and healthy controls. *Psychiatry Research, 134*(2), 169–179.

Beck, A. T. (1964). Thinking and depression: 2. Theory and therapy. *Archives of General Psychiatric, 10*, 561–571.

Biegel, D. E., Johnsen, J. E., & Shafran, R. (1997). Overcoming barriers faced by African-American families with a family member with mental illness. *Family Relations, 46*(2), 163–178.

Christy, A., Poythress, N. G., Boothroyd, R. A., Petrila, J., & Mehra, S. (2005). Evaluating the efficiency and community safety goals of the Broward county mental health court. *Behavioral Sciences and the Law, 23*, 227–243.

The Council of State Governments. (2008). *Mental health courts: A primer for policymakers and practitioners*. Available at: http://consensusproject.org/mhcp/mhc-primer.pdf

Diala, C. C., Muntaner, C., Walrath, C., Nickerson, K., LaVeist, T., & Leaf, P. (2001). Racial/ethnic differences in attitudes toward seeking professional mental health services. *American Journal of Public Health, 91*(5), 805–807.

Dingfelder, S. (2004). Treatment for the untreatable. *Monitor in Psychology, 35*(11), 48–49.

Ditton, P. M. (1999). *Mental health and treatment of inmates and probationers* (Special report NCJ 174463). Washington, DC: U.S. Department of Justice, Office of Justice Programs, Bureau of Justice Statistics.

Feisthamel, K. P., & Schwartz, R. C. (2009). Differences in mental health counselors' diagnoses based on client race: An investigation of adjustment, childhood, and substance-related disorders. *Journal of Mental Health Counseling, 31*(1), 47–59.

Feldman, S. (2003). Reflections on the 40th anniversary of the U.S. Community Mental Health Centers Act. *Australian and New Zealand Journal of Psychiatry, 3*, 662–667.

Folsom, D. P., Hawthorne, W., & Lindamer, L. (2005). Prevalence and risk factors for homelessness and utilization of mental health services among 10,340 patients with serious mental illness in a large public mental health system. *American Journal of Psychiatry, 162*(2), 370–376.

Human Rights Watch. (2006). US: Number of mentally ill in prisons quadrupled: Prisoners ill equipped to cope. Retrieved online at http://www.hrw.org/news/2006/09/05/us-number-mentally-ill-prisons-quadrupled.

Koerber, G. (2005). *Veterans: One-third of all homeless people*. National Alliance on Mental Illness, Issue Spotlight. Retrieved June 1, 2005, from http://www.nami.org/Template.cfm?Section=Issue_Spotlights&template=/ContentManagement/ContentDisplay.cfm&ContentID=26958

Kress, V. E., Hoffman, R. M., & Eriksen, K. (2010). Ethical dimensions of diagnosing: Considerations for clinical mental health counselors. *Counseling & Values, 55*(1), 101–112.

Lowe, J., Pomerantz, A. M., & Pettibone, J. C. (2007). The influence of payment method on psychologists' diagnostic decisions: Expanding the range of presenting problems. *Ethics & Behavior, 17*, 83–95. doi:10.1080/10508420701310141

Marley, J. A., & Buila, S. (2001). Crimes against people with mental illness: Types, perpetrators, and influencing factors. *Social Work, 46*(2), 115–124.

Michaud, C. M., Murray, C. J., & Bloom, B. R. (2001). Burden of disease—implications for future research. *Journal of the American Medical Association, 285*(5), 535–539.

Mowbray, C. T., & Holter, M. C. (2002). Mental health & mental illness: Out of the closet? *Social Service Review, 76*(1), 135–179.

New Freedom Commission on Mental Health. (2003). *Achieving the promise: Transforming mental healthcare in America* (DHHS Publication No. SMA 03-3832). Rockville, MD: Author.

North, C. S., Eyrich, K. M., Pollio, D. E., & Spitznagel, E. L. (2004). Are rates of psychiatric disorders in the homeless population changing? *American Journal of Public Health, 94*(1), 103–108.

Palermo, G. B., Smith, M. B., & Liska, F. J. (1991). Jails versus mental hospitals: A social dilemma. *International Journal of Offender Therapy and Comparative Criminology, 35*(2), 97–106.

Porter, R. (2002). *Madness: A brief history*. New York: Oxford University Press.

Powell, J. (2003). Letter to the editor. *Issues in Mental Health Nursing, 24*(5), 463.

Saleebey, D. (1996). The strengths perspective in social work practice: Extensions and cautions. *Social Work, 41*(3), 296–305.

Shern, D. L., Tsemberis, S., Anthony, W., Lovell, A. M., Richmond, L., Felton, C. J., et al. (2000). Serving street-dwelling individuals with psychiatric disabilities: Outcome of a psychiatric rehabilitation clinical trial. *American Journal of Public Health, 90*(12), 1873–1878.

Substance Abuse and Mental Health Services Administration. (2010). Mental Health, United States, 2008. HHS Publication No. (SMA) 10-4590, Rockville, MD: Center for Mental Health Services, Substance Abuse and Mental Health Services Administration.

Sue, S. (1977). Community mental health services to minority groups: Some optimism, some pessimism. *American Psychologist, 32*, 616–624.

Sue, S., & McKinney, H. (1975). Asian Americans in the community mental health system. *American Journal, 45*, 111–118.

Sullivan, W. P. (1992). Reclaiming the community: The strengths perspective and deinstitutionalization. *Social Work, 37*(3), 204–209.

Swindle, R. W., Cronkite, R. C., & Moos, R. H. (1989). Life stressors, social resources, coping, and the 4-year course of unipolar depression. *Journal of Abnormal Psychology, 98*(4), 468–477.

Terayama, H., Nishino, Y., Kishi, M., Ikuta, K., Itoh, M., & Iwahashi, K. (2003). Detection of anti-Borna Disease Virus (BDV) antibodies from patients with schizophrenia and mood disorders in Japan. *Psychiatry Research, 120*(2), 201–206.

Torrey, E. F. (1995). *Surviving schizophrenia, 3rd ed.* New York: Harper-Perennial.

Torrey, E. F., & Miller, J. (2002). *The invisible plague: The rise of mental illness from 1750 to present.* NJ: Rutgers University Press.

Torrey, E. F., Kennard, A, D., Eslinger, D., Lamb, R., and Pavle, J. (2010). More mentally ill persons are in jails and prisons than hospitals: A survey of the states. Report for the Nation al Sheriff's Association and the Treatment Advocacy Center.

U.S. Conference of Mayors. (2011). *Hunger and homelessness survey: A status report on hunger and homelessness in America's cities.* Washington, DC: Author.

Vourlekis, B. S., Edinburg, G., & Knee, R. (1998). The rise of social work in public mental health through aftercare of people with serious mental illness. *Social Work, 43*, 567–575.

CHAPTER 9

Homelessness

Bruce Ayres/Getty Images

The Nature of Homelessness: A Snapshot of Homelessness in America

For as long as there have been established residential settlements, there have been those within the population who have either by choice or by life circumstances been homeless. To address the problem of homelessness, it is important to first understand the nature of this social condition, including developing an understanding of the extent of the homeless problem, determining who is most vulnerable to becoming homeless, as well as discovering the root causes of homelessness. It is only through understanding the demographic nature and common reasons for homelessness that social programs can be developed to assist members of society in obtaining permanent, stable housing, as well as developing preventative measures to protect against homelessness in the future. Many homeless advocates believe that significantly reducing the homeless population is a reasonable goal, and in fact it truly does seem plausible to assume that one of the wealthiest countries in the world would have enough resources to wipe out homelessness all together.

Homelessness is increasing in the United States, particularly among families with children. Most urban cities have reported that they have experienced a marked increase in the number of families seeking assistance, and some cities have cited the need to turn away homeless individuals and families due to a lack of resources. In fact, in the most recent annual U.S. Conference of Mayors report (2011), 42 percent of cities reported an annual increase of almost 20 percent in the homeless population. The rate of homelessness began to increase between 1970 and 1980 due to a decrease in affordable housing and an increase in poverty (National Coalition for the Homeless, 2006), but the 2007 global financial crisis has exacerbated this trend of financial vulnerability, as indicated by most cities in the United States, which have reported a marked increase in requests for emergency shelter and food assistance from 2008 through 2011. In fact, for the first time in years, unemployment leads the

Learning Objectives

- Become familiar with the current demographics of the homeless, including the various ways in which the homeless population is counted
- Understand the root causes of homelessness, including the changing demographics of the homeless population and how this is impacting treatment interventions and practice setting structure and policies
- Become aware of the existence and nature of the stigma of homelessness, including what variables increase this stigma and negative stereotypes of the homeless population in general, and with particularly vulnerable populations
- Develop an understanding of the current challenges facing homeless single mothers, both in gaining self-sufficiency as well as in contending with policies and practices employed by most homeless shelters
- Become familiar with current legislation designed to address the homeless problem in the United States, and the strengths and deficits of the legislation and associated policies

list of causes of homelessness, ahead of a lack of housing and poverty (U.S. Conference of Mayors, 2011). The majority of U.S. city mayors expect the trend of homelessness to continue to increase, and yet do not expect funding for services to keep pace.

Most scholars and homeless advocates agree though that the problems of homelessness are multifaceted and are not caused by one or two factors (i.e., unaffordable housing, lack of employment). Homelessness is caused by many factors, including economic, social, and psychological dynamics, and yet one problem in addressing the problem of homelessness is that federal definitions of homelessness often focus on only one variable: housing (Gonyea, Mills-Dick & Bachman, 2010).

The Difficult Task of Defining Homelessness: The HEARTH Act

To confront the problem of homelessness, it must first be determined *who* is homeless. This is a challenge due to the difficulty in defining homelessness as well as the transient nature of the homeless population. There is currently no universally agreed-upon definition of homelessness, although the 1994 reauthorization of the Stewart B. McKinney Homeless Assistance Act of 1987 (now referred to as the McKinney-Vento Act) defines homelessness as

1. an individual or family who lacks a fixed, regular, and adequate nighttime residence;
2. an individual or family with a primary nighttime residence that is a public or private place not designed for or ordinarily used as a regular sleeping accommodation for human beings, including a car, park, abandoned building, bus or train station, airport, or camping ground;
3. an individual or family living in a supervised, publicly or privately operated shelter designated to provide temporary living arrangements (including hotels and motels paid for by federal, state, or local government programs for low-income individuals or by charitable organizations, congregate shelters, and transitional housing);
4. an individual residing in a shelter or place not meant for human habitation and who is exiting an institution where he temporarily resided. (42 U.S.C. § 11302, et seq., 1994)

When homelessness is defined using the federal definition, there were on average 636,017 individuals who experienced homelessness (sheltered and unsheltered) on any given night in 2007 in the United States (U.S. Department of Housing and Urban Development, 2011). Although this represents a slight decline from prior years, this definition is often criticized because its narrow parameters omit the majority of the homeless population who are difficult to count either because they are not living in traditional emergency shelters or because they do not want to be counted. The "hidden homeless" may include those living in motels, automobiles, and abandoned buildings; those who frequently double up with friends or relatives on a temporary basis; and those who for whatever reason do not want to be counted. When homelessness is defined in a more inclusive manner, estimates jump significantly, to between 2.5 and 3.5 million individuals nationally (National Alliance to End Homelessness, 2009).

> The "hidden homeless" may include those living in motels, automobiles, and abandoned buildings; those who frequently double up with friends or relatives on a temporary basis; and those who for whatever reason do not want to be counted.

Another problem in determining the scope of the homeless problem relates to the transient nature of the homeless population, which the McKinney-Vento Act definition did not address. Most individuals who have experienced homelessness have done so on an intermittent basis, where homelessness occurs in an ongoing cycle of temporary or tenuous housing, leading to eventual homelessness due to economic instability. Thus, as aggressive as any count might be, the number of homeless individuals will range dramatically on any given day.

In response to criticism as well as the quick pace at which the nature of homelessness is changing, the McKinney-Vento Act was significantly amended under the Obama administration in 2009. The reauthorization of the McKinney-Vento Act was signed into legislation as Homeless Emergency Assistance and Rapid Transition to Housing (HEARTH) Act. Among many significant changes to the act, the definition of homeless underwent a much-needed expansion to include those previously considered homeless as well as:

5. an individual or family who—
   A. will imminently lose their housing, including housing they own, rent, or live in without paying rent, are sharing with others, and rooms in hotels or motels not paid for by Federal, State, or local government programs for low-income individuals or by charitable organizations, as evidenced by—
      i. a court order resulting from an eviction action that notifies the individual or family that they must leave within 14 days;
      ii. the individual or family having a primary nighttime residence that is a room in a hotel or motel and where they lack the resources necessary to reside there for more than 14 days; or
      iii. credible evidence indicating that the owner or renter of the housing will not allow the individual or family to stay for more than 14 days, and any oral statement from an individual or family seeking homeless assistance that is found to be credible shall be considered credible evidence for purposes of this clause;
   B. has no subsequent residence identified; and
   C. lacks the resources or support networks needed to obtain other permanent housing; and
6. unaccompanied youth and homeless families with children and youth defined as homeless under other Federal statutes who—
   A. have experienced a long-term period without living independently in permanent housing,
   B. have experienced persistent instability as measured by frequent moves over such period, and
   C. can be expected to continue in such status for an extended period of time because of chronic disabilities, chronic physical health or mental health conditions, substance addiction, histories of domestic violence or childhood abuse.
      (42 U.S.C. § 11302, et seq., 2009)

The HEARTH Act goes a long way in addressing deficits in the previous act by now including many of the hidden homeless previously excluded under the former legislation, such as those who have a long history of housing instability living in motels and

> ## Professional History
>
> Understanding and Mastery of Professional
> History: Historical and current legislation
> affecting services delivery
>
> ---
>
> Critical Thinking Question: The McKinney-
> Vento Act and the HEARTH Act made
> great strides toward defining who is
> "homeless" and gaining an accurate count
> of how many homeless persons there are
> in the United States. Why are a definition
> and a count important steps toward ad-
> dressing the issue of homelessness? What
> else is needed to facilitate the next steps
> toward eradicating homelessness?

on others' couches, and specifically addresses the growing issue of homelessness among families with children and unaccompanied youth. It also increases the focus on prevention of homelessness, by identifying risk factors that often lead to housing insecurity such as domestic violence and child abuse. For instance, a report summarizing several studies across the United States found that about half of all homeless single mothers surveyed (across the country) had experienced child abuse, including child sexual abuse, during their childhoods, and almost all had experienced domestic violence at some point in their lives (National Law Center, 2006).

The U.S. Homeless Population: Gauging the Extent of the Problem

Because of the methodological challenges involved in attempting to accurately count the homeless population, most recent demographic studies now use homeless estimates based on indirect counts obtained by surveying professionals working with the homeless population. This reporting method is wrought with problems, though, including underreporting since this method, as other reporting methods, omits those who are not seeking housing or shelter assistance. One of the most significant concerns with the underreporting of the homeless population due to these methodological challenges relates to the fact that government grant money is often directly linked to census numbers; thus, underreporting leads to less money, which in turn leads to fewer services. For this reason as well as others, in 2004 Congress directed the Department of Housing and Urban Development (HUD) to collect comprehensive data on the homeless population in the United States. This mandate represents the first federal attempt to conduct a direct count of the homeless population. HUD responded to this mandate by developing the Housing Management Information System (HMIS), which provides a computerized method for collecting data on the use of shelter and transitional housing programs within each state. The most recent report published in 2010 (reflecting 2009 data) included a "point-in-time" count of homeless persons and the number and demographic characteristics of homeless individuals and homeless families (U.S. Department of Housing and Urban Development, 2010).

The report revealed that on a single night in 2009, 643,037 people were homeless nationwide (living both in shelters and on the streets). This number reflects relative stability in the number of homeless in recent years, with a decrease in the people living on the streets (about 37 percent of all homeless people) compared to those living in shelters (about 60 percent). The authors note that this increase may be due more to better head counts of people living on the streets versus an increase in the unsheltered homeless population. The report also revealed that approximately two-thirds of the homeless population consisted of individuals, and about one-third consisted of families, which is consistent with prior years. About 50 percent of solitary individuals were sheltered and about 47 percent were unsheltered, while most families were living in some type of shelter (78 percent) on any given night. One disconcerting trend is the increase in homeless families, which has risen in the past two years.

The typical homeless person in 2009 was a middle-aged man who was a member of an ethnic minority group, who was alone. Over two-thirds of the homeless population has some type of disability. There has also been a consistent increase in the overall age of the homeless population, which the authors attribute to the aging cohort of homeless individuals who became vulnerable to homelessness when they were younger.

At any given night in 2009 there were approximately 124,135 chronically homeless individuals living in shelter. Of these, about 13 percent were veterans (down from 15.5 percent in 2006), 4 percent were persons living with HIV/AIDS, 1.5 percent were unaccompanied youth (down from 4.8 percent in 2006), 12 percent were victims of domestic violence (down slightly from 13 percent in 2007), 25 percent were suffering from severe mental illness (down from 27.6 percent in 2007), and 34 percent had a chronic substance abuse problem (down from 39 percent in 2007). The average length of stay in emergency shelter was anywhere from a week to a month, with most staying about two weeks. The average length of stay in a transitional housing program was just under 100 days. Of homeless families, the majority are younger single mothers with two children. Most tend to enter transitional housing programs, are members of minority groups (African American, Latin American, and Native American), and became homeless after leaving someone else's home.

It must be noted that these statistics do not capture the full picture of what most experts are predicting will be a significant increase in financial and subsequent housing insecurity brought on by the 2007 global financial crisis. With a dramatic increase in housing foreclosures, the tightening of the credit market, and mass layoffs, the picture of homelessness in the United States will most assuredly change for the worse. Although it is still too soon to capture specific statistics reflecting these changes, a 2009 study conducted by the Urban Institute exploring the impact of the foreclosure crisis on U.S. families showed that many families who experienced foreclosure found it difficult to rent due to damaged credit ratings and instead were forced to live with family members or friends. This trend is troublesome because the typical path toward homelessness often involves a pattern of moving from self-sufficiency, to living in someone else's housing unit, to ultimately moving into emergency shelter. States with the highest foreclosure rates, such as Nevada, Arizona, California, and Florida, also experienced the highest jump in request for social services such as food assistance. Older adults appear to be particularly vulnerable due to the risk of physical health problems that create barriers to finding employment in a very tight job market (Kingsley, Smith, & Price, 2009).

The Causes of Homelessness

Determining the root causes of homelessness is not only as challenging as determining who is homeless, but also essential, particularly for human service professionals who are committed to advocating for and assisting those who experience poverty and homelessness. Equally important is the task of identifying common biases against and negative stereotypes of the poor and homeless population that dramatically influence the general perception of the poor and homeless, which in turn influences the types of assistance programs that will be supported by state and federal policy makers as well as the voting public.

In general, most people's attitudes toward the poor and the homeless are negative, and the stigma that has always been associated with poverty seems to increase when the poor become homeless. The reasons for this negative bias are likely related to the public nature of homelessness, where those without permanent homes are forced to live out in the open, such as on the streets or alleyways, or in parks or automobiles, where good hygiene is virtually impossible and begging for money and food is often the only means of survival (Phelan, Link, Moore, & Stueve, 1997).

The common association of mental illness and substance abuse with poverty and homelessness also contributes to the negative stigma associated with being homeless, and many experts suspect that the general public assumes that virtually all homeless individuals abuse drugs and alcohol, do not shower, live on the streets, and aggressively beg for money (to buy drugs and alcohol), adding to a sense of perceived dangerousness of the homeless population, particularly those believed to be mentally ill.

This increased negative attitude toward those who are poor and homeless is reflected in several studies and national public opinion surveys. Generally, it appears as though most people blame the poor for their bad lot in life. For instance, one older national survey conducted in 1975 found that the majority of those in the United States attributed poverty and homelessness to personal failures, such as having a poor work ethic, poor money management skills, a lack of any special talent that might translate into a positive contribution to society, and low personal moral values. Those questioned ranked social causes, such as poverty, racism, poor schools, and the lack of sufficient employment, the lowest of all possible causes (Feagin, 1975).

More recent surveys conducted in the mid-1990s reveal an increase in the tendency to blame the poor for their poverty (Weaver, Shapiro, & Jacobs, 1995), even though a considerable body of research points to social and structural issues as the primary cause of poverty, such as shortages in affordable housing, recent shifts to a technologically based society requiring a significant increase in educational requirements, long-standing institutionalized oppression and discrimination against certain racial and ethnic groups, and a general increase in the complexity of life (Wright, 2000). A 2007 study comparing attitudes toward homelessness among respondents in seven countries—the United States, the United Kingdom, Belgium, Germany, and Italy—found that the respondents in the United States and the United Kingdom, the only two English-speaking countries, reported higher rates of lifetime homelessness and fewer social programs, yet had lower levels of compassion for the homeless population (Toro et al., 2007).

In general, though, compassion for poverty-related homelessness tends to be greater during difficult economic times and lower during economic booms, and general compassion for homeless individuals such as families, who are unlike the stereotypical skid row alcoholic, tends to be greater as well. Recent studies reflecting attitudes about poverty during the most recent economic crisis, increasingly referred to as the *Great Recession* or the *Global Crisis* reveal this sense of increased compassion, but they also show an increase in class conflict, where lower income individuals express resentment toward the wealthy. Nowhere was this more apparent than in the *Occupy Wall Street* movement, a series of staged demonstrations across the United States, which began on September 17, 2011, in Manhattan's Financial District, and spread to over 100 cities nationwide. The Occupy Wall Street movement website (www.occupywallst.org) states that it

> is a people-powered movement that . . . is fighting back against the corrosive power of major banks and multinational corporations over the democratic process, and the role of Wall Street in creating an economic collapse that has caused the greatest recession in generations. The movement is inspired by popular uprisings in Egypt and Tunisia, and aims to fight back against the richest 1% of people that are writing the rules of an unfair global economy that is foreclosing on our future. (Occupy Wall Street, 2011)

This popular social movement reflects the findings of a 2012 study conducted by the Pew Center, which found that negative perceptions of each class—the poor of the rich and the rich of the poor—commonly referred to as "class conflict" has significantly increased in recent years. In fact, an interesting shift in attitude is that more white people than ever before are noticing this conflict, whereas the majority of African Americans and Latinos have always perceived a conflict between the rich and the poor. Attitudes toward the wealthy, something not explored in earlier attitudinal surveys about income levels, reveal that almost half of all respondents (about 46 percent) believe that "most rich people 'are wealthy mainly because they know the right people or were born into wealthy families'" whereas 43 percent say wealthy people became rich "mainly because of their own hard work, ambition or education" (Taylor, Parker, Morin, & Motel, 2012, p. 3).

Based on these studies it appears as though their remains at the least confusion about the causes of poverty and wealth, whether poverty and homelessness are caused by behavioral factors or social conditions, and whether wealth is a result of privilege and inheritance or hard work. Despite intermittent increases in compassion toward the poor and homeless, the general public does not appear to understand the underlying causes of poverty and homelessness, which may make it easier to jump to incorrect conclusions based upon negative stigmas. Although perceptions of individual homeless individuals are not as negative as perceptions of specific subgroups within the homeless population (e.g., single men, certain racial groups, alcoholics, undocumented immigrants), possible reasons for the overall negative perception of the homeless population may relate to the *fundamental attribution error,* where people tend to attribute their own personal failures or the failures of people they know well and like to situational factors, but attribute the failures of those they do not know or do not like to personal or dispositional factors. Thus, according to the fundamental attribution error, the average person would assume that those whom they did not know were homeless due to their own personal shortcomings. Yet, if someone they knew became homeless, they would attribute the homelessness to situational causes, such as being laid off or abruptly leaving an abusive marriage.

Human service professionals must understand the stigma associated with homelessness because unless these negative attitudes are acknowledged and challenged, human service professionals may even embrace them, significantly influencing their perceptions of their clients suffering from poverty and poverty-related homelessness. Understanding homelessness from a historical perspective is also useful in understanding the nature of this long-standing social problem so that situational forces can be acknowledged.

History of Homelessness in the United States

The types of people who have experienced homelessness and the reasons for their misfortune have changed significantly throughout the years. Prior to the Middle Ages (from about the 14th to the 17th century), the early church was responsible for the care of the poor, including those without homes. The monasteries embraced this responsibility as one given by God. Thus, at least the "deserving poor" (those who were poor through no fault of their own) were considered blessed, and it was considered a blessing to care for them.

Throughout the Middle Ages, the homeless population consisted primarily of the wandering poor—those individuals, most commonly men, who migrated for employment, either working someone's land or selling goods. The English poor laws (discussed in Chapter 2), which were adopted by many of the American colonies, included harsh measures for dealing with the poor and destitute, adding to the overall negative social stigma associated with poverty. For example, most communities enforced strict residency requirements designed to discourage the wandering poor from settling in more affluent districts to collect social welfare intended to serve longtime residents who had contributed to the community before falling on hard times. This policy, as well as others against vagrancy and even unemployment, is reflective of the overall negative sentiment held of the homeless population in general, particularly when it could be assumed that one was homeless either through choice or some personal failing.

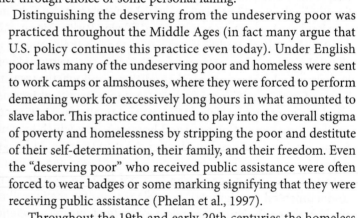

Distinguishing the deserving from the undeserving poor was practiced throughout the Middle Ages (in fact many argue that U.S. policy continues this practice even today). Under English poor laws many of the undeserving poor and homeless were sent to work camps or almshouses, where they were forced to perform demeaning work for excessively long hours in what amounted to slave labor. This practice continued to play into the overall stigma of poverty and homelessness by stripping the poor and destitute of their self-determination, their family, and their freedom. Even the "deserving poor" who received public assistance were often forced to wear badges or some marking signifying that they were receiving public assistance (Phelan et al., 1997).

Throughout the 19th and early 20th centuries the homeless population still consisted of primarily men, either vagrants (men who were unemployed for a variety of reasons, including mental illness or alcoholism) or migrant workers, such as men who were making their way out West to work in the gold mines, the railroads, or the fields. Hobos, for instance, were often counted among the homeless population. Hobos were men of European descent, typically Germany or Scandinavian countries, who were migrating laborers and were often treated with mistrust and contempt despite the fact that they were an integral part of the labor force throughout the 19th century (Axelson & Dail, 1988).

It was not until the Great Depression in the mid-1930s that families began to appear on the homeless scene in significant

The Great Depression resulted in extremely high unemployment and homelessness
Getty Images

numbers. The failure of the financial markets, the closings of many banks, and rampant unemployment resulted in many families losing their homes and wandering the streets in search of sustenance and shelter. Because the Great Depression hit just about everyone in the United States, there was increased compassion for the homeless population and for those suffering from poverty in general. The Great Depression brought most people back to a pre-Protestant ethic time, where people recognized and acknowledged that poverty and homelessness could be caused by circumstances beyond one's control. Thus, although the Protestant ethic and Social Darwinism might have had many people believing that falling on hard times was a result of laziness, the Great Depression reminded everyone that sometimes, no matter how hard one works or is willing to work, circumstances occur that render someone destitute and impoverished. Unfortunately, this spirit of empathy and compassion for society's poor and homeless did not last much past the next economic boom. Apparently, a by-product of personal good fortune may just be a reduction in one's ability to empathize with those less fortunate.

The Contemporary Picture of Homelessness: The Rise of Single-Parent Families

After the Depression, the homeless landscape returned to its former demographic picture, with the majority of the homeless population consisting primarily of single, white men. Yet another significant change was on the horizon. The 1970s and 1980s saw a dramatic increase in the homelessness of families. Yet the difference between the homeless families of the Great Depression era and homeless families of late is that the latter consists primarily of single parents with children. An increase in homelessness families has occurred in the past two consecutive years of available data (2008 and 2009). The typical family consists of a single mother and two children, about 80 percent of whom reside in a shelter of some sort, and 20 percent of whom are unsheltered (U.S. Conference of Mayors, 2010).

Most single-parent families become homeless as a result of a complex set of circumstances.

There is a tendency among policy makers to oversimplify the causes of homelessness, perhaps because a simple cause would warrant a simple solution, and multifactor causes often call for overwhelmingly complex responses. But most single-parent families become homeless as a result of a complex set of circumstances, as illustrated in Case Study 9.1 (p. 211). There are some common themes among single-parent homeless families. The great majority of homeless single mothers are approximately 25 years of age, with two to three children in the preschool to 6 years of age range. The majority of these single mothers are U.S. citizens, native born, and fluent English speakers. Even states that border Mexico have a relatively low

Children are at increasing risk of becoming homeless
Bushnell/Soifer/Getty Images

percentage of homeless immigrants. Families of color are at greatest risk of becoming homeless, although single-parent homelessness among Caucasians is significant as well.

Most homeless single mothers have never been married, and although many are high school graduates, a significant majority of single mothers never established a solid work history for many reasons. Most cite either never having had stability in their housing situations or having experienced unstable housing for several years prior to becoming homeless. Most have experienced homelessness chronically on a cyclical basis, securing housing for a short time only to experience a financial crisis, such as a job loss, which results in a domino effect of negative life events and ultimately another incident of homelessness. Many homeless single mothers are either underemployed or unemployed, with the majority citing the inability to pay for child care as the primary barrier to finding employment, but others cited being undereducated and an inability to secure employment that would pay for market rent. The Great Recession of 2007–2009 has resulted in an increase in single-mother unemployment. For instance, the percentage of single mothers employed in an average month between 2000 and 2009 decreased by about 8 percent, from 76 percent in 2000 to only 68 percent in 2009 (U.S. Department of Labor, 2010). This trend has led to increases in food and housing insecurity among single-mother households as well (Legal Momentum, 2010).

Unfortunately, the safety net in the form of public assistance programs has significantly shrunk since the passage of federal welfare reform legislation in 1996. Welfare reform effectively ended the Aid to Families with Dependent Children (AFDC) program and initiated the Temporary Assistance for Needy Families (TANF), a program that provides assistance at about one-third of the federal poverty level (Nickelson, 2004).

Historically, only about 20 percent of homeless individuals have been on any form of public assistance, even though the majority of these individuals would have qualified for some form of assistance (Shlay & Rossi, 1992). Although single mothers qualify for more aid than single homeless men and single homeless women, as a group they still tend to underutilize public assistance programs. In fact, according to a report evaluating how single mothers and their children have fared between 2000 and 2009, unemployment has increased and welfare utilization has decreased, which has resulted in an increase in extreme poverty among single mothers. What may surprise many is that despite the increases in single-mother unemployment and overall poverty levels in single-mother families between 2007 through 2009, the percentage of these families receiving welfare benefits decreased from 16 percent in 2001 to only 10 percent in 2010 (Women's Legal Defense and Education Fund, 2010). To add to the dilemma of shrinking public assistance programs, aid that is provided in the form of block grants or aid packages may have long waiting lists for certain types of assistance, such as child care.

Another contributing factor to homelessness of single-mother families may relate to the bad childhoods many homeless women experienced. Many single mothers report unstable childhoods filled with physical and sexual abuse. In fact, it appears as though many of these women proceeded to recreate these patterns of abuse in their adult lives because approximately three-quarters of all homeless single mothers who were married prior to becoming homeless cited domestic violence as the primary reason for leaving their marital home and moving into a shelter with their children. Other reasons include having a child early, either during adolescence or early adulthood, which interrupts the development of educational and career goals (Nunez & Fox, 1999). Other personal vulnerabilities include having a substance abuse disorder or mental illness (although some research suggests that the trauma of homelessness is actually a risk factor for substance abuse and mental disorders, such as depression, and not typically the cause of homelessness in single-mother families), having grown up in the state foster care system, and having a poor or absent social support system.

Some structural causes of the recent dramatic increase in homeless single-parent families include the failure of many courts to enforce child support orders, dramatic cutbacks in federal housing programs in the 1980s, and the failure of public welfare to keep pace with inflation and increases in the cost of living. Further increases in homeless families are expected particularly now that welfare benefits are limited to only two to five years (depending on the state), rather than providing long-term benefits on a case-by-case basis.

It is important to discuss the strengths that many of these single mothers exhibit as well, particularly because human service professionals will need to identify such strengths and work with the single-parent client to enhance and build on existing strengths. A 1994 study found that single mothers living in shelters had an amazing amount of determination, a sense of personal pride, and an ability to confront their problems directly. Many of the single mothers interviewed exhibited a strong commitment to the welfare of their children (particularly those who chose homelessness over remaining in an abusive relationship), had strong moral values that acted as a guide in decision making, and had deep religious convictions that provided them with a sense of purpose and meaning. Despite being homeless, many of these single mothers maintained a commitment to helping others in need (Montgomery, 1994). Many also overcame what seemed to be insurmountable odds to keep their children with them rather than have them placed within the state foster care system, despite harsh living conditions. A more recent study evaluating the resiliency of single mothers in general (not homeless) found that respondents were quite resilient, despite all of the challenges they faced. Most stated that they disagreed with the negative stereotypes of single mothers as inadequate, and believed that they had personally grown through the challenges they faced in raising their children alone, many of whom had disabilities. Many found the experience of single parenting transformative and confidence-building (Levine, 2009). Human service professionals can tap into these strengths when assisting single parents' access resources in order to gain self-sufficiency in the face of multiple challenges.

Homeless Shelter Living for Families with Children

The increase in single-mother family homelessness has resulted in the need for significant changes in social welfare policy regarding how homelessness is managed on local,

Human Services Delivery Systems

Understanding and Mastery of Human Services Delivery Systems: Economic and social class systems including systemic causes of poverty

Critical Thinking Question: A substantial number of the homeless are single-mother families with young children. What systemic factors contribute to poverty and homelessness among this population? What changes in policy and/or practice might help to alleviate this?

state, and federal levels. When the homeless population was more homogeneous, consisting primarily of single men living in single-room occupancy (SRO) or on skid row, the community response was less complex focusing on low-cost housing and substance abuse counseling. But this new homeless population presents more complex problems requiring a more multifaceted approach. For instance, the traditional homeless person typically resided on the streets, whereas families often avoid street dwelling opting for shelter living instead. Yet, many emergency shelters are not equipped to serve families.

Elizabeth Lindsey (1998) interviewed single mothers who had lived in homeless shelters with their children and asked them about their experiences in shelters and the impact it had on their family life. Many shelters would not allow boys as young as eight years to sleep in the same area as their mothers, requiring them to stay on the men's side of the shelter alone, stay with relatives, or in some cases, even enter the foster care system.

Other shelters applied the same rules to families as they did to singles, forcing single mothers to leave the shelter at 7:00 A.M., even if they had infants or preschool-aged children, and not allowing them to return to the shelter until 5:00 or 6:00 P.M., regardless of weather conditions or the safety of the community where the shelter was located. Many single mothers complained that there was no way to look for a job when they had to stay out of the shelter with their kids for hours a day. Other complaints included staff who seemed insensitive to children's needs, such as enforcing rules against children running around and playing, which created difficult situations for parents who were mandated to keep their children quiet at all times, with no distractions, such as television or toys, to assist them.

But by far the most difficult aspect of shelter life according to these women involved staff who would override their parenting decisions, such as correcting a parent in front of the child and other shelter residents for how she was disciplining her child. Mothers complained that their authority was often diminished by shelter rules and interfering shelter staff. Other shelter rules that make parenting difficult include rules prohibiting anyone from eating in the shelter at any time other than during designated meal times, including prohibiting mothers from even bringing snacks into the shelter for their young children.

Research studies have shown that such shelter rules and policies, not necessarily created with families in mind, have a powerfully devastating effect on the parent–child relationship, as mothers find themselves no longer the "head of household" with the power to make parenting decisions in the best interest of their children—even basic decisions such as when to bathe and feed their children. Instead, their children are cared for on the shelter's time frame. These issues might seem like minor inconveniences and relatively innocuous in light of the other major crises going on in the lives of homeless mothers, but researchers noted that the disintegration of the mother–child relationship is not just temporarily disruptive, but this disruption essentially further degrades and disempowers parents who are already feeling shamed and powerless by their homeless status leading to an increase in parental distress and depression, which in turn often leads to an increase in child misbehavior and acting out (Lindsey, 1998).

More recent research on single mothers living in homeless shelters with their children showed similar dynamics. In a 2009 study of single mothers' attitudes about the effect of living in a shelter on child-rearing, the mothers cited several challenges to their parenting such as the loss of privacy and the lack of financial resources. They also cited several strengths, such as their ability to persevere through these various challenges, and their faith and optimism. They also recommended that those working with single-mother homeless families maintain a humble attitude and avoid acting like an expert, and avoid negative stereotypes of single mothers (Swick & Williams, 2010).

Homeless Children: School Attendance and Academic Performance

Children are the fastest-growing segment of the homeless population, which creates new challenges for shelters and other social welfare responses, particularly when these children are school-aged. Developing effective programs designed to keep homeless children in school and succeeding academically is essential, otherwise all these homeless children will be at risk for continuing the cycle of homelessness in the next generation, having never experienced physical or emotional security in their own childhoods.

Between the chronic and cyclical nature of homelessness and the fact that most emergency shelters limit the amount of time residents can stay, ranging anywhere from 1 to 30 days, a significant problem for school-aged children was switching schools every time their families were forced to move to a new shelter. I recall when I was working as a school human service professional in the inner city of Los Angeles having several school-aged children who were homeless on my caseload. No sooner did these children get settled and acclimated to their classroom and start the long process of building a trusting relationship with me than they would literally disappear one day. I would typically learn at some point later that the family was forced to move to a different shelter, and even if remaining at their school of origin was a legal possibility, it was not a realistic one because there was no guarantee that the next shelter would be anywhere close to the children's current school.

A 2000 report to Congress stated that only 87 percent of homeless children were enrolled in school, and of these only 77 percent attended school regularly (U.S. Department of Education, 2001). Many school districts attempted to resolve this issue by creating special schools or programs for homeless children, but these programs have been criticized by many because it segregated homeless children, increasing their social stigma and sense of rejection they no doubt already experienced. Federal legislation, discussed later in this chapter, was designed to address this issue and put a stop to poor attendance and student retention and poor academic performance related to homelessness.

Runaway Youth

No one is certain just how many adolescents are homeless and living on the streets, but some estimates put that number as high as 2 million in the United States alone. This is a unique population among the entire homeless population because the reasons, risk factors, and intervention needs are considerably different. Adolescents are far more likely to be living on the streets than in a shelter. They are also far more likely to participate in dangerous behaviors such as drug abuse (including needle sharing), panhandling, theft, and survival sex (sex for food, money, and shelter). These risky behaviors put homeless adolescents at risk for HIV, hepatitis B, hepatitis C, and sexually transmitted diseases (Beech, Meyers, & Beech, 2002). These teens are also at high risk for physical and sexual violence, both by other teens and by adults.

Most homeless adolescents are living on the streets because they have run away from an abusive home, have been kicked out of their homes by parents who no longer wish to take care of them (throw-away youth), or have aged out of the foster care system. The majority of homeless adolescents interviewed in various research studies reported a history of both physical and sexual abuse, which served as a primer for being similarly victimized on the streets (Whitbeck, Hoyt, & Ackley, 1997). Another study of over 600 runaway youth found that sexual abuse was the chief reason adolescents chose to live on the streets rather than remaining in their homes (Yoder, Whitbeck, & Hoyt, 2001). The fact that many of these teens will continue to experience sexual exploitation while living on the streets, whether through outright attacks or through survival sex, is certainly a tragedy, and one that can be addressed by those in the human services field.

One study that involved the surveying of homeless youth found that in most urban cities homeless adolescents often operate as a somewhat cohesive group on the streets, protecting each other and helping one another survive. In fact, this study found that the more seasoned adolescents would often take new homeless teens under their wings, teaching them survival tactics and welcoming them into the "fold." Newer homeless youth who were interviewed talked about what a relief it was to have someone essentially mentor them into the ways of surviving street life. But without glamorizing this life, most teens, both boys and girls, talked of the horrors of having to participate in prostitution to survive. In fact, the adolescents who were interviewed talked about many ways in which they felt exploited, both by older teens and by adults who forced them into drug dealing and prostitution (Auerswalk & Eyre, 2002).

Ironically, many teens reported having a strong belief in God, who they believed watched out for them and kept them alive. One teen stated that when they were not really in need, they would often get no offer of food and little money while panhandling. Yet when they were really in need, having gone without food for a few days, then whatever they needed would just come to them. He attributed this to God knowing what they needed and providing for them when they needed it the most (Auerswalk & Eyre, 2002). In fact, in one study researchers found that over half of all homeless youth interviewed cited faith in God as the primary motivation for survival (Lindsey, Kurtz, Jarvis, Williams, & Nackerud, 2000).

Yet even with this surprisingly high percentage of faith-seeking homeless teens, an estimated 40 percent of homeless youth attempt suicide (Auerswalk & Eyre, 2002). They are also at high risk for post-traumatic stress disorder (PTSD), anxiety disorders, depression, substance abuse, and delinquency (Thrane, Chen, Johnson, & Whitbeck, 2008). Many runway and homeless youth report losing all contact with people in their former lives, even siblings, extended family, and those who had been supportive of them in the past. Most also talked of feeling extremely lonely and distrustful but in desperate need of love and affection. Because the majority of homeless adolescents have run away from abusive homes, it seems likely that many were suffering from some form of emotional disturbance even prior to entering street life (Kidd, 2003).

Unfortunately, many of the adolescents interviewed were highly suspicious of all adults, including outreach workers with human services agencies providing assistance to the homeless adolescent population. The overall perspective of these outreach agencies were negative, and adolescents who accepted assistance from these agencies were considered "sellouts" and foolish. The prevailing belief was that human services organizations would force the teens to return to an abusive home environment, or they'd be turned over to the police. Knowing these attitudes, though, can aid human services agencies in developing outreach efforts and other services designed to overcome these negative perceptions.

Any successful intervention program is going to have to address the issue of the teens feeling like outsiders. In fact, research studies have found that homeless adolescents are acutely aware of their outsider status, and many of them manage this through incorporating this outsider status into their identity. By embracing being an outsider, through multiple piercings, for example, they take control of something that could potentially make them vulnerable (Auerswalk & Eyre, 2002).

Many human service experts strongly recommend that any intervention program be targeted at identifying the adolescents' strengths. But this is challenging when most intervention systems view homeless youth in a deviant manner; first, because they are "runaways," and second, because many of the behaviors they engage in while living on the streets are criminal. Even the classification of their behavior is in pathological terms, such as diagnosing them with conduct disorder. This can be humiliating and shaming to an adolescent who is likely acting out in response to being victimized in her family of origin. Most homeless youth have been both physically and verbally abused and degraded in their homes; thus, in many respects they are living up to their parents' negative expectations of them by dropping out of high school and living on the streets. To then enter into the juvenile justice system that continues to pathologize their behavior and responds in punitive measures rather than supportive ones only adds to their feelings of victimization.

Human service professionals working with this population must provide consistent encouragement, compassionate care, and understanding that promote both self-esteem and self-efficacy (a sense of competence) in these emotionally broken and bruised teens. This can be accomplished while focusing on basic needs such as providing food, shelter, and good healthcare. Yet again the barrier that human services agencies must overcome is significant because so many homeless youth have been so horribly rejected and abandoned by their families, and then further exploited and abused by adults on the streets, that to trust any adult seems foolish and risky.

Developing one-on-one relationships where trust can grow slowly is one method of intervention that may be more successful than more traditional outreach efforts, but the ratio of outreach workers to homeless youth renders this approach challenging. Regardless, any intervention must allow the teen to feel safe and empowered in seeking services.

Single Men, the Mentally Ill, and Substance Abuse

Although single-parent families now comprise a large proportion of the homeless population, just less than 50 percent of the homeless population consists of men, many of whom are single, some of whom are mentally ill, some of whom have substance abuse issues, and most of whom are veterans. Of course these are overlapping categories in many instances. Reasons for homelessness often vary, and some are similar to the causes noted in single-parent families—childhood histories of abuse, growing up in the foster care system, having little or no family or social support, being undereducated and stuck in minimum wage jobs, substance abuse, and mental illness. Social causes include institutionalized racism and oppression, suffering from PTSD after having served in the military during wartime, and changes in the economic infrastructure resulting in fewer well-paying jobs.

> Just less than 50 percent of the homeless population consists of men, many of whom are single, some of whom are mentally ill, some of whom have substance abuse issues, and most of whom are veterans.

Veterans' services address many of these issues in programs designed to meet the complex needs of the homeless population who were enrolled in the armed services. Human service professionals working for the Department of Veterans Affairs (VA) provide both in-house and outreach services and are trained on PTSD recovery and the unique needs of this special population.

Older Adult Homeless Population

Chapter 7 touched briefly on the issue of older adults and homelessness, but it will be explored again somewhat briefly in this chapter because although rates of homelessness among older adults are significantly lower than younger individuals, it is still an important issue worth exploring in some depth, particularly because the number of homeless older adults is expected to increase as the baby-boomer generation ages.

Differences exist between homelessness among younger and older persons, both in terms of the root causes of homelessness and effective responses. Younger homeless individuals report domestic violence and previous incarceration as reasons for becoming homeless far more frequently than older populations. Both groups report equal difficulty in finding affordable housing, and both groups report equivalent rates of alcohol and substance abuse as reasons for homelessness, with 4 percent of younger individuals reporting this as a reason and just over 6 percent of older adults reporting substance abuse as the primary reason for their homelessness. Yet in light of the nature of substance abuse and the tendency for alcoholics and drug addicts to minimize or deny the impact of their addiction, these percentages might be underreported.

Older adult homeless persons report being without shelter for far longer periods than younger individuals, with older adult men reporting an average homeless episode lasting over 60 days, and younger homeless men averaging about 14 days. Older men also reported far longer episodes without a permanent shelter, some reporting homeless episodes of over two years, whereas younger men reported being homeless an average of 11 months (Hecht & Coyle, 2001). This is likely due to fewer social supports and the difficulty in either moving in with a roommate or living with family, often due to caretaking issues related to common age-related physical problems.

Even though there are more similarities than differences between older and younger homeless persons, the response to older adults who are homeless must be vastly different due to all of the variables associated with their advanced age. One variable mentioned in the previous paragraph relates to the diminished capacity of older people in getting back on their feet by finding new employment opportunities or entering a reeducation program to enter a new career; thus, the possibility of regaining financial independence is greatly diminished in the older adult population.

Other issues affecting older adults include their increased vulnerability—both physically and psychologically, leaving them open to physical and financial victimization. Physical disability and illness are also complicating factors in meeting the needs of the older adult homeless population.

Although there is increased funding for services for older adults, most economic support is not available until the age of 65. Self-sufficiency models designed for the general homeless population do not work with the older adult population for the reasons mentioned earlier; thus, some experts suggest responding to the older adult homeless population by developing aid-assisted low-cost housing with social services to assist with financial, physical, and psychological support to deal with the trauma of becoming homeless. As referenced earlier, homeless advocates and policy experts have expressed concern that the recent financial crisis involving the crash of the stock market, loss of retirement funds, mass layoffs, and dramatic increases in foreclosures will have a significantly negative impact on the older adult population due to decreased possibilities to rebound financially.

Human Systems

Understanding and Mastery of Human Systems: Emphasis on context and the role of diversity in determining and meeting human needs

Critical Thinking Question: Specific groups of homeless people—young families, veterans, older adults, and individuals with mental illness or substance abuse issues—have a wide variety of underlying needs. In what ways might a human service professional address the needs of clients from each of these groups?

Current Policies and Legislation

Governmental policies designed to meet the needs of the homeless population are often targeted to subgroups, such as single-parent families and veterans, or toward particular issues that make one more vulnerable to homelessness, such as substance abuse and mental illness. But some legislation has been passed intended to address the homeless problem directly. The McKinney-Vento Homeless Assistance Act of 1987 is probably one of the most important pieces of legislation passed for those suffering from homelessness or who are at risk for homelessness, and prior to the passage of this act, the majority of homeless services were facilitated at the grassroots level. This act guarantees government assistance for the homeless and homeless services and increases in federal funding from passage to the mid-1990s has been significant, from its original appropriation of $180 million in 1987 to $1.8 billion in 1994.

As mentioned earlier in this chapter, the McKinney-Vento Homeless Assistance Act was reauthorized for the first time in two decades in May of 2009 by President Obama as the HEARTH Act (a part of the Helping Families Save Their Homes Act), which provides a variety of remedies focusing on both prevention and response. The federal budget for fiscal year 2009 allotted $2.62 billion of funding for 10 different programs spread across several federal agencies, including the Department of Housing and Urban Development (HUD), the Department of Health and Human Services (HHS), the VA, and the Department of Education (ED), just to name a few. Initially, the increase in homelessness funding did not result in a decrease in homelessness, including a $1.5 billion grant to be spent on homeless prevention. Unfortunately, in response to the Great Recession spending cuts were made essentially across the board in the 2011 federal budget, affecting virtually all social welfare programs, but particularly those focusing on housing security. Cuts to housing programs, such as housing for the elderly, and for people with disabilities averaged about 70 percent in 2011 from prior years. A very creative and interactive tool on the White House website shows the projected 2012 federal budget, and indicates that that housing assistance programs constitute only 1.59 percent of the federal budget and income and housing support programs constitute only 1.44 percent of the federal budget (see http://www.whitehouse.gov/omb/budget).

The McKinney-Vento Education for the Homeless Children and Youth Program is designed to address many of the problems experienced by homeless students. Through this act, states can apply for funding to assist them in managing the many academic challenges associated with a student being homeless. Problems related to enrollment, attendance, and academic achievement are all addressed in this act, and states applying for funds must abide by certain standards and meet various criteria in meeting the complex needs of homeless students. For instance, according to the McKinney-Vento program, schools:

- must provide the same educational opportunities to homeless children and youth that are available to nonhomeless children and youth.
- must not segregate homeless children from the mainstream school environment for reasons based solely on their state of homelessness.
- cannot educate children off-site, such as at a shelter, but must educate them alongside their peers, in a regular classroom setting.
- must make school placement decisions based on the best interest of child, not on the physical location of the shelter. Thus, whenever possible, the child must be allowed to remain in the school of origin, and the school district must make arrangements for the child to be transported to school, if transportation is an issue.
- must designate a liaison to identify homeless students and assist them and their families in addressing barriers to enrollment, attendance, and academic achievement.
- must immediately enroll homeless students, even if they do not have immunization records, birth certificates, or proof of residency. The liaison must then work with the family and the former school in obtaining these records in a timely fashion. Schools are also required to transfer school records immediately when a homeless student transfers to a different school.
- must provide transportation for homeless students so that they may remain in their school of origin.
- must allow "unaccompanied youth" (students who for a variety of reasons, including emancipation, do not have a legal guardian) to enroll in school even if they do not have a parent or legal guardian to sign admittance forms for them.
- must make a determination of homelessness on a case-by-case basis according to the McKinney-Vento definition of "fixed, regular, and adequate nighttime residence."

This legislation goes a long way in addressing the many challenges facing homeless families with school-aged children, yet much more must be done. School human service professionals, for instance, can be utilized to assist in the identification of homeless youth because a great number of families are too overwhelmed and embarrassed to come forward and report their homeless status. In addition, many homeless parents are simply unaware of their children's educational rights, and even though the McKinney-Vento act requires that school liaisons inform students and their families of these rights, school human service professionals are often the link between the families, students, liaison, and school administration and can therefore be extremely instrumental in ensuring that these kids remain in school, without disruption, despite the immense level of instability homelessness causes.

The Role of the Human Service Professional: Working with the Homeless Population: Common Clinical Issues

Working with the homeless population is as challenging as it is meaningful. Whether a homeless client is a grown man, an older adult, a child, or an entire family, being homeless is traumatic, degrading, and for many actually terrifying as one's foundation slips away without any sort of safety net to stop the fall. For many people homelessness is not an isolated incident, but is a way of life, and even when employed and residing in a permanent home, for many people homelessness is only one unexpected financial crisis away.

Many believe that there is a reciprocal relationship between many mental and emotional disorders and homelessness. The process of becoming homeless, which typically comes on the heels of months or years of financial and residential instability, is extremely stressful and often leads to anxiety disorders, depression, loss of self-esteem, substance abuse, and even personality disorders as individuals respond to the harshness of life in various maladaptive and defensive ways.

Research indicates that children who have experienced extreme poverty and homelessness are at risk for higher rates of physical illnesses, depression, anxiety, behavioral problems, learning problems, and low self-esteem (Davey, 2004). Children who live in shelters are often negatively affected as they watch their parent's caretaking roles and responsibilities taken over by shelter staff and human service professionals. Thus, working with the homeless population, whether directly at an emergency or domestic violence shelter, on the transitional housing program, at a school as a liaison or on other human services programs, or indirectly as a school human service professional, general counselor, child welfare worker, or in some other capacity where homeless clients might seek services, will involve working with an extremely wide range of clients and clinical issues.

Human service professionals provide counseling services to homeless adults and children, facilitate support groups, and provide individual counseling. But one of the most significant roles that human service professionals play is advocating for the homeless population, both on a personal case-by-case basis and on a community level by influencing policy and the development of legislation designed to aid the homeless population. Human service professionals also supervise shelter residents and provide case management services for adult and child residents, assisting them in connecting to a wide array of human services that will help them obtain economic and housing stability.

One of the underlying principle values of the human services field is to empower clients by plugging them into a variety of social support systems, moving them toward a state of self-sufficiency. This is particularly important when working with the homeless population and those suffering from severe poverty; thus, networking with other human service providers to provide a comprehensive continuum of care is a powerful intervention tool for human service professionals. The effective human service professional will not attempt to meet all of a homeless client's needs alone, but will depend on the services provided by other governmental and not-for-profit agencies in the area. Even many churches offer services for homeless individuals, including providing respite care for children, job training and networking, and financial assistance.

Many clients facing or experiencing homelessness tend to have multiple problems, which the human service professional might find challenging to address. Single-parent families that are either homeless or on the verge of homelessness are particularly challenging because the human service professional must address the needs of the children as well as the parent, and these needs might conflict with one another. For instance, consider the young, overwhelmed single mother with two young children and absolutely no one to help her with her child care responsibilities. Life in the shelter is depressing and difficult, her children are acting out more than ever because they miss their home and do not understand why they have to live in a shelter with so many strangers and so many odd and confusing rules. It is perfectly understandable for this mother to desperately need some time alone without her children, yet the tremendous amount of instability and the trauma associated with being homeless causes the children to need her more than ever. This dynamic can result in increased frustration on the part of the mother, which in turn creates increased fear and insecurity in the children. The human service professional can work with the mother to help her recognize this relationship interaction and take steps to resolve it through intermittent child care respite and counseling the family so that each member better understands the impact homelessness has on each other as well as themselves.

In light of the burden and stress placed on the single mother, who never enjoys a break from the frightening stressors and responsibilities she experiences living on the streets and in shelters with her children, it is no wonder that many women rush into romantic relationships believing promises of never-ending love and rescue. And although it would be tempting for anyone so completely overwhelmed with life to believe a man's offer to take over the control of one's life and the lives of her children, a relationship that moves too quickly will often result in domestic violence.

Many single mothers make decisions to proceed too quickly in relationships with men believing that such a relationship will provide the stability of an intact family for their children, only to find out a short time later that they have entered into yet another abusive relationship with someone who wants to control them and becomes violent if not successful. The shame these women feel is immense and sometimes results in their choice to remain in the abusive relationship because it seems better than facing homelessness again and having to admit that they made yet another devastating mistake. Yet all this accomplishes is to further lower their self-esteem, and change is rarely possible when one cannot move past the shame. In light of the fact that so many homeless single mothers experienced physical and verbal abuse growing up and then repeat this pattern in adult relationships, it is no surprise that many will eventually believe the horrible things being said to them causing them to further doubt whether they have the ability to make good choices for themselves and their children.

Many people, including human service professionals, become critical and frustrated with single mothers who enter into a string of relationships with abusive men, sometimes becoming pregnant, but I have often challenged people to consider how they might respond if they had no one in the world to help and support them, had no one to share the burdens and difficulties of life with, and did not have the luxury of taking their time to build a truly loving and healthy relationship because they had never enjoyed a solid foundation of love and security in their childhoods, causing them to enter into an adult world desperate for someone to love them and provide for them. I strongly believe this would make anyone impulsive in jumping into a relationship that looked good at first glance, because when you are desperately alone in the world anything looks good—in fact, it is a little like living in a desert with no water and thinking seawater tastes absolutely wonderful, only to find out later that rather than saving you, it will kill you. Consider Case Study 9.1 about Kim, paying particular attention to the complexity of her problems and issues, as well as the "domino effect" occurring in this single mother's life.

CASE STUDY 9.1

Case Example of a Homeless Single Mother

I met Kim when she was homeless and looking for permanent housing and attempting to put the pieces of her life together. Kim was raised in an unstable and abusive home environment where she had been told repeatedly throughout her childhood that she was worthless and that no one would ever love her. Her every move was criticized and served as proof that she was no good. She had the natural need and desire to be loved and accepted, and by the time she was 17 this need peaked to a point that she could not resist the affections of an older man who promised her the world. Although she initially resisted his attempts to become sexual with her, he eventually convinced her that the only way he would know she loved him was if they had sex, and if she refused he would leave her. Kim's immense insecurities and her deep need to be cared for made her vulnerable to his manipulative threats, so she relented and agreed to become sexual with him, believing that she had finally found someone who truly loved and accepted her. Yet when she became pregnant, he became abusive and used many of the same abusive statements she had confided that her father had used to manipulate and control her. She believed that her father must have been right all along, because how else could she explain yet another man seeing such ugliness in her? Ultimately he abandoned her and her unborn child, and when her father learned of her pregnancy he kicked her out of the house and refused to allow her to return.

For the next four years she was intermittently homeless, finding temporary stability through various transitional housing programs that helped her secure employment and an apartment, but any crisis put her on the streets again, such as the time her son got the chicken pox, resulting in her needing to stay home with him for two weeks. Kim was fired even though she had medical verification of her son's illness. This led to yet another financial downward spiral and another episode of homelessness. By the time her son was five he was acting out, considerably adding to her sense of frustration and burden. So when she met a new man who showered her with attention and compliments, all she could think of was that she had finally met the man of her dreams. He said all the right things, offered to let her and her son move in with him, and offered to manage every part of her life. He even told her that she would not have to work and could stay home with her son, so she gladly quit her job and embraced being a stay-at-home mom at last—something she had wanted to do for years.

Kim wanted desperately to believe this was real and accepted his seemingly generous offers because she believed that to do otherwise would mean robbing her son of his only opportunity for a real home and family. When her new boyfriend told her that she was the first woman he ever wanted to have a baby with, she was so flattered she agreed immediately to get pregnant. She believed with all her heart that she finally had it all, and that all the years of suffering were behind her.

Kim became pregnant quickly and dreamed of her new life with her new boyfriend. Although she would have preferred they got married, he claimed to not be ready yet, and because she did not want to create waves in the relationship, she did not push the subject. She talked endlessly to her son about their good fortune in finding this man who was going to take care of them forever. When her new boyfriend hit her for the first time, she convinced herself that it was a one-time incident caused by the stress of having a new family. When she noticed that he drank too much alcohol and seemed impatient with her son, she convinced herself that he needed time to adjust to having an instant family. Then one day he did not come home from work, and when a few days had gone by and he still did not return with her car, she came to the agency where I worked asking for financial assistance because she had no money to pay for the rent due in a few short days.

Unfortunately, we learned that this man had a pattern of treating women in this way, and this was not the first time he had encouraged a single mom to depend on him only to flee when the good feelings ended. Equally unfortunate was the fact that she had absolutely no recourse against him, even for taking her car, because to make insurance matters easier, she had agreed to put his name on the title, a decision that seemed foolish now, but in light of all that he was offering her it seemed the least that she could do. Now she had no money; no job; no car; a devastated, hurt, and angry child; and a baby on the way; and she would be homeless again within the month.

Adding to her burden was the intense sense of humiliation she felt when she realized that she had once again been taken advantage of. She firmly believed that she deserved this treatment and argued that there must be something terribly defective about her because these things kept happening to her. She was devastated that she was so horribly abandoned in the wake of breathing her first sigh of relief in years. She was extremely depressed, which made her at risk for either inadvertently abusing her child or neglecting him in some way, particularly when he expressed anger at her for driving his new daddy away. And, her additional loss of self-esteem left her in no shape to problem solve by gaining employment, finding low-cost housing, and searching out assistance programs, most of which would require her to disclose her reasons for becoming homeless, forcing her to repeat her failures and leaving her vulnerable to the criticisms of others. Although she should have been hospitalized for severe depression and risk of suicide, she refused because it would mean placing her son in temporary foster care.

Ultimately, she managed to piece her life back together, and it was the security of an authentic counseling relationship that enabled her to resist getting into another whirlwind romance and allowed her to see that saying no to a man was not saying no to a secure future, but likely saying no to another abusive and exploitative relationship. Virtually all my guidance meant her acting in a counterintuitive manner. She was desperate for love and companionship, yet I cautioned her to resist getting into a relationship until she was out of crisis. She desperately wanted to avoid revisiting old wounds from her childhood, yet I encouraged her to delve into her early experiences drawing parallels with relationships in her adult life and helping her to see the patterns she seemed helpless to escape. It would be difficult to imagine my client developing the wisdom to respond to her psychological issues and her current life crisis without the benefit of the objective and unconditional support of a human service professional trained to understand and respond to suffering from a social systems perspective—embracing, encouraging, supporting, and guiding in a nonjudgmental manner.

Although Kim's life sounds complicated, it is not at all unique. Understanding the dynamics involved in intergenerational abuse and poverty helps one to understand how and why people repeatedly make what often turns out to be unhealthy choices that when combined with social and structural factors leave them vulnerable and at risk for severe poverty and homelessness. Thus, although it might be easy to sit in the comfort of one's stable and healthy home environment and criticize the immoral lifestyle of single mothers who jump from relationship to relationship getting pregnant along the way, once all the situational factors are known and someone takes the time to truly look at the world through the eyes of someone suffering and alone, it becomes far easier to understand how someone could make the choices my client did. One of the saddest statements of humankind is that it seems as though for every vulnerable and hurting person, there is someone waiting to exploit him or her. Fortunately there are just as many people waiting to lend them an accepting, nonjudgmental, and helping hand as well.

Common Practice Settings for Working with the Homeless Population

Programs designed to aid the homeless population are offered in three levels of service. The first includes *emergency shelters* and *daytime drop-in centers*. Both offer short-term solutions to a long-term problem. Although emergency services are definitely needed, particularly when dealing with a population that might experience a crisis resulting in sudden homelessness, many emergency shelters are sharply criticized for their often unsafe and inflexible environment where residents can stay for as short as one night to as many as 30 days. Another area of criticism is that far greater amounts of funding are appropriated for emergency services rather than for long-term programs and services (Shlay & Rossi, 1992).

The second level of service includes *transitional housing programs*. These programs offer temporary housing for anywhere from six months to two years, with most programs offering a one-year program. Housing is only one part of the program package, though, and residents are typically required to participate in a wide range of adjunct social services such as job training, budgeting classes, adult literacy, substance abuse treatment, and parenting training. Other support services may include child care, job placement, and medical care. Most transitional housing programs focus on a specific target population, such as victims of domestic violence, single-mother families, single men suffering from substance abuse, adolescents, or the aging. These programs tend to be more successful because they provide a wide range of intensive services aimed at addressing the root causes of extreme poverty and homelessness, but they are also challenging to facilitate due to the complexity of the issues being addressed and the cost associated with administering programs offering comprehensive services, particularly because one of the primary root causes of homelessness is the unavailability of low-cost housing. Thus, to secure housing for homeless clients is just as expensive for the administering agency. Unfortunately, transitional housing programs have not garnered the majority of governmental funding.

A type of homeless service that is actually a combination of levels one and two includes domestic violence shelters. Because domestic violence is such a significant issue in the prevalence of single-parent families becoming homeless, shelters specialize in meeting the needs of individuals, most commonly women and children, who are fleeing from dangerous domestic relationships. Although there is some variation, the most common scenario involves a woman with children fleeing from a boyfriend or husband who is physically, emotionally, and verbally abusive. Domestic violence shelters operate on a 24-hour emergency basis, providing safe houses whose locations remain confidential.

Most domestic violence shelters have various homes and apartments spread throughout the community, each shared by a few women and children. Shelter stays range from one month to several months, and residents and their children participate in a broad range of services, including support groups for the mothers and the children. Human service professionals provide counseling, case management, and advocacy services, including assisting clients obtain orders of protection through the court system and advocating for them during any criminal or civil court hearings. Support groups focus on empowerment issues and educating the women on the nature of domestic violence, parenting from a perspective of strength, and developing better boundaries in relationships. Services may also include providing job training skills and job networking, locating child care, referral for substance abuse treatment, and assistance in locating permanent housing. In general, human service professionals provide as many services as are needed by the client.

Issues related to domestic violence will be explored in greater depth in Chapter 12 on violence, but it is important to understand that working with domestic violence victims can be challenging for a variety of reasons, but one of the most difficult aspects of working with this population is the cyclical nature of domestic violence where victims often return to their batterers when promises are made to authentically change.

The third level of service involves the provision of low-cost or *public housing projects* provided by HUD, which theoretically provides a permanent solution for the problem of homelessness. Unfortunately, this solution is the most difficult to provide and does not have a good track record of providing effective resolution to the homeless problem because traditional public housing units mostly built in the 1950s were developed as high-rise units in low-income neighborhoods, essentially creating segregated societies of the poor, not only producing dangerous neighborhoods but also further adding to the general negative stigma associated with poverty. Gang activity, drug dealing, and other crimes often associated with the urban inner city were common in what is often casually referred to as *the projects*.

Once government policy makers realized that housing projects of this type were likely causing more harm than good, an organized attempt was initiated to close the projects down, particularly in large cities such as Chicago and Philadelphia, and to transition residents to new low-rise housing units scattered throughout the city. Yet squatting became a significant problem, with some squatters even using the empty units as drug labs or gang hideouts.

A more current form of permanent low-cost housing includes governmental voucher programs facilitated by HUD. HUD's Section 8 housing voucher program designed for the general population and Section 811 designed for individuals suffering from disabilities (including mental illness) involve qualified individuals or families applying for the program when it is accepting applications (which may only be a few short periods throughout the year) and having their benefits determined. The voucher beneficiary must then locate a landlord who is willing to accept a government rent voucher as rental payment.

Theoretically the voucher can be used with any rental, but either through bias or because of a competitive rental market, many landlords in more expensive communities will not accept Section 8 or 811 rental vouchers. Thus, even though one intention of this program was to avoid the isolation and segregation created by high-rise congregated public housing, in many communities the result is still much the same because it is not the individual landlord of units scattered throughout the city that is most likely to accept a rental voucher, but the owners of large apartment complexes in low-income areas where occupancy rates run high who are the most likely to accept rental vouchers, creating the same sort of isolated high-crime environment experienced with public housing high rises.

Unfortunately, the need for housing has not kept pace with availability, and the recent spike in home foreclosures in response to the Great Recession of 2007–2009 has resulted in an increase in homelessness and a decrease in funding for housing assistance programs, including significant funding cuts in the Section 8 and 811 housing, and hope is limited that funding will significantly improve in the near future. For this program to be successful the federal government must make a firm commitment to subsidized housing that will be reflected in funding these programs appropriately.

Professional History

Understanding and Mastery of Professional History: Historical and current legislation affecting services delivery

Critical Thinking Question: What have been some of the unintended consequences of programs such as public housing projects and Section 8 vouchers? How have these consequences contributed to the institutionalization of poverty?

Concluding Thoughts on Homelessness

Homelessness is a complex social problem with multifaceted causes, including several root causes that lie in the personal domain (such as domestic violence, substance abuse, and teen pregnancy), as well as social causes such as institutionalized racism and oppression and structural causes related to the changing U.S. economy.

Structural issues related to a capitalist society include declining salaries, particularly for the poor, and escalating housing prices, which when combined creates an abundance of low-income renters competing for fewer affordable housing units. The development of affordable housing, although a good idea in theory, is challenging due to the high cost of land and housing in safer areas. In addition, most people who are at risk of homelessness often cannot afford to pay a significant portion of their own rent and many cannot afford to pay any rent at all. Thus, regardless of how the rental subsidies are structured, focusing on affordable subsidized housing as the primary resolution to the homeless problem essentially requires permanent governmental support, and unless adjunct services are provided, some argue that permanent subsidized housing programs may encourage dependency rather than fostering independence (Wright, 2000).

> **Programs offering a wide array of social services focusing on the personal root causes of homelessness, while at the same time addressing structural causes such as declining incomes and escalating housing costs, will have the greatest likelihood of successfully addressing the homeless problem.**

It appears then that programs offering a wide array of social services focusing on the personal root causes of homelessness, while at the same time addressing structural causes such as declining incomes and escalating housing costs, will have the greatest likelihood of successfully addressing the homeless problem with long-term solutions in mind. Human services agencies are on the front lines of developing such programs designed to promote self-sufficiency and personal security.

The following questions will test your knowledge of the content found within this chapter.

1. The rate of homelessness began to increase between 1970 and 1980 due to
   a. a decrease in affordable housing
   b. an increase in poverty
   c. an increase in depression
   d. Both A and B

2. Despite considerable research existing to support the opposite, a 1995 survey revealed that there was an increase in mainstream society's tendency to blame _____ for their poverty.
   a. the poor
   b. racism
   c. poor schools
   d. a poor economy

3. The 1970s and 1980s saw a dramatic increase in the homelessness of
   a. veterans
   b. intact families
   c. female single-parent head-of-households with children
   d. adolescents

4. Many homeless single mothers report having had
   a. unstable childhoods filled with physical and sexual abuse
   b. numerous sexual partners
   c. children with several different men
   d. a history of receiving public assistance

5. The majority of homeless single mothers are
   a. 25 years of age, native-born U.S. citizens with two to three children
   b. over the age of 35, with four to five children and a history of substance abuse
   c. Latino, undocumented immigrants who do not speak English
   d. African Americans who have four to five children from several men, and a history of welfare fraud

6. HUD's Section 8 housing voucher program is designed for
   a. individuals with disabilities (including mental illness)
   b. the general population
   c. single mothers and children
   d. veterans

7. Describe the effect of religious and philosophical ideologies on the perception and treatment of the poor.

8. Describe some of the challenges facing homeless single mothers residing in traditional homeless shelters and how human service professionals might assist clients in managing various challenges.

Suggested Readings

Jencks, C. (1995). *The homeless.* Cambridge, MA: Harvard University Press.

Kozol, J. (1988). *Rachel and her children: Homeless families in America.* New York: Ballantine Books.

Liebow, E. (1995). *Tell them who I am: The lives of homeless women.* East Rutherford, NJ: Penguin Books.

Stephen, B. (2000). *Street crazy: America's mental health tragedy.* Redondo Beach, CA: Westcom Associates.

Internet Resources

Homeless Advocacy Project: http://www.homelessadvocacy project.org

U.S. Department of Housing and Urban Development: http://www.hud.gov

National Coalition for the Homeless: http://www.national homeless.org

References

Auerswalk, C. L., & Eyre, S. L. (2002). Youth homelessness in San Francisco: A life cycle approach. *Social Science and Medicine, 54,* 1497–1512.

Axelson, L. H., & Dail, P. W. (1988). The changing character of homelessness in the U.S. *Family Relations, 37*(4), 463–469.

Beech, M., Meyers, L., & Beech, D. J. (2002). Hepatitis B and C infections among homeless adolescents. *Family Community Health, 25*(2), 28–36.

Davey, T. L. (2004). A multiple-family group intervention for homeless families: The weekend retreat. *Health & Social Work, 29*(4), 326–329.

Feagin, J. R. (1975). *Subordinating the poor.* Englewood Cliffs, NJ: Prentice Hall.

FEANTSA. (2007). *ETHOS—European typology on homelessness and housing exclusion.* Retrieved September 14, 2009, from http://www.feantsa.org/code/EN/pg.asp?Page=484

Gonyea, J. G., Mills-Dick, K., & Bachman, S. S. (2010). The Complexities of Elder Homelessness, a Shifting Political Landscape and Emerging Community Responses. Journal of Gerontological Social Work, 53(7), 575–590. doi:10.1080/01634372.2010.510169

Hecht, L., & Coyle, B. (2001). Elderly homeless: A comparison of older and younger adult emergency shelter seekers in Bakersfield, California. *American Behavioral Scientist, 45*(1), 66–79.

Kidd, S. A. (2003). Street youth: Coping and interventions. *Child and Adolescent Social Work Journal, 20*(4), 235–261.

Kingsley, G. T., Smith, R., & Price, D. (2009, May). *The impacts of foreclosures on families and communities: A report prepared for the Open Society Institute.* Washington, DC: The Urban Institute. Retrieved August 4, 2009, from http://www.urban.org/UploadedPDF/411909_impact_of_forclosures.pdf

Legal Momentum (2010), Single Mothers Since 2000: Falling Farther Down, http://www.legalmomentum.org/our-work/women-and-poverty/resources—publications/singlemothers-since-2000.pdf.

Levine, K. A. (2009). Against all odds: resilience in single mothers of children with disabilities. *Social Work Health Care, 48*(4) 402–419.

Lindsey, E. W. (1998). The impact of homelessness on family relationships. *Family Relations, 47*(3), 243–252.

Lindsey, E. W., Kurtz, D. P., Jarvis, S., Williams, N. R., & Nackerud, L. (2000). How runaway and homeless youth navigate troubled waters: Personal strengths and resources. *Child and Adolescent Social Work Journal, 17*(2), 115–140.

McKinney Homeless Assistance Act, Pub. L. No. 100-77, § 103(2) (1), 101 Stat. 485 (1987).

Montgomery, C. (1994). Swimming upstream: The strength of women who survive homelessness. *Advances in Nursing, 16*(3), 34–45.

National Alliance to End Homelessness. (2009). What we know about housing and homelessness. In *2009 Policy Guide* (pp. 3–6). Retrieved August 10, 2009, from http://www.endhomelessness.org/content/article/detail/2462

National Coalition for the Homeless. (2006). *How many people experience homelessness?* NCH Fact Sheet 2. Washington, DC: Author. Retrieved October 7, 2005, from http://www.ncchca.org/files/Homeless/NCH_How%20Many%20are%20Homeless_06.pdf

National Law Center on Homelessness & Poverty (2006). Some facts on homelessness, housing and violence against women. Retrieved January 20, 2012, from http://www.nlchp.org/content/pubs/Some%20Facts%20on%20Homeless%20and%20DV.pdf.

Nickelson, I. (2004). *The district should use its upcoming TANF bonus to increase cash assistance and remove barriers to work.* Washington, DC: DC Fiscal Policy Institute. Retrieved December 22, 2005, from http://dcfpi.org/?p=69

Nunez, R., & Fox, C. (1999). A snapshot of family homelessness across America. *Political Science Quarterly, 114*(2), 289–307.

Occupy Wallstreet. About. Retrieved July 4, 2012, from http://occupywallst.org/about/

Phelan, J., Link, B. J., Moore, R. E., & Stueve, A. (1997). The stigma of homelessness: The impact of the label "homeless" on attitudes toward poor persons. *Social Psychology Quarterly, 60*(4), 323–337.

Shlay, A. B., & Rossi, P. H. (1992). Social science research and contemporary studies of homelessness. *Annual Review of Sociology, 18*, 129–160.

Swick, K. J. & Williams, R.H. (2010). The voices of single parent mothers who are homeless: Implications for early child education professionals. *Early Child Education Journal, 38*(1), 49–55, DOI: 10.1007/s10643-010-0378-0.

Taylor, P., Parker, K., Morin, R., & Motel, S. (2012). Rising Share of Americans See Conflict Between Rich and Poor. Pew Research Center. Social and Demographic Trends. Retrieved October 23, 2012 from http://www.pewsocialtrends.org/files/2012/01/Rich-vs-Poor.pdf.

Thrane, L., Chen, X., Johnson, K., and Whitbeck, L. (2008). Predictors of post-runaway contact with police among homeless adolescents. *Youth Violence and Juvenile Justice, 6*(3), 227–239.

Toro, P. A., Tompsett, C. J., Lombardo, S., Philippot, P., Nachtergael, H., Galand, B., Schlienz, N., Stammel, N., Yabar, Y., Blume, M., MacKay, L. and Harvey, K. (2007). Homelessness in Europe and the United States: A comparison of prevalence and public opinion. *Journal of Social Issues 63*(3), 505–542.

U.S. Code, Title 42, Chapter 119, Subchapter I, § 11302. General Definition of Homeless Individuals (2005). Available online at: http://uscode.house.gov/download/pls/42C119.txt

U.S. Conference of Mayors. (2010, June). *The 2009 annual homeless report to Congress*. Retrieved January 19, 2012, from http://www.hudhre.info/documents/3rdHomelessAssessmentReport.pdf

U.S. Department of Education. (2001). *Report to Congress fiscal year 2000*. Washington, DC: Author.

U.S. Department of Housing and Urban Development. (2011). 2011 Point-in-time estimates of homelessness: Supplement to the annual homeless assessment report. Washington, DC: Author. Retrieved September 30, 2012, from http://www.hudhre.info/documents/PIT-HIC_SupplementalAHARReport.pdf

U.S. Department of Labor Bureau of Labor Statistics. (2010). Employment characteristics of families – 2010. Retrieved January 22, 2012, from http://www.bls.gov/news.release/pdf/famee.pdf

Weaver, R. K., Shapiro, R. Y., & Jacobs, L. (1995). Trends: welfare. *Public Opinion Quarterly, 59*(4), 606–627.

Whitbeck, L. B., Hoyt, D. R., & Ackley, K. A. (1997). Abusive family backgrounds and later victimization among runaway and homeless adolescents. *Journal of Research on Adolescents, 7*(4), 375–392.

Wright, T. (2000). Resisting homelessness: Global, national and local solutions. *Contemporary Sociology, 29*(1), 27–43.

Yoder, K. A., Whitbeck, L. B., & Hoyt, D. R. (2001). Event history analysis of antecedents to running away from home and being on the street. *American Behavioral Scientist, 45*(1), 51–65.

CHAPTER 10

Healthcare and Hospice

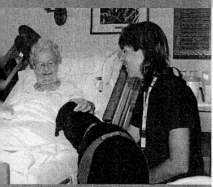

Dennis Sabo/Shutterstock.com

Learning Objectives

- Become familiar with the general nature of trauma counseling often conducted in a hospital setting
- Describe the skills necessary for the rapid assessment of patients in an acute medical situation
- Become familiar with the hospice philosophy including the definition and nature of palliative care
- Explore various theoretical models of grief and mourning, including traditional stage models as well as more contemporary task model approaches
- Understand common ethical dilemmas involved in the care of the sick and dying, including euthanasia, and the inadvertent omission of those individuals who are economically disadvantaged

Human Services in Medical and Healthcare Settings

At 9 A.M. Glenn is called to the labor and delivery department of a large hospital where he tries to talk with a teenage girl who just had a baby. The young girl holds her new infant as Glenn initiates a conversation with her. He explains that he is visiting her because it is hospital policy for the human service professional to visit with all adolescents who have just had babies. He asks the young mother a few basic questions, such as whether she has a place to live once she and her infant leave the hospital, whether her parents knew about her pregnancy, whether the father of the baby is involved, and whether she has a plan for raising her child. The young mother admits that her parents knew nothing of her pregnancy, as she managed to hide it by wearing large clothing and spending a lot of time in her room. She admits she is frightened that if she shares the news with them, they will force her to leave their home. And she admits that the father is no longer in her life and that she has no ability, nor any real desire, to raise a child. Glenn understands that this mother's ambivalence about raising her child is far more related to her age than her character. After some further discussion, Glenn asks her if she'd be interested in talking with a counselor who can assist her sorting through all the options available to her. After learning that her parents were generally supportive and loving people, he offers to call them for her so that they can help her decide how to best manage this unplanned pregnancy. The young mother appears relieved and admits that she considered just leaving the hospital without her baby because she was so desperately frightened and didn't know what else to do. When Glenn returns to his office, he makes the call to the parents; after a 20-minute emotional phone call, he makes plans to meet them in 30 minutes in their daughter's hospital room. After meeting with the entire family, he supplies them with several names of counseling agencies that can assist the young mom in either parenting or placing her infant for adoption.

While walking out of the hospital room Glenn is paged to the emergency room. When he arrives, he finds the entire unit in chaos. Three cars collided, and many people were injured. After talking to the emergency room nurses and physicians, Glenn learns that one of the cars had several children in it, many of whom were seriously injured. Glenn gets to work right away collecting identifying information, making sure that each child's parent is accounted for, and obtaining numbers of parents who need to be notified of the accident and their child's condition. After obtaining all necessary information, Glenn makes himself available to the parents who had children in surgery—parents who were not in the accident, whom he recently called to the hospital—and the children who are not seriously injured but had parents who were. He offers to contact friends and family for support. After contacting spouses and two family pastors, Glenn sat with one family who had two seriously injured children and provided crisis counseling so that they could be calm enough to understand all that was going on with their children. Glenn also offers to be the conduit between the waiting families and the medical team, so for over an hour he goes back and forth between the medical personnel working on the injured and delivering any new information to the family members. Two hours later, all parties were out of crisis and had support systems by their sides, and Glenn was cleared to return to his office.

Next, Glenn began working on several discharge planning cases for various patients who were scheduled to be released from the hospital within the next two days. One was an older patient who was not healthy enough to return home, and it was Glenn's responsibility to assist the family in finding either appropriate alternate housing or in-house services that would enable the patient to remain in his home. Another case involved a survivor of a serious car accident who needed continued therapy, but could no longer remain in the hospital. Glenn's job was to locate a rehabilitation center close to family that would be covered under his insurance plan.

As Glenn's day was coming to a close, he was paged again to the emergency room, where he learned there was a potential victim of sexual assault. Glenn asked the victim if she was comfortable talking to him, but she stated that she was not—she preferred a female counselor. Glenn then called the local county rape crisis center and asked for a volunteer to come to the hospital immediately to counsel and support a sexual assault victim.

On his way out of the emergency room he was asked to consult on a potential child abuse case. Glenn interviewed the parents of a six-year-old boy who suffered a spiral fracture of the arm. Glenn became concerned when he interviewed each parent separately and their stories differed significantly. Because of this and the child's inability to describe in detail how he injured his arm, Glenn felt the case warranted a call to child protective services (CPS). He explained to the parents that he would be making an abuse allegation report and that the child would not be released until a CPS caseworker came to the hospital and interviewed everyone in the family.

Prior to Glenn leaving for the day, he was asked to visit with a patient and her adult son who just learned of her terminal diagnosis. Glenn provided both with some crisis counseling and made a referral to a local hospice agency. He offered to meet with them again tomorrow and to meet with them and the hospice team if they wished. Glenn's last case for the day was to provide counseling to a 60-year-old man who recently underwent a liver transplant and was about to be released from the hospital. Research indicates that transplant patients often experience depression after being released from the hospital, thus Glenn's focus was to help this patient adjust to the realities of being a transplant patient, as well as preparing him for experiencing some depression in the coming weeks. He made sure this patient left armed with names of counselors who had experience working with transplant patients.

This is a typical day of a human service professional working in a medical or healthcare setting, and although this description is realistic, it is probably more realistic to state that there is no such thing as a "typical" day for a human service professional in a healthcare setting! In fact, someone interested in a career in the human services field and looking for structure and predictability would probably not fare well in a healthcare center, where the broad range of patient issues determines the range of issues dealt with by the human service team.

Human service professionals have traditionally worked in hospital settings in a variety of capacities, yet as the healthcare field branched out to other arenas, including community-based healthcare centers, primary care full-service clinics, and specialized health centers serving special populations (such as AIDS/HIV patients, women, or those suffering from cancer), human service professionals can now be found in a variety of healthcare-related practice settings.

> **Healthcare settings are highly regulated fields and as such professionals working in these environments are typically required to have both advanced degrees and state licensing.**

Healthcare is one field where most professionals working in human services are often required to be licensed human service professionals in some capacity. There is some variation from state to state, but healthcare settings are highly regulated fields and as such professionals working in these environments are typically required to have both advanced degrees and state licensing.

Human service professionals working in medical settings are true generalists: they must be flexible and able to deal with a variety of issues, often in a setting wrought with crisis and trauma. But despite their broad generalist functions, the scope of human service professional functions in healthcare and hospital settings can be quite specific. These functions and responsibilities may include

- conducting psychosocial assessments on patients as needed
- providing information and referrals for patients
- preadmission planning
- discharge planning
- psychosocial counseling
- financial counseling
- health education
- postdischarge follow-up
- consultation with colleagues
- outpatient continuity of care
- patient and family conferences regarding health status, care, and future planning
- case management for patients
- facilitation of and referral to self-help and emotional support groups for patients and families

- patient and family advocacy
- trauma response
- assistance in exploring bioethical issues
- outcome evaluations on best practice committees

In addition to performing these various functions, some of the issues addressed by human service professionals working in medical settings include addressing patient problems related to activities of daily living, assisting patients and their families in dealing with illness adjustment, assessing possible physical and sexual abuse, including child abuse and domestic violence, and assessing patients with potential mental health problems (NASW, 1990).

Crisis and Trauma Counseling

A large part of a human service professional's role in a medical or healthcare setting is to provide crisis and trauma counseling to patients and their families. In fact, when the hospital has notified the family of a patient who has been seriously injured either through illness or accident, it is often someone from the human services department who meets the family at the emergency room doors.

A good model for how to approach an individual or family in crisis is one developed by Abraham Maslow. Maslow (1954) created a model focusing on needs motivation. As Figure 10.1 illustrates, Maslow believed that people are motivated to get their most basic physiological needs met first (such as the need for food and oxygen) before they attempt

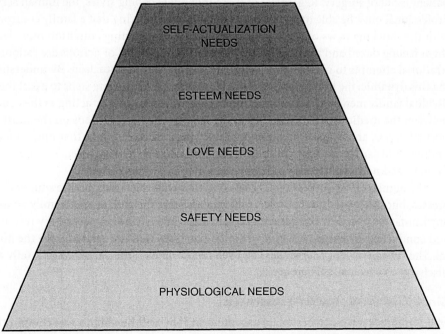

FIGURE 10.1
Maslow's Hierarchy of Needs

Maslow, Abraham H.; Frager, Robert D.; Fadiman, James, *Motivation and Personality*, 3rd Ed., ©1987. Reprinted and Electronically reproduced by permission of Pearson Education, Upper Saddle River, New Jersey

to meet their safety needs (such as the security we find in the stability of our relationships with family and friends). According to Maslow, most people would find it difficult to focus on higher level needs related to self-esteem or self-actualization when their most basic needs are not being met. Consider anyone you know who suffers from low self-esteem and then consider how he might react if a war suddenly broke out and his community was under siege. Maslow's theory suggests that thoughts of low self-esteem would quickly take a back seat as worries about mere survival took hold.

When individuals are facing a significant crisis, they often feel compelled to get their most basic needs met. In situations where family members or close friends have been called to the hospital in response to a loved one having been in a serious accident or suffering from some life-threatening illness, their first priority is often to obtain information about the medical status of the patient and it is very important for the human service professional to avoid escalating in panic or anxiety along with the family. In fact, it is vital that professionalism be maintained in the midst of the crisis so that amidst crying, screaming, and perhaps even misplaced anger, the human service professional can serve as a calming influence that the family can rely on as they attempt to regain their composure.

Each family handles crises differently; thus, it is important for the human service professional to quickly recognize the family's coping style. Some families will focus on mundane details, such as asking how long their loved one will be in the hospital when the patient has not even emerged from emergency surgery, and some families will focus directly on important issues, such as repeatedly asking whether the patient will survive the surgery, even though there might be no way to answer such a question until the patient is out of surgery. Regardless of these individual coping styles, the human service professional must be able to read between the lines recognizing that a family confronted with the shocking news of a loved one having a life-threatening condition often leaves them feeling dazed and powerless, and many of the questions or actions are rational or irrational attempts to recover some sense of control over the situation. By understanding this dynamic, the human service professional can take concrete steps to assist the individual family members in gaining as much control as possible by acting as the conduit between the medical staff and the family, by helping the family focus on the most important issues, and by assisting them in developing a plan of action that might include finding child care for younger children, having someone go to the patient's house to care for pets, and notifying friends and employers on behalf of the patient.

The human service professional's role continues with the family as the situation progresses, but takes on a different role, including assisting the patient and family adjust to any limitations posed by the patient's condition or injury, finding necessary resources, and conducting discharge planning when the patient is well enough to leave the hospital. The human service professional will even follow up with the patient and family after discharge to check on their progress.

Single Visits and Rapid Assessment

Most human service professionals assume that they will be able to work with their clients—regardless of their role in the helping process—over an extended period of time. Yet, this is typically not the case in a hospital setting due to the trend toward significantly shortened duration of hospital stays. In fact, often human service professionals working in a hospital setting will see patients only one or two times. Because of this pattern, there is a growing body of literature on single session encounters with clients, and how human service professionals can develop a set of skills that allow for rapidly assessing the patient and their situation, and assist them effectively, depending upon the role and functions of the human service worker.

Gibbons and Plath (2009) explored this very issue by interviewing several patients and asking what they found helpful in these single sessions and which skills and qualities were helpful and which were not. They isolated seven basic skillsets that medical social workers needed in order to engage successfully with patients during a single session. These included the ability to

1. quickly put the patient at ease
2. establish a rapport and a sense of trust quickly
3. exhibit a sense of competence
4. engage in active listening and exhibit empathy
5. be nonjudgmental
6. provide needed information quickly
7. organize support services

Although the study focuses on medical social workers, the application is appropriate for generalist human service workers as well, at least in some general respects. Since many hospitals hire human service professionals with certificates and bachelor's degrees to conduct discharge planning, case management, and patient advocacy, as well as other functions related to patient care, the skill set discussed by Gibbons and Plath can be applied to a broader range of roles within the human services profession.

Working with Patients with HIV/AIDS

Human service professionals working in a medical setting, particularly in public health, commonly work with various health-related epidemics or pandemics. HIV, which causes AIDS, is an example of such a pandemic. HIV and AIDS were first discussed in the medical literature in 1981 (Gottlieb et al., 1981). Medical treatment during these early years typically occurred in a crisis setting when patients presented in the emergency room with advanced or end-stage AIDS infections, such as *Pneumocystis carinii* pneumonia (PCP) and Kaposi's sarcoma, both opportunistic infections common in end-stage AIDS patients. In August 1981, 108 AIDS cases were reported in the United States by the Centers for Disease Control and Prevention (CDC). By 1986 the CDC reported 16,458 cases of AIDS in the United States and 8,361 deaths, and just two years later those numbers rose to 72,024 cases with an estimate that 1 to 1.5 million Americans were infected with HIV (Centers for Disease Control, 1988).

During the early years of the AIDS crisis, the role of the human service professionals focused almost exclusively on the crisis of receiving a terminal diagnosis and included conducting emergency discharge planning, death preparation, arranging for acute care, and initiating hospice services. By the 1990s education efforts led to earlier diagnoses and better medical treatment for those who could afford it and clinical intervention

Human Systems

Understanding and Mastery of Human Systems: Emphasis on the context and role of diversity in determining and meeting human needs

Critical Thinking Question: Human service professionals working in hospital settings often have only one or two brief contacts with their clients; and they are called on to serve a wide variety of people whose circumstances vary enormously. How might these factors shape the manner in which the professional approaches her or his work?

focused more on the psychosocial issues involved with having a chronic, debilitating, and sometimes terminal disease that carried a stigma with it. These psychosocial issues typically included a fear of discrimination, concerns about receiving quality medical care, job accommodations and other income sources, and housing accommodations when physical health begins to decline (Kaplan, Tomaszewski, & Gorin, 2004).

When HIV/AIDS first emerged in the United States, there were no medical treatments available to address the actual disease process (other than symptomatic relief), but through grassroots efforts (that led to significant fund-raising efforts for medical research) significant medical advances were gained throughout the 1990s until HIV/AIDS is now considered more of a chronic, rather than terminal, disease—for those individuals fortunate enough to have access to expensive antiviral therapy. Despite these medical advances, however, the treatment of HIV/AIDS remains a serious public health concern, particularly for those individuals who have no access to advanced medical treatment or who do not respond positively to the most aggressive antiviral therapies, commonly referred to as the *AIDS cocktail*.

The most recent statistics available from the Centers for Disease Control and Prevention (CDC) (2010) indicate that at the end of 2009, about 1.2 million people in the United States were living with HIV/AIDS (both diagnosed and undiagnosed). The CDC estimates that there were approximately 41,540 new cases of HIV diagnosed in 2009. According to the CDC, the population at greatest risk remains what the CDC refers to as the "Men who have sex with men" or MSM group. This group represents 2 percent of the population, but consists of 61 percent of all new HIV cases diagnosed in 2009 and about half of all people currently living with HIV/AIDS. White MSM and black MSM account for the majority of new diagnoses in 2009 (11,400 and 10,800 cases, respectively) followed by Latino MSM (6,000 cases). Heterosexuals consisted of approximately 27 percent of all new diagnoses in 2009. Young black heterosexual men were the only group that experienced an increase in diagnoses in 2009. Women consisted of about 23 percent of all new diagnoses in 2009, with black women consisting of about 65 percent of those cases.

The ongoing trend in the demographics of AIDS that disproportionately affects people of color, particularly black women, has led to many changes in the psychosocial needs of the HIV/AIDS population, which has had an impact on the roles and functions of medical human service professionals working with this population. There is still considerable social stigma associated with an HIV/AIDS diagnosis, particularly in light of the uninformed belief that it is a disease affecting only the homosexual population. But because HIV/AIDS was a disease that affected primarily Caucasians when it first surfaced in the United States, racial discrimination was not a central psychosocial issue. But now that this disease is affecting many minority communities, racial discrimination has been coupled with the existing social stigmas that often presume immoral behavior, such as sexual promiscuity and drug abuse. Despite aggressive public awareness campaigns in both the general public and the professional community designed to increase general awareness and remove stigma, many individuals with the HIV/AIDS virus are forced to endure numerous barriers to getting basic needs met—some of which are related to the stigma, some related to institutionalized racial discrimination, and some related to a combination of both (Kaplan et al., 2004).

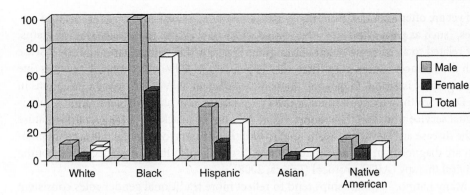

FIGURE 10.2
Bar Graph Comparing Rates (per 100,000 population) of AIDS by Race/Ethnicity, 2004, 50 States, Including District of Columbia
Source: Based on statistics from CDC, HIV/AIDS Surveillance Report, 2004, Vol. 16.

For instance, quality medical care is lacking on most Native American reservations, and native advocates argue that reasons for this relate to racial disparity and historic mistreatment and oppression. When reservations were first confronted with a rapidly increasing incidence of HIV/AIDS, elders complained that the medical neglect experienced on most reservations was yet another form of racial discrimination and oppression, evidenced by the fact that the federal government was not allocating sufficient funding to address this issue on the reservations (Weaver, 1999). Human service professionals working within the medical field must be aware of the various ways that racial prejudice plays out within the community, whether such discrimination be direct and overt or institutionalized (such as where federal monies are allocated). This awareness can then translate into advocacy and outreach as well as increased sensitivity as practitioners challenge their own perception of the HIV/AIDS crisis, including their attitudes about those populations that are currently being most significantly affected by this disease.

HIV/AIDS and the Latino Population

Although the white MSM population still account for the majority of those in the United States diagnosed with HIV/AIDS in 2009 (the most recent data available), according to the CDC (2011), the Latino population is disproportionately affected by the HIV/AIDS virus. While Latinos represent approximately 16 percent of the population, they constituted about 20 percent of all new HIV diagnoses in 2009, an incidence rate three times that of Caucasians (relative to their representation in the population). The largest group among Latinos diagnosed with HIV in 2009 is Latino MSM.

Among the U.S. Latino/a population, those living in the southern states, where the Latino/a population has grown by over 200 percent since 1990, are particularly vulnerable for a variety of reasons. A report focusing on a "two-year fact-finding and cooperation program" facilitated by the *Latino Commission on AIDS* explored the extent and nature of the HIV/AIDS problem within the Latino/a population living in the Deep South, which includes Alabama, Georgia, Louisiana, Mississippi, North Carolina, South Carolina, and Tennessee (Frasca, 2008).

The program, also referred to as the "Deep South Project," found evidence that Latinos/as/ are being infected with HIV at disproportionate and rising rates in the south, and yet are often excluded from the healthcare system, as well as HIV/AIDS-related services, (such as prevention and educational programs) due to their immigration status, fear related to an increase in anti-immigrant hostility, the stigmatization of the disease with the Latino/a community. They also lack access to HIV/AIDS-related services due to geographic isolation. In general, there are insufficient HIV/AIDS-related programs in the Deep South focusing on the Latino/a population, and an insufficient number of bilingual service providers. The report found that due to a lack of awareness of the nature of the disease and lack of access to quality healthcare, many within the Latino/a population are diagnosed in the later stages of the disease, which limits the success rate of the antiviral therapy (ART) protocol (Frasca, 2008).

Many Latino/a relationships tend to reflect more traditional gender roles consistent with machismo culture, which increases the risk of contraction and transmission of the HIV/AIDS virus. For instance, within the machismo culture, men commonly engage in high-risk sexual behaviors in order to prove their manhood, such as having multiple sex partners, despite being married, and not wearing a condom. Latina women often cite an awareness that their male partner's sexual behavior places them at greater risk for contracting the HIV/AIDS virus, but they do not believe they have enough power in the relationship to make demands for safe-sex practices, such as fidelity and wearing a condom (Acevedo, 2008).

Human service professionals working within the Latino/a population must develop a level of cultural competence in working with this population, becoming aware of the many culturally related risk factors affecting this population. They must also be aware of how racial prejudice, social exclusion based upon immigration status (or perceived status), and various stigmas impact the Latino/a's access to educational and prevention services, as well as access to quality and timely healthcare (Acevedo, 2008).

The Deep South project report makes several recommendations, including public health departments conducting needs assessment of the Latino/a population, increasing outreach efforts in high Latino/a communities, increasing the number of bilingual service providers, increasing the cultural competency of service providers working with the Latino/a population, and increase HIV/AIDS research on Latino/a populations so that the literature more accurately reflects the nature and needs of the Latino/a population (Frasca, 2008).

Concluding Thoughts on Working with the HIV/AIDS Population

When confronting the HIV/AIDS crisis, human service professionals engage in a four-pronged approach to psychosocial care, including prevention and educational awareness (such as the practice of safe sex), client advocacy, and case management/counseling. Human service professionals are actively involved in both practice and policy aspects of the HIV/AIDS pandemic, including meeting the psychosocial needs of those diagnosed with HIV/AIDS, as well as being on the front lines of prevention efforts, community and patient educational and awareness campaigns, advocacy for increased funding of intervention and treatment programs, and participating in lobbying efforts, advocating for the passage of laws designed to protect the privacy and legal rights of those diagnosed with HIV/AIDS.

Human service professionals working in a medical or healthcare setting assist those with HIV/AIDS in obtaining necessary medical services, obtaining the necessary funding for treatment, and providing counseling for those infected individuals, their families, and caregivers. The nature of the counseling will change depending on the progression of the virus. Clients newly diagnosed will need counseling focusing on acceptance of a potentially terminal disease, whereas other clients will need counseling focusing on living with a chronic illness, accepting a life of potential disability, accepting a life that includes multiple medications taken on a daily basis, and learning to live with the consequences of stigmatized disease.

Depending on the demographic nature of the patient, the human service professional may help secure child care; help the patient apply for financial assistance; obtain home healthcare; maintain or obtain employment, housing, and medical care, including care for other health-related issues, such as substance abuse; and finally help the patient and family contend with the various stressors involved with having a stigmatized illness (Galambos, 2004).

> **Human Services Delivery Systems**
>
> Understanding and Mastery of Human Services Delivery Systems: Range of populations served and needs addressed by human services
>
> Critical Thinking Question: Why is it critical for human service professionals to demonstrate cultural sensitivity and cultural competence in their HIV/AIDS prevention, education, and treatment work with Latino/a, African American, and Native American populations?

The Hospice Movement

Hospice care is a service provided to the terminally ill that focuses on comprehensive care addressing their physical, emotional, social, and spiritual needs. Although hospices have existed since about the 4th century, the biblical and Roman concepts of hospice involved providing refuge for the poor, sick, travelers, and soldiers returning from war. Hospice as a refuge or service for the terminally ill was not developed until the mid-1960s.

The modern hospice movement emerged from the general dissatisfaction with how dying individuals were being treated by the established medical community. Western medicine is curative by design with a focus on restoring individuals back to a state of healthy functioning. This model left the majority of the traditional medical community at a loss as to how to treat those who were beyond the hope of recovery. Dying patients often felt neglected and isolated in depersonalized hospital settings where they were typically subjected to needless and futile medical interventions. The hospice movement challenged the treatment provided by the traditional medical community that often failed to address pain management effectively and often neglected the psychosocial and spiritual needs of the dying patient.

> The hospice movement challenged the treatment provided by the traditional medical community that often failed to address pain management effectively and often neglected the psychosocial and spiritual needs of the dying patient.

The History of Hospice: The Neglect of the Dying

Dame Cicely Saunders, the founder of the modern hospice movement, recognized this lapse of appropriate care for the dying and set about to make significant changes that would affect how the world viewed the dying process. Originally trained as a nurse,

Saunders eventually earned her degree in medicine and quickly challenged what she saw as the medical community's failure to address the comprehensive needs of terminally ill patients. Saunders was passionate about the care of the terminally ill and in 1958 wrote her first paper, entitled "Dying of Cancer," addressing the need to approach dying as a natural stage of life. Through her work with the terminally ill, Saunders recognized that dying patients required a far different approach to treatment than the traditional one that tended to see death as a personal and medical failure.

In Saunders's personal letters, she describes in detail her discussions with terminally ill patients in the hospice where she worked, as well as her dedication to the prospects of developing a system of care committed to a dying process without pain, while enabling terminally diagnosed patients to maintain their sense of dignity throughout the dying process (Clark, 2002). Saunders founded St. Christopher's Hospice of London in 1967. Her model of care used a multifaceted approach, where dying patients were treated with compassion so that their final days were spent in peace rather than undergoing invasive and futile medical treatments and where they were free to attend to the business of dying, such as saying good-bye to their loved ones.

The Connecticut Hospice, Inc., was the first hospice opened in the United States in 1974 in New Haven, Connecticut, funded by the National Cancer Institute (NCI). The hospice was created for many of the same reasons noted by Saunders—the belief that good end-of-life care was severely lacking within the U.S. hospital system and the belief that the dying process was a meaningful one worthy of honor and respect (Stein, 2004). When the HIV/AIDS crisis first began in the 1980s, and prior to the development of antiviral treatment, hospices took on a significant role in the end-of-life care of those dying of the AIDS virus. Although there are some freestanding hospices, hospice is not a "place," but rather it is a concept of care and can be provided anywhere a patient resides (Paradis & Cummings, 1986).

The hospice movement has grown immensely in a relatively short period of time, and what began as a grassroots effort of trained volunteers supported by philanthropic agencies, such as the United Way, has become a highly regulated and profitable industry staffed by a team of professional service providers. Although the core goals and philosophy of hospice remain the same, the professionalization and governmental regulation of this field has influenced its service delivery model. For instance, although hospice care was originally developed as an alternative to hospital care, many hospices in the United States are now in some way affiliated with a hospital or other healthcare organization, most are accredited, and almost all are Medicare certified (National Hospice and Palliative Care Organization, 2003; Paradis & Cummings, 1986).

The Hospice Philosophy

The hospice philosophy employed today is similar to the one envisioned by Saunders. Dying is seen not as a failure, but as a natural part of life, where every human being has the right to die with dignity. Hospice care involves a team approach to the care and support of the terminally ill and their family members. A core value of the hospice philosophy is that each person has the right to die without pain and that the dying process should be a meaningful experience. Because Western culture often perceives accepting death as synonymous with giving up, individuals battling illness are often inadvertently encouraged to fight for their survival to the "bitter end"; thus, the hospice philosophy is counterintuitive to Western cultural wisdom.

Hospice treatment involves palliative care rather than curative care. The hospice movement is highly supportive of patients remaining in their homes, but when that is not possible, hospice service is provided in hospitals, nursing homes, and long-term care facilities and can be an adjunct to other medical services provided. The only stipulation of most hospice agencies is that the patient has stopped pursuing curative treatment and that the patient received a terminal diagnosis of six months or less.

THE HOSPICE TEAM   The hospice team is interdisciplinary by design, and although there is considerable overlap in many of the roles of the various service providers, the hospice human service professional serves a unique purpose on the team, emanating from the distinct values underlying the human services and social work discipline (MacDonald, 1991).

The hospice team typically consists of a *hospice physician* who makes periodic visits and monitors each case through weekly reports from other team members; a *nurse* who visits patients wherever they reside at least three times per week; a *human service professional* who provides case management services, counseling to the patient and family, including helping the patient say good-bye to friends and family, help resolve any past conflict, and assistance with end-of-life issues such as preparation of legal documents such as wills, and advance directives (which will be explored in the next section); a *chaplain* who provides spiritual support; a *home health aid* who provides daily care such as personal hygiene; a *trained volunteer* who provides companionship including reading to patients or taking them for strolls in a wheelchair, and a *bereavement counselor* who provides counseling and support to surviving family members after the death of the patient. One might question whether the interdisciplinary team model works, when so many varied professions are involved; yet, research indicates that the hospice interdisciplinary team model is effective as long as there is good communication, trust, and mutual respect among team members, as well as administrative support (Oliver & Peck, 2006).

The Role of the Hospice Human Services Worker

The hospice human service professional provides numerous services to hospice patients and their families, including providing *advocacy* for patients, particularly in regard to obtaining services and financial assistance; *crisis intervention* when emergencies arise; *case management* and *coordination of services* for the comprehensive care of patients and their family members; *case consultation* services among hospice and other healthcare staff; assisting the patient and family in *planning* for the patient's eventual death; and *bereavement counseling* to assist patients in accepting their terminal illness and in saying good-bye to loved ones, as well as counseling surviving family members after the patient's death.

The Psychosocial Assessment

Prior to providing these services, a hospice human service professional must complete a thorough psychosocial assessment to evaluate the strengths and deficits of the patient and family members. How are the client and family accepting the reality of the terminal illness? Can the family realistically provide for current and future needs of the patient? Family members who are still reeling from the news that their loved one is dying are often unrealistic in their expectations of the rigors involved with caring for a terminally ill person and will need help to recognize their limitations and need for outside assistance.

Conducting a thorough psychosocial assessment is the first step in making these clinical determinations and ascertaining what services are needed. Chapter 4 outlines the basic criteria of a psychosocial assessment, but the information sought and the focus of the assessment change depending on the issues at hand. Thus, a psychosocial assessment of a hospice patient will focus more on the patient's current living conditions, whether they are appropriate in relation to the patient's declining health, as well as end-of-life issues. Other dynamics explored may include the state of the patient's current relationships, and whether there are any unresolved issues that need to be resolved before the patient's passing.

Intervention Strategies

Once a thorough psychosocial assessment has been conducted, the human service professional can determine the nature and level of intervention necessary to meet the needs of the patient and family members. In fact, the psychosocial assessment in many respects acts as the blueprint for the human service professional, determining the course of case management and counseling intervention strategies for the patient and family.

For instance, if the psychosocial assessment reveals that the patient is older and has an aging spouse and no adult children in the immediate area, plans might need to be made for the patient's eventual placement into a facility for full-time care once the illness has progressed to a point beyond the spouse's caregiving ability. Thus, even though the patient's spouse might currently be managing the daily rigors of caring for the patient, plans will need to be made for the patient's care once the illness progresses and care requirements become more complex. This can occur through either placement in a residential facility, contracting with a home healthcare agency, or utilizing a day respite center (depending on the nature of the illness or condition). If the psychosocial assessment reveals that the patient has insufficient health insurance benefits, the human service professional will assist the patient and family with applications for governmental assistance such as Medicare.

If the psychosocial assessment reveals a mental health history of depression or anxiety, then an intervention involving a course of antidepressant or antianxiety medication might be in order. Finally, if the psychosocial assessment reveals a history of conflict within the family, then the human service professional can plan an intervention strategy designed to help the family work out their issues so that they might move toward a place of resolution before the patient dies.

CASE MANAGEMENT AND COUNSELING SERVICES   One of the most common roles for hospice human service professionals includes providing case management and counseling services to patients and their family members that address the issues noted in the psychosocial assessment. For instance, issues related to how the patient and family are dealing with the terminal illness, the loss of control because of increasing debilitation, and the impending death are all explored and counseling provided as necessary.

Yet, because each family is different, the counseling will vary dramatically from patient to patient. For instance, if the patient is a five-year-old child dying of cancer, the human service professional will need to assess the needs of the parents and siblings involved. Yet, if the patient is 85 years old with an ailing spouse and adult children in their sixties, the clinical issues will be different, and although it would be incorrect to automatically assume that the level of grief is lessened simply because this death is expected in the natural course of life, the needs of the different parties involved are obviously going to vary significantly. Thus, the actual nature of the illness or condition, the age of the patient, and the specific demographics and characteristics of the family members all combine to determine the nature of the counseling.

I recall working with one client who was dying of amyotrophic lateral sclerosis (ALS), also known as *Lou Gehrig's disease*. She was suffering from almost complete paralysis and was unable to communicate once hospice was hired, thus I worked primarily with her husband. This couple was in their early eighties and had been married for over 50 years. The surviving spouse was heartbroken at the prospect of losing his wife who was also his best friend. Our counseling relationship lasted for months and consisted primarily of him talking about his wife, their relationship, and how agonizing it was for him to watch his once capable, articulate wife, who was a leader both in the community and within their family, become slowly imprisoned and paralyzed by ALS. During our initial sessions he shared some wonderful memories of their life together and of his wife's strengths and accomplishments (attending seminary after raising their children), but would then become emotionally upset when sharing the pain and powerlessness he felt as he watched her struggle to communicate, at that point by blinking. My role was not to put a "happy face" on his suffering, nor was it to reframe this tragedy in some positive light, as might be appropriate in another type of counseling in another practice setting. Rather, my role was to remain comfortable when in the presence of his emotional expressions of grief and sadness, which in some sense gave him permission to have these necessary feelings. I did my best to provide comfort and a forum for his sadness, but I never gave him the impression that his feelings were in any way wrong.

Well-meaning but misguided counselors are often uncomfortable when confronted with a client's intense emotions of sadness and anger and, in an attempt to alleviate this pain and their own discomfort, try to make the client feel better by pointing out the positive side of a crisis or by encouraging the client to not dwell on feelings of sadness and anger. This approach often leaves grieving clients feeling as though their intense feelings are somehow unacceptable, or at the least burdensome, which in turn results not in them feeling any better, but as they shut off communication, they ultimately risk suffering in isolation.

Hence, one of the greatest challenges facing hospice workers lies in their ability to increase their comfort level for intense and unpleasant emotions. Those who are grieving can intuitively sense when those around them are comfortable with their emotions, and many hospice clients report that hospice counselors are the only people with whom they feel safe and comfortable sharing their deepest and most painful feelings of loss, sadness, anger, and mourning.

One of the greatest challenges facing hospice workers lies in their ability to increase their comfort level for intense and unpleasant emotions.

RESISTING THE REALITY OF THE DEATH   Another challenge facing human service professionals working in hospice is resistance on the part of the patient and/or family members in directly dealing with the realities associated with a terminal diagnosis. As mentioned earlier, embracing death often feels all too much like letting go of life, and North American culture is far more comfortable embracing life. Many people are fearful that if they accept the reality of the terminal diagnosis, they are essentially letting go of their loved one, which not only sends the wrong message, but also feels far too much like giving up. This attitude has helped to create a sort of taboo surrounding death where many people are resistant to even think about their own deaths, let alone the impending death of a loved one.

In some families, to accept the reality of the terminal diagnosis is synonymous with losing hope, thus resisting the acceptance of a terminal diagnosis can feel like fighting for life. A human service professional might be seen as someone who will attempt to rob the patient and family of their hope, thus many times families make the decision to either reject social work services when first signing up for hospice care or prohibit the human service professional from talking about the terminal diagnosis in front of the patient. Yet, because many of the issues addressed by human service professionals working in hospice are designed to also deal with problems that will confront the family at some point in the future—perhaps even years after their loved one has died when social work services are not available to assist them, it is important that the human service professional be able to confront the family's denial and assist them in understanding that to accept the impending death of their loved one is not synonymous with hastening the death or with losing hope.

Counseling can be particularly challenging when the patient is asking for information and the family does not want the information about the terminal diagnosis to be shared. In this situation, the human service professional must be sensitive, but clear that the patient is the identified client, and what is in the best interest of the patient will also eventually be in the best interest of the family, even if they do not initially recognize it as such. A human service professional must delicately assist the family with the task of accepting the terminal illness, facing this approaching loss, and addressing each emotional complication that arises.

Human service professionals working in hospice then must be comfortable confronting the realities of death within themselves before they can ever hope to be comfortable dealing with this taboo with patients and families. Knowing how to respond effectively and compassionately when a family accepts social work services, but prohibits any discussion of the terminal illness, requires clinical skills based not only on good training and education, but also on the human service professional's self-awareness and comfort level in dealing with these difficult issues.

PLANNING FOR THE DEATH   The human service professional working in hospice also assists the patient and family with the practical aspects of planning for increased disability and eventual death. Such practical planning may include something as specific as assisting the patient and family prepare *advanced directives* or as broad as helping the patient and family sort through their feelings of sadness and even anger in response to the reality of the impending death. Generally, advanced directives include the spelling out of one's end-of-life wishes. Legal documents such as do-not-resuscitate (DNR) orders, living wills, and medical powers of attorney are designed to clearly define a patient's wishes regarding the nature of their medical care if and when they reach a point where they are no longer able to make decisions for themselves. Preparing advanced directives is an emotional process, though. Imagine sitting with a patient who recently learned he is terminally ill and will likely die in less than six months and discussing whether or not the patient and his family want extraordinary measures taken to save his life when a point is reached in his disease process where he is unresponsive and stops breathing. Making a decision that essentially will mean allowing a family member to die without intervention, either through the removal of a feeding tube or not using cardiopulmonary resuscitation (CPR) to revive their loved one, often generates feelings of immense guilt at the prospect of abandoning their family member. Such emotional turmoil has the potential to create significant conflict and rifts within a family system that is already buckling under the emotional strain of their impending loss. A human service professional's role then is not simply to assist the patient and family with the practical matters involved with preparing advanced directives, but to help the family navigate this emotionally rocky path as well.

Another role of the human service professional is to assist the patient with the preparation of *funeral arrangements*. The thought of planning one's own funeral might seem rather morbid to some, but it can actually be rather therapeutic for someone who is facing a terminal illness or other life-limiting condition. Consider experiencing a life event that stripped you of all control—you can no longer plan for your future because you have only six months to live, you can no longer bound out of the door for a morning jog or even to run errands whenever the mood strikes. A terminal illness robs its victims of their hopes for the future, but it also robs them of their control in all respects, particularly in their everyday lives, and patients—even aging patients—often struggle with the reality of their increasing dependence on others. Planning their funeral, such as selecting scriptures, music to be played, whether it will be a celebration of life, or a more traditional and formal funeral, a graveside service, or a memorial service with no coffin, gives patients a sense of control in the midst of their increasing powerlessness.

The hospice human service professional can utilize what might initially appear to be a practical matter (making funeral arrangements) to facilitate discussions and elicit feelings about the patient's increasing debilitation and resultant confinement and dependence. I recall working with a hospice patient who at the age of 93 years shared heartfelt grief at the thought that he could no longer take his dog for a walk or run to catch up with a friend. In his confinement to a bed, he recalled how he had taken his physical freedom for granted and felt powerless and hopeless in response to the realization that his body could no longer cooperate with what his mind wanted to do. Planning his funeral was the one thing he felt he still had control over in the midst of the powerlessness he felt in every other aspect of his life.

The Spiritual Component of Dying

Hospice care has its roots in the caring of the dying by religious orders, because religious leaders recognized the spiritual component of facing one's mortality and eventual death. Even though religious issues and spiritual concerns may technically fall under the purview of the hospice chaplain, every professional on the hospice team will likely be asked by a patient or family member to pray with them, and human service professionals, including bereavement counselors must be comfortable in doing so, even if they do not happen to share the same faith as the patient. Facing one's mortality can be a frightening experience for many, and relying on or reconnecting to the faith of one's youth is a common experience for those dying of a terminal illness.

Counseling commonly takes on a spiritual tone as hospice patients attempt to make sense out of their terminal diagnosis. Patients might experience anger, confusion, and a loss of hope and may seek answers from God, yet pose these questions to the human service professional. Although no one expects someone in human services to be an expert in theology, it is important that the human service professional feel comfortable enough to help the patient sort through these questions, and even if questions cannot be answered, the human service professional can then direct a pastor or other religious leader to the patient.

Death and Dying: Effective Bereavement Counseling

Several research surveys have noted that whereas about 60 percent of human services and social work programs at both a bachelor's and a master's level offered courses related to death and dying, these courses were primarily offered as electives, and only about 25 percent of students actually took them. Related studies found that over 60 percent of new human service professionals felt as though their educational program did not adequately prepare them for counseling clients dealing with end-of-life issues (for a complete discussion of these surveys, see Kramer, Hovland-Scafe, & Pacourek, 2003). This is unfortunate because many human service professionals work directly or indirectly with death and dying issues, including loss and bereavement. In light of this, it is essential that those in the human services field obtain the necessary education and training so that they feel competent in providing services to clients dealing with death and dying.

The Journey Through Grief: A Task-Centered Approach

Several theoretical models are available for dealing with bereavement related to death and dying. Traditional grief models, including Elisabeth Kübler-Ross's (1969) model of grief, depict grieving in terms of distinct, but overlapping stages, where a mourner meets a loss with a sense of *denial* and disbelief, then moves on to the *anger* stage, where the mourner often feels a sense of injustice and rage in response to the loss. The object of the anger varies depending on the circumstances surrounding the loss, but might include being angry with God, the loved one who died, or everyone in general. The next stage is marked by the mourner *bargaining* to avoid the loss. Individuals whose loss is due to a death will often bargain with God—perhaps promising a sinless life if their loved one can be returned to them. The stage of *depression* follows the bargaining stage. During this stage mourners experience deep melancholy, often citing a sense of hopelessness and despair. The final stage of grieving involves the mourner's *acceptance* of the loss. Although Kübler-Ross's stage theory has dominated the field of grief and loss for many years, there has been a recent turn away from perceiving the mourning process as one where the bereaved progress through distinct emotional stages.

Many contemporary theorists have recently focused more on task theories, which suggest that mourners are confronted with tasks or challenges they need to conquer as they make their way on their grief journey. Alan Wolfelt, a *thanatologist* (an expert on death and grieving), has developed a task-based theory of grief and loss. Wolfelt (1996) cites seven reconciliation needs that both adults and children need to face and tackle to find healing. It is interesting to note that Wolfelt does not discuss healing in terms of acceptance, which he believes may put too much pressure on the bereaved, particularly those mourning a significant loss, such as the death of a child.

Wolfelt's seven reconciliation needs include acknowledging the reality of the death, embracing the pain of the loss, remembering the person who died through memories, developing a new self-identity in the absence of the loved one, searching for some meaning in the loss, receiving ongoing support from others, and reconciling the grief (reconciling is different than acceptance).

Bereavement counseling can be facilitated by human service professionals from various disciplines, including a human service generalist with a bachelor's or master's degree, a licensed therapist or social worker, or even hospice volunteers. In fact, it is typically a volunteer who follows up with family members after the death of the patient to explore how the surviving family members are faring, as well as to determine the need for ongoing bereavement counseling. Human service professionals who conduct bereavement counseling may do so on an individual basis, but will commonly facilitate support groups focusing on a particular loss. Groups for children surviving the loss of a parent or groups for widows or widowers are examples of grief-specific bereavement support groups. Most hospices offer free bereavement counseling for up to one year after the death of the patient as a part of the full continuum of care. Knowing that their loved ones will be cared for after their death often provides a sense of comfort for dying hospice patients; thus, bereavement counseling is an important aspect of hospice care.

> ## Human Services Delivery Systems
>
> Understanding and Mastery of Human Services Delivery Systems: Major models used to conceptualize and integrate prevention, maintenance, intervention, rehabilitation, and healthy functioning
>
> ---
>
> Critical Thinking Questions: Kübler-Ross's stages of grief and Wolfeldt's task-centered approach to grieving provide two models that human service professionals can draw on to assist the families of clients with terminal diagnoses. How might a human service professional use these with clients and their families to help them prepare for the client's death? How might they be used following the death to help the family cope?

Multicultural Issues

In general, individuals from many ethnic minority and migrant groups tend to underutilize hospice care. The reasons for this underrepresentation appear to relate to numerous factors, including lack of awareness of hospice care; Medicare regulations, which create barriers for immigrant, low-income, and minority groups; a lack of diversity within the hospice staff leading to a general mistrust and discomfort with hospice services; and a lack of knowledge of hospice care on the part of many physicians who serve minority populations. Many ethnic groups maintain values that are inconsistent with hospice values and perceive acceptance of death negatively, and although this attitude is not significantly different from Western values in general, many within the majority culture have slowly adopted new cultural values that espouse acceptance of death as an important part of life.

A 1999 study that examined barriers to hospice service for African Americans found that many African Americans held religious beliefs that conflicted with the hospice philosophy. Subjects stated that they did not feel it was appropriate to talk about, plan for, or accept their death. In addition, a majority of the subjects interviewed stated that they felt more comfortable turning to those within their own community, particularly their church, for support during times of crisis, rather than to strangers within the healthcare system (Reese, Ahern, Nair, O'Faire, & Warren, 1999).

Researchers involved in this study acknowledge the importance of not pushing a service on the African American culture if it is truly unwanted and perhaps even unneeded, but they cite leaders within the African American community who argue that members of the community would in fact benefit from hospice care, stating that a chief reason why hospice care is often rejected lies more in the lack of knowledge about the services provided. Thus, rather than accepting these differences in philosophy, the principle investigators suggest that hospice agencies adapt their services to meet the needs of the African American community (Greiner, Perera, & Ahluwalia, 2003; Reese et al., 1999).

> Although there may be multiple barriers facing some populations in receiving hospice care services—some financial and some cultural—one of the foundational values of the hospice philosophy is that hospice care will be available to every dying individual.

No research has been conducted to date on usage patterns or barriers to service for Asian Americans, Latino/a Americans, or Native Americans, but similar issues are likely to emerge within these communities as well. It is imperative that hospice agencies remain flexible enough to meet the needs of all cultural groups and that policies that either directly or inadvertently discriminate against ethnic minority groups, such as various admittance requirements, be challenged and if possible changed so that all individuals who desire hospice care can benefit from this service. Although there may be multiple barriers facing some populations in receiving hospice care services—some financial and some cultural—one of the foundational values of the hospice philosophy is that hospice care will be available to every dying individual.

Certainly hospice administrators are responsible for developing admittance policies that do not directly or inadvertently discriminate against low-income patients while protecting the financial status of the hospice. But human service professionals who are professionally committed to advocating for low-income and underserved populations are in the unique position of securing financial assistance in the form of private and government assistance through effective case management.

Another ethical dilemma faced by hospice staff involves the issue of euthanasia, or physician-assisted suicide. Dr. Jack Kevorkian made national headlines in the 1990s for assisting numerous terminally ill patients in the ending of their lives and is now serving a prison sentence. Because euthanasia is illegal in all states except Oregon, patient requests for physician-assisted suicide create an ethical dilemma complicated by the illegal nature of such an act. Requests for physician-assisted suicide present a particularly challenging ethical dilemma for conservative faith-based hospice agencies that believe that issues related to death and dying fall under the sole dominion of God (Burdette, Hill, & Moulton, 2005).

Those who believe that euthanasia should be legalized typically cite an argument based on the inalienable human right to choose death when pain and suffering robs them of a meaningful life. Although a counterargument could be based on the meaningful nature of suffering, a better argument might be based on the hospice philosophy that dying persons have a right to die without physical, emotional, and spiritual pain. In fact, several studies examining similarities among terminally ill patients expressing a desire to hasten their deaths found that the chief reasons cited included (1) depression and a sense of hopelessness, (2) poor symptom management, (3) poor social support, (4) fear of becoming a burden on family members, and (5) a poor physician–patient relationship (Kelly et al., 2002; Leman, 2005). Thus, the question is: If these issues could be addressed effectively, would these same patients still seek physician-assisted suicide?

Although the hospice philosophy advocates for neither hastening nor postponing death, hospice agencies have more in common with supporters of physician-assisted suicide than one might initially think. In fact, the leading reasons among terminally ill patients for requesting a quicker end to their lives listed previously include the very issues hospice care is designed to manage. Hospice workers can respond to this ethical dilemma by advocating for the meaningful nature of the dying process from spiritual, psychological, and social perspectives, made possible when patients are helped to confront feelings of sadness and hopelessness, when symptoms are well managed, when social support is bolstered, when families are assisted with the care of the patient, and when the hospice physician maintains a close relationship with patients based on a palliative care model. In fact, one human service professional working in hospice explained that if a choice is made to cut the dying process short, then many opportunities for growth and even last-minute resolution may be lost, as it is often the last weeks, days, hours, or even minutes of a person's life that many lifelong problems are resolved. Hospice advocates cite the value of every life experience and remind us how these types of end-of-life realizations and resolutions also benefit surviving family members and friends (Mesler & Miller, 2000).

Concluding Thoughts on Human Services in Hospice Settings

Human service professionals perform a valuable service to hospice patients and their family members and serve an important function on the hospice team. Although other members of the hospice team may perform case management and counseling services as a function of their role as hospice team members, neither the nurses nor the chaplains have the same approach to service provision as do professionals in the human services field. Unfortunately, with the increasing reliance of hospices on Medicare benefits, the psychosocial component of hospice care has eroded. This is primarily due to Medicare's (and managed care in general) cost-containment efforts, and because each service provider is billed separately in many hospice agencies, social work services have come to be seen as an "optional" service unless otherwise prescribed by law (Reese & Raymer, 2004).

Some hospice experts are concerned that this attitude has led to a "turf war," particularly among some nurses who are in the position of determining the family's needs. Reese and Raymer (2004) caution that although nurses often provide some psychosocial care, they are not trained to perform services in the same manner and with the same focus as human service professionals. In fact, Reese and Raymer's research study was borne out of this concern among social work leaders. The authors' recommendations include that hospices work toward the goals of human service professional involvement in all intake interviews and that social work involvement not be solely on a crisis or as-needed basis, because ongoing social work intervention will likely prevent many of these crises in the first place.

Finally, the authors challenge the common notion that social service involvement increases and strains budgets, suggesting that although budgets might increase initially with social work involvement, consistent social work intervention from case inception reduces financial outgo in the long run as expensive and time-consuming crises are avoided. This contention is based on the well-researched connection between many psychosocial and physical crises, where many medical emergencies requiring costly intervention have their origin at least in part in the psychosocial realm, such as patient depression and anxiety (Reese & Raymer, 2004).

Another challenge facing hospice agencies is the well-established pattern of patients being referred for hospice far too late for any of the meaningful work to be effectively accomplished. Despite the immense growth of the hospice movement and the general assumption that hospice care is a wonderful concept, only 22 percent of dying individuals are actually referred for hospice services, and of these about three-quarters are referred within three weeks of their death (Stein, 2004). Lorenz, Asch, Rosenfeld, Lui, and Ettner (2004) cited numerous barriers to hospice admission including patients being rejected for hospice admittance because they were still seeking curative medical treatment such as chemotherapy. Lorenz et al. recommended that hospices re-examine their enrollment policies that might inadvertently exclude appropriate patients from receiving services. They suggested that there might be a link between the general knowledge that the majority of hospices deny enrollment to patients still undergoing curative treatment and the fact that the majority of dying patients are either not referred at all to hospice or are referred so late in their disease process.

It seems clear that hospices must take responsibility for developing educational programs focusing on the nature of hospice care and the importance of early referral. As experts in the psychosocial dynamics commonly at play in end-of-life care, those within the human services field can lead these educational efforts both with the hospice administrators who determine enrollment policies and within the medical community and general public. A family's willingness to forgo curative treatment immediately on learning of the terminal diagnosis (necessary for hospice referral) is likely an unrealistic expectation on the part of hospice administrators. Deciding to pull a feeding tube or stop chemotherapy are psychosocial issues that evoke considerable emotional turmoil within families and could be considered a psychosocial goal of hospice counseling. Thus, although continuing to actively seek a cure is clearly contrary to the hospice philosophy, perhaps the transition from curative to palliative care could be one that occurs as a part of hospice care, not a condition of it.

Human service professionals are an integral part of the hospice team and must remain so for hospice care to remain true to its original goals and philosophy. But human service professionals must also be on the front lines of effecting change within the hospice field, which will ensure that hospice care is flexible in meeting the needs of a changing society.

Professional History

Understanding and Mastery of Professional History: Historical and current legislation affecting services delivery

Critical Thinking Question: Both Medicare and private insurance tend to list the services of human service professionals as "optional" forms of care in a hospice setting. How does this affect the ability of human service workers to adequately perform their jobs? How does it impact the dying patient and her or his family?

The following questions will test your knowledge of the content found within this chapter.

1. Since everyone handles a medical crisis differently, it is important for the human service professional to
   a. educate the family on the best way to manage the crisis
   b. match the level of emotion exhibited by the patient's family
   c. quickly recognize the family's coping style
   d. None of the above

2. During the early years of the HIV/AIDS crisis, the role of the medical human service professional or human service professional focused almost exclusively on
   a. the crisis of receiving a terminal diagnosis
   b. the chronic care needs often associated with an AIDS diagnosis
   c. the discrimination often endured by AIDS sufferers
   d. Both A and C

3. What groups are the most significantly impacted by the HIV/AIDS virus?
   a. Women
   b. Homosexuals
   c. Ethnic minorities
   d. Heterosexual females

4. The first hospice, St. Christopher's hospice of London, was founded in 1967 by
   a. Dorthea Dix
   b. Jane Addams
   c. Clifford Beers
   d. Dame Cicely Saunders

5. A core value of the hospice philosophy is that each person has the right to
   a. keep fighting for life, even when all seems hopeless
   b. die without pain
   c. continue curative medical treatment even after receiving a terminal diagnosis in order to keep hope alive
   d. Both B and C

6. A 1999 study that examined barriers to hospice service for African Americans found that many African Americans
   a. held religious beliefs that conflicted with the hospice philosophy
   b. did not feel it was appropriate to either talk about, plan for, or accept their death
   c. felt more comfortable turning to those within their own community, particularly their church, for support during times of crises, rather than to strangers within the healthcare system
   d. All of the above

7. What are some ways in which human service professionals can assist patients in an emergency room setting?

8. Describe the hospice philosophy, including some of the tasks a human service professional working in hospice may engage in with terminally ill clients, and their family members.

Suggested Readings

Byock, I. (1997). *Dying well*. New York: Riverhead Books.

Callanan, M., & Kelley, P. (1997). *Final gifts: Understanding the special awareness, needs, and communications of the dying*. New York: Bantam Books.

Klaas, D., Silverman, P. R., & Nickman, S. L. (1996). *Continuing bonds: New understandings of grief*. Washington, DC: Taylor & Francis.

Lord, J. H. (1992). *Beyond sympathy: What to say and do for someone suffering and injury, illness or loss*. Ventura, CA: Pathfinder Publishing.

McCracken, A., & Semel, M. (1998). *Broken heart still beats after your child dies*. City Center, MI: Hazelden.

Internet Resources

Hospice Foundation: http://www.hospicefoundation.org
Hospice.net: http://www.hospicenet.org/

The National Hospice and Palliative Care Society: http://www.nhpco.org/templates/1/homepage.cfm

References

Acevedo, V. (2008). Cultural competence in a group intervention designed for Latinos living with HIV/AIDS. *Health & Social Work, 33*(2), 111–120.

Burdette, A. M., Hill, T. D., & Moulton, D. E. (2005). Religion and attitudes toward physician-assisted suicide and terminal palliative care. *Journal for the Scientific Study of Religion, 44*(1), 79–93.

Centers for Disease Control. (1988). Quarterly report to the domestic policy council on the prevalence and rate of spread of HIV and AIDS—United States. *Morbidity and Mortality Weekly Report, 37*(36), 551–554.

Centers for Disease Control and Prevention. Table 5a. Estimated numbers of cases and rates (per 100,000 population) of AIDS, by race/ethnicity, age category, and sex, 2004—50 states and the District of Columbia. Retrieved July 12, 2012, from http://www.cdc.gov/hiv/surveillance/resources/reports/2004report/pdf/table5.pdf

Centers for Disease Control and Prevention. HIV Surveillance Report, 2010; vol. 22. http://www.cdc.gov/hiv/topics/surveillance/resources/reports/. Published March 2012. Accessed June 12, 2012.

Frasca, T. (2008). Shaping the new response: HIV/AIDS and Latinos in the Deep South. Latino Commission on AIDS. Retrieved January 28, 2012, from http://img.thebody.com/press/2008/DeepSouthReportWeb.pdf

Galambos, C. M. (2004). The changing face of AIDS. *Health & Social Work, 29*(2), 83–85.

Gottlieb, M. S., Schroff, R., Schanker, H. M., Weisman, J. D., Fan, P. T., Wolf, R. A., et al. (1981). *Pneumocystis carnii* pneumonia and mucosal candidiasis in previously homosexual men: evidence of a new acquired cellular immunodeficiency. *New England Journal of Medicine, 305*(24), 1425–1431.

Greiner, K. A., Perera, S., & Ahluwalia, J. S. (2003). Hospice usage by minorities in the last year of life: Results from the National Mortality Follow Back Survey. *Journal of the American Geriatrics Society, 51*, 970–978.

Kaplan, L. E., Tomaszewski, E. S., & Gorin, S. (2004). Current trends and the future of HIV/AIDS services: A social work perspectives. *Health & Social Work, 29*(2), 153–159.

Kelly, B., Burnett, P., Pelusi, D., Badger, S., Varghese, F., & Robertson, M. (2002). Terminally ill cancer patients' wish to hasten death. *Palliative Medicine, 16*, 335–339.

Kramer, B. J., Hovland-Scafe, C., & Pacourek, L. (2003). Analysis of end-of-life content in social work textbooks. *Journal of Social Work Education, 39*(2), 299–320.

Kübler-Ross, E. (1969). *Living with death and dying.* New York: Macmillan Publishing Co.

Leman, R. (2005). *Seventh annual report on Oregon's death with dignity act.* State of Oregon, Department of Human Services, Office of Disease Prevention and Epidemiology. Retrieved March 2, 2004, from http://oregon.gov/DHS/ph/pas/docs/year7.pdf

Lorenz, K. A., Asch, S. M., Rosenfeld, K. E., Lui, H., & Ettner, S. L. (2004). Hospice admission practices: Where does hospice fit in the continuum of care? *Journal of Geriatrics Society, 52*, 725–730.

MacDonald, D. (1991). Hospice social work: A search for identity. *Health & Social Work, 16*(4), 274–280.

Maslow, A. (1954). *Motivation and personality.* New York: Harper.

Mesler, M. A., & Miller, P. J. (2000). Hospice and assisted suicide: The structure and process of an inherent dilemma. *Death Studies, 24*, 135–155.

National Association of Human service professionals. (1990). *Clinical indicators for social work and psychosocial services in the acute care medical hospital.* Washington, DC: Author.

Oliver, D., & Peck, M. (2006, September). Inside the interdisciplinary team experiences of hospice human service professionals. *Journal of Social Work in End-of-Life & Palliative Care, 2*(3), 7–21.

Paradis, L., & Cummings, S. (1986). The evolution of hospice in America toward organizational homogeneity. *Journal of Health and Social Behavior, 27*(4), 370–386.

Reese, D., & Raymer, M. (2004). Relationships between social work involvement and hospice outcomes: Results of the National Hospice Social Work Survey. *Social Work, 49*(3), 415–422.

Reese, D. J., Ahern, R. E., Nair, S., O'Faire, J. D., & Warren, C. (1999). Hospice access and use by African Americans: Addressing cultural and institutional barriers through participatory action research. *Social Work, 44*(6), 449–559.

Saunders, C. (1958). Dying of cancer. *St Thomas's Hospital Gazette, 56*(2), 37–47.

Stein, G. (2004). Improving our care at life's end: Making a difference. *Health & Social Work, 29*(1), 77–79.

Weaver, H. N. (1999). Through indigenous eyes: Native Americans and the HIV epidemic. *Health & Social Work, 24*(1), 27–34.

Wolfelt, A. (1996). Healing the bereaved child: Grief gardening, growth through grief, and other touchstones for caregivers. Fort Collins, CO: Companion Press.

CHAPTER 11

Substance Abuse and Treatment

Andy Sotiriou/Andy Sotiriou/
Getty Images

Learning Objectives

- Understand the history of substance abuse in the United States, including usage trends, and the various ways in which society has responded to those who abuse substances
- Become familiar with the different theoretical models of substance use and abuse
- Become familiar with the various types of practice settings where individuals with substance abuse problems seek treatment
- Explore the most effective treatment interventions and modalities most commonly utilized by human service professionals in response to various types and levels of substance abuse problems in the United States.
- Become familiar with the importance of incorporating cultural sensitivity when addressing substance abuse issues, including identifying vulnerable groups and treatment obstacles facing certain ethnic minority groups.

Over 22 million people in the United States suffer from either a substance abuse or a substance dependence problem, with the greatest number of people being addicted to marijuana, and a growing number developing addictions to prescription drugs (Substance Abuse & Mental Health Services Administration [SAMHSA], 2012). Every day, human service professionals are intricately involved in prevention efforts and in providing treatment services for individuals and families in over 11,000 substance abuse treatment programs in the United States (SAMHSA, 2005).

Despite the widespread nature of the substance abuse problem in the United States, specialized treatment is often viewed as a part of human service practice set apart from the mainstream, seen as operating completely independently from all other services. Many human service professionals express an aversion to working with substance-abusing clients, and some believe that one must be a recovering addict to effectively counsel others with this problem. Until fairly recently, most human service and mental health providers did not receive specific training in substance abuse issues as a part of their normal course of studies.

In practice, however, all human service professionals are affected by the issue of substance abuse. Although only a small percentage may work directly in specialized substance abuse treatment programs, all will find that the issue of substance abuse frequently touches the lives of the clients with whom they work. All human service professionals need to be familiar enough with the dynamics of substance use, substance abuse, substance dependence, and addiction to be able to recognize when it may be a primary or secondary problem for their clients. Human service professionals also need to be aware of their own feelings and attitudes that may help or hinder their ability to work effectively with both clients who have substance abuse problems and those whose lives have been affected by the substance abuse of others.

Those who do choose to work directly in substance abuse treatment will encounter a diverse field with many practice settings. Human service professionals may focus on prevention and voluntary treatment with chemically dependent clients and their families, or even with mandated clients within the criminal justice system. In this chapter, we will examine the history and evolution of substance abuse treatment in the United States. We will then explore the many meaningful roles that human service professionals fulfill in this challenging area of practice.

> **In practice, all human service professionals are affected by the issue of substance abuse.**

History of Substance Abuse Practice Setting

Throughout recorded history, people have used psychoactive substances to change how they feel. Evidence from the earliest prehistoric and ancient civilizations indicate the use of fermented grains and honey to produce alcoholic beverages and the use of plants containing psychoactive substances in medicinal and religious rituals. The particular substance of choice has varied with time and from one society to another, but the use and abuse of substances have been so prevalent as to be routinely regarded as part of the human condition.

Most societies sanction some use of psychoactive substances. In the United States, it is legal for adults to consume alcohol, nicotine, and caffeine, which are all drugs that affect the central nervous system. The use of other psychoactive drugs in the United States is either prohibited or regulated. Many uses, such as the medical use of marijuana or the use of the peyote cactus in religious ceremonies by some Native Americans, remain controversial and the subject of ongoing legal and public policy debates at the state and federal level (Inaba & Cohen, 2004).

Societies have also developed ways of responding to individuals whose use of substances "cross the line" of what is considered acceptable by creating problems for the individual and the society as a whole. How a society has responded to this problematic use has varied according to that society's beliefs about the nature of the problem. For example, societies that view substance abuse as the result of personal misconduct or moral failure tend to focus on a call to repentance and/or punishment for the offender. Societies that regard substance abuse as an illness are more likely to focus on providing treatment.

History of Use and Early Treatment Efforts Within the United States

Attitudes and practices regarding substance use and abuse in the United States have undergone significant changes over time and continue to evolve. William White (1998) traced the history of addiction treatment and recovery in the United States, focusing on the development of the professional field that has emerged in response to the problem of substance abuse. This historic review provides perspective on the prevalence of the substance abuse problem from the very beginning of U.S. history. In exploring social attitudes, White noted that

> [a]lcohol use and occasional drunkenness were pervasive in colonial America, but it wasn't until per capita alcohol consumption began to rise dramatically between the Revolutionary War and 1830 that Americans began to look at excessive drinking in a new way and with a new language. (p. xiii)

The term *alcoholism* was first introduced by physician Magnus Huss in 1849, but it took another 100 years, and the birth of Alcoholics Anonymous (AA), for the term to become fully accepted (White, 1998).

Early efforts to provide treatment for substance abuse began in the United States in the mid-1800s, prompted by public concern over the problems resulting from increased levels of public drunkenness. White (1998) traced the roots of this increase back to colonial America, describing the variety of attitudes and practices regarding drug and alcohol use held by the diverse cultural groups that immigrated to colonial America. Many immigrant groups had previously used drugs or alcohol only in moderation and often in the context of social, religious, or medical practices. Wine may have been used to celebrate a wedding, partake in a communion service, or deaden the pain of an injury, but excessive use of alcohol was often condemned.

Coming to colonial America, immigrants were affected by what White described as "the utter pervasiveness of alcohol," which was consumed throughout the day by virtually everyone: man, woman, and child. Alcohol was commonly integrated into everyday social and political life, often in the form of more concentrated distilled liquor such as whiskey and rum. Native Americans, who previously used only weak forms of alcohol ceremonially, were also affected by the introduction of distilled liquor.

A number of laws were passed in an effort to combat public drunkenness and vagrancy, but drinking itself was not yet perceived as a problem. Other psychoactive substances in common use included laudanum, opium-laced alcohol used for many medical problems, and tobacco, a major crop for both domestic use and export (Inaba & Cohen, 2004).

By the end of the colonial period, there was a shift in societal attitudes about the use of alcohol in the United States. Instead of being seen as a blessing of God, it was increasingly seen as a curse. This shift in thinking birthed the temperance movement, which initially focused on encouraging moderate use of alcohol (thus the term *temperance*), but eventually came to advocate total abstinence from alcohol when it became clear that problem drinkers were frequently unable to maintain moderate drinking. This shift in thinking coincided with the rise of medicine as a profession. Dr. Benjamin Rush suggested that chronic drunkenness represented a "progressive medical condition" rather than a moral failure, thus introducing the disease concept of alcoholism (White, 1998).

The Prohibition Movement

Attempts to eliminate drug and alcohol problems through legal prohibition lead to the passage of several pieces of federal legislation. In 1906, the Pure Food and Drug Act established the Food and Drug Administration (FDA) and gave it authority to approve all drugs meant for human consumption, to establish that certain drugs required a prescription, and to mandate warning labels on drugs that were potentially habit forming. (Prior to this time, drugs such as opium and cocaine were freely available and not regulated.) In 1914, the Harrison Act was passed, which regulated the *medical* use of certain drugs such as opium, morphine, cocaine, and their derivatives and, at the same time, criminalized the *nonmedical* use of these same drugs (Whitebread, 1995).

The temperance movement was successful in establishing alcohol prohibition laws in many states, and eventually the ratification of the Eighteenth Amendment in 1919 made alcohol manufacture, transportation, and sale illegal in the United States. Musto (1999) noted that the Eighteenth Amendment, like earlier state prohibition laws, enjoyed wide public support and reflected societal fear that even small amounts of alcohol posed a danger both to the individual and to society as a whole.

Prohibition, described by President Hoover as a "noble experiment," proved to be short-lived. The Twenty-First Amendment repealed the Eighteenth Amendment in 1933, ending Prohibition and thereby legalizing the manufacture and sale of alcohol once again in the United States. Several factors provided the impetus for this change, including the widespread disregard for the law and the rise of organized crime in the production and distribution of bootleg liquor. Inaba and Cohen (2004) concluded, however, that the widespread belief that Prohibition was a failure is incorrect. "An examination of medical records concerning diseases caused by excess alcohol consumption as well as criminal justice records shows that Prohibition did reduce health problems, domestic violence, crime, and consumption" (p. 323).

The perceived failure of Prohibition to rid society of drug and alcohol problems, the closing of specialty addiction treatment programs, and the financial hardships of the Great Depression combined to create an atmosphere in the 1930s that offered little help or hope for those with drug and alcohol problems (White, 1998). This combination of factors made the climate right for the birth of the mutual aid society of AA, "a fellowship of men and women who share their experience, strength and hope with each other that they may solve their common problem and help others to recover from alcoholism" (n.d.). The growth of AA from two men (known simply as Dr. Bob and Bill W.) meeting in Akron, Ohio, in 1935 to a worldwide organization with over 50,000 meetings (Abadinsky, 2004) is indeed remarkable and represents a major component in the development of the current treatment of addictions.

The Rise of Modern Addiction Treatment in the United States

Several factors shaped the course of addiction treatment in the United States during the second half of the 20th century up until the present time. The growth of AA played a major role in the broad (but by no means universal) acceptance of the *medical model* of addiction treatment. The establishment of private health insurance provided increased access to treatment for a greater percentage of the population; this in turn led to a significant increase in the number of substance abuse treatment programs. After initially operating as separate entities, alcohol treatment and drug treatment services combined at both the public and private level in favor of substance abuse treatment that serviced both populations. With this change came further professionalization of the field.

Finally, the development of *managed care* as a means of controlling rising health care costs led to a shift from inpatient hospital treatment to outpatient services as the treatment setting most frequently authorized and approved by insurance carriers. Each of these factors has a significant impact on how human service professionals provide substance abuse treatment today (White, 1998). Before examining the various treatment settings available

▶ **Professional History**

Understanding and Mastery of
Professional History: Creation of
human services profession

Critical Thinking Question: Over time,
societal attitudes about drug and alcohol
consumption and abuse have changed,
driving policies and practices related
to the availability and use of these substances. How do our culture's current
attitudes about the use of alcohol and
drugs shape the field of substance abuse
treatment, and vice versa?

· ·

today, it is important to understand the scope of the problem, the professional vocabulary used to define the problem, and the ongoing effect that societal attitudes and perceptions have on the availability and
utilization of services.

Demographics, Prevalence, and Usage Patterns

Over the years that I have worked in addiction treatment, I have
spoken to many community groups. I often begin by asking them to
describe to me their picture of an alcoholic or a person addicted to
drugs. There is always a wide range of responses. As we begin, someone will usually mention the man on skid row, drinking out of a bottle
concealed in a brown paper bag. Others think of the image of a "drug
bust" on a television crime show, police breaking down the door as the
people inside scramble to flush drugs down the toilet. As the discussion progresses, some brave soul will bring the examples closer to home. They may say,
"I remember my father, drunk and passed out on the couch every night" or "My favorite
aunt is in detox right now…I've lost count of how many times she's been there." The next
person may add, "My brother is in jail right now for drug possession" or "I've been in AA
for five years now." Invariably, what begins as a discussion that focuses on someone else's
problems "out there" in society becomes personal to the group. When I have this same
discussion with students, they are often surprised to realize how many of their classmates'
lives are affected by substance abuse.

Although it is certainly true that substance abuse is a problem that exists within all
levels of society, it is usually this type of facilitated discussion that brings home this very
point. As you continue to read this chapter, I encourage you to consider how substance
abuse affects your life at both the personal and professional level. Because of the prevalence of the problem, and because each person with a substance abuse problem affects
the lives of the people around them, most can identify a direct link to this issue.

SAMHSA, a division within the Department of Health and Human Services (HHS),
conducts an annual survey on the prevalence of substance use in the United States
and the problems associated with that use. In 2011, an estimated 133 million North
Americans (12 years of age or older) were current drinkers of alcohol. This represents
just over one-half of the population. Just over 58 million people over the age of 12
(22.6 percent of the population) had engaged in binge drinking. During this same time
period about 16 million people over the age of 12 engaged in heavy drinking and illicit
drug use (about 6.3 percent of the population). Marijuana was the most commonly used
illicit drug, followed by psychotherapeutics (nonmedical use of prescription drugs),
cocaine, hallucinogens, and inhalants (SAMHSA, 2011).

What do these numbers mean to the human service professional? At a minimum, they
alert us to the reality that a significant number of the clients with whom we work in any
practice setting already have a primary substance abuse problem with illicit drugs or alcohol and that many others are using alcohol in a way that may complicate their current problems and affect their ability to utilize or benefit from any services we may offer to them.

These statistics also reinforce the need for all human service professionals to have a working knowledge of addictions so that they can accurately assess the needs of their clients.

The consequences of drug and alcohol abuse in the United States are enormously costly. Although the costs can be evaluated in dollars, they are more readily understood in human terms: family discord, neglect and/or abuse of children, personal misery, financial straits, medical problems, fetal alcohol syndrome, HIV infection, lower work productivity, and job loss—and the list goes on. Combating and reducing the source of these problems have proven to be difficult indeed, but one of the most straightforward and least controversial ways is to provide effective treatment to drug abusers (Boren, Onken, & Carroll, 2000).

Defining Terms and Concepts

Thus far, we have used the terms *substance use, substance abuse,* and *alcoholism* in a general way, without providing detailed definitions. It is important to understand how these terms are understood in the professional community. As noted earlier, during much of the 20th century, treatment for alcohol problems was conducted separately from treatment for problems with other drugs (White, 1998). The term *alcoholism* came into common use with the acceptance of the medical model and the understanding of alcoholism as a disease. Alcoholism and drug dependence has been defined in many different ways, but most experts describe alcohol and drug dependence as a chronic and progressive disease that is influenced by one's environment. Individuals who suffer from alcohol and drug dependence may struggle constantly with their addictions, or cyclically, but remain preoccupied with alcohol and drugs even though their use has very negative consequences—both psychologically and related to their lifestyle. In extreme cases, dependence on alcohol and drugs can be fatal (Morse & Flavin, 1992).

Gradually, during the second half of the 20th century, the treatment community focused less on the differences between alcohol abuse and abuse of other drugs and more on the similarities that existed between them. Most treatment programs are now designed to meet the needs of clients with alcohol and/or other drug problems. Currently, treatment professionals use the broad term of *substance abuse disorders* with many subtypes of the disorder, depending on the substance being used. In keeping with the medical model, these disorders are defined in the *Diagnostic and Statistical Manual of Mental Disorders,* fourth edition, text revision (*DSM-IV-TR*) of the American Psychiatric Association (2000).

In general terms, individuals are described as abusing a substance when they continue to use the substance despite experiencing negative consequences from their use. These negative consequences can include health problems; difficulties in their family, work, and social life; and financial and legal problems. Individuals are said to be *dependent* on the substance when, in addition to these negative consequences, they build tolerance and experience withdrawal if they stop using the drug. Tolerance occurs when a person's body has become accustomed to the drug and thus needs to use more in an attempt either to regain the pleasurable effects of the drug or merely to feel normal. Withdrawal symptoms occur when individuals become physically dependent, meaning that if they stop using the drug, their body will experience uncomfortable symptoms.

These symptoms vary depending on the nature of the drug use. If individuals have been using a central nervous system depressant, such as alcohol or tranquilizers, they will experience symptoms associated with their central nervous system speeding up when they stop their use. These symptoms typically include anxiety and agitation, but may be severe enough to cause grand mal seizures. Conversely, if individuals have been using a central nervous system stimulant, such as cocaine or amphetamines, they will likely experience a "crash" of exhaustion and depression when drugs are withdrawn. In severe cases, this can include suicidal thoughts and behaviors. The range of severity of withdrawal symptoms varies with the individual and with the amount of use. It is important to note that the withdrawal experienced from some drugs can be life threatening and therefore require medical supervision (Inaba & Cohen, 2004). For these reasons, addiction treatment programs must include appropriate medical professionals, either on their direct staff or available for consultation.

Theoretical Models of Use and Abuse

Although it might be ideal to present a single theoretical model that explains the nature of addictions and how they should be treated, no such model currently exists. In fact, there continues to be significant controversy over the best way to understand and to treat addictions. There are also significant advances in the knowledge of how the brain works and responds to drugs that inform and modify current treatment models.

Throughout history, there have been many theoretical models for understanding the nature and cause of substance abuse and addiction. For thousands of years, addiction was primarily seen as the result of an individual's moral failure. More recently, theories have been developed that incorporate new knowledge from psychology, biology, and medicine. Inaba and Cohen (2004) identified three prevalent models of understanding addiction: the addictive disease model (also known as the medical model), which focuses on the influence of heredity; the behavioral/environmental model, which focuses on the influences of environment and behavior; and the academic model, which focuses on the physiological effects of psychoactive drugs.

- *Addictive disease model:* We have already introduced the medical model and the related "disease concept" of addiction. Disease is defined as impairment of health or a condition of abnormal functioning. This model stresses that addiction, like other diseases, has identifiable symptoms, a predictable course, and a likely outcome if left untreated; it further understands that genetic influences may result in a predisposition, making the development of the disease more likely.

  Inaba and Cohen (2004) explain that the 'medical model' views addition as a disease that is enduring, will continue to progress (particularly without treatment), and is ultimately incurable, in fact fatal if left untreated. This model posits that at the root of all addictions is genetic irregularity within the brain's chemistry and anatomy, which is likely activated when a certain drug is abused.

- *Behavioral/environmental model:* This developmental model describes the possible progression of substance use through six stages:
  - Abstinence, meaning no use of alcohol or drugs

- Experimentation, marked by curiosity that leads to limited use
- Social/recreational use, marked by seeking out drugs/alcohol in these settings
- Habituation, meaning repeated use without negative consequences
- Abuse, defined as continued use despite negative consequences
- Addiction, meaning abuse plus the presence of tolerance and withdrawal

This model examines how factors in a person's environment, such as peer pressure or easy access to drugs, can foster the progression from one level to the next. Although abstinence is the only stage that can be seen as "risk free," note that it is not until one reaches the stages of abuse and addiction that the hallmark behaviors of continuing to use despite negative consequences, obsession with drug taking, and loss of control are seen (Inaba & Cohen, 2004).

- *Academic model:* This model understands addiction from the standpoint of the changes that occur in people's bodies over time as they use drugs. These changes occur at the cellular level and result in the development of *tolerance,* meaning that as persons become resistant to the drug's effects, they will need increasing amounts of the drug to achieve the desired effects. *Tissue dependence* occurs when the body has become so accustomed to the drug that it needs the drug "to feel normal." Even where tissue dependence does not occur, the memory of the pleasurable effects of the drug and the ongoing desire for that feeling may result in *psychological dependence.* If use is interrupted, the person may experience uncomfortable physical and psychological symptoms known as *withdrawal;* the fear and dread of withdrawal symptoms plays a major role in the addict continuing to use (Inaba & Cohen, 2004).

> ## Human Services Delivery Systems
>
> Understanding and Mastery of Human Services Delivery Systems: Major models used to conceptualize and integrate prevention, maintenance, intervention, rehabilitation, and healthy functioning
>
> Critical Thinking Question: There is still no consensus regarding a model for explaining substance abuse, although the medical, behavioral/environmental, and academic models currently all have proponents. What are each model's implications for treatment, and how important is it for a human service professional working in the field of substance abuse to remain familiar with current scholarly literature on the topic?

Inaba and Cohen (2004) propose that it is actually an integration of these models that best explains the predisposition and process by which addiction develops over the course of one's life. Each provides a type of lens through which an individual's substance abuse problem can be understood and solutions explored; they do not need to be seen as mutually exclusive.

Consider the case example about Jack in Case Study 11.1.

CASE STUDY 11.1

Case Example of an Alcoholic

Jack is a 45-year-old married man with two teenage daughters. Jack is the manager of a busy restaurant located in a shopping mall. Over many years, Jack has developed a pattern of eating his lunch in the restaurant's bar in the quieter time between the busy lunch and dinner hours. He initially drank a beer with lunch, but that has increased to three or four beers over the years. He finds that he looks for opportunities to offer a drink to regular customers and has another drink along with them. After the dinner rush, he will sit at the bar and have several more drinks as his employees do the cleaning before he locks up for the night. Jack is aware that many of his food servers use speed (amphetamines) to get through a busy shift, and he finds that he is doing this more and more himself. When he uses speed, he finds he needs to have a few extra drinks so that he can fall asleep at night. Because he still sleeps poorly, he increasingly needs the speed to get going the next day. The employees, who used to be happy to give him speed once in a while, now want him to pay for those pills, creating some financial problems. Jack is starting to feel uncomfortable that his employees know about his use and worries that it undermines his authority with them. Because Jack is drinking more, he is getting home later and is less involved with his family. Initially this caused arguments, but his wife and daughters have grown accustomed to his being either at work or passed out on the couch. They have learned to plan their life without much involvement from Jack.

Given the progressive nature of his use, Jack is now likely to experience some of the many predictable problems that could bring him into contact with a human service professional. Jack is a likely candidate for getting fired when the owner learns about his drug and alcohol use at work, for a drunk-driving arrest, for escalating family problems, or for a major health problem such as a heart attack. Any of these events could create a crisis for Jack and his family that could lead to their entering substance abuse treatment.

Jack's problem would be understood somewhat differently depending on the theoretical model held by the treatment professional assessing it. Those working from the addictive disease model would identify factors that predispose Jack for substance abuse, such as a family history of alcoholism and a work environment with easy access to drugs. They would see his increasing sleep problems, financial problems, and family tension as symptoms of an escalating disease. Those working from the behavioral/environmental model would trace how Jack's use has progressed from habitual use to abuse and likely addiction. The academic model would explain how Jack's body has developed tolerance for alcohol, needing more drinks to achieve the same results, and how Jack has begun to attempt to counteract the negative depressant effects of alcohol with stimulants. No matter which model seems most helpful to the human service professional in understanding his use, it is clear that each model provides some relevant information in conceptualizing Jack's problem. Conceptual models such as these also assist in treatment planning. If, instead of considering Jack, we examined the history of a 15-year-old cheerleader who is using speed to lose weight and is partying on the weekends or a 30-year-old homeless veteran addicted to heroin since returning from the war, we would find both similarities and differences in their substance abuse that would inform the type of treatment they need.

Types of Substances Abused

Many categories of drugs are subject to abuse because they create effects that are desirable, at least to some users. Because the psychoactive qualities of drugs differ, different people find different drugs attractive. It is extremely important that human service professionals understand the effects of these drugs, so that they are able to recognize the signs of substance use in their clients as well as to understand how they may affect how their clients perceive and utilize services. Inaba and Cohen (2004) defined a psychoactive drug as "any substance that directly alters the normal functioning of the central nervous system" (p. 32) and divided psychoactive drugs into these three broad categories.

Uppers are central nervous stimulants, increasing chemical and electrical activity. Drugs in this category include cocaine, amphetamines (such as methamphetamine), caffeine, and nicotine. Note that this category includes both legal and illicit drugs. Some of the reasons people are drawn to the use of stimulants are to increase attention and energy, to suppress appetite, and to feel more confident. These effects are the result of the forced release of the brain's "energy chemicals": norepinephrine and epinephrine, two neurotransmitters. Because tolerance builds rapidly with stimulant drugs, abuse and addiction can develop quickly.

Many physical and psychological problems are associated with the abuse of central nervous system stimulants. The depletion and imbalance of neurotransmitters can lead to depression, paranoia, and psychosis. The ongoing speeding up of the central nervous system (without time to recover) may result in insomnia and the problems associated with lack of sleep, cardiovascular problems, and weight loss. In fact, with the use of stronger stimulants, the brain "does not signal the need for food, drink, or sexual stimulation, resulting in malnutrition, dehydration, or a reduced sex drive" (Inaba & Cohen, 2004, p. 131).

A very serious substance abuse problem relates to what many experts are referring to as the methamphetamine epidemic. Methamphetamine (also called "meth," "crystal," and "crank") is a highly addictive synthetic stimulant that acts on the central nervous system, wreaking havoc on the body, particularly the brain and heart. Meth was first developed in Germany in the late 1800s, as a cure for diseases. The drug appeared on the illicit drug scene in the United States around 1979 and exploded as one of the most highly manufactured and used drugs in the United States. In 2009, 1.2 million people ages 12 and older had used meth at least once in the prior year (National Institute on Drug Abuse [NIDA], 2010). Although the drug initially gained popularity among males living in the Western part of the country, it has since grown in popularity and is now a problem across the country among both genders, and across a variety of lifestyles.

Methamphetamine use has declined considerably, from an all-time high of 731,000 active users in 2006 to 439,000 active users in 2011 (SAMHSA, 2012). First-time use is also declining, with first-time users of methamphetamine in 2011 estimated to be approximately 133,000 (12 years and older), compared to first-time users in 2006, which was estimated to be approximately 318,000. Despite this decrease, use of methamphetamine remains a considerable problem, with far-reaching consequences.

Symptoms of methamphetamine use include inability to sleep; increased sensitivity to noise; nervous physical activity, such as scratching, irritability, dizziness, or confusion; extreme anorexia; tremors or convulsions; and increased heart rate and blood pressure. Long-term affects include dependence, addiction psychosis, paranoia, hallucinations, mood disturbances, repetitive motor activity, stroke, and weight loss (NIDA, 2010). Methamphetamine is similar in structure and affect to other psychostimulants, such as cocaine, but unlike cocaine, meth remains in the body far longer and is not as easily metabolized, leading to a more sustained stimulant effect. For instance, the high from cocaine lasts about 30 minutes, whereas the high from meth can last for up to 24 hours. The increased potency of meth is one of the reasons why it is so dangerous—not only with regard to its highly addictive nature, but also because the sustained state of stimulation can cause a number of serious cardiovascular problems.

Methamphetamine use has become a significant social problem because of how highly addictive it is, and because of how it ravages the body and the mind. In fact, it not only causes serious damage to the heart and brain but also destroys users' physical appearance, including leading to rapid weight loss, accelerated aging, severe facial blemishes, and the rotting of the teeth (commonly referred to as "mouth rot"). Meth use has had a significant impact on various social systems within the country, including the criminal justice system, public health, and child welfare agencies. For instance, emergency department visits involving meth use increased 54 percent from 1995 to 2002 (SAMHSA, 2010). Meth use has also led to an increased risk of contracting HIV, and Hepatitis B and C due to increased high-risk sexual behavior and the sharing of needles (Centers for Disease Control and Prevention, 2011). Meth use has also had a significant effect on law enforcement due to a dramatic rise in meth-related robberies and domestic violence (National Association of Counties, 2006a). Also, the increasing prevalence of meth use has brought public attention to the additional dangers posed to children when their meth-abusing parents neglect their children's needs or place them at risk by creating in-home meth labs; responding to these forms of child endangerment creates challenges for human service agencies, specifically child welfare agencies, as well as law enforcement (National Association of Counties, 2006b). Human service professionals within all of these areas will no doubt be affected by the methamphetamine epidemic in one way or another; thus it is vitally important that they remain abreast of meth-use trends and ongoing treatment options.

Downers are central nervous system depressants, slowing down its overall functioning. Depressant drugs include painkillers (such as morphine, Darvon, Demerol, Vicodin, and OxyContin), sedative-hypnotics (such as Valium, Xanax, and Seconal), and alcohol (beer, wine, and hard liquor). Depressants slow heart rate and respiration, relax muscles, dull the senses, diminish pain, and induce sleep. Because they depress or lower inhibitions, the initial effect of these drugs may seem like a stimulant; someone who is drinking alcohol may feel increasingly social or sexually disinhibited, however the long-term effect is that of a depressant. As with the stimulants, tolerance builds with repeated use. As people need more of the drug to feel high, they experience more of the negative side effects of the drug: loss of coordination, impaired judgment, memory problems, and the development of physical dependence.

All Arounders is the term used by Inaba and Cohen to describe psychedelics. This category includes marijuana, LSD, phencyclidine, MDMA (Ecstasy), and mescaline. Hallucinogens distort sensory perceptions and can create altered or intensified sense of sight, touch, and hearing. Users may experience auditory and visual hallucinations or distorted thinking (delusions). Side effects from hallucinogens vary, but include increased appetite and respiratory damage (with marijuana); "bad trips" and flashbacks (with LSD); and increased blood pressure, amnesia, and combativeness (with phencyclidine). Because these drugs are generally manufactured and processed illegally, users run the risk of taking stronger doses than anticipated or even getting a different drug than anticipated. These drugs may present even greater risks for individuals with preexisting mental disorders (Inaba & Cohen, 2004).

Other drugs commonly abused include inhalants (such as glue, metallic paints, and nitrous oxide), anabolic steroids, and other "performance-enhancing" drugs. All these drugs are associated with serious health consequences that can be life threatening.

Abuse of Prescription Drugs

A growing area of concern in the United States is the abuse of prescription drugs. Much media attention has been given to the problem of "street sales" of drugs used as pain-killers, such as OxyContin, those used to treat anxiety, such as Valium and Xanax, and those used to treat attention deficit/hyperactivity disorder (ADHD) such as Ritalin. Drug addicts have long attempted to deceive and manipulate physicians into giving them prescriptions for pain medication and tranquillizers by creating or exaggerating symptoms or by altering the number of pills authorized on the prescription form. The National Center on Addiction and Substance Abuse at Columbia University (NCASAC, 1998) conducted a three-year study of the abuse and diversion of prescription medications including opioids, central nervous system stimulants and depressants, and steroids. The study found that from 1992 to 2003, the number of Americans who abuse controlled prescription drugs had nearly doubled from 7.8 million to 15.1 million. Nearly one-half of physicians surveyed reported that patients commonly try to pressure them into prescribing controlled drugs. CASA places these figures in the context of the widespread acceptance of the use of prescription medication in the United States in general and the growing acceptance of the use of psychotropic medications. A more recent study found that in the past five years the abuse of prescription drugs has remained relatively stable among females, has declined slightly for adolescents but has increased among males (SAMHSA, 2009).

> From 1992 to 2003, the number of Americans who abuse controlled prescription drugs nearly doubled, from 7.8 million to 15.1 million.

Problems with prescription drugs include those who intentionally abuse and those who inadvertently become addicted to legally prescribed medication. The CASA study suggests that this is a problem that has not been adequately addressed. In assessing for substance abuse problems, human service professionals are therefore encouraged to explore use of prescription drugs with their clients in addition to their use of any "street drugs."

Common Psychosocial Issues and the Role of the Human Service Professional

The Presence of Substance Abuse across All Practice Settings

Although some human service professionals might assert that they have no interest in working with individuals who have substance abuse problems, it is important to note that, because alcohol and drug abuse are so prevalent in the United States, it is virtually impossible to entirely avoid working with this issue.

Many human service professionals do not begin their careers with the intention of specializing in substance abuse, but quickly encounter the issue in the lives of their clients. I began my career over 30 years ago in a county public assistance office. I soon realized that many of the clients applying for General Relief were alcoholics whose long-term use of alcohol had led to loss of employment, family, and health. When I worked in hospital settings, I again found that many of the patients needing treatment were suffering from conditions that resulted from or were complicated by their use of alcohol or other drugs. I later chose to work directly in substance abuse treatment programs, eventually providing treatment in outpatient, residential, hospital inpatient, and partial hospitalization settings with substance-abusing clients and their families.

Acceptance of Problem

One of the most common practice issues human service professionals must address with substance-abusing clients is helping the client acknowledge that the substance abuse is in fact a problem. It can be perplexing for a professional to listen to clients describe various incidents occurring in their lives that clearly seem to be negative consequences of their substance abuse yet know that clients are either unable or unwilling to make that connection. Such clients may forcefully maintain that their problems have nothing to do with the substance use. Clients may describe, for example, a recurrent pattern of getting drunk (or high), followed by getting into fights with their spouse. They may even acknowledge that the fights only happen when they are using drugs, yet they still maintain that there is no connection between the two. Clients who have lost relationships, jobs, and money because of their use may still defend their alcohol or drug consumption, asserting that "with all the problems I have right now, it is the only thing that is keeping me going . . . the only friend I have left."

This *denial* of the problem is more than a psychological defense mechanism. It reflects the learned experience of most substance abusers that, at the outset of their use, the substance was giving them positive effects. A common phrase in treatment programs is *what starts out as the solution becomes the problem.* In other words, the drinking that initially provided a mild relaxation of inhibitions to feel more relaxed and sociable at a party now with increased use results in inappropriate and aggressive behavior at the party. Hence, the solution has now become the problem, but the persons using the substance are often the last to recognize this reality; they have learned to believe that it is the solution to their problems and are resistant to changing this belief. An additional consideration is that the psychoactive nature of the substance being used alters the users' thoughts and perceptions in ways that may hinder their recognition of the problem. Human service professionals who understand this dynamic are less likely to become frustrated with their clients' statements and thus are more likely to be effective in their attempts to help clients accept their problem. Figure 11.1 provides a comparison of those who perceived that they had the need for substance abuse treatment with those who actually entered a treatment program. This graphic clearly indicates the tendency of those suffering from substance abuse problems to avoid seeking treatment.

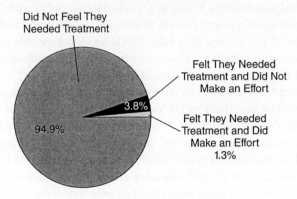

Past Year Perceived Need and Effort Made to Receive
Specialty Treatment among Persons Aged 12 or Older
Needing But Not Receiving Treatment for Illicit Drugs
or Alcohol: 2003

Did Not Feel They
Needed Treatment

Felt They Needed
Treatment and Did Not
Make an Effort

3.8%

Felt They Needed
Treatment and Did
Make an Effort
1.3%

94.9%

20.3 Million Needing But Not Receiving
Treatment for Illicit Drugs or Alcohol

FIGURE 11.1

Perceived Need and
Effort Made to
Receive Substance
Abuse Treatment

Source: SAMSHA,
Office of Applied
Studies (http://www.oas.
samhsa.gov/nhsda/2k3
nsduh/2k3overview.htm).

Hitting Bottom

Traditionally, addiction treatment professionals have thought it necessary for those addicted to drugs or alcohol to "hit bottom" before they recognize their problem and the need for treatment. Although some individuals were described as having a "high bottom" because they reached this point of recognition with relatively minor consequences such as spilling a drink on an expensive rug or one relatively minor verbal outburst, the common wisdom was that substance abusers could not be helped until they were ready to help themselves. Hitting bottom was seen as the starting point, much to the dismay of concerned family members, friends, and employers who were tired of waiting for this recognition to occur because from their perspective, their loved one "hit bottom" long ago.

Generalist Practice Interventions

There have, however, been many approaches utilized to help the substance abuser "hit bottom" more quickly. Those expressing concern for the substance abuser have been advised by treatment professionals to "stop enabling" and to instead allow the individuals to suffer the natural consequences of their use. Abusers are advised not to call in sick for the person when they are hungover, not to put them to bed when they pass out, and not to bail them out when they are in jail. Friends and family members are advised that although these enabling behaviors are well intended, they actually help substance abusers to continue to deny or minimize their problem. If, instead of waking up in bed, the drinkers wake up on their front lawns in full view of the neighbors, they experience the negative consequences of their drinking rather than having to trust the description provided by a spouse the next morning. They are less likely to believe that a spouse is "just exaggerating" or lying about events that they may be unable to remember.

This, of course, creates true dilemmas for family members. Allowing a loved one to wake up on the front lawn, lose a job, or remain in jail may lead to serious consequences for the individual as well as the family as a whole. Human service professionals can help concerned family members and friends to identify their options and to think through the implications of the actions they take. Clinicians who are able to listen nonjudgementally and communicate their understanding of the difficulty of the decisions to be made are more likely to be truly effective in this helper role.

Among the options available to families seeking help are *interventions.* Formal interventions were first described by Dr. Vernon Johnson, a pioneer in alcohol treatment, in his 1973 book *I'll Quit Tomorrow.* Although many variations have been developed in addition to the original Johnson model, interventions typically bring together all concerned individuals in the lives of the substance abusers in order to confront them with the negative consequences of their substance abuse. They meet as a group, joined by an intervention specialist, and share ways that their own lives have been negatively affected by the substance abusers' use of drugs and/or alcohol and rehearse ways to present this information to the substance abusers in the hope of breaking down their denial. They then meet with the substance abuser to share their concerns and encourage the person to enter treatment, often immediately after the meeting. The goal is to precipitate a crisis that will result in change. Styles for conducting interventions vary from collaborative to highly confrontational. Interventions have sometimes been subject to criticisms of conflict of interest when the intervention specialist is part of the staff at a treatment facility where the person is being encouraged to enter treatment.

Motivational Interviewing

Another common way to help the substance abusers recognize their need for treatment is through the use of motivational interviewing. This approach differs from methods that use confrontation or coercion to attempt to engage substance abusers in treatment. As defined by Rollnick and Miller (1995), "Motivational interviewing is a directive, client-centered counseling style for eliciting behavior change by helping clients to explore and resolve ambivalence." Hettema, Steele, and Miller (2005) explain that, in creating a safe atmosphere, motivational interviewing allows individuals to confront their ambivalence by examining both the positive and the negative aspects of changing their current behaviors. This approach acknowledges that it is normal for people to have "mixed feelings" about change and invites them to explore all sides of their ambivalence. In recognizing the cost that they pay for maintaining their substance use, individuals may become more likely to willingly choose to make a change.

Motivational interviewing (and the related motivational enhancement therapy) stresses that people vary in how ready they are to make changes. Drawing on the work of Prochaska, DiClemente, and Norcross, an individual's willingness to change is understood as occurring in stages: *precontemplation, contemplation, determination,* and *action* (Miller, 2000). Although not limited to use in substance abuse, it is seen as a helpful model that allows the human service professional to build an alliance with the client toward change. *Precontemplation,* as the name implies, is the stage people are in before they ever give any thought to the need to change their behavior (although it may be clear to others that a problem exists). If family members or clinicians try to convince substance abusers at this stage that they should take *action,* such as stopping their substance use or entering a treatment program, it is not likely that the suggestion will be positively received or even make sense to the abusers. Instead, human service professionals would focus their efforts on assisting substance abusers to become more ready to change by exploring with them the impact of their use. As substance abusers explore what they see as both the positive and the negative aspects of their use, they may begin to *contemplate* a need for change. Clinicians may help clients look at various ways to solve the problems associated with substance abuse, increasing the clients' *determination* to make a change. Persons who have reached these higher levels of readiness to change are more likely to respond positively to treatment suggestions made by human service professionals.

In discussing the relative popularity of these different approaches to dealing with the clients' denial among treatment professionals, White (1998) described "four overlapping stages in its view of the role of addict motivation in addiction recovery" (p. 291). He sees an evolution over the second half of the 20th century from a "baseline position" that people must first hit bottom before they are willing to change, to a focus on teaching those in the substance abusers' life to stop "rescuing," and allow them to experience the consequences of their behavior (so that they can hit bottom), to a focus on "raising the bottom" through formal intervention. These three stages share an emphasis on increasing pain as a motivation to enter treatment. White believes that the treatment community is currently more accepting of the idea that for some substance abusers it is not an absence of pain but rather a lack of hope that change is possible that keeps them from entering treatment. This has lead to more treatment programs providing "pretreatment" services that assist clients in becoming more ready for change, as described earlier. Here human service professionals may utilize their generalist practice skills to assist clients in this stage of the process.

Cultural Sensitivity

As with all areas of human services, gender and culture play a significant role in individuals' perception of a problem and their attitude about receiving help. It has long been recognized that different cultures have different patterns of alcohol and drug use. For example, among Western cultures, those that socialize children to drink responsibly by establishing patterns of when and where to drink, while at the same time discouraging drunkenness, tend to have lower rates of alcohol abuse than those that forbid their drinking altogether (Vaillant, 1995).

Yet, current research stresses the difficulty of describing any cultural group as fitting a stereotyped pattern of use. Caetano, Clark, and Tam (1998) found that ethnic minorities are underrepresented in alcohol research in the United States and that existing studies often fail to take into consideration the differences that exist between subgroups within a given cultural group. It is therefore wise to assess cultural attitudes with each client and avoid assuming that stereotypes apply. Because clients may well feel loyal to their culture, clinicians need to listen nonjudgementally to the information shared by their clients. Cultural sensitivity also involves practitioners being aware of how their own attitudes and beliefs have been influenced by their cultural background (Corey, 2005).

Human service professionals are routinely encouraged to consider those things that might present obstacles to their clients receiving help. It is important to keep in mind that even the concept of seeking professional help outside one's family reflects a Western worldview that is open to the idea that outsiders are appropriate sources of assistance. Many cultures reject the focus on the individual or the belief that it is helpful to share one's feelings, a theme that is central to professional counseling. Given the prevalence of group forms of therapy, this may be particularly difficult in substance abuse treatment and calls for sensitivity and understanding on the part of the treatment staff. Clients who have difficulty sharing feelings with members of the opposite sex, looking directly at another group member when they speak to them, or confronting an older group member may be reflecting their cultural norms rather than resistance (Corey & Corey, 2006).

Defining Treatment Goals

Abstinence

In most treatment programs utilizing the medical model, abstinence is seen as the necessary first step in treatment. This means that the person commits to completely eliminating the use of alcohol and all illicit drugs; in some programs this includes eliminating even medically prescribed psychoactive drugs and pain medication. Abstinence is seen as the necessary beginning point before other problems can be accurately assessed and addressed. It is not, however, generally seen as the only goal of treatment.

Harm Reduction

Although abstinence is the goal in most treatment programs, some argue that harm reduction may also be an appropriate goal (Inaba & Cohen, 2004). Harm reduction can include a variety of goals designed to limit the negative consequences (for both the individual and society) of substance use for those unwilling or unable to achieve abstinence. Those who favor harm reduction may see abstinence as the eventual goal, but believe that it may be appropriate to first focus on smaller/intermediate goals such as using less-dangerous drugs, decreasing the frequency or quantity of drug use, or limiting the health risks associated with drug use. For example, those who argue for needle exchange programs (for intravenous drug users) to reduce the transmission of AIDS base their position on the concept of harm reduction. As both a public policy issue and a treatment philosophy, harm reduction continues to be very controversial.

This approach is more likely to be advocated by programs working with clients dealing with other problems in addition to their substance abuse disorder such as homelessness or mental illness. Advocates for harm reduction argue that clients must overcome many obstacles to enter treatment and that if they are required to be abstinent prior to entering a treatment program, they may never do so. They draw the parallel that doctors do not ask patients with other illnesses to eliminate their symptoms before they can be treated, but that treatment programs often require some period of abstinence before people can enter their program. Like motivational interviewing, harm reduction approaches favor "meeting people where they are," not where we would like them to be.

Mode of Service Delivery

Availability of Treatment

SAMHSA (2005) lists over 11,000 addiction treatment programs on its online resource directory, including outpatient, residential, hospital inpatient, and partial hospitalization/day treatment programs. The services provided by these programs include rehabilitation, counseling, behavioral therapy, medication, and case management (NIDA, 1999). In order to be included in the directory, programs must be approved by the substance abuse agency for the state in which they are located. Also included are those programs administered by the Department of Veterans Affairs, the Indian Health Service, and the Department of Defense. From the earliest days of treatment in the United States, addiction treatment has been funded by both public (government) and private sources (payment by private insurers, out-of-pocket payment by the person receiving treatment, or payment by charitable sources).

Public Programs

Federal and state governments currently provide the majority of funding for substance abuse programs. Although this is a source of ongoing public policy debate, there has been a general consensus that money invested in providing substance abuse treatment is well spent. One study found that for every dollar spent on substance abuse treatment, seven dollars are saved in reduced health care, crime, lost productivity, and the like. Studies have also established that it is significantly less expensive to provide treatment to substance abusers than to incarcerate them. However, in an era where all levels of government face increasing budget deficits, providing treatment funds despite the benefits continues to be controversial (Scanlon, 2002).

Private Programs

Even those individuals who have health insurance that provides coverage for substance abuse treatment are likely to find that their insurance plans provide strict guidelines that limit how they can utilize their benefits. Beginning in the late 1980s, in an effort to control rising health care costs, employers increasingly opted for offering managed behavioral health (mental health and substance abuse) care as a part of their group health insurance plans.

The American Society of Addiction Medicine issued a report on the impact of *managed care* on addiction treatment in the United States. Their study, conducted by the Hay Group, found that from 1988 to 1998, the value of insurance coverage for addiction

Human Systems

Understanding and Mastery of Human Systems: Emphasis on context and the role of diversity in determining and meeting human needs

Critical Thinking Question: The concept of harm reduction is gaining popularity among human service professionals, particularly in fields such as substance abuse treatment. What are some of the benefits of adopting a harm reduction approach? What are some of the risks? How might a human service professional address the ethical implications of a harm reduction policy?

treatment had declined by 75 percent for employees who participated in group health plans offered by mid- to large-size companies. During the same time period, the report found a much smaller decrease (11.5 percent) in the value of overall health insurance coverage. A major factor in the decrease of the value of addiction benefits comes from a reduction in the authorization of inpatient hospital treatment in favor of less-expensive outpatient treatment options (Galanter, Keller, Dermatis, & Egelko, 2001).

For those working specifically in addiction treatment, it has often meant the loss of a job when their inpatient unit closed. In working with clients, it has meant becoming well versed in the criteria used by managed care companies. Human service professionals are often called on to provide referrals for their clients; clinicians who are able to provide direction to their clients in navigating the managed care system can be of great assistance to those needing to arrange for treatment. Many people still associate substance abuse treatment with the 30-day inpatient programs common in the 1980s. Because this is no longer a realistic option for most people, human service professionals need to be aware of other treatment options and should be familiar with the types of programs, both public and private, available locally to meet their clients' needs.

Continuum of Care

The currently accepted goal is that communities provide a *continuum of care* so that individuals, groups, and families can receive the form of substance abuse treatment most appropriate for their needs. This concept acknowledges both that different people have different treatment needs and that an individual person's needs vary over his or her course of treatment.

> **The currently accepted goal is that communities provide a continuum of care so that individuals, groups, and families can receive the form of substance abuse treatment most appropriate for their needs.**

Prevention services are generally targeted to populations known to be at higher risk for substance abuse. Although it is possible for anyone to develop a substance abuse problem, the Center for Substance Abuse Prevention (2004) identifies six risk factors that may lead to substance abuse or addiction:

- Substance use by parents of other family members
- Substance use by peers and the perception that "everyone is doing it"
- Preadolescent use of alcohol, tobacco, or other drugs
- Being a victim of physical or sexual abuse
- Abusive or violent environment at home or in school
- Economic deprivation

Many well-known prevention programs focus on providing youth with information about the risks associated with using drugs and alcohol and the skills to "just say no."

In addition to their role in providing services that help prevent substance abuse, human service professionals are often in a position to recognize warning signs of drug use in their clients. For example, school social workers should be alert to the possibility that a student's attendance problems, declining grades, poor physical appearance, or change of peer group may be an indication of substance use. Other behaviors that may be signs of substance use with clients in any practice setting include declining work performance, financial problems, dramatic mood changes, attempts to cover the smell of alcohol, always wearing sunglasses to hide dilated or constricted pupils, wearing long sleeves to hide needle marks, or stealing from family and friends (Hepworth, Rooney, & Larsen, 2002). Although any of these behaviors may be due to issues other than substance abuse, human service professionals should be alert to the possibility. It is often the recognition of these changes in behavior that lead concerned family members and friends to seek out assistance, initiating a request for assessment of the problem.

The skills needed to assess a substance abuse problem vary according to the practice setting. For the human service professional in a practice setting outside substance abuse treatment, screening may be a part of the normal intake procedure. The agency's intake form may prompt the clinician to ask about current and past drug and alcohol use, family history of substance abuse, and any negative consequences associated with substance use. In some settings, this may be all that is indicated to screen for problems that may require additional referrals or may affect the client's ability to utilize services.

In agencies where the prevalence of substance abuse is more likely to be high, more extensive screening may be needed. For example, human service professionals in settings dealing with domestic violence, the homeless, or families at risk for child abuse may determine that even alcohol use that does not meet the *DSM-IV-TR* criteria for abuse might interfere with treatment efforts. Agencies may reasonably set policies that require clients to refrain from drinking alcohol prior to attending parenting classes, anger management sessions, or any other agency service. The intake clinician at a shelter for the homeless may need to have more sophisticated screening skills to determine if a client presenting for services can be safely housed in a program or instead needs a referral for detoxification.

Likewise, clinicians who deal directly with substance abusers need to be able to assess if a client is in need of medically supervised detoxification and, if so, whether it can be provided on an outpatient basis or requires hospitalization. Clinicians involved in this level of assessment would normally be part of a treatment team including immediate access to medical services. Here assessment would require not just the recognition of a problem, but the ability to refer to the appropriate level of treatment. Those skills needed to screen for a possible problem and initiate a referral for further assessment and possible treatment would appropriately be considered part of *generalist* human services practice. The ability to assess or treat specific substance abuse problems would generally be considered a *specialist* skill requiring specific training.

Many assessment tools have been developed to assist clinicians and health care providers in the assessment of substance abuse problems. The brief (four-item) CAGE questionnaire and 25-question Michigan Alcoholism Screening Test are designed to identify the presence of negative consequences of alcohol use that may indicate a need for intervention. The more comprehensive Addiction Severity Index is more likely to be used in substance abuse specialty programs where it may be used as a part of treatment planning, outcome evaluation, or in conducting research. As previously described, diagnosis of substance abuse or dependence to the range of psychoactive drugs of abuse is performed according to criteria set forth in the *DSM-IV-TR* (Inaba & Cohen, 2004).

Treatment Modalities

Once an assessment is completed indicating that a problem exists, treatment options can be explored. Treatment of a substance abuse problem is a complex process that occurs in stages over a period of time. Recovery from substance abuse is often described as being *a process, not an event.* Depending on the treatment setting, staff may be drawn from a variety of disciplines with different levels of training. The multidisciplinary team may include psychiatrists and other medical doctors, nurses, psychologists, social workers, addiction counselors, family therapists, recreation therapists, occupational therapists, and chaplains. All members of the treatment team may assess the client and participate, along with the client, in developing treatment goals and plans.

The Role of the Human Service Professional

Human service professionals generally referred to as counselors working in substance abuse programs come from a wide range of experiences and training backgrounds. During the period of rapid growth of alcohol and drug treatment programs from the 1960s to the 1980s, few professionally trained counselors had specialty training in addiction treatment. In most programs, "frontline" counselors, who conducted much of the individual and group counseling, came to the field by way of personal experience in recovery; such counselors were often described as *paraprofessionals.* In some longer-term residential settings, it was common for individuals to successfully graduate or "phase out" of treatment and return almost immediately as a member of the treatment team. This had the advantage of providing staff who knew the program, were dedicated to its mission, and were often willing to work for low wages.

Common problems arose if the counselors relapsed, had difficulty separating their own treatment experience from that of their clients, or became overwhelmed by the demands of attending to their own recovery while providing emotionally intense counseling for their clients. Making the transition from being a resident (or patient) in a program to being a counselor was generally not easy. There was considerable controversy as to whether personal recovery experience was a help or hindrance to working in the field (White, 1998). Over time, most programs developed policies to address the common problems that arose. For example, programs might require a graduate of their program to have a minimum of one to two years of sobriety after completing the program before they could be hired as an employee.

Many steps have been taken to advance the training of substance abuse counselors. Certificate programs were added to the curriculum of many community and four-year colleges that gave recovering individuals an opportunity to build on their life experience with academic and professional training. Specialty programs can prepare counselors who are themselves in recovery to deal with ethical issues that are unique to the field, such as how to manage interaction with one of their clients if they attend the same AA meeting (Bissell & Royce, 1994).

Other advances in the professionalization of the field are the development of standards for professional certification and the growth of professional organizations at the state and national level. At the same time, college programs (for human service professionals, marriage and family therapists, social workers, and psychologists) have added substance abuse training to their normal course of studies, as evidenced by the inclusion to this text of the chapter you are reading.

Stages of Recovery

Many models have been developed that describe the process of recovery in terms of the stages one must complete to arrive at health. These incorporate basic understandings that problems that develop over a long period of time will take time to heal (in other words, "You didn't get sick in a day, you're not going to get well in a day either"). Here it is helpful to distinguish between the concepts of *abstinence* and *sobriety* as used in recovery. Terence Gorski (1989), a pioneer and leader in the area of relapse prevention, regards abstinence from mood-altering chemicals as "a necessary first step in learning what to do to get and stay healthy in all areas of life" (p. 4). Sobriety, as described by Gorski, involves more: "abstinence plus a return to full physical, psychological, social, and spiritual health" (p. 4). Recall that when individuals first begin to use a psychoactive substance, it is for them "a solution." Whether it provides the "liquid courage" to ask someone to dance, the energy to stay up all night to complete a paper or clean the house, or a means to feel accepted by one's peers, the substance used has provided some positive reinforcement for continued use.

When people stop using the substance, at a minimum, they must determine what functions their use provided for them and how they will go about meeting these needs in healthier ways in the future. Often this journey will involve painful psychological work dealing with issues of past trauma or abuse. For example, veterans, who have used painkillers as a way of numbing their memories of war, will have to deal with the emergence of these memories in recovery. For most clients, their substance use has led to multiple losses: often family, friends, job, and health. Grieving these losses is another significant treatment issue.

Most clients also become increasingly aware of the ways that their use has harmed others and must deal with the associated feelings of guilt and shame; this is often particularly painful work for parents who realize they have abused or neglected their children. The timing of this work requires sophisticated skill on the part of the counselor to decrease the likelihood of precipitating a relapse.

Relapse Prevention

Throughout the stages of treatment and recovery, counselors increasingly introduce the concept of *relapse prevention*. Although not limited to substance abuse treatment, relapse prevention draws on cognitive-behavioral strategies to help clients build skills to maintain abstinence and to address relapse should it occur (NIDA, 1999). Individuals are taught to recognize potential triggers for relapse such as being in neighborhoods where they once used, sights or smells associated with use, or experiencing difficult emotions. Counselors may help clients develop a list of coping strategies such as calling a friend, attending a support meeting, or "thinking through" the consequences if they should relapse. Clients may carry a list of such possible strategies with them in their wallet so that they will have to see the list if they try to buy a drink or drugs. Counselors also encourage clients to plan their response should they relapse. Rather than telling themselves that "I've blown it now, I'll never be able to stop," they are encouraged to tell themselves, "Get back to treatment." Clients are educated to understand that addiction is a disease prone to relapse, and they are encouraged to be active in their efforts to prevent relapse.

Common Treatment Settings

As previously noted, since the 1980s, there has been a shift away from inpatient treatment programs as the standard for care of substance abuse in favor of outpatient programs. In part, this has been the result of managed care efforts to control rising health care costs. Others, however, argue for outpatient treatment on philosophical grounds. The choice of a treatment program is best made based on determining the individual needs of a specific client. However, it is important for human service professionals to familiarize themselves with all the types of programs generally available and with the specific resources available in their community. As you read about the various types of treatment programs, keep in mind that human service professionals are employed in each of these settings, generally providing the core treatment services of counseling and case management.

All treatment programs will begin by assessing the needs of the individual (or family) requesting treatment to determine if the individual is an appropriate candidate for that program. In the event that the program is unable to provide the indicated treatment, or the client rejects the services offered, it is the ethical responsibility of the human service professional to provide the person seeking help with appropriate referrals. Most agencies keep up-to-date resource directories to aid in this process. SAMHSA maintains an online national directory of substance abuse programs; state and local directories are also available for most communities.

Detoxification Programs

As previously noted, clients who have become physically addicted to drugs or alcohol need detoxification for the medical management of their withdrawal. Although many substance abusers have stopped using abruptly (often referred to as "quitting cold turkey"), this can be both uncomfortable and dangerous, depending on the drugs involved. Recall that addicts using downers (such as alcohol, barbiturates, and tranquillizers) that depress the central nervous system will experience a speeding up of their nervous system when in withdrawal. This can result in life-threatening seizures and therefore requires medical supervision. Although "medically necessary detoxification" has been a common criterion for inpatient treatment, in most cases, this can be accomplished on an outpatient basis, a practice that is becoming more common.

Although detox is generally regarded as necessary before treatment can begin, some clients will seek detox as an end in itself, as a way to either find housing or reduce their tolerance so that they can reduce the cost of their drug intake (Doweiko, 2006). In this setting, human service professionals play a key role in encouraging clients to remain in treatment despite the discomforts of withdrawal and the urges to leave and resume their substance abuse.

Inpatient Treatment Programs

Traditionally, the term *inpatient* was used to refer both to programs located in hospitals or freestanding programs (such as the Betty Ford Center) that were staffed to provide medical services, including detoxification. Inpatient units existed in both general hospitals and psychiatric hospitals. Although once common, the 30-day inpatient programs often associated with substance abuse treatment are now relatively rare. The treatment focus of these programs, however, continues to shape much of outpatient treatment that has become more common.

Most inpatient programs utilized what is known as the Minnesota Model of treatment, which has its roots in the 1940s and 1950s in three treatment programs in that state: Pioneer House, Hazelden, and Willmar State Hospital. Developing over time, a defining concept of the Minnesota Model was an understanding of addiction as a primary, progressive disease (rather than a symptom of other problems) that would be the focus of treatment, with lifetime abstinence as the goal. Seeing addiction as affecting all areas of a person's life, treatment was provided by a multidisciplinary team including doctors, nurses, psychologists, social workers, and clergy.

Recovered alcoholics were also part of the counseling staff. Each discipline completed an assessment of the patient, giving input into an overall treatment plan. The principles of AA were incorporated into the treatment, and patients attended meetings as a part of their treatment program. Other treatment activities included educational lectures, group and individual counseling, family treatment, reading and written assignments, and informal discussions with other patients, which combined to make a highly structured program (White, 1998). Some programs offered specialized units for adolescents, "impaired professionals" (doctors and nurses), dually diagnosed patients who suffered from an additional mental illness, or patients who wanted their treatment integrated with their faith (most commonly Christian). Although all these treatment activities and areas of specialization continue to be available, they are now more likely to be provided in an outpatient setting.

Partial Hospitalization Programs

Partial hospitalization allows patients to attend all the day activities provided at an inpatient program, while returning to their home to sleep. For patients who have a relatively stable home environment, this can allow them to integrate what they are learning in treatment into their family and home life. If problems arise at home, they can deal with it in treatment the next day. Because the costs are reduced, insurance companies may authorize more treatment days for partial hospitalization than for inpatient care.

Residential Treatment Programs

Although inpatient programs may also be referred to as residential, the distinction made here is that residential treatment is more likely to occur in a homelike setting, over a longer period of time, providing less medical care. Like hospital-based programs, residential programs provide 24-hour supervision so that the residents can focus on their treatment, free of the stresses and responsibilities of their outside life and (at least theoretically) free of opportunities to use chemicals. Historically, residential programs (known as therapeutic communities) worked with drug addicts who had generally exhausted all resources. Many utilized a more confrontational approach designed to "tear down" the street image and "build up" a new, healthy identity. The resident's day was highly structured with active involvement in the needs of the house, such as cleaning and cooking, in addition to group and individual counseling. Over time, residents gave up their addict identity in favor of being a member of the program community, often referred to as family (NIDA, 1999, 2002). Human service professionals serve in all treatment roles in residential programs using titles such as case manager, counselor, clinical director, or program manager.

Some residential programs provide a "step-down" or transition from inpatient treatment or detox. In these programs, such as halfway houses or sober living facilities, residents experience living in a supportive community free of drugs and alcohol, but may continue their employment during the day. Residents are generally required to attend a set number of mutual aid meetings each week in addition to house meetings. The inclusion of additional on-site counseling, provided by human service professionals, varies from program to program.

Outpatient Treatment

Intensive outpatient treatment (IOT) provides community-based treatment for substance abuse. Programs vary in intensity, but include psychoeducational and therapeutic efforts such as lecture, group, and individual counseling and activities designed to enhance life skills. Programs vary in format, but generally involve the client in a minimum of 10 hours per week of treatment activities. To accomplish this, IOT uses many of the same principles as described for inpatient treatment including a multidisciplinary treatment team and individualized treatment planning.

IOT has grown in popularity as inpatient treatment has become less common. In many ways, it bridges a gap between the 28-day medically managed programs once prevalent and the traditional outpatient counseling where the client was seen only once a week. In most programs, the number of hours patients are involved in treatment decreases as their length of sobriety increases. Stepped-down aftercare services may be available for a year or longer. At this point, the client may be attending treatment services only once a week, in addition to his 12-step participation. Staffing for IOT increasingly includes licensed therapists along with other human service professionals.

Traditional outpatient counseling, where a client sees a counselor once a week, is likely to be inadequate for the client with a serious substance abuse problem. In the past, mental health counselors frequently attempted to provide such counseling, often treating the substance abuse as a symptom of underlying problems. It was the failure of this approach that birthed current addiction treatment. Today, educational programs that train human service professionals such as counselors, social workers, and psychologists should include training on recognizing substance abuse problems.

At a minimum, clinicians should be aware that clients' substance abuse will severely affect their ability to participate in counseling and thus should consider referring their clients to appropriate substance abuse treatment.

Pharmacological Treatments

The use of medication to treat substance abuse and substance abusers has been a source of ongoing debate. Much of substance abuse treatment has been provided in "drug-free" programs, stressing the need for abstinence from all psychoactive substances, including medication prescribed for the treatment of psychiatric disorders such as antipsychotics and antidepressants. This meant that substance abusers with psychiatric disorders were often told that they were not appropriate candidates for substance abuse treatment programs.

Hospital-based inpatient programs were more likely to include psychiatric medications as part of treatment, but there was controversy even in those settings. Some argued that a substance abuser must be drug free for some period of time before being accurately diagnosed with a mental illness. Others maintained that providing medication for mental illness would serve to enhance the success of the substance abuse treatment. "Recent epidemiologic studies have shown that between 30 percent and 60 percent of drug abusers have concurrent mental health diagnoses including personality disorders, major depression, schizophrenia, and bipolar disorder" (Leshner, 1999, p. 1). Because the co-occurrence rate of substance abuse and mental disorders is high, there has been a growing emphasis on the importance of clinicians in both substance abuse and mental health treatment being aware of the special needs of *dual-diagnosis* patients who suffer from both disorders. Generally, there has been increased acceptance of the need for psychiatric medication for these patients, although drug-free programs may still decide that they do not have the medical services available to accept such patients into their program.

Self-Help

Earlier in this chapter, we discussed the birth of AA from the perspective of the history of addiction treatment. Now we will look further at AA and other 12-step programs (such as Narcotics Anonymous and Cocaine Anonymous) from the perspective of treatment. Twelve-step programs play a significant role in the treatment of addiction, both as a primary source of support and as an adjunct to professional treatment. Because they are free, well known, and widely available, self-help groups represent a major resource to human service professionals and their clients.

Twelve-step programs provide a setting in which members can share their "experience, strength, and hope" with other members. Although commonly referred to as self-help programs, the term *mutual aid society* may more accurately reflect the belief that one who suffers from addiction, but has received help, is in the

---

Human Services Delivery Systems

Understanding and Mastery of Human Services Delivery Systems: Range and characteristics of human services delivery systems and organizations

Critical Thinking Question: Currently, the trend in substance abuse services is to provide a community-centered "continuum of care," including prevention programs and a range of treatment options spanning detox programs, inpatient care, partial hospitalization programs, and outpatient treatment. To what extent is the trend away from intensive inpatient treatment driven by financial considerations? What are the benefits and risks of this new approach for clients?

Twelve-step programs play a significant role in the treatment of addiction, both as a primary source of support and as an adjunct to professional treatment.

best position to help a fellow sufferer. Providing this help to newcomers helps the older member to stay sober.

Family Involvement

Many substance abuse programs include a component for family participation such as a multifamily group, "family night," or separate groups for family members. These groups play a particularly important role in programs for adolescent substance abusers, where the need for family work is immediate. Support groups such as Al-Anon, Alateen, and Co-Dependents Anonymous also provide ongoing support for family members and friends. These groups help individuals to identify the ways in which their own life has become negatively affected by the substance abuse of another, and how to make healthy changes.

Commonly, family members come to understand that in focusing too much on the substance abuser, they have neglected taking care of themselves. Some behaviors that were intended to help the substance abuser, such as covering for them or taking over their responsibilities, may in fact have enabled the substance abuser to continue their use. Support groups for family members help them to determine clearer boundaries between "what my responsibility is and what it is not" and to make necessary changes in their own behavior. Typically, family members come to realize that all attempts to control the substance abuser have been futile and that they have only the power to control their own actions. Family members can, therefore, benefit from treatment even if their chemically addicted member never participates in treatment.

Concluding Thoughts on Substance Abuse

The use and abuse of psychoactive substances has been present from the earliest known societies and continues to be a major health problem in the United States today. Efforts to address this problem in the United States have included legislation to regulate or prohibit the manufacture and sale of drugs and alcohol, prevention programs designed to decrease risk factors and increase protective factors, and treatment for those with substance abuse problems. These efforts have evolved over time, influenced by societal attitudes about substance abuse and, more recently, by scientific research.

Human service professionals play a major role in the provision of prevention and treatment services. Because substance abuse affects all areas of an individual's life, human service professionals will encounter this issue in every practice setting. Research has established that prevention and treatment are effective and are increasingly utilized in guiding program development and provision of treatment. In a variety of roles and settings, human service professionals can assist substance-abusing clients in recognizing the negative effects of their use, in obtaining necessary treatment, as well as in working with them throughout the entire treatment process. Skilled human service professionals routinely find this practice setting both challenging and rewarding.

The following questions will test your knowledge of the content found within this chapter.

1. Early efforts to provide treatment for substance abuse began in the United States in the mid-1800s, prompted by:
   a. a dramatic increase in opium addictions
   b. a religious revival where the use of any alcohol was deemed inappropriate and "sinful"
   c. public concern over the problems resulting from increased levels of public drunkenness
   d. the dawn of psychiatry

2. What of the following factors is commonly considered contributors to the repeal of Prohibition, which once again legalized the manufacture and sale of alcohol in the United States:
   a. widespread disregard for the law
   b. a significant rise of organized crime in the production and distribution of "bootleg" liquor
   c. the first medical studies that revealed and benefits of moderate consumption of alcohol
   d. All of the above

3. The behavioral/environmental model describes the possible progression of substance use through five stages:
   a. abstinence, experimentation, social/recreational use, habituation, abuse, addiction
   b. social/recreational use, experimentation, abuse, addiction, habituation
   c. addiction, abuse, habituation, social/recreational use, experimentation, abstinence
   d. social/recreational use, experimentation, addiction, abuse, habituation, abstinence

4. The academic model focuses on:
   a. the influence of a collegiate environment on the drinking behavior, particularly the fostering of binge-drinking behavior
   b. the study of addictive behavior
   c. the physiological effects of psychoactive drugs
   d. Both A and B

5. Motivational interviewing is directive, client-centered counseling style for eliciting behavior change by helping clients to explore and resolve:
   a. past hurts
   b. past loss
   c. ambivalence
   d. unresolved anger

6. Many of the human service professionals who conduct individual and group counseling within substance abuse came to the field by way of personal experience in recovery are often called:
   a. "frontline" counselors
   b. recovery counselors
   c. graduated counselors
   d. paracounselors

7. Describe the harm reduction treatment model, including its goals and rationale for utilization with the substance abusing population seeking treatment.

8. Describe various treatment modalities for substance abuse disorders, including the strengths and deficits of each, treatment goals, and efficacy levels suggested by research.

Suggested Readings

Abbott, A. A. (2000). *Alcohol, tobacco, and other drugs: Challenging myths, assessing theories, individualizing interventions*. Washington, DC: NASW Press.

Beattie, M. (1987). *Codependent no more: How to stop controlling others and start caring for yourself*. New York: Harper & Row.

Black, C. (1981). *It will never happen to me*. New York: Ballantine.

Johnson, V. E. (1980). *I'll quit tomorrow*. New York: Harper & Row.

Miller, W. R., & Munoz, R. (1982). *How to control your drinking: A practical guide to responsible drinking*. Albuquerque: University of New Mexico Press.

Philleo, J., Brisbane, F. L., & Epstein, L. G. (1997). *Cultural competence in substance abuse prevention*. Washington, DC: NASW Press.

Vogler, R. E., & Bartz, W. R. (1982). *The better way to drink: Moderation and control of problem drinking*. Oakland, CA: New Harbinger.

Woititz, J. G. (1983). *Adult children of alcoholics*. Deerfield Beach, FL: Health Communications.

Internet Resources

Adult Children of Alcoholics (ACOA): http://www.adultchildren .org

Al-Anon/Alateen: http://www.al-anon.alateen.org

Alcoholics Anonymous: http://www.alcoholics-anonymous .org

Narcotics Anonymous: http://www.na.org

National Center on Addiction and Substance Abuse at Columbia University: http://www.casacolumbia.org/absolutenm/templates/ article.asp?articleid=287&zoneid=32

National Institute on Drug Abuse: http://www.nida.nih.gov

SAMHSA's National Clearinghouse for Alcohol and Drug Information: http://www.health.org

References

Abadinsky, H. (2004). *Drugs: An introduction* (5th ed.). Belmont, CA: Wadsworth.

Alcoholics Anonymous. (n.d.). *Information on A.A.* Retrieved December 22, 2009, from http://www.aa.org/lang/en/subpage.cfm

American Psychiatric Association. (2000). *Diagnostic and statistical manual of mental disorders* (4th ed., Text Revision). Washington, DC: Author.

Bissell, L., & Royce, J. E. (1994). *Ethics for addiction professionals* (2nd ed.). Center City, MN: Hazelden Foundation.

Boren, J. J., Onken, L. S., & Carroll, K. M. (Eds.). (2000). *Approaches to drug abuse counseling* (NIH Publication No. 00-4151). Bethesda, MD: National Institutes of Health.

Caetano, M. D., Clark, C. L., & Tam, T. (1998). Alcohol consumption among racial/ethnic minorities: Theory and research [Electronic version]. *Alcohol Health and Research World, 22*(4), 233–241.

Centers for Disease Control and Prevention. (2011). *Diagnoses of HIV infection and AIDS in the United States and dependent areas, 2009* (HIV Surveillance Report, Vol. 21). Atlanta, GA: U.S. Department of Health and Human Services. Retrieved from http:// www.cdc.gov/hiv/topics/surveillance/resources/reports/

Center for Substance Abuse Prevention. (2004). Risk factors for substance abuse. In *It won't happen to me: Substance abuse-related violence against women for anyone concerned about the issues (module 2)*. Retrieved December 9, 2009, from http:// pathwayscourses.samhsa.gov/vawc/vawc_4_pg2.htm

Corey, G. (2005). *Theory and practice of counseling and psychotherapy* (7th ed.). Pacific Grove, CA: Brooks/Cole.

Corey, M. S., & Corey, G. (2006). *Groups: Process and practice* (7th ed.). Belmont, CA: Thomson Brooks/Cole.

Doweiko, H. E. (2006). *Concepts of chemical dependency* (6th ed.). Belmont, CA: Thomson Brooks/ Cole.

Galanter, M., Keller, D. S., Dermatis, H., & Egelko, S. (2001). The impact of managed care on substance abuse treatment: A problem in need of solution: A report of the American Society of Addiction Medicine. In *Recent Developments of Alcoholism*. NY: Springer Publishing.

Gorski, T. T. (1989). *Passages through recovery: An action plan for preventing relapse*. New York: Harper & Row.

Hepworth, D. H., Rooney, R. H., & Larsen, J. A. (2002). *Direct social work practice: Theory and skills* (6th ed.). Pacific Grove, CA: Brooks/Cole.

Hettema, J., Steele, J., & Miller, W. R. (2005). Motivational interviewing. *Annual Review of Clinical Psychology, 1*, 91–111.

Inaba, D. S., & Cohen, W. E. (2004). *Uppers, downers, all arounders: Physical and mental effects of psychoactive drugs* (5th ed.). Ashland, OR: CNS Publications.

Johnson, V. E. (1973). *I'll quit tomorrow*. New York: Harper & Row.

Leshner, A. I. (1999). Drug abuse and mental disorders: Comorbidity is reality. In *National Institute on Drug Abuse: A collection of NIDA notes articles that address drug abuse treatment, 14*(4), 98–99. (NIH Publication No. NN0026). Bethesda, MD: National Institutes of Health. Retrieved on December 22, 2009, from http://archives.drugabuse.gov/NIDA_Notes/NNVol14N4/ DirRepVol14N4.html

Miller, W. R. (2000). Motivational enhancement therapy: Description of counseling approach. In J. J. Boren, L. S. Onken, & K. M. Carroll (Eds.), *Approaches to drug abuse counseling* (pp. 89–93). Bethesda, MD: National Institute on Drug Abuse.

Musto, D. F. (1999). The impact of public attitudes on drug abuse research in the twentieth century. In M. D. Glantz & C. R. Hartel (Eds.), *Drug abuse: Origins and interventions* (pp. 63–78). Washington, DC: American Psychological Association.

Morse, R. M., & Flavin, D. K. (1992). The definition of alcoholism. The Joint Committee of the National Council on Alcoholism and Drug Dependence and the American Society of Addiction Medicine to Study the Definition and Criteria for the Diagnosis of Alcoholism. *Journal of the American Medical Association, 268*(8), 1012–1014.

National Center on Addiction and Substance Abuse at Columbia University. (1998, January 8). *CASA releases report: Behind bars*. Retrieved July 11, 2005, from http://www.casacolumbia .org/templates/PressReleases.aspx?articleid=167&zoneid=49

National Association of Counties. (2006a). *The meth epidemic in America: The criminal effect of meth on communities: A 2006 survey of U.S. counties.* Retrieved March 31, 2012, from http://www.in.gov/cji/files/NAC0_Meth_Survey_Report_-_Jul_2006.pdf

National Association of Counties. (2006b). *The meth epidemic in America: Two new surveys of U.S. Counties: "The effect of meth abuse on hospital emergency rooms" and "the challenges of treating meth abuse".* Retrieved September 15, 2009, from http://www.csam-asam.org/sites/default/files/pdf/misc/NatAssocCoMethER.pdf

National Institute on Drug Abuse. (1999). *Principles of drug addiction treatment: A research-based guide* (NIH Publication No. 00-4180). Bethesda, MD: National Institutes of Health.

National Institute on Drug Abuse. (2002). *Research report series: Therapeutic community* (NIH Publication No. 02-4877). Bethesda, MD: National Institutes of Health.

National Institute on Drug Abuse. (2010). *Methamphetamine.* Bethesda, MD: National Institutes of Health. Retrieved online at http://www.drugabuse.gov/publications/drugfacts/methamphetamine.

Rollnick, S., & Miller, W. R. (1995). What is MI? [Electronic version]. *Behavioral and Cognitive Psychotherapy, 23,* 325–334. Retrieved November 13, 2004, from http://motivationalinterview.org/Documents/1%20A%20MI%20Definition%20Principles%20&%20Approach%20V4%20012911.pdf

Scanlon, A. (2002). *State spending on substance abuse treatment.* Retrieved November 27, 2004, from National Conference of State Legislatures website: http://www.ncsl.org/programs/health/forum/pmsas.htm

Substance Abuse and Mental Health Services Administration, Office of Applied Studies. (2005). *The NSDUH Report: Methamphetamine use, abuse, and dependence: 2002, 2003, and 2004.* Rockville, MD: Substance Abuse and Mental Health Services Administration. Retrieved online at http://www.oas.samhsa.gov/2k5/meth/meth.pdf

Substance Abuse and Mental Health Services Administration. (2005). *Substance abuse treatment facility locator.* Retrieved September 28, 2005, from http://findtreatment.samhsa.gov

Substance Abuse and Mental Health Services Administration, Office of Applied Studies. (February 5, 2009). *The NSDUH Report: Trends in Nonmedical Use of Prescription Pain Relievers: 2002 to 2007.* Rockville, MD.

Substance Abuse and Mental Health Services Administration, *Results from the 2011 National Survey on Drug Use and Health: Summary of National Findings,* NSDUH Series H-44, HHS Publication No. (SMA) 12-4713. Rockville, MD: Substance Abuse and Mental Health Services Administration, 2012.

Substance Abuse and Mental Health Services Administration, Office of Applied Studies (2010). *Drug Abuse Warning Network, 2007: National Estimates of Drug-Related Emergency Department Visits.* Rockville, MD.

Vaillant, G. E. (1995). *The natural history of alcoholism revisited.* Cambridge, MA: Harvard University Press.

White, W. L. (1998). *Slaying the dragon: The history of addiction treatment and recovery in America.* Bloomington, IL: Chestnut Health Systems/ Lighthouse Institute.

Whitebread, C. (1995). *The history of the non-medical use of drugs in the United States.* Retrieved December 2, 2004, from the Schaffer Library of Drug Policy website: http://www.druglibrary.org/schaffer/History/whiteb1.htm

CHAPTER 12

Human Services in the Schools

Mary Kate Denny/PhotoEdit

Learning Objectives

- Become familiar with the role and function of the school social worker, including gaining an understanding of the roots of the various helping models
- Understand the various activities the human service worker engaged in within a school setting, including individual and group counseling, crisis intervention, and case management
- Explore the nature of the multidisciplinary team within the public school system, identifying the role and function of each member, including understanding how student issues are addressed by each member
- Become familiar with the broad range of psychosocial and academic issues facing school-aged children and adolescents, including the difference between urban, suburban, and rural school environments
- Develop an understanding of the broad range of psychosocial interventions most commonly utilized by human service professionals work

Where there are children, there will be counseling, and the U.S. public school system is no exception. The field of human services has had a strong presence in the U.S. public school system for over 100 years, and this presence continues to grow, particularly in urban areas, where crime and poverty continue to flourish.

Counseling on public school campuses is primarily conducted by three types of professionals: school social workers, who are typically trained professionals with a Master of Social Work (MSW) degree; school counselors, who have a master's degree in school counseling and have a background in teaching; and school psychologists, who have a master's degree or doctorate in school psychology and, in addition to instruction in educational counseling, are trained to conduct specialized educational and psychological testing of students. Together, these human service professionals comprise what is often called *student services* or *pupil support services*.

Although each of these providers conducts counseling in some respect, they use somewhat different approaches to counseling and student support, have different standards of practice, and even have different service and treatment goals. And although there are significant differences in practice guidelines between various states, regions (urban, rural, etc.), and districts, school social workers tend to focus more on the psychosocial aspects of students' lives, providing counseling and case management that focus on traditional social work concerns such as the students' overall mental health, violence on campus and at home, the risk of suicide among the student population, and the need for advocacy on behalf of vulnerable students, including the homeless student, students of color, and a range of other students who fall into various "at-risk" populations. School counselors tend to focus more on academic counseling, career guidance, and emotional or psychological issues that directly pertain to student achievement. School psychologists focus on testing, particularly in response to numerous federal and state mandates that require the academic testing of students to place them in the proper educational setting, but may also provide counseling for students who are experiencing emotional difficulties affecting their academic achievement.

Regardless of a counselor's designated role, when one works with human beings experiencing strife, one immediately becomes a generalist having to deal with a broad range of issues and serving in several different roles. Thus, although a school counselor might initiate a counseling session with a student regarding academic performance, study skills, or career planning, the session can take a quick detour focusing on the student's recent breakup, a bullying incident, a friend's suicide, or a parent's alcohol abuse. A school psychologist charged with the responsibility of facilitating all the school district's educational testing might easily find her- or himself spending extra time with a student who breaks down during testing because she or he is living in a homeless shelter and knows no one at school. In a similar vein, school social workers whose goal is to focus on students' psychological and emotional issues that are creating a barrier to learning might find themselves conducting a study skills workshop or helping students explore where they want to attend college or what they want to do for a career. Despite the overlap in the functions of these three school-based careers, each of these fields has unique professional standards and roles that, in many respects, delineate them from one another.

School Social Work

School social work has its roots in the settlement house movement discussed in Chapter 2. Settlement house workers in the late 1800s and early 1900s, all of whom were women, recognized the poor job urban schools were doing in keeping in touch with and connecting to the parents of many of their students. Because settlement houses were designed to provide services and relief primarily to low-income immigrant populations in large urban areas, most of the children who were the original focus on these early efforts to connect school with home were from families who had recently emigrated from non-English-speaking countries. Settlement house communities frequently suffered from overcrowding, both within neighborhoods and within the classroom, where some schools had as many as 50 students per class (McCullagh, 1993, 1998). Thus, these early school social workers served an important support function enabling teachers to focus on the task of teaching academics.

Mass urbanization was also occurring during this time with scores of families moving from agricultural lifestyles to the city in search of factory jobs. With them came children, many of whom were not adjusting well to city life, particularly when it involved living in cramped quarters with parents who worked long hours. Of chief concern among school districts that were the first to use school social workers were child "maladjustment," child "handicaps," and erratic school attendance, and it was the school social worker's primary goal to address these concerns by ensuring that children's adjustment needs were met, children with handicaps received necessary services, and children attended school regularly (McCullagh, 1993).

These school social work pioneers had many different titles: visiting teachers, home visitors, special visitors, and visiting social counselors (McCullagh, 1998), and they often lived in the settlement houses acting as a liaison between the school, child, and home. This early work, often referred to as the *Visiting Teacher's Movement*, tended to focus on school-related matters such as irregular attendance issues, various health problems, searching the city for children who were not attending schools (such as deaf children and orphans living on the streets), and various other home-centered matters affecting students. The guiding philosophy of home and school visiting committees was that the child was to be viewed from a holistic perspective—not solely as a student causing problems for the district (McCullagh, 1993).

Through the development of numerous committees, and the creation of a governing and organizing association called the Public Education Association (PEA), these visiting teachers or counselors gained in popularity and quickly became an integral part of many school districts throughout New York, Boston, and Philadelphia over the next several decades (McCullagh, 1993).

By the early part of the 20th century, teachers in low-income, high-need urban neighborhoods had begun to look to these home visitors for advice and assistance on several issues related to their students, including those concerning inappropriate behaviors, potential problems at home, lack of attendance, and general issues related to school functioning. This reliance and general appreciation of the services provided by these early school social workers reflected the teachers' and school administrators' increasing respect for this support service. In fact, by about 1910, many larger school districts were lobbying to have school social workers become paid members of the school district and board of education, rather than being contracted volunteers of the settlement houses supported by philanthropic organizations (McCullagh, 1993).

Schools were also seen as the chief means for "Americanizing" foreign children (and, it was hoped, their families), thus government interest remained high in social work efforts to connect schools with families because it was believed that through such connections more effective assimilation of the immigrant families would occur. This focus on expanding the purpose of schools to include both the education *and* the social needs of the child is still widely reflected in today's public school systems that not only offer solely academic education and services, but also offer counseling, case management, food programs, and on-campus health services. But even if the goal of government was social control, the focus of the school social worker remained on the individual child; the commonly cited goal of these early social workers involved making certain that the individuality of each child did not get lost in the chaos of the overcrowded classroom (McCullagh, 1993).

School social work continued to expand and professionalize over the next 40 years, along with social work in general, and although originally more aligned with teachers and the field of education, by the 1940s visiting teachers and counselors were wholly aligned with the social work profession, and the PEA officially changed its name to the American Association of School Social Workers; later in the decade the name was again changed to the National Association of School Social Workers (NASSW) (McCullagh, 1998). By 1955, several different social work-related committees merged to create the National Association of Social Workers (NASW), and although the NASSW still exists, it is now under the auspices of the NASW. The role of the school social worker continued to grow and expand through the 1960s, fueled by the social turbulence that marked this era. This awareness led to many universities developing school social work degree programs (McCullagh, 2001).

Finally, in 1975 Congress passed the Individuals with Disabilities Education Act (Pub. L. No. 94-142), requiring that public schools provide "free and appropriate" public education to all school-aged children between the ages of 3 and 21 years, regardless of their disability. This law has required school districts to provide increased funding for social work services for students with special needs, when deemed appropriate.

Currently, school social work remains a growing field that offers excellent practice opportunities for those wanting to work with school-aged children. Issues such as international academic competition, concerns about increasing violence in schools, and continued reliance on social work services for the regular as well as special education students have continued to propel school social work forward into the 21st century and helped to offset periodic reductions in education budgets due to cyclical economic downturns. During difficult economic times though, it is not uncommon for school districts to consider cutting back social work services. This is unfortunate because research consistently shows how school social workers make a significant difference in the lives of students and in the levels of their academic success. It is vital then that school social workers consistently communicate their practices and effective interventions to their administrators who appear to lower the possibility of cuts in school social work personnel (Bye, Shepard, Partridge, & Alvarez, 2009; Garrett, 2006).

The School Social Work Model

The traditional model of school social work involves the social worker providing school-based social work services as an employee of the school district and as a part of a multidisciplinary team. Although some districts utilize school-based social workers employed by outside agencies (primarily as a cost-saving measure), most school districts in the United States still employ the traditional model. Regardless of the school social worker's actual employer, the roles and functions of the social worker are typically generalist in nature, but have become increasingly specialized as managed care has forced many school districts to seek government reimbursement for services (such as Medicaid or Medicare), which in turn has prompted an increase in specialized credentials beyond licensing (Lewis, 1998).

> The traditional model of school social work involves the social worker providing school-based social work services as an employee of the school district and as a part of a multidisciplinary team.

Most states require that school social workers have an MSW with a specialization in school social work, have accrued several hundred hours in an internship at a public school, and have passed a state content-area test. Some states still require only a bachelor's degree from an accredited social work program, but there is a national push toward master's level education.

School Social Work Roles, Functions, and Core Competencies

School social workers perform a variety of tasks, serve numerous functions, and operate within several different roles depending on the demographics of the school population, the type of children served, and the capacity in which the social worker is functioning. In general, school social workers exist to assist children in managing any psychosocial issues that are creating a barrier to learning. These could include physical barriers in the form of a disability, cognitive barriers such as intellectual or learning disabilities, or behavioral barriers such as students who are depressed, anxious, or acting out. School social workers also work to develop, enhance, or maintain a close working relationship between student families and the school, advocating for the family in a variety of situations.

According to the NASW, school social workers should be competent in providing individual, group, and family counseling; should be well versed in theories of human behavior and development; and should have knowledge of and be sensitive to the demographic makeup of the school population with which they work, including relevant issues related to socioeconomic status (SES), gender, race, sexual orientation, and any community stressors that might affect a student's ability to perform (such as a high crime rate or gang infiltration). School social workers must also have competencies in the areas of assessment, must be familiar with local referring agencies, and must be committed to the values and ethics of the social work profession, including those relating to social justice, equity, and diversity (NASW, 2003).

School social workers may work with the general school population or may be hired to work within the special education department either with physically or mentally handicapped children or with students who are behavior disordered. Direct practice will often include individual counseling and group counseling, as well as some family counseling, if necessary. In most school settings, for a child to receive social work services either they must be designated as a required service per the Individualized Education Plan (IEP), which serves as a sort of contract between the school and family for students identified for special education services, or the student's emotional or psychological problems must in some way be interfering with academic performance. Thus, if a student was experiencing depression, but his academic performance was not affected, this student might not be an appropriate candidate for social work services and would likely be referred out for mental health services.

Individual counseling might include psychological counseling for a high school student, or it might involve play therapy, including drawing, therapeutic games, or doll play, for an elementary school–aged student. Likewise, *group counseling* might involve getting six or eight students together whose parents recently divorced, or who recently moved from another school, and providing them with an opportunity to talk about their struggles and feelings. Yet group counseling might also have a structured and specific curriculum focusing on issues such as anger management or social skills training. School social workers also conduct home visits to obtain vital information about the student's life outside school as well as to ensure a strong link between home and school.

Case management is also provided and can include the organization and coordination of numerous services received by a student. For instance, a student's case might involve an outside therapist who is providing psychological counseling, a psychiatrist who supervises psychotropic medication such as antidepressants, a truancy officer, the police department, a child welfare agency, the family, all the student's teachers, and the school principal. Thus, depending on the actual issues of the student receiving services, the social worker will likely be involved in the coordination of services and the appropriate dissemination of information of a number of involved parties. For instance, new medications or medication changes in students who are suffering from clinical depression would be vital information for school social workers.

Crisis intervention is also an important role of a school social worker. Whether the crisis involves a natural tragedy, such as a tornado or earthquake, the crisis surrounding a student suicide, or the crisis of on-campus violence such as student-on-student assaults, school social workers provide crisis counseling to the entire student population, families, and even the school staff. Crisis counseling might include helping students face the initial shock of some tragedy, but also often involves implementing a safety plan. For instance, the suicide of a student often elicits emotional distress in other students and can lead to an increased chance of other students committing suicide. A school social worker will be involved in creating awareness (through classroom presentations or staff meetings), maintaining a visible presence on campus, and conducting outreach services to vulnerable students.

School social workers may also facilitate *conflict resolution* and *violence prevention* programs. For instance, a school social worker might conduct a structured violence prevention workshop or presentation in a classroom or manage a peer-led conflict resolution program, training students to conduct resolution counseling sessions with students who are engaged in some conflict.

Most social workers are assigned to more than one school, thus they might spend only a few days per week at any one school site. They typically have a caseload of students they must see on a weekly or biweekly basis either on an individual basis or in a group, and perform these various other tasks on an as-needed basis. Because the range of student population types is so wide, it is difficult to describe precisely what a school social worker does on a daily basis, but as with most human services positions, school social workers must be generalists to effectively manage the variety of issues with which they are confronted. Case Study 12.1 provides a wonderful example of some of the issues a school social worker might encounter, but again the specific nature of the work depends in great part on the demographics of the student population, the age of the students, and the capacity in which the social worker was hired.

CASE STUDY 12.1

Case Example—A Day in the Life of a School Social Worker

Mario is a junior at a public high school in a large urban school in a state bordering Mexico. He does not have a behavior problem, and does relatively well in his academic studies, but has come to the attention of school social workers due to excessive absences. His teachers also report that he seems particularly "stressed out" lately, and not himself. There is concern that he may be withdrawing emotionally and socially due to an increase in anti-immigrant sentiment exhibited among some students and school personnel. A psychosocial evaluation reveals that Mario is the oldest of four children. Mario's parents are undocumented immigrants from Mexico, who have been living in the United States for approximately 15 years, having been recruited to the United States by a large agricultural company. Mario's parents do not speak English, and Mario disclosed that he often misses school so that he can translate for his parents, or intercede on behalf of his parents who are often scared to seek out services themselves in light of anti-immigration legislation recently passed in the state. Mario also disclosed that he has in fact been the target of anti-immigrant sentiment in the form of derogatory statements, and scapegoating. For instance, on several occasions while walking down the halls in school he has heard random students shout out to him asking for proof of his legal status. He has also experienced negative statements directed toward all Hispanic immigrants, including a few teachers and some office assistants making statements appearing to scapegoat the Latino population for everything from escalating violence in the drug war, to scapegoating Latinos for high regional unemployment rates. The school social worker, Kate, responds to Mario and his parents reassuringly, and explains that Mario can receive supportive services—both from government human services and from programs within the school. At this point in the session, Mario admits that he just learned that he does not in fact have legal status. Mario grew up believing that he was born in the United States, but after a recent meeting with a state human services agency, he was informed that his Social Security number was not valid. His parents then told him that he was six months old when they emigrated from Mexico, and they used false papers provided to them by men from the agricultural company that recruited them. Mario became extremely distraught when sharing this secret, expressing discouragement and fear that he would not be able to attend college and receive financial aid, despite having lived in the United States almost his entire life, and working so hard to do well in school, or worse, that he could be legally deported to a country he has never visited, and where he knows no one.

Before Kate can competently provide guidance, services, and referrals to Mario and his family she must be aware of several areas of law that impact the migrant population—both those who are documented and undocumented. These overlapping areas include federal and state immigration laws (much of what changed significantly post-9/11), changes in public assistance policies in response to 1996 welfare reform (that barred the majority of residents, documented and undocumented, from receiving any public assistance), differences in legislation and policies on various levels (federal, state, county, and school), as well as having an awareness of pending legislation that may have an impact on Mario and his family, such as the Dream Act, federal legislation that would make it possible for students like Mario to attend college, under certain circumstances. Gaining this level of awareness of macro issues affecting the Hispanic students at Kate's school is a vital part of providing cultural competent social work services. One way to learn more about current issues affecting undocumented students is to attend workshops and conferences focusing on this issue, as well as seeking out resources identifying key issues and dynamics published by advocacy organizations, or other authoritative organizations. For instance, the National School Boards Association and the National Education Association jointly published an online report in 2009 in cooperation with several professional organizations, including School Social Work Association of America, entitled "Legal Issues for School Districts Related to the Education of Undocumented Immigrants" (Borkowski & Sorensen, 2009). This publication would be a great place for Kate to start in learning about a public school's obligations and responsibilities regarding the education of undocumented students.

School Counseling

Historical Roots of School Counseling

The professional school counselor often has an overlapping role with the school social worker, but typically focuses more on academic concerns and career guidance. School counseling also has a history reaching back to the late 1880s and early 1900s, with roots in the vocational guidance counseling movement (Schmidt & Ciechalski, 2001). In fact, early school counselors focused primarily on matching male high school graduates with an appropriate vocational or job placement.

In the 1920s, theories of intelligence and cognitive development became popular, influencing the work of school guidance counselors who, with the advent of intelligence and aptitude testing, now had new tools with which to do their jobs. The 1930s saw advancements in the areas of personality development and motivation, which directly influenced the field of school counseling, enabling counselors to further assist students in identifying areas of aptitude, as well as developing motivational techniques. Social trends and political movements were chief among various influences that led to a gradual shift from a primary focus on the vocational needs of students to a more comprehensive focus where school counselors proactively meet various developmental needs of students (Schmidt & Ciechalski, 2001). Many school counselors working in a secondary school setting not only continue to provide general guidance and academic counseling, but also continue to strive to meet the needs of the whole student.

As with school social work, the Education for All Handicapped Children Act of 1975 (Pub. L. No. 94-142)—which required, among other things, that children with special needs receive all support services necessary to their academic success—led to school counselors becoming involved in special education departments. In addition, governmental committee reports, such as "A Nation at Risk" (1983), and federal legislation, such as the No Child Left Behind Act of 2001 (now referred to as *the Elementary and Secondary Education Act* [ESEA]) (U.S. Department of Education, 2001), have meant an increase in funding in many school districts' budgets for school counseling, because concern for academic achievement (or, in some districts, concerns about academic decline) has countered budgetary concerns.

School Counselors: Professional Identity

Although school counseling programs have continued to grow within most school districts, one challenge consistently plaguing the field is role definition. A review of the literature relating to the school counseling field clearly reveals a long-standing struggle to define the role and function of school counselors. This is perhaps due to the overlap—and even some professional territorial struggles—with school social workers and school psychologists, all of whom are concerned with psychosocial counseling and intervention with students.

School districts that have made budgetary decisions to hire only one mental health provider may employ a school counselor to provide all counseling to students, including

---

Professional History

Understanding and Mastery of Professional History: Historical and current legislation affecting services delivery

Critical Thinking Question: Most states now require school social workers, counselors, and psychologists to hold master's degrees in their fields, with several hundred hours' worth of internship experience, and often with a special certification, as well. Why is it important for human services provider positions in schools to be so highly professionalized?

guidance, career, and mental health. In this instance, the role of the school counselor is similar to that of a school social worker. Yet in many schools that employ both school social workers and school counselors, the latter commonly will provide more academically related counseling and even be responsible for many administrative functions, including maintaining school records and monitoring attendance. For instance, Lambie and Williamson (2004) complained that in many school districts school counselors are working as assistant principals. Lambie and Williamson cite this practice as an example of role confusion within the school counseling field, suggesting that the American School Counselor Association (ASCA), the professional organization for school counselors, continue its quest to outline and define the professional identity of school counselors.

Challenges Facing Urban "Inner-city" Schools

The plight of urban schools has received considerable attention in the past several years, from both educators and the federal government. In response to these concerns, ASCA and the Education Trust (a not-for-profit agency committed to working for high academic achievement among all children) have made numerous recommendations regarding school counseling programs, including developing systematic programs designed to address many of the issues currently confronting urban schools, such as gang activity, poverty, homelessness, child abuse, violence on and off campus, increasing rates of clinical depression, unplanned pregnancy, and low academic performance (Baggerly & Borkowski, 2004; Holcomb-McCoy, 2005; Lee, 2005).

In addition, urban schools face what is referred to as an *achievement gap* when compared to suburban youth. Urban youth are far more likely to drop out of high school and are less likely to meet the minimum standard on national standardized tests. Urban schools have far greater difficulty retaining quality teachers, must contend with political issues often not confronting suburban schools, and are often located in high-crime areas of concentrated poverty (Olson & Jerald, 1998).

Other issues facing urban schools and school counselors working in these settings include dealing with high student absenteeism, unstable family systems, including a high percentage of students living in foster care, and high student transience, where students often transfer in and out of school frequently (Green, Conley, & Barnett, 2005; Lee, 2005). Each of these issues is far more complex than one might think initially. For instance, consider the issue of high student mobility: One might think that this issue would not necessarily affect the school the student is leaving, yet students who leave schools suddenly due to family instability often fail to return their textbooks, which can lead to significant financial losses for schools, many of which are already suffering serious budgetary shortfalls.

Urban schools are often overcrowded and located in high-crime neighborhoods.
Will Hart/PhotoEdit

California is one state that has a significantly higher incidence of student mobility than many other states, due in part to the immigrant population. In a 1999 study of the impact of student transience on school districts, school researchers made several suggestions including utilizing school counselors to reach out to departing and incoming students to coordinate transfers and minimize disruptions (Rumberger, Larson, Ream, & Palardy, 1999).

Common Roles and Functions of School Counselors

School counseling programs generally focus on three basic areas: *academic counseling, career development,* and *personal–social development* (Dahir, 2001). What form this counseling takes depends in large part on whether the counselor is working at an elementary school, middle school, or high school. Other issues influencing the nature of the counseling include the size of the student population, whether the school is in an urban or rural area, and the nature of surrounding community. A school counselor who works at a high-crime, overcrowded high school in inner-city Chicago will certainly have a different role and perform different functions than a school counselor working in a high-income suburban elementary school.

> School counseling programs generally focus on three basic areas: academic counseling, career development, and personal–social development.

According to the ASCA website (see http://www.schoolcounselor.org), school counselors may engage in the following activities:

- individual student academic program planning
- interpreting cognitive, aptitude and achievement tests
- counseling students who are tardy or absent
- counseling students who have disciplinary problems
- counseling students as to appropriate school dress
- collaborating with teachers to present guidance curriculum lessons
- analyzing grade point averages in relationship to achievement
- interpreting student records
- providing teachers with suggestions for better management of study halls
- ensuring that student records are maintained as per state and federal regulations
- assisting the school principal with identifying and resolving student issues, needs, and problems
- working with students to provide small and large group counseling services
- advocating for students at individual education plan meetings, student study teams and school attendance review boards
- disaggregated data analysis (American School Counselor Association, 2005, p. 1).

In general, school counselors provide individual student guidance, such as helping students develop good study skills, do some preliminary career planning, develop effective coping strategies, and foster good peer relationships through the development of prosocial skills, such as exhibiting empathy, showing kindness to others, and managing anger appropriately. School counselors also develop and facilitate programs on substance abuse awareness and multicultural awareness. School counselors assist students with goal setting, academic planning, and planning for college. They facilitate crisis intervention with individual students, the student body, families, and the school as a whole. They collaborate with parents, teachers, and school administrators and provide community referrals as necessary. They may also facilitate programs focusing on making the transition to the next level in school or to work. School counselors identify and work with at-risk students, managing behavioral and mental health issues such as substance abuse, suicide threats, classroom disruptions, student–teacher conflicts, and other issues as they arise.

Among school counseling competencies, Lee (2005) lists *cultural competence,* the ability to advocate for students in an attempt to remove barriers to academic success, a willingness to be leaders in educational reform, and the ability to effectively communicate with and collaborate with other educational professionals.

Common Ethical Dilemmas Facing School Counselors

As with many other human service-related disciplines, school counselors face ethical dilemmas on a daily basis that require them to not only be acutely aware of the ethical standards of the school counseling profession, but also be aware of common dynamics they may face that could result in sliding down the "slippery slope" from genuine caring about students to the egregious violation of ethical boundaries. Some of these challenges are pretty straightforward, such as maintaining confidentiality of student counseling and related records or reporting child abuse in accordance with mandatory child abuse laws. But there are other areas of ethical concern that are not so clear cut, and involve far more of that ethical "slippery slope," where appropriate responses to complex situations are very much in the gray area.

For instance, consider the Latina school social worker who is passionate about advocating for Latina students because of what she endured in school, or the white school counselor who, without awareness, seems to automatically show bias toward other white students, and against students of color. At what point does passionate advocacy become excessive single-minded bias toward one subpopulation, and directly or indirectly, against another? School counselors deal with very complex situations on a daily basis, and make decisions about how to handle these situations not solely upon their professional training, but also on their own personal experiences—often those very experiences that brought them to this career to begin with. Thus, it is important for a school counselor to be aware of how easy it is for unethical behavior to be rooted in a sincere desire to show care and concern for students, particularly those students who are particularly vulnerable.

In the ASCA online website, there is a section devoted to legal and ethical issues for school counselors. Several articles posted in this section of the website pose ethical dilemmas, exploring the nature of these dilemmas in such a way as to assist school social workers in recognizing the ethical and unethical nature of various approaches to student problems and situations. For instance, one article entitled "Boundary Crossing: The Slippery Slope," features a vignette of a student from a particularly chaotic and neglectful home who develops a strong attachment to the school counselor, popping into her office spontaneously whenever he needs some additional support. The student then invites the school counselor to a wrestling match a considerable distance from the school on a Saturday evening. The school counselor attends the match, and then drives the student home, stopping for dinner on the way home as a gesture of congratulations for a job well done. Readers are asked whether any aspect of this scenario violated ethical boundaries and why. Most respondents were somewhat mixed on whether the fluidity of the office visits and traveling a long distance to attend a student's school-related event were ethical, but all respondents perceived the school counselor driving the student home from the match and stopping for dinner, as clearly representing an ethical boundary violation (Stone, 2011).

Stone (2011) goes on to explain each level of boundary violations and risks involved, including the violation of boundaries regarding *roles* (confusing the role of school counselor with a caregiver), the violation of boundaries regarding *time* (allowing the student to so frequently make impromptu office visits, an arrangement that cannot be maintained for all students), the violation of boundaries regarding *place* (attending an event outside of school hours and such a long distance away, driving the student home, and stopping for dinner). Such boundary violations, while coming from good intentions on the part of the school counselor, can lead to confusion on the part of the student who may develop unrealistic expectations of the school counselor, and can also show bias toward one particular student, when many students have similar backgrounds and needs. In closing, the author summarizes how many ethical boundary violations come out of good intentions on the part of the school counselor, and seemingly innocuous initial events:

> Boundary violations do not necessarily arise from bad character. When school counselors do not recognize boundary crossings, innocent acts merely intended to be supportive can spiral downward to boundary violations such as counter-transference or worse. Egregious boundary violations are usually preceded by relatively minor boundary excursions. (Stone, 2011, p. 12)

Concluding Thoughts about School Counselors

Although it may be true that the school counseling profession is still struggling to assert a strong professional identity, establish the roles and functions of school counselors, and maintain a presence among other student services professionals, as educational reform movements continue to grow, schools will benefit most from a multidisciplinary team that addresses the comprehensive needs of all students. Although school social workers and school counselors often have overlapping roles and missions, a school with a student body of about 3,000 that employs two school social workers and four school counselors will certainly have enough student issues to keep all student services personnel busy!

School Psychologists

The National Association of School Psychologists (NASP, n.d.) includes the following statement on its website: "School psychologists help children and youth succeed academically, socially, and emotionally. They collaborate with educators, parents, and other professionals to create safe, healthy, and supportive learning environments for all students that strengthen connections between home and school."

If you think this explanation is similar to the description of school social workers and school counselors, you are correct! As with the other two student services positions discussed in this chapter, school psychologists have a broad range of responsibilities and functions that depend on the actual school environment. But one significant difference between a school psychologist and a school social worker and/or school counselor is that a school psychologist conducts academic testing on students to evaluate and assess their academic abilities and deficits and often is the only student services professional who is trained in evaluating intervention programs.

Most school psychologists have a master's degree in educational psychology, have completed a lengthy internship at a school, and have a special credential designating them as a school psychologist. Those with master's degrees in social work and counseling who want to become a school psychologist can earn an EdS (specialist in education), which will enable them to obtain a school psychologist credential.

Common Issues and Effective Responses by Human Services Personnel

Due to the overlap that exists in the roles and functions of school social workers, school counselors, and school psychologists, any of these professionals will encounter similar clinical issues while working in a public school. Thus, although some of the information contained in this section might appear to be oriented more toward one discipline or the other, it is important to remember that all human service professionals working in a school setting could conceivably confront these same issues, depending on their role within their assigned school.

I mentioned earlier that the nature of work performed by school social workers, school counselors, and school psychologists can vary significantly, depending on a wide range of variables and circumstances; yet certain issues will arise on virtually every public school campus, and human service professionals working on school campuses must be trained to both recognize and respond to them when they occur.

Depression and Other Mental Health Concerns

The National Institute of Mental Health (1999) states that approximately 3 percent of children and 8 percent of adolescents suffer from some form of depression. These statistics underscore the importance of human service professionals having the tools necessary to both recognize and respond to depression in the school environment.

Symptoms of depression in children and adolescents are similar to that of adults, except that oftentimes children exhibit symptoms of irritability rather than melancholy. Another important consideration is that it is often the quiet children, sitting in the back of the classroom bothering no one, suffering silently, are often the most in need of help, yet likely to be overlooked by school personnel because they are not acting out in any visible way.

Abrams, Theberge, and Karan (2005) recommended that school counselors (and other mental health providers) use an ecological model (discussed in Chapter 1) as a lens through which a depressed student is assessed. Students who are identified as suffering from depression are evaluated from a perspective that considers a student's "contextual map" to truly grasp the reciprocal nature in the relationships between the depressed students and their environment, including their families, close friends, neighborhoods, and school (microsystem); to truly grasp the reciprocal nature in the relationship between depressed students and their broader community (mesosystem); and finally to allow the counselor to evaluate the impact of the broadest aspects of the student's world, including the effect of cultural mores, various social reforms, political policies, and the impact of natural tragedies (exosystem).

For instance, in assessing and evaluating a potentially depressed student, the clinician would evaluate the relationship the student has with peers, family members, and even teachers. Is the student experiencing conflict with one or both parents? Has the student recently experienced fights with peers? The counselor will then evaluate the relationship the student has with the broader community. Is the student involved with the legal system? Does the student have involvement with a truancy officer? Finally, the clinician will evaluate how anything in the broader society might be affecting the student.

For instance, the terrorist attacks of September 11, 2001, had a devastating impact on virtually everyone. The evaluation of any student for depression in the months subsequent to September 11 was likely assessed in the context of these devastating events. Did the student have any friends or family members who were directly affected by these attacks? Does the student have a parent or close family member who was deployed to Iraq or Afghanistan in response to these attacks? Similarly, any significant changes in governmental social policy have the potential to affect students, particularly those who are living in government-subsidized housing and who have parents who are subsidized by public assistance. Do these changes in policy affect the student's family in a way that consequently puts pressure on the student because of increased stress within the household?

In general, the human service professional not only evaluates anything that might be a contributing factor to the student's current mental health status, but also evaluates strengths and support within the student's world (Abrams et al., 2005). Does the student belong to a church body or faith community that offers or has the potential of offering support? Does the student have any extended family members who might come forward and offer to support the student during a difficult time? A student who is experiencing depression because his father was deployed to Afghanistan might have an untapped support system in a support group for children sponsored by the U.S. Army.

The value of this model is that it is complementary with the overall model of human service professionals who are trained to consider the entire context within which the student is operating. This approach also enables the social worker and counselor to provide more effective case management once contributing factors and support systems are identified. This model also encourages the involvement of the student's family system. In fact, research so strongly supports the positive impact of parental involvement in the student's mental health on academic achievement that Vanderbleek (2004) suggested that school social workers and school counselors identify and address any barriers to families becoming involved in the counseling process. These barriers might be cultural or racial, such as a less-than-welcoming environment toward non-English-speaking parents or parents who do not feel well treated by school personnel. Barriers can also be more concrete, such as a parents' lack of transportation or a work schedule that makes meeting with school personnel impossible. Flexibility on the part of support services personnel, including a willingness to conduct home visits—after school hours, if necessary—will help to reduce the majority of these barriers.

Auger (2005) wrote in favor of a multifaceted approach to depression intervention within the school system and suggested that school counselors collaborate with school personnel, families, and other mental health practitioners, challenge the student to address any pessimistic or negative thinking, encourage the development of greater insight into feelings and their connection with behavior, help the student develop better social skills, and create opportunities for the student to succeed. Auger even advocated encouraging the student to increase physical activity because there appears to be a relationship between physical activity and positive mental health.

There is, of course, a limit to the amount of mental health services a school can provide to its students. Student support personnel must learn to recognize when a student's mental health problems have evolved past the purview of the school social worker's or counselor's area of expertise. Many students experience a level of depression that can successfully be addressed within the school, but human service professionals' training may not extend to the level necessary to deal with a student who is profoundly depressed and/or whose family system is so desperately impaired, that outside referral—and possibly hospitalization—is the only viable option.

Human Services Delivery Systems

Understanding and Mastery of Human Services Delivery Systems: Major models used to conceptualize and integrate prevention, maintenance, intervention, rehabilitation, and healthy functioning

Critical Thinking Question: Many human service professionals employ a tool called an "ecogram"—a graphic model that represents clients' relationships with other people and institutions, as well as stressors and sources of strength and support. How might this model be used to assist a school child who is suffering from depression?

Diversity and Race

In virtually every school, some students fit into the mainstream and others do not. It is often the student who does not fit in who is most likely to be vulnerable to scapegoating, bullying, and violence. Students who do not feel safe in school, who are subject to bullying, and who are made to feel as outcasts because of race, sexual orientation, or any reason that seems to set them apart from other students will be at risk for academic failure or at least academic difficulty.

Although the responsibility for keeping students safe lies with all adults associated with the student—teachers, all school personnel, and even parents—human service professionals are in a unique position to identify potential problems related to diversity and intervene by advocating for diverse students.

Racial and ethnic diversity can be a wonderful asset to any school environment leading to a richness in experiences for students and teachers alike. But in some school environments, racial prejudice and discrimination can lead to violence and conflict among many within the student population. Students who comprise a part of a racial minority either within the school or within the broader society are at risk for academic failure for many reasons including social, economic, and political conditions such as poverty, racial intolerance, and higher rates of violence often associated with the urban school environment discussed earlier in this chapter. A school environment that is hostile to racial minorities contributes to an environment where students feel unwelcome and possibly where school policies either directly or indirectly discriminate against students of color. The target of racial discrimination is not limited to people of color though, particularly on a school campus where any student who is a racial minority can be a ready target for bullying and violence.

Human service professionals can assist teachers and school administrators in recognizing and addressing discrimination and prejudice on campus. They can also assist in the development of cultural diversity training focusing on racial sensitivity and respect for diversity. Equally important is the cultural competence of the counselors themselves. It is vital that human service professionals undergo training, focusing on the nature of counseling from a multicultural perspective (Holcomb-McCoy, 2004).

> **It is vital that human service professionals undergo training, focusing on the nature of counseling from a multicultural perspective.**

An examination of traditional counseling theories and interventions reveals a bias against racial minorities, particularly African Americans. Fusick and Charkow (2004) discussed the tendency of traditional Euro-American theories to pathologize racial minorities rather than recognizing the social oppression that contributes to violence, gang activity, and juvenile delinquency. For those who disregard the power of long-standing racism and its resultant oppression, one must ask whether they believe that certain racial groups are simply more violent than others. If not, then credence must be given to the possibility that dysfunctional behavior is not solely a result of individual pathology, but can be the result of social causes as well.

Because of a history of abuse by child welfare agencies (as discussed in Chapter 5) and a court system that has often not recognized the long-standing effect of generations of racial discrimination, certain minority groups may be mistrustful of counseling and mental health treatment (Horejsi, Craig, & Pablo, 1992; Surbeck, 2003). Thus, counselors should not assess a student's or family's wariness of mental health personnel as paranoia or as a sign of deception, but should recognize and understand the roots of such mistrust—one that can often be overcome through the development of an authentic helping relationship and student-centered advocacy.

Fusick and Charkow (2004) also discussed the effect of biased assessment tools that were created for assessment and evaluation of the majority culture, with Caucasian middle-class values and mores. African American students in particular are far likelier to be referred for counseling and social work services for behavioral problems, further exacerbating the hostility often felt toward mental health professionals. Fusick and Charkow recommend that social workers and counselors be neither too directive nor too appeasing in counseling sessions, but instead focus on developing a truly authentic relationship.

I worked as a school social worker in an urban school that was primarily African American. Not only was I one of very few Caucasians on campus, but my caseload consisted primarily of boys, and as a woman, there was a natural discomfort with me on the part of my students. A fellow social worker suggested that I try to speak to the students using some of the slang used by many of the African American or Latino American youth. Knowing that I could never get away with this, I decided to just be myself and express my desire to get to know each of them. I spent time getting to know them and their interests, and I quickly learned that many of the boys loved athletics. I purchased packs of sports cards as encouragement and rewards. In time, the majority of my students recognized my sincere desire to understand and help them. And although I never judged them or their feelings, I was never afraid to jump in and make suggestions for either perceiving or handling situations in a different way.

Lesbian, Gay, Bisexual, Transgendered, and Questioning Youth

Students who are in the sexual minority, such as lesbian, gay, bisexual, transgendered, and those students who are questioning their sexuality in some way (LGBTQ), are often the victims of violence, both verbal and physical. Many of these children spend a considerable amount of time feeling different and isolated, often believing that no one will understand their feelings and accept them unconditionally. Such individuals have an alarmingly high rate of suicide attempts, with over 30 percent admitting to having attempted suicide at some point in their lives. Approximately 75 percent of gay and lesbian students admit to having been verbally abused at school, and over 15 percent have been physically abused (Pope, 2003).

Most of the youth in Pope's study reported that the violence they experienced was a direct result of their sexual orientation, with boys being abused more often than girls. Pope discussed this type of abuse in terms of the pressure on most high school students to conform to the norms of their peer group. When faced with the overwhelming demands to be just like everyone else, students who stand out, either because they look different or, as is the case with gay and lesbian students, when their sexual orientation is different, they can quickly become outcasts.

In 2009, the advocacy organization Gay, Lesbian and Straight Educational Network (GLSEN) conducted a national survey of LGBTQ students on their experiences with the following issues:

- hearing biased and homophobic remarks in school
- feeling unsafe in school because of personal characteristics, such as sexual orientation, gender expression, or race/ethnicity;
- missing classes or days of school because of safety reasons; and
- experiences of harassment and assault in school

The results of the study found that a significant majority of LGBTQ students experience verbal and physical harassment on a daily basis in school, with little to any intervention or advocacy on the part of school personnel. For instance, between 75 percent and 90 percent of LGBTQ students surveyed heard homophobic terms used in a derogatory manner, such as "gay," "dyke," and "faggot" in school, and most respondents reported feeling distress in response. Almost 85 percent reported that they had been verbally harassed at school due to their sexual orientation, and almost as many reported that they'd been verbally harassed because of their gender expression (not being femi-nine or masculine enough). About 40 percent of respondents reported that they had been victims of physical harassment at school because of their sexual orientation, and about 20 percent were physically assaulted. Over 50 percent of respondents were vic-tims of cyberbullying and harassment through text messaging, emails, and social media. In most of these cases, there was little to no response on the part of school person-nel, leaving the majority of these students feeling very unsafe in their respective school environments.

The report details the most frequent consequences of these various types of bully-ing related to a student's sexual orientation and gender expression, including higher-than-average absenteeism, lower educational achievement, and a negative impact on their psychological well-being (higher rates of depression, anxiety, and lower levels of self-esteem). The authors of the report recommend the following solutions: gay-straight alliance clubs (GSAs), inclusive curriculum (course curriculum that includes positive representations of LGBTQ people and events, currently and historically), supportive educators (training educators in LGBTQ awareness and advocacy), and incorporation of strict bullying and harassment legislation and policies. Schools that had incorporated these remedies shows marked reductions in LGBTQ bias–based bullying (Kosciw, Grey-tak, Diaz, & Barkiewicz, 2010).

It is vital that school personnel address the harassment that most gay and lesbian students experience and develop a plan for combating this response to students in the sexual minority. The first step is to establish a *zero-tolerance policy,* where teachers, school administrators, and student services professionals make it clear to the student population through policy and action that harassment will not be tolerated in any re-spect. Developing a plan for making school safe for all vulnerable students begins with the education of school personnel.

School social workers, counselors, and psychologists are the ideal candidates to edu-cate both school staff and students on the importance of tolerating diversity. Such a pro-gram must begin with the school staff, particularly the teachers, who are most likely to be present when the abuse of gay and lesbian students occurs. Teachers do not need to be convinced that homosexuality is an acceptable orientation. In fact, regardless of how strongly the student support professionals feel about wanting to create a consensus of acceptance, it is probably unrealistic to assume that everyone on the campus is going to perceive alternate sexual orientations as a positive, albeit alternative, lifestyle choice. What needs to be emphasized is that regardless of one's personal beliefs about the issue of sexual orientation, no human being should be subjected to verbal and physical ha-rassment and abuse. Nor should people be solely defined by their sexual orientation or any other singular aspect of their personhood. School personnel should be taught that personal feelings should be set aside and the focus should be placed instead on teaching students to respect human dignity and everyone's basic right to self-determination.

A particularly effective program facilitated by school social workers, counselors, and psychologists across the nation is called the Making Schools Safe project (Otto, Middleton, & Freker, 2002). This program was developed by the American Civil Liberties Union (ACLU) and was designed to combat antigay harassment on school

Human Systems

Understanding and Mastery of Human Systems: Emphasis on context and the role of diversity in determining and meeting human needs

Critical Thinking Question: Lesbian, gay, bisexual, transgendered, and questioning (LGBTQ) youth are at extremely high risk for bullying and harassment and the myriad emotional wounds that result from such treatment. How might a school social worker, counselor, or psychologist best work to meet the needs of this vulnerable population?

campuses. The ACLU recommends that all teachers and administrators use this curriculum, which focuses on the vital importance of creating a safe learning environment for all children.

The Terrorism Threat and the Impact of 9/11

On September 11, 2001, members of a terrorist organization called Al-Qaeda hijacked four commercial airliners and crashed two of them into the World Trade Center towers in New York City and one into the Pentagon in Arlington, Virginia. The fourth airplane, allegedly intended for the White House, crashed in Somerset County, Pennsylvania, after passengers temporarily overpowered the hijackers. This series of terrorist attacks was followed by a month-long bioterrorism attack with letters sent through the post office laced with anthrax (Baggerly & Rank, 2005). The media was filled with reports of feared future attacks.

Many school districts scrambled to develop programs to address students' feelings and concerns in the wake of the September 11 attacks. The most common psychological response was post-traumatic stress disorder (PTSD), a disorder that often occurs in the wake of a traumatic event. Individuals with PTSD continue to experience fear, hopelessness, and horror long after the event (American Psychiatric Association, 2000). Vicarious victimization was also prevalent on many school campuses. The events of September 11 were difficult for adults, but were particularly hard on children who lack the ability to think abstractly and who often lack the ability to communicate their feelings.

A 2004 study found that 65 percent of respondents reported that students experienced moderate to high levels of distress in the weeks following the attacks (Auger, Seymour, Roberts, & Waiter, 2004). The most frequently reported symptoms included fear, worry, anxiety, sadness, anger, and aggression. Students who were personally affected by these terrorist attacks or who already suffered from some mental health issues, such as depression, were the most at risk for developing PTSD symptoms.

Auger et al. (2004) also noted that although most schools surveyed took appropriate action in responding to the attacks, 12 percent took no responsive action. The majority of the schools surveyed took no action to assist school personnel in dealing with their own feelings. Over one-third of school counselors stated that they did not feel prepared to respond to a serious trauma, suggesting that ongoing training of all school personnel is essential.

There have been many longer-term consequences to the September 11, 2001, attacks but one particularly troubling reaction is a marked increase in what is called Islamophobia, the irrational fear and hatred of Muslims (or those perceived to be Muslims). The Runnymede Trust—a social policy think tank organization—has identified eight components of Islamophobia:

Seeing Islam as

1. a "monolithic bloc," static and unresponsive to change
2. "separate and other" with values that are dissimilar to other cultures
3. inferior to the West—"barbaric, irrational, primitive and sexist"

4. violent, aggressive and threatening, and in support of terrorism
5. having a political ideology used for military advantage

Responded to by non-Muslims by

1. summarily rejecting any criticisms made by Islam of the West
2. justifying discrimination and social exclusion of Muslim populations based upon this hostility
3. perceiving anti-Muslim hostility as normal (Conway, 1997).

A 2011 policy brief published by the Institute for Social Policy and Understanding states that bullying of Muslim children in school environments is on the rise since the September 11, 2001, terrorist attacks (Britto, 2011). While the increase in bias-based bullying is on the rise in general, this brief identifies the chief reason why Muslim children are being bullied is due to "American mainstream's limited knowledge, pervasive misperceptions, and negative stereotypes about Muslims" (p. 1). Britto cites the influence of media on the attitudes of non-Muslim youth, which frequently depicts Muslims as potential terrorists, and ideological extremists. She recommends using the media to counteract these negative and incorrect stereotypes, such as creating YouTube videos depicting accurate reflections of Muslim culture.

The Learning Channel (TLC) attempted to do just that with its new series called *All-American Muslims*—a reality-based show featuring Muslim families in their everyday lives. The show features the lives of several families living in Michigan, including a high school football coach and his family, and a young newlywed couple expecting their first child. The purpose of the show according to TLC and its producers is to educate the non-Muslim American population about the range of Muslim culture by illustrating how Muslim-Americans are often concerned about the same ordinary issues as everyone else. Yet, despite the positive intention of the show's producers, significant controversy ensued, leading to most of the show's advertisers pulling their ads during the show. The majority of the criticism came from a conservative Evangelical Christian organization called Florida Family Association (FFA), whose founder, fundamentalist David Caton, is better known for attacking GSAs (according to one parent he once compared gays to murderers) (Freedman, 2011). Caton claimed that the show, *All-American Muslims*, had an "Islamic agenda," which was a threat to American traditional values. On FFA's website, a statement about the show reads as follows:

> Florida Family Association urged advertisers to stop supporting the Learning Channel's new show All-American Muslims because it appeared to be propaganda designed to counter legitimate and present-day concerns about many Muslims who are advancing Islamic fundamentalism and Sharia law. The show profiled only Muslims that appeared to be ordinary folks while excluding many Islamic believers whose agenda poses a clear and present danger to liberties and traditional values that the majority of Americans cherish. (FFA, 2012)

So essentially, Canton and his small fundamentalist organization had a problem with the show because it *did not* feature Muslims who were extremists, had values contrary to American culture, or who embraced Sharia law! Canton claims that he and his organization were successful in influencing the majority of the show's advertisers, citing that 100 out of 112 of them did not advertise on future episodes in response to Canton's email campaign. What concerns social justice advocates is that such a small fundamentalist organization was able to successfully utilize social media to spread Islamophobic propaganda to such an extent that it influenced numerous major advertisers into pulling support (although many have since reinstituted advertising support in response to strong public criticism).

This is just one example of what many fear is becoming mainstream Islamophobia being projected at attempts to dispel stereotypes and myths regarding the American-Muslim population. Other examples, particularly those affecting school children, include making constant references to Muslim children as being terrorists, and making jokes about Muslim children and their families making bombs (Abdelkader, 2011). Such bias-based bullying should not be tolerated and school social workers, in coordination with school counselors, school psychologists, teachers, school administrators, parents and other students, can counteract Islamophobia through the implementation of educational programs designed to increase awareness of the range moderate belief systems embraced by the mainstream Muslims, both in the United States and abroad. Yet, as TLC found, social workers would be wise to expect controversy on some level, particularly by parents and those in the community who may be threatened by attempts by any marginalized group to assert its collective right to enter the mainstream of America.

Substance Abuse

Substance abuse both on and off campus continues to be a growing problem across the United States, primarily in high schools, but also in some middle schools. School social workers, counselors, and psychologists must be able to identify the signs of substance abuse as well as be prepared for the various ways of intervening when substance abuse is suspected. (See Chapter 11 for more on the issue of the lack of training in the area of substance abuse.) Although many graduate programs in the mental health–related fields are addressing this issue by including more courses on substance abuse, the majority of programs still only offer substance abuse courses as electives. Many mental health professionals in student services are unprepared to deal with substance abuse issues or the complexity of adolescent substance abuse, particularly with regard to complicated family systems (Lambie & Rokutani, 2002). The reality is that 74 percent of high school seniors in suburban high schools have reported using alcohol, and 40 percent of high school seniors in suburban high schools have reported using illegal drugs (Greene & Forster, 2004), making substance abuse one of the most significant issues confronting school personnel.

> **74 percent of high school seniors in suburban high schools have reported using alcohol, and 40 percent of high school seniors in suburban high schools have reported using illegal drugs.**

School counselors need to be able to identify adolescent substance abuse and respond with an intervention strategy. That strategy must include a response from the school as well as from outside referral sources that will involve the entire family system. The model most often used to describe the nature of adolescent substance abuse is similar to an adult model and does not take into consideration factors related to adolescent development. Adolescents tend to be egocentric, often acting and feeling in ways that tend to be self-focused. They also tend to display behavior that is impulsive, appearing to lack any real sense of consequences. This seeming sense of omnipotence, coupled with developmental egocentrism, often complicates traditional models of substance abuse.

Lambie and Rokutani (2002) suggested using a systems perspective in evaluating substance abuse in the adolescent population. Rather than viewing substance abuse in the adolescent as an individual problem, a systems perspective views the substance abuse as a sign of something amiss within the family system. The substance-abusing adolescent often serves some purpose within the family system, such as enabling the parents to focus on the adolescent's dysfunctional behavior rather than on problems in the marriage. The substance-abusing adolescent sometimes serves as an apparent symptom of deeper problems within the family system that are purposely hidden from view. For instance, the family who works hard to appear "normal" and healthy will be compelled to deal with underlying dysfunction when one or more of the children begin acting out in ways that require outside attention and intervention, such as abusing drugs and alcohol.

Another issue to consider when using a systems perspective is whether the adolescent's substance abuse is mirroring a parent's substance abuse. A parent's abuse of alcohol or drugs has been shown to influence an adolescent's decision to begin drinking (Lambie & Sias, 2005; Piercy, Volk, Trepper, Sprenkle, & Lewis, 1991). In general, families that have system problems such as parental substance abuse and other forms of maladaptive behavior tend to be rigid closed-family systems and lack the ability or capacity to handle the increased stressors associated with children entering the adolescent years. Adolescents demanding changes to long-standing rules, pushing for more privileges, developing a far wider circle of peers, and questioning family rules can often leave a family that is wary of outsiders and rigidly adheres to rules and discipline with few effective coping skills to adapt to these changes. In addition, problems that have their roots in early childhood most often manifest during adolescents. In fact, I have worked with adolescents for years and cannot think of a single adolescent who did not act out in response to an issue or condition with roots in his or her childhood.

A school social worker, counselor, or psychologist working with substance-abusing adolescents must first be able to identify the common signs of abuse, including erratic behavior, mood swings, red eyes, and slurred speech. They must then be able to provide support to both the student and the family, acting as a liaison between student, family, school, and community-based treatment programs.

On a broader level, student services personnel can institute prevention programs in the school, such as the Drug Abuse Resistance Education (DARE) program that involves police and other community agencies coming into the schools and creatively (through plays, dance, and songs), and in an age-appropriate manner, enlighten students about the dangers of drug abuse and encourage students to avoid substance use and abuse.

Child Abuse and Neglect

School social workers, counselors, and psychologists are often in the position of having to report child abuse to their local child welfare agency. (See Chapter 5 for a discussion of child protective services' involvement in child abuse cases.) School social workers, counselors, and psychologists are often in the precarious position of having to decide what should constitute a "hotline" call. For instance, a child showing up to school with bruises, who discloses she has been physically abused by her mother, clearly mandates a call to child protective services, but frequently a counselor might not have such a clear indication of abuse and must make a determination based on suspicion. It is important for student services personnel to understand that they do not need to be certain of abuse; if there are indicators of any type of abuse, it is their legal obligation to file a report and allow child protective services to conduct an investigation.

It is important that the school social worker, counselor, or psychologist remain composed when a student discloses abuse, but express compassion, support, and encouragement. It is equally important that promises are not made that cannot be kept. For instance, the counselor should not promise not to tell anyone, because the student will feel betrayed when report of child abuse is made (Lambie, 2005). There might also be reticence on the part of the counselors to make a report of child abuse if they know the parents and are suspicious of the student's disclosure, but the counselor must adhere to the law, which requires that any abuse disclosure be reported as required.

Teenage Pregnancy

A newspaper article in 2005 reported that 13 percent of the female students at an Ohio high school were pregnant, causing serious concern about why this high school's pregnancy rate was nearly double the national average (Garvey, 2005). Although teenage pregnancy has been on the decline in recent years (Karraker, 2004), it remains a serious concern, with over 60 percent of high school seniors reporting they were sexually active (Greene & Forster, 2004). Various research studies have pointed to many factors that might influence pregnancy. Beyond sexual activity in the adolescent population, other factors include early alcohol use (Stueve & O'Donnell, 2005) and poverty (Young, Turner, Denny, & Young, 2004).

Research on prevention points to religiosity (Rostosky, Regnerus, & Wright, 2003), peer influence, appropriate parental supervision, good and direct parental communication, SES, race (Corcoran, Franklin, & Bennett, 2000), and involvement in sports that is correlated with remaining abstinent in high school or at least becoming sexually active later in adolescence.

Sex and pregnancy prevention programs have been included in school curriculums for several decades with mixed reviews. *Abstinence-only* programs, although somewhat controversial, have shown to be surprisingly successful (Toups & Holmes, 2002). In fact, Toups and Holmes reviewed several studies that revealed marked reductions in teenage pregnancy after experiencing a school-based abstinence-only program. In fact, one cited study evaluated all 5,000 teenagers who participated in an abstinence program in one year. Not only did few of these teenagers become sexually active, but also over 50 percent of the students who had been sexually active stopped having sex. Proponents of abstinence-only sex education cite the decrease in adolescent sexual activity as evidence that these programs work. Yet others have questioned whether these programs are as successful as some of these studies indicate, citing poor study designs and a wide range of abstinence programs with some defining abstinence as postponing sex until early adulthood and some more religiously based programs sending the message that premarital sex should always be avoided. Without a clear definition of "abstinence," critics claim that it's impossible to determine the success of these programs (Kirby, 2002).

Some educators are concerned, though, that abstinence-only programs will not work for all teenagers, particularly those who have any of the complicating factors mentioned earlier. Teenagers living in poverty, who have poor communication with their parents, and who are not supervised well by their parents may not respond positively to abstinence-only programs because of the other forces pushing them in the direction of sexual activity. Based on the belief that some adolescents will have sex no matter who tells them not to, education programs focus on safe sex practices, such as using condoms during sexual intercourse. Many of these programs focus among other things on HIV/AIDS education, which is often later cited as a chief reason among adolescents for using condoms. Although there has been some concern that educating teenagers to use contraception and even making contraception available is sending a mixed message (i.e., "You should not have sex during adolescence, but just in case you do, use a condom!"), which in essence promotes sexual activity during adolescence, a review of 28 studies examining this issue clearly shows that such programs do not increase sexual activity among teenage participants, nor do they lead to sexual activity at an earlier age. In fact, many studies indicated that safe sex programs increase the usage of contraception (Kirby, 2002).

One of the most popular programs currently used in high schools across the nation is called the Baby Think It Over (BTIO) program, which uses a computerized doll programmed to cry and fuss intermittently throughout the day and night to educate teenagers on the realities of having a baby. This program has been successful in educating teenagers about the hardship and burden of having a child at such an early age (Somers, Johnson, & Sawilowsky, 2002).

Another issue commonly noted by school social workers, counselors, and psychologists who work with female high school students is a pervasive tendency for girls who are sexually active to report that they had not considered the possibility that they could have said no to a boyfriend's sexual advances. Developing "empowerment" support groups where girls can have a safe place to talk about their feelings about sex, support each other in their right to say no, and consider the positive consequences of doing so can be a successful tool in encouraging better boundary setting, which is likely to result in a reduction of sexual activity.

Attention Deficit Disorder and Attention Deficit/Hyperactivity Disorder

In the past 20 years, diagnoses of attention deficit disorder (ADD) and attention deficit/hyperactivity disorder (ADHD) have literally skyrocketed, with school personnel being on the leading edge of those referring children for evaluation and assessment. According to the *DSM-IV-TR* (*Diagnostic and Statistical Manual of Mental Disorders*, fourth edition, text revision), individuals with ADD suffer from inattention, have difficulty following directions, have difficulty maintaining a sense of organization, and are reluctant to engage in activities that require sustained mental effort (American Psychiatric Association, 2000). Additional symptoms that might warrant a diagnosis of ADHD include hyperactivity, impulsivity, and poor self-control (Kos, Richdale, & Jackson, 2004).

Children diagnosed with ADD/ADHD often present significant challenges in the classroom due to difficulty in paying attention to the teacher and sitting still for extended periods of time. Classroom management with such children can often be difficult because many children with ADD/ADHD symptoms have difficulty with social skills as well, making peer relations a problem.

The most common treatment for ADD/ADHD involves the use of medication, most commonly Ritalin, which is a stimulant that has a calming effect on the child. But many schools have instituted behavioral plans that include token rewards for children who are able to remain focused for increasing amounts of time and who display prosocial behaviors. School social workers, counselors, and psychologists are often called on to work with children exhibiting ADD/ADHD symptoms, both in the classroom and outside the classroom. Many schools utilize a therapeutic group model, bringing several such students together to work on issues such as impulse control, social skills, and maintaining attention. Many therapeutic board games on the market are designed to encourage these skills by engaging the students in play while teaching them how to delay gratification and control their impulses as a winning strategy.

Despite the prevalence in the diagnosing of these disorders, many are concerned that too many referrals for ADHD are coming from educational circles. With most disorders, such as depression or anxiety, referrals for evaluation and counseling might come from one's employer, spouse, or friend. Yet schools tend to be by far the largest source of referrals for ADD/ADHD, presumably because children with these symptoms can cause serious disruption in the classroom, and with class sizes increasing, it can be taxing on a teacher to contend with students who are not paying attention and are acting on their every impulse.

Another concern is that the *DSM-IV-TR* criteria are too broad and in many respects self-fulfilling. For instance, criteria number 1-d states: "Often does not follow through on instructions and fails to finish schoolwork, chores, or duties in the workplace (not due to oppositional behavior or failure to understand instructions)." This criterion is extremely broad and in many respects could be used to describe just about any child at one point or another during his academic career. Certainly most clinicians would not diagnose a child with ADD simply due to one or two incidences of procrastination, but rather they would look for a pattern of behavior. But, a recent study that compared a group of school-aged children assessed using *DSM-IV-TR* criteria and a group of school-aged children using neuropsychological criteria found that the *DSM-IV-TR* group had an 18 percent prevalence rate and the neuropsychological group had a prevalence rate of only 3.5 percent (Guardiola, Fuchs, & Rotta, 2000). This seems to support the criticism that the *DSM-IV-TR* criteria may be too broad.

Another criticism of what some claim is the overdiagnosing of ADD/ADHD includes a concern that boys are disproportionately referred for and diagnosed with ADHD for what many consider to be typical "boy" behavior, including being more naturally active than girls (Sciutto, Nolfi, & Bluhm, 2004). In addition, several studies have found that several other factors might account for ADHD-like behavior including lack of sleep (Brown & Modestino, 2000) and even gifted intelligence with high creativity, leading to a concern that many children are being misdiagnosed with ADHD when other issues might better account for the child's inability to focus, such as fatigue or, in the case with the intellectually gifted child, a need for increased mental stimulation (Hartnett, Nelson, & Rinn, 2004).

One of the most significant concerns of all, though, involves concern among medical personnel, therapists, and parents about the wisdom of giving children a stimulant with cocaine-like properties throughout their developmental years. Historically the medical community did not believe that Ritalin caused any permanent brain changes, yet many recent studies seem to contradict this belief. A 2001 study on rats found that Ritalin use did cause permanent neurological brain changes (Andersen, Arvanitogiannis, Pliakas, LeBlanc, & Carlezon, 2002), and although rats are certainly not humans, they are amazingly similar to humans in the sense that they often respond chemically in ways similar to humans. The results of this study were surprising to researchers because the initial goal of the study was to evaluate whether long-term Ritalin use made subjects *more* vulnerable to drug abuse later in life. What they found, though, was that the rats who were on Ritalin desired cocaine *less*. Further research discovered that this was due to Ritalin causing an increase in a certain protein that affects the pleasure centers of the brain. So although this result might be good with regard to decreasing one's desire for drugs, it is not so good if it makes other activities less rewarding, such as eating and sexual activity.

Another 2001 study also indicates that Ritalin, commonly thought to have a short-term life in the body, has long-term effects, many of which are permanent and most of which remain unknown. This study found that Ritalin may affect or even alter gene expression, which may lead to enduring changes in brain cell structure and function (Acheson, Thompson, Kristal, & Baizer, 2001; Brandon & Steiner, 2003). Basically what this means is that the Ritalin may actually be turning a certain gene on that then turns other genes on, a reaction also found in the brains of those who abuse cocaine. Because these studies are still in the animal model phase, it is impossible to conclude anything other than the implication of the results, which at this point seems to clearly indicate that long-term use of Ritalin in developing children will likely have a permanent effect on their brains. Even at this preliminary stage it seems clear that Ritalin should be used only in cases of serious hyperactivity and perhaps only with neurological testing to determine if Ritalin is medically necessary.

Finally, a 2009 longitudinal study on medical and behavioral intervention for ADHD in children found that while there is some short-term benefit to taking medication, such as Ritalin and Adderal in controlling ADD/ADHD symptoms, there appears to be no long-term benefit. In fact, the researchers noted that at an eight-year follow-up of children diagnosed with ADD/ADHD, children who had stopped taking medication functioned as well as children who were still medicated. In light of this and other research, highlighting the risk and limited benefits of medicating children exhibiting ADD/ADHD symptoms, there is wisdom in approaching the "Ritalin Revolution" with some healthy skepticism.

Human Services Delivery Systems

Understanding and Mastery of Human Services Delivery Systems: Range of populations served and needs addressed by human services

Critical Thinking Question: Increasing numbers of students are being diagnosed with ADD/ADHD, often as a result of referrals from school personnel, and treated with medications such as Ritalin. What are some of the ethical considerations that a school human service professional must take into account as part of a student's treatment team?

Human service professionals may find themselves going against conventional wisdom, by advocating for behavioral interventions with short-term medication protocol (or none at all), only if the child's symptoms warrants taking the risk. Despite these criticisms and concerns, children who exhibit behaviors that are not conducive to the classroom environment need assistance to learn to adapt to a structured world. School social workers, counselors, and psychologists can work with the students in a manner that both respects different learning and personality styles and at the same time encourages children to work effectively in a structured environment.

Concluding Thoughts on Human Services in the Schools

Human service professionals are an integral part of the public school system providing emotional guidance and academic counseling to thousands of students every year. School counselors, school social workers, and school psychologists work within their respective specialties as a part of a multidisciplinary team meeting student needs and increasing student success.

The role of all of these human service professionals is expected to continue to expand in the future in response to a projected increase in many of the social trends experienced today, including an increase in poverty, homelessness, and single-parent families. Teams of human service professionals work together to remove barriers to learning, paving the way for teachers to do what they do best—teach students in their designated academic discipline.

The following questions will test your knowledge of the content found within this chapter.

1. School social work has its roots in the
   a. intelligence testing movement
   b. settlement house movement
   c. English Poor Laws
   d. the Vocational Testing movement

2. Most of the children who were the original focus of the early efforts of early school social workers were:
   a. from families who recently emigrated from non-English-speaking countries
   b. identified for services due to consistent acting out and rebellious behavior
   c. from upper-income families whose children were often left with caregivers
   d. from area orphanages

3. The Disabilities Education Act (Public Law 94-142) required that public schools provide:
   a. "free and appropriate" public education to all school-aged children between the ages of 3 and 21, regardless of their disability
   b. increased funding for social work services for students with special needs, when deemed appropriate
   c. before- and after-school care for all children with "Individualized Education Plans"
   d. Both A and B

4. The field of school guidance counseling was influenced by:
   a. the advent of developmental psychology
   b. the advent of intelligence and aptitude testing
   c. the need for increased vocational counseling to fill the gap left by young men going to war
   d. A

5. Urban schools face what is referred to as:
   a. the urban dilemma
   b. the "drop out" phenomenon
   c. an "achievement gap" when compared to suburban schools
   d. the "funding paradox"

6. Therapeutic board games often used in school settings are designed to encourage which of the following skills?
   a. Self-sufficiency skills
   b. Delaying gratification and managing impulse control
   c. Competitive skills
   d. Both A and C

7. Describe the nature of teen pregnancy, including demographics, associated risk factors, and effective intervention strategies used by human service professionals.

8. Compare and contrast the functions and roles of the various human service professionals working in a school environment, including origins of the different school-based professions, respective professional identity including role overlap, and treatment goals.

Suggested Readings

Brock, S. E., Lazarus, P. J., & Jimerson, S. R. (2002). *Best practices in school crisis prevention and intervention.* Washington, DC: NASP.

Huxtable, M., & Blyth, E. (2002). *School social work worldwide.* Washington, DC: NASW Press.

Sprick, R. S., & Howard, L. (1995) *The teacher's encyclopedia of behavior management.* Longmont, CO: Sopris West.

Torrey, E. F. (2001). *Surviving schizophrenia: A manual for families, consumers, and providers* (4th ed.). New York: Collins.

Tourse, R. W. C., & Mooney, J. F. (Eds.). (1999). *Collaborative practice: School and human service partnerships.* Westport, CT: Praeger Publishers.

Internet Resources

American School Counselor Association: http://www.schoolcounselor.org

International Network for School Social Work: http://internationalnetwork-schoolsocialwork.htmlplanet.com

National Association of School Psychologists: http://www.nasponline.org

School Social Work Association of America: http://www.sswaa.org

References

Abrams, K., Theberge, S. K., & Karan, O. C. (2005). Children and adolescents who are depressed: An ecological approach. *Professional School Counseling, 8*(3), 284–292.

Acheson, A. W., Thompson, A. C., Kristal, M. B., & Baizer, J. S. (2001). Methylphenidate induces c-fos expression in juvenile rats. *Society of Neuroscience Abstracts, 27*, 223–224.

Abdelkader, E. (2011, October 24). Islamophobic bullying in our schools. *Huffington Post*: Religion.

American Psychiatric Association. (2000). *Diagnostic and statistical manual of mental disorders* (4th ed., Text Revision). Washington, DC: Author.

American School Counselor Association (2005). *The ASCA National Model: A Framework for School Counseling Programs, Second Edition*. Alexandria, VA: Author. Retrieved January 27, 2012, from http://www.schoolcounselor.org/files/appropriate.pdf

Andersen, S. L., Arvanitogiannis, A., Pliakas, A. M., LeBlanc, C., & Carlezon, W. A. (2002). Altered responsiveness to cocaine in rats exposed to methylphenidate during development. *Nature Neuroscience, 5*(1), 13–14.

Auger, R. W. (2005). School-based interventions for students with depressive disorders. *Professional School Counseling, 8*(4), 344–352.

Auger, R. W., Seymour, J. W., Roberts, W. B., & Waiter, B. (2004). Responding to terror: The impact of September 11 on K–12 schools and schools' responses. *Professional School Counseling, 7*(4), 222–230.

Baggerly, J., & Borkowski, T. (2004). Applying the ASCA national model to elementary school students who are homeless: A case study. *Professional School Counseling, 8*(2), 116–124.

Baggerly, J. N., & Rank, M. G. (2005). Bioterrorism preparedness: What school counselors need to know. *Professional School Counseling, 8*(5), 458–465.

Borkowski, J. W. & Sorensen, L. E. (2009). *Legal issues for school districts related to the education of undocumented students*. Alexandria, VA: National School Boards Association. Retrieved January 29, 2012 from http://www.nsba.org/SchoolLaw/COSA/Search/AllCOSAdocuments/Undocumented-Children.pdf

Brandon, C. L., & Steiner, H. (2003). Repeated methylphenidate treatment in adolescent rats alters gene regulation in the striatum. *European Journal of Neuroscience, 18*(6), 1584–1592.

Britto, R. (2011). Global battleground or school playground: the bullying of America's Muslim children, Policy Brief #49. Washington, D.C.: Institute for Social Policy and Understanding. Retrieved January 29, 2012, from http://ispu.org/pdfs/ISPU_Policy%20Brief_Britto_WEB.pdf

Brown, T. E., & Modestino, E. J. (2000). Attention-deficit disorders with sleep/arousal disturbances. In T. E. Brown (Ed.), *Attention-deficit disorders and comorbidities in children, adolescents, and adults* (pp. 341–362). Washington, DC: American Psychiatric Association.

Bye, L., Shepard, M., Partridge, J., & Alvarez, M. (2009, April). School social work outcomes: Perspectives of school social worker and school administrators. *Children & Schools, 31*(2), 97. Retrieved September 15, 2009, from MasterFILE Premier database.

Conway, G. (1997). Islamophobia: A challenge for us all. London: Runnymede Trust.

Corcoran, J., Franklin, C., & Bennett, P. (2000). Ecological factors associated with adolescent pregnancy and parenting. *Social Work Research, 24*(1), 29–39.

Dahir, C. A. (2001). The national standards for school counseling programs: Development and implementation. *Professional School Counseling, 4*(5), 320–327.

Florida Family Association. (2012). *Emails to All-American Muslim advertisers made a difference. 101 out 112 companies did not return to the show*. Retrieved January 29, 2012, from http://floridafamily.org/full_article.php?article_no=108.

Freedman, S. (2011, December 16). *Waging a one man war against Muslims*. New York Times. Retrieved from http://www.nytimes.com/2011/12/17/us/on-religion-a-one-man-war-on-american-muslims.html?_r=0

Fusick, L., & Charkow, B. (2004). Counseling at-risk Afro-American youth: An examination of contemporary issues and effective school-based strategies. *Professional School Counseling, 8*(2), 102–116.

Garrett, K. (2006, April). Making the case for school social work. *Children & Schools, 28*(2), 115. Retrieved September 15, 2009, from MasterFILE Premier database.

Garvey, M. (2005, September). Preggo high school; kids are readin', writin' & reproducin'. *New York Post*, p. 19.

Green, A. G., Conley, J. A., & Barnett, K. (2005). Urban school counseling: Implications for practice and training. *Professional School Counseling, 8*(3), 189–195.

Greene, J. P., & Forster, G. (2004). *Sex, drugs, and delinquency in urban and suburban public schools* (Education Working Paper 4). New York: Center for Civic Innovation, Manhattan Institute. (ERIC Document Reproduction Service No. ED483335)

Guardiola, A., Fuchs, F. D., & Rotta, N. T. (2000). Prevalence of attention-deficit hyperactivity disorders in students: Comparison between *Diagnostic and Statistical Manual of Mental Disorders-IV (DSM-IV-TR)* and neuropsychological criteria. *Arquivos de Neuro-Psiquiatria, 58*(2b), 401–407.

Hartnett, D. N., Nelson, J. M., & Rinn, A. N. (2004). Gifted or ADHD? The possibilities of misdiagnosis. *Roeper Review, 26*(2), 73–76.

Holcomb-McCoy, C. C. (2004). Assessing the multicultural competence of school counselors: A checklist. *Professional School Counseling, 7*(3), 178–186.

Holcomb-McCoy, C. C. (2005). Investigating school counselors' perceived multicultural competence. *Professional School Counseling, 8*(5), 414–423.

Horejsi, C., Craig, B. H., & Pablo, J. (1992). Reactions by Native American parents to child protection agencies: Cultural and community factors. *Child Welfare, 71*(4), 329–343.

Karraker, M. W. (2004). Adolescent pregnancy: Policy and prevention services. *Family Relations: Interdisciplinary Journal of Applied Family Studies, 53*(1), 115.

Kirby, D. (2002). Effective approaches to reducing adolescent unprotected sex, pregnancy and childbearing. *Journal of Sex Research, 39*(1), 51–57.

Kos, J. M., Richdale, A. L., & Jackson, M. S. (2004). Knowledge about attention-deficit/hyperactivity disorder: A comparison of in-service and preservice teachers. *Psychology in the Schools, 41*(5), 517–526.

Kosciw, J. G., Greytak, E. A., Diaz, E. M., and Bartkiewicz, M. J. (2010). *The 2009 National School Climate Survey: The experiences of lesbian, gay, bisexual, and transgender youth in our nation's schools.* New York: GLSEN.

Lambie, G. W. (2005). Child abuse and neglect: A practical guide for professional school counselors. *Professional School Counseling, 8*(3), 249–258.

Lambie, G. W., & Rokutani, L. J. (2002). A systems approach to substance abuse identification and intervention for school counselors. *Professional School Counseling, 5*(5), 353–359.

Lambie, G. W., & Sias, S. (2005). Children of alcoholics: Implications for professional school counseling. *Professional School Counseling, 8*(3), 266–273.

Lambie, G. W., & Williamson, L. L. (2004). The challenge to change from guidance counseling to professional school counseling: A historical proposition. *Professional School Counseling, 8*(2), 124–131.

Lee, C. C. (2005). Urban school counseling: Context, characteristics, and competencies. *Professional School Counseling, 8*(3), 184–188.

Lewis, M. R. (1998). The many faces of school social work practice: Results from a research partnership. *Social Work in Education, 20*(3), 177–190.

McCullagh, J. G. (1993). The roots of school social work in New York City. *Iowa Journal of School Social Work, 6,* 49–74.

McCullagh, J. G. (1998). Early school social work leaders: Women forgotten by the profession. *Social Work in Education, 20*(1), 55–64.

McCullagh, J. G. (2001). NASW and school social work: Selected events, developments and publications, 1947–2001. *Journal of School Social Work, 12*(1–2), 5–35. (ERIC Document Reproduction Service No. ED467859)

National Association of School Psychologists. (n.d.). *What is a school psychologist?* Retrieved January 5, 2005, from http://www.nasponline.org/about_sp/whatis.aspx

National Association of Social Workers. (2003). *NASW standards for school social work services.* Washington, DC: Author.

National Institute of Mental Health. (1999). *Depression research at the National Institute of Mental Health* (NIH Publication No. 00-4501). Bethesda, MD: Author.

Olson, L., & Jerald, C. D. (1998). *Quality counts '98: The urban picture.* Retrieved June 18, 2004, from http://rc-archive.edweek.org/sreports/qc98/challenges/tables/ta-n.htm

Otto, N., Middleton, J., & Freker, J. (2002). *Making schools safe: An anti-harassment program from the Lesbian & Gay Rights project of the American Civil Liberties Union.* New York: Lesbian & Gay Rights Project, American Civil Liberties Union. (ERIC Document Reproduction Service No. ED475274).

Piercy, F. P., Volk, R. J., Trepper, T., Sprenkle, D. H., & Lewis, R. (1991). The relationship of family factors to patterns of adolescent substance abuse. *Family Dynamics of Addiction Quarterly, 1*(1), 41–54.

Pope, M. (2003). *Sexual minority youth in the schools: Issues and desirable counselor responses.* Information Analysis. (ERIC Document Reproduction Service No. ED480481).

Rostosky, S. S., Regnerus, M. D., & Wright, M. L. C. (2003). Coital debut: The role of religiosity and sex attitudes in the add health survey. *Journal of Sex Research, 40*(4), 358–367.

Rumberger, R. W., Larson, K. A., Ream, R. K., & Palardy, G. J. (1999). *The educational consequences of mobility for California students and schools.* Berkeley, CA: Policy Analysis for California Education. (ERIC Document Reproduction Service No. ED441040).

Schmidt, J. J., & Ciechalski, J. C. (2001). School counseling standards: A summary and comparison with other student services' standards. *Professional School Counseling, 4*(5), 328–333.

Sciutto, M. J., Nolfi, C. J., & Bluhm, C. (2004). Effects of child gender and symptom type on referrals for ADHD by elementary school teachers. *Journal of Emotional and Behavioral Disorders, 12*(4), 247–253.

Somers, C. L., Johnson, S. A., & Sawilowsky, S. S. (2002). A measure for evaluating the effectiveness of teen pregnancy prevention programs. *Psychology in the Schools, 39*(3), 337–342.

Stone, C. (2011, July). Boundary crossing: The slippery slope. ACSA School Counselor. Retrieved January 27, 2012, from http://www.ascaschoolcounselor.org/article_content.asp?edition=91&section=140&article=1221.

Stueve, A., & O'Donnell, L. N. (2005). Early alcohol initiation and subsequent sexual and alcohol risk behaviors among urban youths. *American Journal of Public Health, 95*(5), 887–893.

Surbeck, B. C. (2003). An investigation of racial partiality in child welfare assessments of attachment. *American Journal of Orthopsychiatry, 73*(1), 13–23.

Toups, M. L., & Holmes, W. R. (2002). Effectiveness of abstinence-based sex education curricula: A review. *Counseling and Values, 46*(3), 237–240.

U.S. Department of Education, *A nation accountable: Twenty-five years after a nation at risk*, Washington, DC, 2008.

U.S. Department of Education. (2001). *No Child Left Behind Act of 2001 (H.R.1)*. Washington, DC: Author.

Vanderbleek, L. M. (2004). Engaging families in school-based mental health treatment. *Journal of Mental Health Counseling, 26*(3), 211–224.

Young, T., Turner, J., Denny, G., & Young, M. (2004). Examining external and internal poverty as antecedents of teen pregnancy. *American Journal of Behavior, 28*(4), 361–373.

CHAPTER 13

Faith-Based Agencies

Robin Nelson/PhotoEdit

Historically, the mental health and medical communities in Western society have had a tendency to divide human beings into biological, intellectual, social, emotional, and spiritual parts, with minimal recognition of how each of these dimensions interacts with the other. But in recent years there has been a growing interest both within professional circles and within the general public in moving away from such a dichotomous view and toward regarding humans from a *holistic* perspective, where a person is considered as a whole with each aspect of the person being inextricably linked with the other.

Essentially, a holistic approach to mental health involves the process of acknowledging, addressing, and evaluating the mind, the body, and the spirit (or soul) when considering any potential issue affecting one's psychological health. In other words, rather than attempting to determine whether depression is a biological disorder with psychological manifestations or a psychological disorder with biological implications, depression would be seen as a disorder or condition having a reciprocal impact on the whole person: mind, body, and soul.

Although mental health providers in the past have had a tendency to shy away from integrating spirituality within the counseling relationship, recent studies have revealed the dramatic ways in which religion or personal spirituality affect people's physical and mental health. In fact, several recent research studies have focused on the mind–body–soul connection in an attempt to understand the reciprocal relationship of each, with a specific focus on how spirituality affects an individual's physical and mental health (Idler & Kasl, 1992; Koenig, George, et al., 1998; Koenig, Larson, & Weaver, 1998; McLaughlin, 2004; Powell, Shahabi, & Thoresen, 2003).

For instance, many of these studies have shown that personal religiousness and spirituality have been linked to a decrease in depression, an increase in greater social support, an increase in cognitive functioning

Learning Objectives
- Understand the various ways in which a faith-based agency is defined
- Become familiar with the various ways different faith traditions provide human services to those in need
- Identify the benefits and potential risks of human services provided by faith-based agencies
- Become familiar with the basic traditions of giving in the "Big Three" faith traditions: Judaism, Christianity, and Islam
- Become familiar with recent legislation and government policy that influences faith-based agency practices

A holistic approach to mental health involves the process of acknowledging, addressing, and evaluating the mind, the body, and the spirit (or soul) when considering any potential issue affecting one's psychological health.

(Koenig, George, & Titus, 2004), an improvement in the ability to cope with crises (McLaughlin, 2004), and better ability to cope with substance abuse problems (Fallot & Heckman, 2005). Although there is some confusion about the difference between *religiousness* and *spirituality,* religiousness is commonly defined as a social experience grounded in traditional religion, whereas spirituality is often defined as having an independent relationship with God (Miller & Thoresen, 2003). Of course, people can be religious *and* spiritual (having a personal relationship with a deity that is grounded in a particular religious faith), religious without being particularly spiritual (a cultural or secular involvement in a religious faith), or spiritual without being grounded in a particular religious tradition. Thus, it is important for practitioners to explore what religiousness and spirituality mean to their individual clients.

The issue of faith is one of importance to all clinical practitioners, particularly mental health practitioners and those within the human services field. Several recent studies have revealed that the majority of U.S. Americans (between 80 and 90 percent) identify themselves as being either religious or spiritual, stating that their faith is an important aspect of their lives (Gallup & Lindsey, 1999; Grossman, 2002). In fact, several of these research studies suggest that clinicians should both acknowledge and address the spiritual dimension of mental and emotional disorders within the counseling relationship, particularly if clients identify themselves as being "spiritually grounded" (Fallot, 2001; Kliewer, 2004; Miller, Korinek, & Ivey, 2004).

Yet human service professionals (or any service provider) must realize that as much as incorporating spirituality into the counseling relationship may be helpful for some clients, there is also the potential for harm, particularly when the religion of a provider is pushed onto a client in a directive or aggressive manner.

Incorporating spirituality into a counseling relationship requires cultural competent counseling skills because many religious traditions are rooted in cultural tradition. Thus, religious abuse is similar in many ways to culture abuse where a provider of a dominant culture is inappropriately directive in forcing the values of the dominant culture onto a client of a different culture. For instance, prior to the recent surge of interest in holistic health, practitioners in the West were often dismissive of Eastern philosophy, which acknowledged the mind–body–soul connection for centuries (Tseng, 2004), rendering many in the human services profession ill-equipped to provide effective services to Asian clients from Buddhist or Hindu traditions (Hodge, 2004).

Religious abuse can be avoided through the sensitivity of the human service professional who not only recognizes the value of addressing matters of faith but also recognizes that the issue of spirituality is only helpful if this exploration is client driven and client centered (Hall, Dixon, & Mauzey, 2004).

Faith-Based Versus Secular Organizations

Human service professionals can incorporate matters of spirituality in virtually any practice setting in response to their clients' disclosure that faith is an integral part of their lives or something they wish to explore. Thus, it is important that providers receive training on the nature of various religious traditions, particularly those they might have an opportunity to encounter in their clinical practice. But there are also numerous human services agencies that operate within a particular religious faith that reach out to those within and outside the tradition.

Such faith-based agencies are often ignored in discussions of human services practice settings, but any review of helping agencies should include an exploration of faith-based agencies because of the long history of religious traditions offering help to those in need, and the fact that matters of spirituality are now recognized as being integral to many people's lives.

What is a faith-based organization? And what makes it different—in nature and service delivery—from a secular agency? It's easy to identify a faith-based organization when it's a synagogue, church, or mosque filled with religious symbols and a mission statement that identifies serving God as a primary function and purpose of the organization. But what about agencies that might be considered parachurch organizations that do not function as churches but more as a human services agencies? Or human services agencies that have their roots in a particular religious tradition but don't integrate religion or faith into practice? Would those agencies be considered "faith based"?

These are more challenging questions than they might appear. Even the courts are not particularly clear on what makes an organization religious in nature (Ebaugh, Pipes, Chafetz, & Daniels, 2003). The difficulty lies in the fact that many secular agencies provide almost identical services as faith-based agencies and there is often no distinguishable difference between the two. Ebaugh et al. discussed the various ways in which policy makers, social scientists, and historians have defined faith-based organizations, with most criteria relating to an organization's *dependency* on religious entities or denominations for support, whether the *mission statement* identifies agency goals that reflect *core values* that are religious in nature, and whether the employees of the organization are religious and adhere to a *statement of faith*.

However "faith based" is defined, it is important to remember that "faith based" does not necessarily mean *Christian*, as might be presumed in some Western countries, such as the United States. In fact, a number of religiously oriented organizations provide faith-based human services grounded in faiths other than Christianity. Thus, although it is true that the majority of faith-based organizations are Christian in nature, many are not. Faith-based agencies may be Jewish, Muslim, Mormon, and Buddhist, each serving communities either broadly or choosing to serve individuals of that particular faith.

Faith-based human services can be facilitated as a ministry of a house of worship, or they can be facilitated as a program within a religious organization that functions as a human services agency, such as the Salvation Army. Such organizations might have the goal of converting clients to that particular faith, believing that conversion is the first step toward wholeness, or they might deliver human services in a manner similar to secular agencies, but operating in a manner consistent with the values of its religious roots. It's important to be aware of the church's or agency's mission because it will have a significant impact on how human services are delivered.

Federal Faith-Based Legislation

Historically, it has been difficult, if not impossible, for a faith-based agency to receive government monies in support of services. The 4th Amendment to the U.S. Constitution, which guaranteed freedom of worship, had been interpreted by the courts to require separation between religions and the government. Thus, unless the agency operated as a secular organization and did not incorporate faith into practice, it could not receive government funding. The government remains sensitive to those members of society who do not share the same faith as the majority culture and, as such, attempts to protect these individuals by passing laws that ensure that they will not be placed in positions where they are either directly or indirectly coerced into praying to a God in which they do not believe.

But in 2001 former president George W. Bush passed the Faith-Based Community Initiatives Act, also known as Charitable Choice, or Care Services Act (CSA), which made it legal for faith-based organizations to receive federal funding as long as these organizations were not involved in religious worship, instruction, or proselytization, at least within the aspect of the organization seeking federal funding. Many saw this as a positive step toward reengaging religious organizations in the care of those in need. Their belief is that it was unfair to exclude faith-based organizations from government funding, as well as a belief that churches and other faith-based organizations can often provide human services more proficiently than government agencies. Yet others express concern that faith-based organizations may enforce arbitrary conditions on service delivery based on religion-based morality that either directly or indirectly discriminates against certain groups, such as gays and lesbians, single parents, the poor, or individuals who embrace different values than the majority population (National Association of Social Workers [NASW], 2002).

But does making services contingent on the performance of some behavior or act rob the client of self-determination and risk forcing cultural and moral values on those who do not share these same social mores? Take, for example, single women in the 1940s and 1950s who had children out of wedlock. It was not uncommon for these women to have services denied to them unless they agreed to place their infants for adoption—a practice based on the cultural and moral belief that premarital sex was wrong and that it would be immoral for a single mother to raise an out-of-wedlock child (Edwards & Williams, 2000). The goal of this chapter is not to determine which side of this debate has a stronger argument. Certainly each side has merit, and a meaningful debate must continue.

NASW expressed concerns about former president Bush's faith-based initiative. These concerns relate primarily to issues of forced morality, the value of self-determination, and the importance of keeping services *voluntary* for all members of society regardless of their race, gender, religion, and sexual orientation. All one needs to do is conduct a cursory review of history to recognize how easy it is to confuse faith with cultural values. For example, slavery was once considered a practice sanctioned by God, and scriptural support was even offered in support of a Christian man's right to have a slave. In fact, a host of issues once considered sinful (e.g., divorce, homosexuality, women in the ministry, single parenting) are now considered appropriate within many mainstream religious denominations, indicating that the interpretation of biblical scripture—and thus God's intent—is influenced by the current moral climate of society.

Most critics of former president Bush's faith-based initiative are not necessarily critical of faith-based agencies' ability to provide effective human services; rather, they feel that faith-based agencies should not become the primary human services providers in the United States. The NASW advocates for government remaining responsible for providing comprehensive human services programs to the public to guarantee equal and available access to human services that encourages utilization on a voluntary basis, human service delivery that is accountable to the public and professional community in all respects, and a guarantee that service providers have appropriate levels of education and are professionally licensed in their field (NASW, 2002).

Potentially in response to these concerns, on February 5, 2009, President Obama signed Executive Order 13199, which established the White House Office of Faith-Based and Neighborhood Partnerships. After signing the order, President Obama pledged to not favor one religions group over another—changing how decisions on funding practices are made from his predecessor. According to a February 2009 White House press release, the Office of Faith-Based and Neighborhood Partnerships will focus on four key priorities:

1. The Office's top priority will be making community groups an integral part of our economic recovery and poverty a burden fewer have to bear when recovery is complete.
2. It will be one voice among several in the administration that will look at how we support women and children, address teenage pregnancy, and reduce the need for abortion.
3. The Office will strive to support fathers who stand by their families, which involves working to get young men off the streets and into well-paying jobs and encouraging responsible fatherhood.
4. Finally, beyond American shores this Office will work with the National Security Council to foster interfaith dialogue with leaders and scholars around the world. (White House, 2009, para. 5–6)

The shift in priorities will likely alleviate some fears of NASW and other human service professionals who recognize the value and long-term contributions of faith-based agencies but advocate for distribution of funding of agencies from a wide range of religious traditions and approach social issues from a perspective representing a wide range of views and perspectives.

Many agree that the arbitrary exclusion of all religious organization from federal funding is neither fair nor in the best interest of clients, and human service professionals must advocate for fairness, equity, and objectivity in the dissemination of federal funding, avoiding the politicization of this issue so that clients of all faith

Professional History

Understanding and Mastery of Profession History: Historical and current legislation affecting services delivery

Critical Thinking Question: Recent changes in policy have opened up more opportunities for faith-based organizations to receive government funding. What are some of the potential advantages of this shift? What are some potential dangers or drawbacks?

traditions have similar opportunities to seek assistance from agencies that share their religious views.

Methods of Practice in Faith-Based Agencies

The counseling methods used in faith-based counseling are also sometimes debated, with some expressing concern that certain behaviors are moralized in some faith-based counseling, which can be hurtful to the client. Certainly, some faith-based counseling techniques may incorporate a moralistic style, a method some will agree with and some will not. But faith-based agencies can address issues of immorality such as marital unfaithfulness or child maltreatment with grace and forgiveness *as well as* a measure of accountability. What many human service professionals in faith-based agencies may argue is that too often secular practitioners assume that clients are "hit over the head with their sin" in a faith-based practitioner's office, when many times clients who are buried in shame for past poor choices are taught to approach their past mistakes and the mistakes of others with a sense of grace, forgiveness, and mercy. Thus, faith-based counseling can be less about theology and more about grace, forgiveness, mercy, and love—concepts that are universal to nearly every religion in the world.

Such debates regarding the appropriateness of how faith-based human services are provided arise even within religious circles, with some religions or denominations focusing more on social justice, where issues related to social oppression, racism, and classism are addressed in the same manner as secular agencies, and other religions and denominations professing a belief that problems in life are solved by having a relationship with God, thus bringing someone into relationship with God is the necessary first step toward healing. Even if a consensus could be obtained on this issue, evangelism in a counseling relationship outside a ministry setting remains inappropriate in most circumstances if for no other reason than it would violate the foundational principles underlying the human services profession.

The Benefits of Faith-Based Services

The majority of Americans not only identify spirituality as being an important part of their lives, but also identify themselves as being members of faith communities (groups of individuals who share similar religious beliefs and come together for a time of worship and fellowship). Many members of faith communities rely on their congregations when going through a difficult time. Faith communities provide individuals with a valuable support system during difficult times, providing both guidance and emotional support. One goal of human services is to connect people to a broad support system, and a faith community can easily provide this for its active members. Religious coping has also been found to provide more benefits over other coping methods such as general social support and other counseling methods (Pargament, Tarakeshwar, Ellison, & Wulff, 2001).

A recent study questioned individuals within a church congregation who had recently experienced a crisis. The subjects were asked to rank various resources that they found helpful during their crisis. Factors included family, friends, religious beliefs, praying, reading scripture, and professional services, including counseling, legal services, and psychological services. The researchers were surprised to learn that most people ranked professional services last as far as helpfulness and ranked religious beliefs and praying the highest (Stone, Cross, Purvis, & Young, 2003).

Another study conducted after the September 11 terrorist attacks on the World Trade Center and the Pentagon revealed that of 560 adults questioned in a national telephone survey, 90 percent sought positive religion often in the context of a faith community as a way of coping with this tragedy. Examples of positive religion include seeing God as a source of strength and support and perceiving God and a faith community as supportive rather than a source of judgment (Meisenhelder & Marcum, 2004). These studies confirm what many therapists would likely say: that in times of crisis, many people draw strength and support from their faith communities, which provide them with a sense of comfort and familiarity while providing a sense of being a part of a larger whole and reminding them that they are not alone.

Religious Diversity in Faith-Based Organizations

Understanding the distinction in theology and ideology between the various faith-based organizations, whether that includes interfaith differences or variations among various denominations within the same faith, is important because a religious organization's theology and underlying ideology about human nature will likely serve as a reflection of the types of interventions utilized in the delivery of human services.

Many non-Christian faith-based organizations provide many of the same services as Christian-based services. Jewish Family Services (JFS), which acts as an umbrella agency for Jewish community centers, offers comprehensive human services to Jewish and non-Jewish communities across the nation. Islamic human services agencies focus primarily on the Muslim community both within the United States and overseas, such as Bosnians and Palestinians, but also support causes outside the Muslim faith.

In fact, a recent Associated Press article discussed the outpouring of Muslim support for victims of Hurricane Katrina, the devastating natural disaster that hit New Orleans and surrounding states in August 2005 and left thousands of people homeless and with absolutely nothing. Faith-based organizations such as the Muslim American Society, the Council on American-Islamic Relations, Islamic Relief USA, and the Muslim American Society all participated in the Muslim Hurricane Relief Task Force, which took turns manning relief shelters and feeding those left homeless by Hurricane Katrina. This is an example of how various religious faiths and houses of worship often come together to offer assistance to the poor and destitute through donations and assistance during times of crisis and natural disaster (Associated Press, 2005).

Muslim charity work.
ZUMA Press/Newscom

Faith-Based Agencies: Services and Intervention Strategies

In this section we'll look at several different types of faith-based agencies offering human services and examine their success in both identifying and addressing the needs of their target population. We'll also explore the role of the human service professional working in these faith-based organizations, noting any significant differences between their role and those played in secular agencies. Most of the agencies featured in this section operate separately from any church or religious entity but are either supported by a particular faith or operated as an arm or branch of a particular denomination. All featured agencies operate in a manner consistent with the commonly accepted definition of a faith-based agency discussed earlier in this chapter.

It is important to have a basic working understanding of the values held by these different religious faiths in the event that a human service professional has a client who practices a different faith or if a human service professional coordinates services with a faith-based agency of a different faith. Having more than a superficial understanding of different faiths will enhance the human service professional's experience by enabling them to move beyond common negative stereotypes and see the value of diversity within a service delivery context.

Jewish Human Services: Agencies and the Role of the Human Service Professional

> If one of your countrymen becomes poor and is unable to support himself among you, help him as you would an alien or a temporary resident, so he can continue to live among you. (Leviticus 25:35)

The Jewish faith is rich in admonitions and examples of charity and general provision of the poor. The Torah, the Jewish holy book called the *Tanakh,* is what Christians call the Old Testament. The Talmud is the transcribed collection of oral tradition handed down from generation to generation, guiding the interpretation of the Tanakh. Charity, as referenced in both the Tanakh and the Talmud, is defined as giving to the poor and is a requirement for the Jewish people. According to Jewish law, forgiveness of sins is granted with prayer, repentance, and charity.

As with the Christian faith, the Jewish faith has different denominations called *movements,* including Orthodox, Conservative, Hasidic, Humanist, Reform, Sephardic, Ashkenazi, and Reconstructionist. Some of these movements evolved through geographic divisions and some through philosophical divisions. Nevertheless, all Jewish movements hold that charity and benevolence (kindness and compassion) are an integral part of righteousness. Good financial stewardship is highly valued in many faith traditions, and the Jewish faith is no exception. Unlike some Christian denominations that consider giving all of one's earthly possessions to the poor a blessed act, giving 5 to 10 percent of one's income to charity is considered an obligation among all Jewish denominations. Charity is not solely related to duty, though, but also reflects the value of community and the commitment to remain connected to all Jews worldwide. This sense of community is based on shared experiences of both current and historical persecution, which binds the Jewish people together in a communal determination of self-sufficiency and survival. The Talmud specifies different levels of giving, with the lowest level involving giving begrudgingly and the highest levels including giving anonymously to a stranger and helping someone attain self-sufficiency by giving them work (Babylonian Talmud, Chagigah 5a; Maimonides, Hilchos Matnos Aniyim 10:7–14).

Jewish human services agencies are coordinated into a national umbrella organization that serves as a network of support for smaller human services agencies that provide direct service. Human services are directed toward Jewish and non-Jewish communities as well as targeting domestic and international causes.

THE JEWISH FEDERATIONS OF NORTH AMERICA   The Jewish Federations of North America (JFNA) is an international umbrella humanitarian organization that represents over 100 Jewish federations in North America alone. The JFNA provides humanitarian relief and human services worldwide to those in need. The goals of social justice and strengthening Jewish community are a reflection of the scriptures in the Talmud that command giving to the poor, sick, widows, and orphans. The JFNA exists to provide financial support and educational services to Jewish federations and Jewish community centers; it also funds the rescue and resettlement of Jews living in high-conflict or unsafe areas worldwide.

A component of the JFNA is the Human Services and Social Policy Pillar (HSSP), which is responsible for social lobbying action on local and national levels in an attempt to influence social policy. Whether it's lobbying for increased funding for geriatric services, homeless resources, or refugee programs, the HSSP, or *the pillar* as it is commonly called, relies on human service professionals and volunteers to coordinate services of human services agencies within and without the Jewish community.

ASSOCIATION OF JEWISH FAMILY AND CHILDREN'S AGENCIES   The Association of Jewish Family and Children's Agencies (AJFCA) acts as the umbrella organization for JFS and Jewish community centers across the United States and Canada. The AJFCA also acts as an information clearinghouse for local JFS agencies, which provide comprehensive human services to the Jewish community. AJFCAs also provide funding for local federations of Jewish human services agencies, advocate for social justice, and provide information on education and training opportunities.

Local JFS agencies offer a number of different services including individual and family counseling, marital counseling, substance abuse counseling, AIDS counseling and awareness programs, anger management courses, employment services, parenting workshops, children's camps, teen programs, and geriatric programs including Kosher Meals on Wheels and hospice. No one is denied services due to an inability to pay, and payment for services is typically on a sliding scale.

One program that is relatively unique to this organization includes refugee resettlement programs, which assist individuals and families who have legally entered the United States having fled from persecution. Refugees of either Jewish or non-Jewish descent come from various countries, including Russia and other former Soviet-bloc countries, the Middle East, and Africa. Services typically include providing short-term housing on arrival, emergency financial support, case management, medical care, assistance with school enrollment, job placement, and language courses. These agencies have excellent reputations in assisting refugees to gain financial independence rapidly, particularly in light of the often tragic circumstances the refugees have faced prior to coming to the United States.

Services focused exclusively on the Jewish community include Holocaust survivor services to Jews who lived in European countries under Nazi rule between 1933 and 1945. In addition to providing counseling services related to post-traumatic stress disorder (PTSD), in-home services related to geriatric care are also provided. Other Jewish-related services offered include counseling and case management services for Jewish armed services personnel, Jewish chaplaincy services, family services, and outreach focusing on assisting families reconnect with their Jewish roots by learning how to incorporate Jewish traditions and values into their family systems. Premarital and marriage services are also offered to Jewish couples and interfaith couples, focusing on marriage and parenting in a Jewish context.

Human service professionals within these agencies provide a wide range of services because JFS agencies typically offer comprehensive human services similar to those discussed through this entire text. In fact, many of the JFS agencies offer just about every type of human services one could imagine! The primary difference between the manner in which human services professionals deliver services at a JFS agency versus a secular agency is the focus on connecting Jewish clients to the broader Jewish community, both domestically and worldwide, as well as the incorporation of Jewish values throughout the various programs. Counselors and case managers are also primarily Jewish and well connected to the Jewish community, including being familiar with local synagogues and other Jewish services within the local community.

Virtually, all JFS programs are eligible for federal funding as long as proselytizing does not occur as a function of any program receiving funding. Even synagogues offering human services programs are eligible to receive government funding as long as the programs are operated separately from any religious functions.

The Jewish human services agency network provides a network of comprehensive services designed to address human needs on all levels. They provide invaluable services to the Jewish community, as well as those outside of the Jewish faith, both within the United States and abroad.

CASE STUDY 13.1

Case Example of a Client at a Jewish Faith-Based Agency

Raisa, a 77-year-old Jewish widow, began counseling at a local Jewish community center about one year ago for depression. Her initial psychosocial assessment revealed a long history of mild depression with mild anxiety that had escalated in recent years to a point where intervention was necessary. Raisa shared that her normal sadness increased dramatically when she lost her husband four years ago and did not abate even when she found herself feeling more at peace with her husband's death. Raisa and her husband were married for 45 years, both having immigrated from Europe shortly after World War II. They were unable to have children of their own and thus adopted one child, a daughter, who resides in a different state about three hours away by car. Raisa's daughter is married and has one child, also through adoption. Sarah, her counselor, presumed that Raisa may have been a Holocaust survivor, and if so that some of the earlier trauma and grief issues were likely at play in her current depressive state, but Sarah chose not to address this possibility in counseling, choosing to wait until Raisa was ready to share her experience. Despite weekly counseling sessions and several courses of antidepressant medication, Raisa's depression and anxiety continued to worsen. During one session approximately nine months into their counseling relationship, Raisa was discussing the difficult early years of her marriage when she and her husband first moved to the United States. Raisa became extremely emotional as she shared that they were both orphans because of the war and thus had no family to help or guide them, either in their transition or in their marriage. Sarah immediately not only recognized the grief Raisa was reexperiencing, but also noted that once Raisa became obviously distressed, she became equally uncomfortable, apologized for her "outburst," and then quickly changed the subject and regained her composure. Sarah did not push Raisa, understanding that Raisa's decision to share her distant but obviously still-powerful memories was just that—Raisa's decision. As the months progressed Raisa began to pensively share more stories of her early marriage, which seemed to be marked by considerable loss and struggle. She was 18 when the war ended. She met her husband, Reuben, one year later, although they had met once or twice several years before. They became inseparable almost immediately, likely out of sheer loneliness, Raisa suspected, rather than any type of love at first sight, although in retrospect Raisa wasn't sure there was a difference—both were emotions that encompassed a significant amount of passionate intensity. Raisa and Reuben spent two years searching for family members. Her husband located an aunt and uncle in the United States. Raisa learned that her brother had escaped to Israel. They never located any other surviving family members. After some thought and consideration, they decided to move to the United States in the hope of connecting with her husband's relatives. When they first arrived in New York, they experienced a long-overdue measure of relief, but this was to be short lived when Reuben's aunt and uncle announced plans to move to California. Deciding not to follow, Raisa and Reuben were left to survive on their own in a big city that offered as much risk as opportunity. Although Raisa spent most of her time focusing on the physical and financial hardships of her early life, she appeared to avoid any discussion of her feelings. In fact, Sarah noted that whenever Raisa risked becoming emotionally grieved such as when Sarah asked any question that required Raisa to reflect on her childhood (even positive aspects of her youth), Raisa became emotionally and physically rigid, as if she were talking herself out of the "nonsense" of her feelings to regain composure. Sarah became increasingly concerned about Raisa's psychological stability, particularly in light of her very recent increase in anxiousness. In fact, there were two recent occasions where Raisa was so anxious she did not feel comfortable leaving her home to attend her counseling session. In light of Raisa's worsening condition and her fear that Raisa might be at risk of suicide, Sarah made the decision to have a session with Raisa where she would more assertively address Raisa's Holocaust experience, believing that to be the root of her unresolved grief and the source of complicated mourning related to many of the losses she experienced after the war. Sarah went to Raisa's house for this session so that Raisa could remain in the safety of her surroundings if the session became too difficult. Sarah also had implemented a safety plan for Raisa, including collecting a list of emergency numbers and the number of a local geriatric outreach center that Raisa had been involved with intermittently for several years. Sarah began her session with Raisa by gently expressing her concern about her emotional well-being, as well as sharing her belief that Raisa may be suffering long-term effects from being a Holocaust survivor, and that unless she faced her past, her depression and anxiety might not get better and may, in fact, continue to worsen. Raisa was immediately uncomfortable, but Sarah reassured her that although she wanted to push Raisa a bit, she had made sure she could remain with Raisa for the entire afternoon, thus Raisa could take her time. Although Sarah had spent considerable time in counseling sessions with Raisa conducting "psychoeducation"—teaching Raisa about the normal stages of grief and the common psychological responses to trauma—Sarah reiterated this information now in the hope that Raisa would begin to accept that her feelings were normal. During this session Raisa shared that her early childhood was one of constant happiness. Her father was a professor at a local university in Holland. Although they were not very religious, they attended synagogue weekly and observed the Sabbath. Without realizing it at the time, Raisa's family was quite immersed in the Jewish culture, which in her family meant close ties to extended family and friends within the community who had a shared culture, customs, and life perspective. Raisa recalled the emergence of a different feeling in her neighborhood when she was about 11 years old. She is not sure if this marked the slow invasion of the Nazi party into her small town, but she does recall that it was about this time that her family could no longer protect her brother and her from the fact that their lives were about to change forever. Raisa shared that her family started closing the front door and drawing the shades more frequently and that various neighbors suddenly began to disappear. She recalls the day, at the age of 13, when almost everyone in her neighborhood was forced to wear yellow stars on their sleeves, and she marked this as the day she realized that some of her favorite neighbors were apparently not Jewish, because they did not have to wear the yellow star. Raisa emotionally shared the night she and her brother, two years older than her, were awakened in the middle of the night by their parents and told to dress quietly in the dark. They were going on a long trip but had to remain quiet. She shared that she did not recall thinking much about what was happening. Perhaps she was too scared, or maybe she had experienced so much change and shock in the past year, she simply accepted this as one more confusing event in a long line of bewildering experiences. Months earlier Raisa's father had told her that it was important for her to obey him without asking questions because not obeying him might have serious consequences. She recalled crying when he said this to her because he was so firm, an emotion she rarely saw in her father. He responded by telling her that tears were useless now—they would not help, and that she needed to be strong. She obeyed him now as she folded one change of clothing into a small dark knapsack, confused and afraid, but resolved not to cry. The next thing Raisa remembers is that she and her family were crouching down outside in the dark and running along the hedge line. She recalled that there was no moon, and the night was so dark she was certain she would lose her brother, who was directly in front of her. She kept running though, trusting that someone would come back for her eventually if she lost her way. They arrived at a stranger's house, and her father knocked on a back door that appeared to lead to a basement. A young woman opened the door and hurried Raisa and her brother through the door. Raisa had only a quick moment to look back and see her mother and father, who to her horror were not following behind her and her brother. Instead, her father and mother were crying, peering into the dark basement with a look of dread and horror on their faces. Raisa immediately recalled her mother telling her earlier that she loved her very much, yet Raisa could not recall having said it in return. This was something that would torment Raisa for years. Did she tell her mother that she loved her? She would never be sure that she had. That was the last time that Raisa and her brother saw their parents. Raisa learned after the war that their parents were forced to leave their home shortly after arranging to smuggle their children out of Holland, and after a short stay in what became known as a Jewish ghetto, they were sent to a concentration camp. Although she was never able to obtain exact information, Raisa learned that both of her parents were executed likely sometime in early 1943. Raisa and her brother remained in the dark basement with little food or water for about three days before being driven, during the middle of the night, to another home. Raisa recalled crying sometimes but her brother, like her father, told her to stop and to be strong, and she complied. This time period was particularly difficult for both Raisa and her brother, who were tempted to escape and return home to their parents. She is not sure whether it was fear or wisdom that kept them from this course, but she realizes now that had they returned home, their fate would have been the same as their parents'. The next trauma for Raisa occurred when she learned that she would be separated from her brother. Although her parents had arranged for them to remain together, increased risk led her rescuers to conclude that two children suddenly showing up in a home was far riskier than one; thus in the middle of one night several weeks into their frightening journey, Raisa's brother was hurried into one car, and she into another. This, too, would remain a source of considerable pain for Raisa, as she realized that once again she was denied a proper good-bye. Her last memory of her brother was his surprised face looking out the car window as he realized that she was being escorted into a different car. Raisa fled to Italy, where she lived in a converted attic, and although enjoying some measure of freedom, she had to remain relatively hidden until the war was over. Her foster parents were nice, but stern. They were not Jewish, thus Raisa was compelled to live a lifestyle very different from the one she had enjoyed in Holland. She dressed differently, attended church rather than synagogue, and ate food very different from what she was used to. It did not occur to Raisa until she was much older that there wasn't any possibility of seeing her family again. Her attitude during the balance of her childhood was one of "waiting it out" until the war was over and she could go home and resume life as she had known it before the war. But of course that was a dream that would never come true. When the war ended, her host family wished her good fortune, and at 17 years of age Raisa was completely on her own and alone in the world. Although God had never played much of a role in her life before, she found herself praying to the God of her childhood that her family was safe and waiting for her at home. Raisa got a job in town so that she could earn enough money to return to Holland. She met Reuben on her first day of work. He was employed at the same shop, but for different reasons. It was Reuben who told her there was nothing to return to—that his family, and likely hers, were dead, and the only choice Jews had was to immigrate somewhere safe. Raisa had been sheltered by her host family and had heard nothing of the concentration camps and the unchecked slaughter of millions of Jews. She had difficulty describing the way she felt once she learned that her entire family was likely dead. She described it as both surreal and numb. She had no idea where her brother had been taken, and she had fantasized for years about finding him walking down an Italian street or shopping in the town center, but he was all she could think of now. She had to find him. She and Reuben made the singular goal of finding whatever family they had left. At some point in their planning, they became a couple and decided to marry. Raisa learned through a charitable organization that her brother was living in Israel. She shared earlier that their decision to immigrate to New York to join Reuben's family was a practical one. She shared now Reuben's fear that if they immigrated to Israel, they might find themselves in the same situation as in Holland—in the center of a war—and he could not risk becoming involved in another war ever again. Raisa let go of her hope to return to her brother when Reuben decided it would be wiser for them to move to the United States. Raisa did reconnect with her brother again, but they never enjoyed the closeness of their childhood. When she and Reuben visited her brother in Israel many years later, it felt to Raisa as if she were visiting a complete stranger. Her brother had become quite religious, embracing the faith of their youth—a choice antithetical to Raisa's, who in response to their earlier losses chose to distance herself from her Jewish roots. Raisa shared all these stories with emotion, but no tears; she was still being strong. Although Sarah decided to hold off on approaching the subject of Raisa and Reuben's infertility resulting in the adoption of their daughter, she made a mental note that she would visit this issue in a later session. Sarah knew this too would likely be a very difficult subject for Raisa and a source of great pain—both from a generational perspective (issues related to infertility were typically not discussed in earlier generations) and from a loss perspective. Sarah assumed that Raisa and Reuben looked forward to having their own children not simply as a way of starting their own family as so many couples do, but as a way of *replacing* the family that had been taken from them both. Sarah would learn later that Raisa's first child was a stillbirth, that the loss was almost too much for Raisa to bear, and that this was likely when Raisa's melancholy transitioned into a clinical depression. Even when Raisa and Reuben experienced the joy of adopting their daughter, Raisa shared that a sense of sadness remained hidden within her. After this intense and very long session, Sarah developed a treatment plan for Raisa—one that involved both trauma and grief counseling. Sarah suspected that in addition to depression and anxiety Raisa also suffered from PTSD, thus she incorporated aspects of treatment designed to help her deal more effectively with being a survivor of trauma. Sarah suspected that Raisa was in many ways still operating with a survivor mentality, which compelled her to obey her father's distant admonition to resist crying and remain strong. Raisa's tendency to equate crying with weakness could be addressed through cognitive behavioral therapy, where Raisa would be encouraged to recognize that such rules about emotion may have been necessary in wartime, but were no longer needed and were actually damaging. The challenge for Raisa would likely lie in a fear that to change her perspective on crying might indicate a betrayal of her father. One of Sarah's ultimate treatment goals for Raisa was to help her develop a more realistic and timely definition of authentic strength that did not dishonor her father's guidance. Another treatment goal involved helping Raisa learn to grieve all her past losses and finally to rebuild the community she lost so many years ago. Although Raisa had a daughter, she had avoided ever getting too involved in her community, perhaps out of a fear that she might lose again what she had lost as a child—a close-knit community of neighbors who shared a culture and a faith and who operated in many respects as an extended family. Although Sarah suspected that Raisa might have some objections to getting involved in the local Jewish community, Sarah planned to explore this possibility with her to reconnect her to the faith and culture of her childhood. A significant portion of Raisa's healing came from a pilgrimage of sorts that Sarah helped her plan involving returning to Holland with her daughter and her brother. During this long-overdue visit Raisa and her brother tearfully revisited their childhood home, as well as other places of nostalgia, and although things had changed significantly since their youth, Raisa and her brother found great healing in their trip "home." The final leg of their trip involved creating a memorial for Raisa and her brother's parents and all her lost family and friends. Raisa's last session with Sarah prior to her trip involved writing a poem that they would leave at the site where the Chelmno concentration camp once stood. The trip not only helped her to create meaning surrounding the death of her parents, but it also helped her to reconnect emotionally with her brother and involve her daughter in a part of her life she had previously kept hidden. In succeeding years Raisa's debilitating depression lifted, and her anxiety receded. She learned how to genuinely grieve her past losses and learned to recognize how her early trauma and loss impacted virtually every area of her life. She did ultimately become involved in her community, and in the years preceding her death at the age of 84, she even resumed attending synagogue. Sarah's relationship with Raisa involved more than counseling. It involved incorporating aspects of faith and culture into sessions, case management that involved connecting Raisa to a community from which she had been generally estranged. It also involved Sarah drawing on her own Jewish faith, which enabled her to understand much of what Raisa experienced both in her past and in her current life.

Christian Human Services: Agencies and the Role of the Human Service Professional

> For I was hungry and you gave me something to eat, I was thirsty and you gave me something to drink, I was a stranger and you invited me in, I needed clothes and you clothed me, I was sick and you looked after me, I was in prison, and you came to visit me . . . I tell you the truth, whatever you did for one of the least of these brothers of mine, you did for me. (Matthew 25:35–36, 40)

Because a fair amount of faith-based organizations in the United States are Christian in nature, it is valuable to have an understanding of the range of theologies and ideologies within the Christian church. The historic role of the Catholic Church discussed earlier in the chapter reflects Catholicism's strong commitment to caring for the poor. This commitment is reflected in today's Catholic Church in ministries such as Catholic Charities, which facilitates numerous human services programs throughout the United States.

Mainstream Protestant denominations such as Methodist, Presbyterian, and Lutheran often embraced the "social gospel," the Old Testament mandate to provide for those in society in need, but these denominations did not necessarily link charity to evangelism. Rather, the predominant view among these mainstream denominations was to show the love of Christ through giving as well as through addressing social concerns for the poor and the oppressed.

Conservative Christians, such as evangelicals, fundamentalists, and Pentecostals, tend to focus on evangelism as the initial priority, addressing social causes and the needs of the poor through winning souls for Christ. If one truly believes that the only path toward wholeness is by surrendering one's life to Christ, repenting of one's sins, and becoming a new creation through a personal relationship with God, then it makes sense to want this experience for anyone who is suffering. The conflict arises when such evangelism occurs in the counseling office or anywhere else where social services are being provided, without the client understanding that this is the goal of the service provider. As mentioned earlier in this chapter, professional standards of the human services field, whether social work, counseling, psychology, or psychiatry, discourage proselytizing to clients. Critics of evangelical practitioners who do attempt to evangelize clients might suggest that as worthy as this act might be perceived, it is more appropriately conducted in the vein of pastoral counseling or ministry efforts (Belcher, Fandetti, & Cole, 2004).

This ethical dilemma is worth exploring in both secular and religious circles and can be addressed in a variety of ways. For instance, there is nothing inherently unethical in talking about matters of faith and spirituality as long as it is client driven. In fact, it is the human service professional's comfort level in talking about such issues and willingness to allow the client to determine the depth and direction of the discussion that is important. For instance, consider the client who enters a counseling session utilizing negative religious coping strategies such as perceiving God as punishing, abandoning, and distant, particularly when tragedy occurs. A human service professional in a faith-based agency can comfort the client by reframing the client's punitive view of God by teaching the client to use positive religious coping methods where God is perceived in a positive manner and a source of guidance, strength, and support. Because research supports the mental health benefits of positive religious coping, this intervention strategy can be used with the understanding that it is truly in the best interest of clients who are being hurt by their negative views of God.

Although evangelizing clients is not appropriate in a secular setting or even in a faith-based organization receiving federal funding, it is appropriate if the human service professional works for a religious organization that makes clear its goal is to evangelize the client so that the client enters the counseling relationship with full disclosure and equal participation. For instance, many outreach ministries provide emergency services such as food pantries or homeless shelters, but do not hide the fact that the ultimate goal of the agency is to lead one down the path of greater religious commitment, which may

Human Services Delivery Systems

Understanding and Mastery of Human Services Delivery Systems: Range and characteristics of human services delivery systems

Critical Thinking Question: How might human service professionals, especially those working for faith-based organizations, walk the fine line between supporting clients' self-determination and forcing their own (or their organizations') religious viewpoints on clients?

involve a deepening relationship in a client's existing faith or may involve a complete conversion to a new faith.

RURAL COMMUNITIES AND THE BLACK CHURCH   Rural communities, typically those with high minority populations, tend to be significantly underserved with regard to mental health services. Yet research in the last 10 to 15 years has revealed that African American churches, particularly those within rural communities, have picked up the slack by offering significantly more human services than White churches (Blank, Mahmood, Fox, & Guterbock, 2002).

> African American churches, particularly those within rural communities . . . , offer significantly more human services than White churches.

There are several potential reasons for including the long-held conflict between secular mental health providers and clergy. It has been addressed in recent years but remains a point of contention with clergy not necessarily endorsing the "medical model," or secular approaches to mental health concerns, and secular mental health providers not readily perceiving mental disorders in spiritual terms. Yet as mentioned earlier in this chapter the church has a long history of providing for the social and mental healthcare needs of individuals within society, and the fact that African American churches tend to offer far more human services programs than White churches may be a reflection of the African Americans' general sense of distrust of the mainstream mental health community and having greater trust of African American providers (Blank et al., 2002; Thomas, Quinn, Billingsley, & Caldwell, 1994).

Regardless, one variable often neglected in research on human services in the church is that for many generations the African American church has been the center and backbone of the African American community, thus these clergy might be more willing to engage deeply in the lives of their parishioners and those within their community. It appears that in many respects African American churches, particularly those in rural communities, have acted in some respects as the Catholic Church in the Middle Ages, taking responsibility for the mental health concerns and basic needs of those within the community.

CATHOLIC CHARITIES   Catholic Charities USA is a network of Christian human services agencies that has a long tradition of caring for those in need. Services are provided to all individuals seeking assistance regardless of religious affiliation. Currently, there are approximately 1,600 local Catholic Charities agencies across the United States offering a wide variety of human services designed to meet the needs within the particular community served. According to the Catholic Charities website services provided at most of its local agencies focus on advocacy and direct services related to reducing poverty, supporting families, and empowering communities. They do this by facilitating programs that focus on adoption, disaster operations, housing counseling, disaster case management, racism and diversity, human trafficking, and climate change (Catholic Charities USA, 2010).

Catholic Charities USA claims to have provided services to over 10 million individuals in the year 2010 alone, making it one of the largest networks of human services agencies in the world, similar in nature to the Jewish federations. The majority of funding comes from federal and state sources, with only a small percentage coming from the Catholic Church. Catholic Charities has not had significant problems obtaining federal funding because providing services directly linked to religious ministry is not typically an aspect of services provided, thus Catholic Charities has not been particularly affected by the faith-based initiative.

In many respects Catholic Charities provides similar services as secular agencies except that most local Catholic Charities agencies also provide support to archdiocesan schools and parishes. Adoption and children's services are also provided, but remain consistent with the values of the Catholic faith, thus option counseling of women experiencing an unplanned pregnancy would not include referrals to abortion services. Most agencies also provide Catholic Youth Organizations (CYO), an after-school and weekend athletic program focusing on the development of sportsmanship-like behavior and ethical values consistent with the Catholic faith.

Other services include child care, domestic and international adoption, domestic violence, employment and job training, gang intervention, health care, HIV/AIDS services, immigration and naturalization services, nutrition counseling, refugee resettlement services, senior services, homeless assistance and emergency housing, senior housing, and substance abuse counseling. Most local Catholic Charities agencies also offer community centers that focus on providing comprehensive human services for those who are homeless or at risk of becoming homeless.

Human service professionals are not required to be Catholic, but many of those in leadership positions are due to the close and supportive relationship with local Catholic parishes. Human service professionals include social workers with Master of Social Work (MSW) degree and therapists, psychologists, and caseworkers and practitioners with Bachelor of Social Work (BSW) degree. Services provided are generalist in nature, and as is the case within the Jewish federation agencies, human services interventions and clinical issues depend on the actual services being provided. Human service professionals working at Catholic Charities have the benefit of working within a broad network of agencies that provide extensive support and educational opportunities.

PRISON FELLOWSHIP MINISTRIES   Chuck Colson reached the peak of his political career as President Richard Nixon's aide, or as many referred to him, President Nixon's "hatchet man." In 1973 Colson became a Christian, and in 1974 he pleaded guilty to obstruction of justice charges in association with the Watergate scandal. Colson served seven months of a three-year sentence and on his release founded Prison Fellowship Ministries (PFM) in 1976, based on his own dramatic religious conversion and his belief that no one is beyond hope. His ministry is now one of the largest prison ministries in the world, reaching out to prisoners, ex-prisoners, their families, and victims. PFM is also involved in criminal justice reform through a PFM affiliate, Justice Fellowship, which focuses on numerous social justice issues including prison safety and eliminating prison rape.

PFM is now one of the largest prison ministries in the world, reaching out to prisoners, ex-prisoners, their families, and victims.

Such social advocacy is particularly important for groups of individuals who do not evoke sympathy in the average person, and prisoners certainly fall into this category. Yet, it is essential for people to realize that prisoners are not a uniform group of evil pedophiles and serial rapists who deserve whatever hardship the prison system can dish out. Most prisoners have had childhoods marked by poverty and abuse, many serve longer sentences because they could not afford adequate legal services, and some are innocent. PFM is committed to stop the intergenerational cycle of crime and poverty by offering prisoners hope for a second chance through the Christian faith.

Citing the difference that this ministry can make in the lives of prisoners as well as in society in general, Colson references the dramatic shift in climate experienced at Angola Prison in Louisiana, once touted as the most dangerous prison in the United States, but now considered the most peaceful under the leadership of Burl Cain, a Christian who invoked the services of local seminaries to minister to inmates. In a similar vein, PFM trains volunteers to counsel and minister to prisoners in virtually every prison across the country.

PFM facilitates a number of ministries including training volunteers to visit prisoners, many of whom receive no other visits. The ministry does not receive federal funding because PFM volunteers focus extensively on the evangelism of prisoners and their family members. The goal of PFM is to bring the gospel of Christ to every prisoner incarcerated in the United States. PFM also facilitates a "pen pal" program linking prisoners with volunteers who are willing to minister to them in writing. PFM also provides services to the family members of prisoners, particularly those with children. An example of such services includes the Angel Tree program, which collects Christmas presents for these children and also facilitates a camp and a mentoring program.

Human service professionals, who are primarily volunteers, working with PFM provide markedly different services than those working in secular agencies. Because evangelism is the primary intervention tool, volunteers facilitate Bible studies and in-prison seminars, mentor at-risk youth, counsel prisoners and crime victims, serve in youth camps, organize Angel Tree programs, visit prisoners regularly, counsel ex-prisoners and crime victims, and write letters to prisoners in the pen pal program. Human service professionals also hold paid positions with PFM, including field director positions, which manage and provide support of ministry teams, including recruiting and training volunteers and reaching out to local churches for assistance and financial support.

CASE STUDY 13.2

Case Example of a Client at a Christian Faith-Based Agency

Castle Christian Counseling Center (CCCC) is a not-for-profit, ecumenical counseling center contracted by the county to provide mandated counseling services, including anger management and alcohol counseling for individuals who have been charged with an alcohol offense. Julie was required to attend anger management as a part of her probation for a domestic battery charge. Julie's initial psychosocial assessment recommended that she participate in both group and individual counseling. The group counseling consisted of a

26-week program focusing on anger management and personal accountability. Her individual counseling was designed to help Julie deal with the underlying reasons for intense anger and inappropriate behavior. Julie was 24 years old when she was charged with domestic battery against her husband of three years. When Julie began counseling she was both emotionally needy and defensive. Her counselor, Dana, suspected that beneath Julie's defensiveness lay a tremendous amount of shame, so she chose not to confront Julie until much later in their counseling relationship. During the first several months of counseling Julie expressed much anger and frustration with her husband, who she perceived as being quite passive. In response to his seeming inability to make decisions or take the lead in any aspect of their life, Julie expressed extreme disappointment and at times rage. It became clear to Dana that Julie's husband was in many respects being set up for failure by Julie. For instance, Julie often expressed to her husband that she wished he would be more proactive in their social life, but if he did forge ahead and make plans without checking with her first, she would become irate that he chose an activity he knew she would not like. Yet if he checked with her first before making plans, she would become angry that he did not have the confidence to make plans without her, and she would accuse him of ruining the surprise for her. The incident that resulted in the charge of domestic battery involved a fight that escalated over their finances. Julie had decided to quit her job to try to get pregnant, even though her husband had expressed concerns that he did not make enough money to be the sole provider. He ultimately supported her decision, and Julie quit her job, but after a few months, when money got tight, and they ultimately did not have enough money to pay bills, Julie lost her temper. During her tirade she accused her husband of not caring about their finances and of sabotaging their plans to start a family. Julie became physically abusive toward her husband when he attempted to stand up to her by telling her that he had not in fact wanted her to quit her job because he feared this very thing. Julie became hysterical, accusing her husband of hating her and of just looking for an excuse to leave her. Dana recognized Julie's tendency to change the facts to support whatever theory she was attempting to prove at the moment. She also recognized Julie's all-or-nothing thinking—people either loved her or hated her, were for her or against her. According to secular psychology Julie would have met the criteria for borderline personality disorder, but Dana recognized her behavior as indicative of a contemporary form of idol worship. Julie was expecting her husband to be God, yet there was only one God who could meet all Julie's needs. Dana knew that over the next several months she would be Julie's representative of God—showing her unconditional love as well as truth. She made a commitment to Julie that she would always be honest with her, and there would be nothing that Julie could do that would lead Dana to end their relationship. She trusted that Julie could handle the truth if it were delivered in love, not shame. It was only a few days later that Julie seemed to test Dana's commitment. Julie called Dana and left a frantic message, stating that she was very upset and needed to talk immediately. When Dana had not returned her call within the hour, Julie called again and but this time was enraged. She accused Dana of being like everyone else—making promises but then abandoning her when she was most in need. Before returning Julie's call, Dana prayed for wisdom and insight. She immediately had an image of truth as light, and for Julie, any truth at all was like a flashlight blaring into her eyes, causing Julie to have to bat the light away to avoid the pain. Dana knew immediately from then on that she would have to be gentle not only in the amount of truth she shared with Julie, but also in the way she shared her wisdom. In the face of Julie's intense and abrasive defensiveness, Dana resisted the natural tendency to force truth on her. Instead she indulged Julie a little, suspecting that Julie's initial feeling when she made a mistake was intense shame, but before she could respond to this emotion she reacted by flipping her shame outward into anger against anyone who represented the source of shame—anyone who made her feel guilty in some way, who exacted accountability, and even who cried in response to one of her rages. Dana's intuition told her that if she could relieve some of Julie's shame—take her off the hook in some manner—this might give Julie the emotional space to explore her feelings of intense shame and guilt. When Dana did call, she suspected that Julie would already be feeling immense shame and guilt, regretting her tirade. Dana also suspected that Julie would not be able to emotionally manage these feelings, thus would have a need to rationalize her behavior by escalating Dana's "sin" to match her own reaction. Dana knew that if she admonished Julie for her tantrum, this would set this process in motion, so she did something different; she took Julie off the hook and rather than admonishing her, she praised her for her ability to communicate her feelings! Julie was so taken off guard that it actually enabled her to experience feeling a small amount of guilt. After Dana had finished complimenting Julie on her willingness to communicate, Julie admitted that she should have handled her feelings differently, that she should have been more patient, and that in some respects she believed she was expecting to be let down by Dana, thus she didn't even give her a chance to meet her needs. Success! By taking this counterintuitive approach and lifting the burden of shame, Julie was able to actually recognize her internal process without rationalizing her feelings away. During the course of their counseling Dana addressed Julie's negative feelings about God. Julie shared that she felt very insignificant whenever she thought of God. She then shared new elements of her childhood. She had already disclosed a childhood wrought with abuse and emotional humiliation at the hands of both her father and her mother, but during this particular session, Julie shared that whenever she made a mistake as a child, her father would tell her she was going to hell, that she was a disappointment to God, and that she could not hide from God—he could see her wherever she was and he knew what she was doing and what she was doing the majority of the time was bad. Julie's father would often physically abuse her, sometimes using a Bible to beat her on the head. When Dana asked Julie to draw a picture of her relationship with God, Julie drew a picture where she was quite small, crouched down and running, and God, a large presence on the page, was looking down on her with a stern scowl on his face. Dana asked Julie if she ever turned to God when going through a difficult time. Julie looked shocked, expressing her belief that if she was in trouble, God would be the last she would consider turning toward. In fact, Julie shared that she believed that the only time God paid any attention to her was when she had messed up. She imagined God saying, "There you go again—I knew you would blow it eventually!" Dana told Julie that she would like to spend some time sharing a different type of God with her, not a punishing God, but a loving God who acted as a father to his children—guiding his children when they were walking down the wrong path, like any good father, and applauding when they did well. Dana shared about her own feelings toward her young son. She found herself chuckling even when he got himself into a bit of trouble, like the time he wrote his name in purple crayon all over his closet door, only to deny his culpability when Dana came upon his artwork. Dana was not harsh, nor punishing, but she did want to teach her son that defacing property was not the best choice. She did this in love, extending grace and forgiveness because she understood that at this age her

Human Systems

Understanding and Mastery of Human Systems: Emphasis on context and the role of diversity in determining and meeting human needs

Critical Thinking Questions: In this case example, how did the human service professional utilize her own faith to guide her practice? Did she get the same results that a similarly skilled secular practitioner treating the client for borderline personality disorder would have gotten?

son did not know any better. She also smirked as she admired her son's artwork, knowing that drawing on the wall with crayon was a perfectly normal thing to do. Julie could not fathom a God who was anything but condemning but she was very interested in learning about the concepts of grace and forgiveness. Once Dana was confident that Julie trusted her, she began to respond to each of Julie's rage episodes by first empathizing with Julie's emotions—her disappointment, her fear, her anger—but then followed by gently sharing truth. When Julie asked if Dana thought she was wrong to have such high expectations of her husband, Dana said yes, but that did not mean that Julie should have no expectations. Rather, Dana explained that once Julie developed a more solid emotional base within herself, including having a more solid relationship with God, her expectations of her husband would likely be more realistic. Julie's counseling also consisted of a significant amount of grief counseling, mourning her lost childhood, gaining insight and understanding of the abuse she had endured, and learning her emotional triggers and ways to avoid them. Dana taught Julie to contain her emotions, so that she wouldn't have to react the moment she experienced an intense emotion, such as the intense fears that she was going to be abandoned, which would often turn toward anger. Dana used guided imagery directing Julie to imagine Jesus holding her firmly, but lovingly. Imagery exercises of this type also helped Julie make God more real in her life. Dana also encouraged Julie to read one new scripture per week. Julie's favorite was Romans (8:28), "And we know that in all things God works for the good of those who love him who have been called according to his purpose." For Julie this meant that even the abuse she endured would be used for good—like making lemonade out of lemons. Another favorite scripture that brought great comfort to Julie was Jeremiah (29:11–13), "For I know the plans I have for you," says the Lord. "They are plans for good and not for disaster, to give you a future and a hope. In those days when you pray, I will listen. If you look for me in earnest, you will find me when you seek me. I will be found by you." Julie felt that this scripture meant that God had good intentions for her, not evil ones. He wanted the best for her, not the worst. He would not hide from her, and she did not have to hide from him. Julie continued counseling even after she met her mandated requirement. In her second year of counseling Dana shifted focus from Julie's childhood to her current relationships, including the relationship with her husband. Julie's intense fear of abandonment often led her to be so self-focused that she was blinded to the damage she caused other people. As her fear of abandonment subsided and her shame diminished, Dana was able to coach Julie into looking through the eyes of her husband. This process would have been impossible a year ago because the shame would have paralyzed her, but with her increasing internal strength, Julie was able to accept her behavior and the pain it caused. Once she saw herself as deserving of forgiveness, she could address her own abusive behavior. Within the second year of therapy, Julie's anger receded significantly, and she was able to talk through her feelings rather than act them out. She remained in counseling intermittently for years to maintain her program of faith building, emotional containment, and extending forgiveness to self and others.

Islamic Human Services: Agencies and the Role of the Human Service Professional

> It is not righteousness that you turn your faces towards East or West; but it is righteousness to believe in Allah and the Last Day and the Angels and the Book and the Messengers; to spend of your substance out of love for Him, for your kin, for orphans, for the needy, for the wayfarer, for those who ask; and for the ransom of slaves; to be steadfast in prayers and practice regular charity; to fulfill the contracts which you made; and to be firm and patient in pain (or suffering) and adversity and throughout all periods of panic. Such are the people of truth, the God fearing. (Qur'an 2:177)
>
> And those in whose wealth is a recognized right; for the needy who asks and those who are deprived. (Qur'an 70:24–25)

Islam is a religion that is often misunderstood and mischaracterized, both by the general public and by the media. This mischaracterization is due in part to the differences between more liberal Western values and the more conservative values held by many in the Islamic community. The terrorist acts of September 11, 2001, and the subsequent increase in xenophobia (an unreasonable fear, dislike, or hatred of foreigners, or people who are different) and Islamophobia have further exacerbated the tendency to view the entire Muslim world as one that endorses violence, extremist dogma, and female oppression. In truth, every culture and every religious faction has its peaceful members and its violent ones. A domestic batterer who uses the Christian fundamentalist concept of submission to justify the oppression and abuse of his wife does not define Christianity any more than does a terrorist bent on destruction define the Muslim religion.

The word *Islam* means submission, and followers of Islam submit themselves to the monotheistic God, Allah. The Muslim holy book is called the Qur'an (sometimes referred to as Koran, but because this is the Anglicized spelling, most Muslims prefer the spelling included previously because it most accurately reflects the correct pronunciation in Arabic). The Qur'an is considered by Muslims to be the recited words of God revealed to the Prophet Muhammad in the 7th century. Islam recognizes and relies on the holy books of Judaism and Christianity (the Old and New Testaments), but Muslims consider the Qur'an to be God's final revelation to humankind.

There are approximately one billion followers of Islam, which makes it the second-largest religion in the world. The majority of Muslims live in Southeast Asia, Northern Africa, and the Middle East. There are two primary sects within Islam due to an early dispute over who should have been Muhammad's successor. The Sunnis tend to be more religiously and politically liberal (for instance, they believe that Islamic leaders should always be elected). Approximately 90 percent of all Muslims are Sunnis. Shiites, on the other hand, tend to be more orthodox in their religious beliefs and political philosophies, having developed a more strictly academic application of the Qur'an. They believe that all successors to Muhammad (Imams) are infallible and sinless. They appoint their clergy and hold them in high regard.

The majority of Muslims who live in the United States are Sunnis, 75 percent of whom are foreign born. The Muslim community tends to be both college educated and middle class, thus Muslims tend not to rely on government-sponsored human services to meet basic needs, and much of the focus of charity is directed toward Muslims in other parts of the world who are suffering, either because of war or some other form of oppression, or is focused on concerns related to marriage and family.

Because Muslims hail from many different countries there is considerable diversity within the Muslim community, particularly in the United States. Yet despite the variability of cultural beliefs and practices, the House of Islam shares five basic pillars of faith:

- *Shahada:* Faith in one God
- *Salat:* Ritual prayer five times a day while facing Mecca
- *Zakat:* Charitable giving to the poor with the understanding that all wealth belongs to God
- *Sawm:* Fasting from sunrise to sunset during the month of Ramadan
- *Hajj:* Pilgrimage to Mecca

According to the Qur'an (9:60), there are eight categories of people who qualify to receive zakat. These include the poor, the needy, those who collect zakat, those who are being converted, captives, debtors, and travelers. The three foundational values within the Islamic community include community, family, and the sovereignty of God. Family is often defined as the joining of two extended families, thus what might be considered *enmeshment* in North American society is often seen as a sign of respect as extended families are drawn close and remain an active part of the immediate family's life. Men and women typically adopt traditional roles with men working outside of the home and women caring for the home and children, although this trend is changing, just as it is in other cultures within U.S. society. Modesty is seen as an important ingredient necessary for keeping order within society, and women often wear clothing (*hijab*) that covers the greater percentage of their bodies (Hodge, 2005).

Hodge (2005) pointed out the areas of obvious conflict between Islamic values and liberal North American values. For instance, Western culture values individualism, self-expression, and self-determination, whereas Islamic culture values community, self-control, and consensus. Thus, whether working with an Islamic human services agency, coordinating services with one, or directly serving the Islamic community, Hodge cautions human services workers not to view Islamic values through the eyes of Western culture.

Western culture values individualism, self-expression, and self-determination, whereas Islamic culture values community, self-control, and consensus.

For example, it is common for Westerners to view the Islamic tenet of modesty as primitive and oppressive to women, which for most Westerners is a "hop, skip, and a jump" away from endorsing domestic violence. Yet the Qur'an states that husbands and wives must express respect and compassion toward one another, and domestic violence is not endorsed. To truly understand the values of modesty and traditional roles embraced within the Islamic culture, one must take the time to understand what these values mean to the men and women within the Islamic culture itself.

Hence, although a human service professional might not share the traditional values held within the Islamic community, working in association with Islamic human services agencies provides human service professionals with an opportunity to display their respect for cultural diversity.

There has been a recent surge in interest in developing human services programs within mosques and Islamic centers across the United States in response to growing concerns about social issues and demonstrated needs within the Muslim community, particularly related to marriage, family, and general hostility often expressed toward this community in the post–September 11 climate. The discipline of human services is relatively new to the Islamic community, but charity is not new and has been practiced within Islamic and broader communities for generations. Islamic human services professionals include social workers, counselors, and psychologists, but these services can also be offered by an Imam, a Muslim religious leader. Islamic human service agencies provide services to those within Muslim and non-Muslim communities, and are increasingly relied upon to serve as a liaison for Western aid agencies in Muslim communities experiencing a crisis (De Cordier, 2009).

Islamic charities have suffered since the September 11 terrorist attacks, though, because many Muslims in the United States are afraid that monies they donate in good faith to Islamic charities may be frozen by the U.S. government and not directed to humanitarian causes as planned. Muslims are also giving less because they are afraid that they might be held in suspicion if a charity they donate money to is later investigated for diverting funds to terrorist causes. Mosques and Islamic centers across the nation are reaching out to legislators in a campaign called Charity without Fear, asking them to establish a list of Islamic charities in good standing, so that devout Muslims can give to charity without fear of being accused of supporting terrorist organizations (Council of Islamic Organizations, 2005).

Although there are not as many Muslim human service agencies as there are Christian organizations, there are several that make valuable contributions to the human services field on a national and international level. The following agencies are a few of these:

ISLAMIC SOCIAL SERVICES ASSOCIATION   Although human services agencies are not yet prolific within the Muslim community, they are increasing in numbers. The Islamic Social Services Association (ISSA) (http://www.issausa.org) acts as an umbrella organization for all Muslim human services agencies in the United States and Canada. The ISSA provides training and educational services, acting as a network linking and equipping Muslim communities.

INNER-CITY MUSLIM ACTION NETWORK   One group of agencies is called Inner-City Muslim Action Network (IMAN) (http://www.imancentral.org), which focuses on meeting the needs of those in the inner city in Chicago by operating food pantries, health clinics, and prayer services. The agency's offices, which are located in a storefront on Chicago's South Side, offer a free computer lab with free Internet service, General Educational Development (GED) courses, and computer training classes. IMAN is also involved in community activism such as lobbying against the granting of liquor licenses in high-crime areas, community development, and coordination of outreach events with other community agencies both Muslim and non-Muslim.

MUSLIM FAMILY SERVICES   There is considerable concern within the Islamic faith community that Muslim marriages are being negatively affected by the casual nature of divorce in the United States. Muslim Family Services (MFS), which is sprinkled throughout the United States, focuses on divorce prevention. MFS (http://www .muslimfamilyservices.org/home) is a division of the Islamic Circle of North America (ICNA), an organization designed to assist Muslims live a more devout life. MFS offers human services to families and couples, teaching them how to have a marriage according to Islamic principles.

MFS provides education, such as workshops for married couples and training for Imams; premarriage, marriage, and parenting counseling; emergency services; foster care; and advocacy in court and with social services, particularly in relation to Muslim family values. Islamic values are stressed, including the belief that marriage is the foundation of society and the pillar on which family is built. Human service professionals working for MFS understand that Muslim couples living in the United States are often caught between two cultures, thus many are influenced by the more liberal Western values. This has led to increased divorce rates and also many parenting challenges as adolescents in particular challenge traditional Islamic values such as modesty and male–female relationships.

ISLAMIC RELIEF USA   Poverty-related crises exist all over the world. Poverty alleviation depends upon the coordination of human service agencies, including government and nongovernmental organizations (NGOs). Islamic Relief USA (http://www.irusa.org) engages in poverty alleviation, disaster relief, and development work throughout the United States and throughout the world.

A similar organization that coordinates services with Islamic Relief USA is Islamic Relief Worldwide (www.islamic-relief.com) provides services on a worldwide basis, also enabling communities to deal with disasters, provides disaster relief and recovery services, and protects vulnerable and marginalized populations by confronting poverty. Islamic Relief Worldwide and Islamic Relief USA both engage in six types of aid work, including poverty alleviation in the form of sustainable livelihoods, education, health and nutrition, child welfare, water and sanitation, and emergency relief and disaster preparedness.

Although there are not an abundant number of human services agencies such as MFS, human services professions working within these agencies are Muslim and must be familiar with Islamic family values and the Qur'an, particularly in matters related to marriage and raising children. Many human service providers use similar counseling methods as do providers in secular agencies, but case management and generalist services are not as widely practiced because a human services network is not as well developed within this community.

The Muslim community within the United States will continue to be confronted with issues related to acculturation, modernization, and the eroding of traditional values, and problems within the family will no doubt continue to rise. Competing marital roles, adolescent rebellion, and at times social isolation, including the internalization of the majority culture's negative views of the Muslim faith, will continue to add stress to the Muslim family system. Human services agencies can assist Muslim families feel less isolated, can provide much-needed education and support, and can provide a sense of connectedness among Muslims who are feeling unsupported within their communities.

CASE STUDY 13.3

Case Example of a Client at a Muslim Faith-Based Agency

Maya is a 42-year-old Muslim woman who was referred to an Islamic women's center for advocacy and counseling. She has been married to Asad, a 44-year-old physician, for 18 years. Maya is the stay-at-home mother of their three children, aged 10, 12, and 14. Both Maya and Asad are originally from Egypt, having immigrated to the United States shortly after getting married. Maya reports that she and her husband have always been devout Muslims, being very involved in their local mosque. They have had what she considers a traditional Muslim marriage, where her husband is the leader of the home and provides for the family financially, and Maya takes care of the home and the children. For the majority of their marriage Maya believes that their marriage has been a good one. She believes that her husband was always very respectful of her and relied on her wisdom and input in making decisions impacting the family, particularly with regard to the children. Because Maya was an accountant prior to getting married, Asad has relied on her to help with financial matters related to his medical practice. Maya reported that about five years ago Asad began to "bring his work home with him," which led to an increase in his general irritability and frustration. In the last two years Maya noted that he began to become more controlling of her whereabouts, getting angry with her if he could not reach her at a moment's notice. She did not reach out then because she believed Asad when he said that it was his right to control her in this manner. Although Maya's father did not behave in this manner, she began to believe that perhaps she needed to endure Asad's behavior in order to be a good Muslim wife. Maya shared that in the past few months his aggression had escalated to the point of screaming at her, both at home and in public, backing her into corners. His drinking has escalated as well. The incident that prompted Maya to finally reach out for help occurred after she refused to sleep with Asad because he was extremely intoxicated and verbally abusing her. Asad became irate and began beating her, citing his right per the Qur'an (4:34–35). Maya initially went to the Imam at her mosque, who supported her completely and also explained that her husband's use of the Qur'an was a misinterpretation. He explained that Islam did not in any way condone abuse. He provided her with a considerable amount of information regarding the "cycle of violence" and services in the community for victims of domestic violence, including support groups for both adults and children. Maya contacted the Muslim women's center that day and saw a counselor later in the week. During Maya's first counseling center she expressed relief that her community was so supportive of her, but she expressed sadness as well because the information and resources she received seemed so fatalistic and hopeless. Her counselor explained that her husband was acting in a manner inconsistent with the will of Allah and if he was truly committed to following Islam and being a good Muslim husband and father, then perhaps he would be open to receiving counseling as well. Domestic violence, the counselor explained, not only destroyed everyone in the family but also affected the entire community, thus the Muslim community was as concerned about Asad as it was about Maya. During counseling Maya began to understand the underlying dynamics of her husband's behavior and gained wisdom regarding the difference between a husband who led his family with respect, as described by Muhammad, and the controlling and abusive behavior exhibited by her husband. As Maya gained confidence in herself and her decisions, she felt strongly that Allah was leading her to be strong for the sake of her family. Strength, according to her counselor, meant that she could not tolerate abuse. Asad met with the Imam for several weeks and then reluctantly agreed to attend a one-year anger management program that was led by an Imam at the community Islamic center, and Maya agreed not to make any decisions about whether to consider a divorce until after Asad had finished his program. Both the Imam and the counselor agreed that family counseling should not occur until after Asad had received enough counseling to recognize that the root of the family and marital problems lay within him. As Maya continued counseling, she began to realize the intergenerational cycle of abuse that existed in her husband's family and how important it was, particularly for the sake of her children, that she become strong enough to break the cycle. The most difficult aspect of this process for Maya was maintaining good boundaries with Asad and realizing that he had the choice not to change, which would force her hand in a sense, forcing her to leave the marriage to avoid repeating the patterns of abuse.

Human Systems

Understanding and Mastery of Human Systems: Processes to effect social change through advocacy

Critical Thinking Questions: What roles are Muslim faith-based organizations playing in changing societal attitudes about Islam? How are these organizations supporting Muslim individuals and families in coping with the increase in xenophobia and anti-Islamist sentiment that has grown in the United States since 9/11?

Concluding Thoughts on Faith-Based Human Services Agencies

As the field of human services evolves and matures, the scope with which this discipline is viewed is broadened and the value of services provided by those not within the mainstream mental health community will be increasingly recognized. Whether these services are delivered informally through church-sponsored programs or through highly organized faith-based human services agencies, recognizing that human services delivery can occur through a variety of systems acknowledges the reality that different people seek help in different ways.

The following questions will test your knowledge of the content found within this chapter.

1. Recent interest in faith-based counseling is based upon a:
   a. holistic approach to mental health
   b. religious revival within the United States
   c. recognition that religious organizations often provide human services more effectively
   d. Both B and C

2. Religiousness is commonly defined as _____, whereas spirituality is often defined as _____.
   a. a personal faith expressed within the structure of religious tradition/faith and general spirituality without a specific belief in a deity
   b. a social experience grounded in traditional religion/an independent relationship with God
   c. adherence to a traditional religious faith/an intense feeling of faith
   d. All of the above

3. Citing biblical support for slavery is an example of:
   a. the danger of organized religion
   b. how easy it is to confuse faith with cultural values
   c. the hypocrisy inherent in most religious traditions
   d. Both A and C

4. The Association of the Jewish Family Services and Children's Agencies acts as a(n):
   a. service agency for all Jewish agencies
   b. referral agency for all Jewish agencies in North America

   c. information clearing house for local Jewish Family Service agencies
   d. an advocacy organization that works at an international level

5. Mainstream Protestant denominations such as the Methodist, Presbyterian, and Lutheran denominations embraced caring for the poor, but these denominations did not necessarily link charity to _____:
   a. giving money (tithing)
   b. social welfare
   c. a biblical mandate
   d. evangelism

6. Islamic human service agencies:
   a. provide services to those within Muslim and non-Muslim communities
   b. reject the notion of human services due to a clash in values
   c. are increasingly relied upon to serve as a liaison for Western aid agencies in Muslim communities experiencing a crisis
   d. Both A and C

7. What are the benefits and challenges of faith-based counseling? Why is it important for human service professionals to develop competency in this area? Cite the rationale for incorporating spirituality into the counseling relationship, including exploring the difference between positive and negative religious coping mechanisms.

8. Compare and contrast the basic tenets of the three religious traditions cited in the text, describing how each approaches the use of human services within the respective traditions. Cite the various perspectives on the strengths and challenges of relying on faith-based agencies to provide human services.

Suggested Readings

Allender, D. B., & Longman, T. (1993). *Bold love.* Colorado Springs, CO: Navpress Publishing Group.

Bloom, J. H. (Ed.). (2006). *Jewish relational care A-Z: We are our other's keeper.* Binghamton, NY: Haworth Judaica Practice Press.

Cloud, H., & Townsend, J. (1994). *Boundaries.* Grand Rapids, MI: Zondervan.

Cnaan, R. A., Wineburg, R. J., & Boddie, S. C. (1999). *The newer deal: Social work and religion in partnership.* New York: Columbia University Press.

Derezotes, D. (2005). *Spiritually oriented social work practice.* Boston: Allyn & Bacon.

Donaldson, D., & Carlson-Thies, S. W. (2003). *A revolution of compassion: Faith-based groups as full partners in fighting America's social problems.* New York: Baker Books.

Ellor, J. W., Netting, F. E., & Thibault, J. M. (1999). *Religious and spiritual aspects of human service practice.* Columbia: University of South Carolina Press.

Martin, E. P., & Martin, J. M. (2003). *Spirituality and the black helping tradition in social work.* Washington, DC: NASW Press.

Yarhouse, M. A., Butman, R. E., & McRay, B. W. (2005). *Modern psychopathologies: A comprehensive Christian appraisal.* Downers Grove, IL: Inter-Varsity Press.

Internet Resources

Islamic Social Welfare Association: http://www.livingislam.co.za/companion/profiles/497-islamic-social-welfare-association.html

Jewish Federations of North America: http://www.ujc.org

References

Associated Press. (2005). *Muslim groups help Katrina victims on 9/11 anniversary.* Retrieved December 31, 2009, from http://www.amvoice-two.amuslimvoice.org/html/body_katrina_relief.html

Belcher, J. R., Fandetti, D., & Cole, D. (2004). Is Christian religious conservatism compatible with the liberal social welfare state? *Social Work, 49*(2), 269–276.

Blank, M. B., Mahmood, M., Fox, J. C., & Guterbock, T. (2002). Alternative mental health services: The role of the Black church in the South. *American Journal of Public Health, 92*(10), 1668–1672.

Catholic Charities USA. (2010). Catholic Charities at a glance. Available online at: http://www.catholiccharitiesusa.org/document.doc?id=2853

Council of Islamic Organizations. (2005). *Charity without Fear.* Retrieved December 31, 2009, from http://www.ciogc.org/Go.aspx?link=7654625

De Cordier, B. (2009). Faith-based aid, globalisation and the humanitarian frontline: An analysis of Western-based Muslim aid organisations. *Disasters, 33*(4), 608–628. doi:10.1111/j.1467-7717.2008.01090.x

Ebaugh, H. R., Pipes, P. F., Chafetz, J. S., & Daniels, M. (2003). Where's the religion? Distinguishing faith-based from secular social service agencies. *Journal for the Scientific Study of Religion, 42*(3), 411–426.

Edwards, C. E., & Williams, C. L. (2000). Adopting change: Birth mothers in maternity homes today. *Gender and Society, 14*(1), 160–183.

Fallot, R. D. (2001). Spirituality and religion in psychiatric rehabilitation and recovery from mental illness. *International Review of Psychiatry, 13*, 110–116.

Fallot, R. D., & Heckman, J. D. (2005). Religious/spiritual coping among women trauma survivors with mental health and substance use disorders. *Journal of Behavioral Health Services and Research, 32*(2), 215–226.

Gallup, G., & Lindsey, D. M. (1999). *Surveying the religious landscape: Trends in U.S. beliefs.* Harrisburg, PA: Morehouse.

Grossman, C. L. (2002, March 7). Charting the unchurched in America. *USA Today,* p. D01.

Hall, C. R., Dixon, W. A., & Mauzey, E. D. (2004). Spirituality and religion: Implications for counseling. *Journal of Counseling & Development, 82*, 504–507.

Hodge, D. R. (2004). Working with Hindu clients in a spiritually sensitive manner. *Social Work, 29*(1), 27–38.

Hodge, D. R. (2005). Social work in the house of Islam: Orienting practitioners to the beliefs and values of Muslims in the U.S. *Social Work, 50*(2), 162–173.

Idler, E. L., & Kasl, S. (1992). Religion, disability, depression and the timing of death. *American Journal of Sociology, 97,* 1052–1079.

Kliewer, S. (2004). Allowing spirituality into the healing process. *Journal of Family Practice, 53*(8), 616–624.

Koenig, H. G., George, L. K., Hays, J. C., Larson, D. B., Cohen, H. J., & Blazer, D. G. (1998). The relationships between religious activities and blood pressure in older adults. *International Journal of Psychiatry Medicine, 28*, 189–213.

Koenig, H. G., George, L. K., & Titus, P. (2004). Religion, spirituality, and health in medically ill hospitalized elderly patients. *Journal of American Geriatrics Society, 52*(4), 554–562.

Koenig, H. G., Larson, D. B., & Weaver, A. J. (1998). Research on religion and serious mental illness. *New Directions in Mental Health Surveys, 80*, 81–95.

McLaughlin, D. (2004). Incorporating individual spiritual beliefs in treatment of in-patient mental health consumers. *Perspectives in Psychiatric Care, 40*(3), 114–119.

Meisenhelder, J. B., & Marcum, J. P. (2004). Responses of clergy to 9/11: Post-traumatic stress, coping and religious stress. *Journal for the Scientific Study of Religion, 43*(4), 547–554.

Miller, M. M., Korinek, A., & Ivey, D. C. (2004). Spirituality in MFT training: Development of the spiritual issues in supervision scale. *Contemporary Family Therapy, 26*(1), 71–81.

Miller, W. R., & Thoresen, C. E. (2003). Spirituality, religion, and health: An emerging research field. *American Psychologist, 58,* 24–35.

National Association of Social Workers. (2002, January). *NASW priorities on faith-based human services initiatives.* Retrieved September 13, 2005, from http://www.naswdc.org/advocacy/positions/faith.asp

Pargament, K. I., Tarakeshwar, N., Ellison, C. G., & Wulff, K. M. (2001). The relationships between religious coping and well-being in a national sample of Presbyterian clergy, elders, and members. *Journal for the Scientific Study of Religion, 40*(3), 497–513.

Powell, L., Shahabi, L., & Thoresen, C. E. (2003). Religion and spirituality: Linkage to physical health. *American Psychologist, 58,* 36–52.

Stone, H. W., Cross, D. R., Purvis, K. B., & Young, M. J. (2003). A study of the benefit of social and religious support on church members during times of crisis. *Pastoral Psychology, 51*(4), 327–340.

Thomas, S. B., Quinn, S. C., Billingsley, A., & Caldwell, C. (1994). The characteristics of Northern Black churches with community health outreach program. *American Journal of Public Health, 84*(4), 575–579.

Tseng, W. S. (2004). Culture and psychotherapy: Asian perspectives. *Journal of Mental Health, 13*(2), 151–161.

White House, Office of the Press Secretary. (2009, February 5). *Obama announces White House Office of Faith-based and Neighborhood Partnerships* [Press Release]. Retrieved March 8, 2012, from http://www.whitehouse.gov/the_press_office/ObamaAnnouncesWhiteHouseOfficeofFaith-basedandNeighborhoodPartnerships/

Andrew Lichtenstein/Corbis

CHAPTER 14

Violence, Victim Advocacy, and Corrections

······································

Learning Objectives

- Develop a basic understanding of the nature of violence in its various manifestations, including its impact on both victims and perpetrators
- Become familiar with the current legislation and government policy having an impact on victims of crime, including the Victim Bill of Rights
- Explore the key roles and functions of human service professionals working in various forensic-related practice settings
- Understand the dynamics involved in gang activity and other similar forms of criminal activity
- Develop an awareness of the multicultural issues involved in forensic human services, including how institutional racism influences the punishment of certain crimes and treatment of prisoners

The field of forensic human services includes areas in which the human services discipline intersects with the legal system. Thus, human service professionals who work in practice settings dealing with domestic violence, sexual assault, gang activity, and criminal justice agencies such as police departments, probation, state, and county prosecutors, and within correctional facilities such as jails and prisons are considered forensic human service providers. The role and function of these practitioners will vary dramatically depending on the legal or criminal issues at play, but most forensic service providers require specialized training in areas such as crime victimization and crisis counseling, as well as developing a thorough understanding of the legal and criminal justice system.

Violence has always been a part of human history. In fact, violence exists in all segments of life among living creatures. A lion's survival is dependent on its killing of a wildebeest or zebra, which involves an act of violence. Yet although biologists would likely argue that violence is a natural aspect of survival in the animal kingdom, controversy abounds when this theory is applied to humankind. Does a review of history reveal that war is necessary? Certainly war has always existed, but is our existence dependent on competition for resources won through violent means? At what point does the act of war become the act of genocide? How can ordinary people live side by side peaceably for years and suddenly commit heinous acts, such as was the case during the Holocaust or the more recent genocide in Rwanda, and somehow justify their actions? Perhaps having the ability to respond in intense anger that manifests in violence is necessary when one is defending oneself, but doesn't the unjust use of violence make this defensive response necessary in the first place?

Determining the answer to these questions lies at the heart of violence research within the domain of social scientists such as sociologists, social psychologists, anthropologists, and criminologists, as well as those who work in the "applied" fields such as human services and those working within the criminal justice system. In this chapter the various types of violence will be explored, such as domestic violence, sexual assault, battery, and murder. Ways in which society and those within the human services fields most often intervene to reduce violence that affects not only its victims but also society as a whole will be explored.

Intimate Partner Violence

Intimate partner violence (IPV) (a more inclusive term than *domestic violence*) involves the physical, sexual, and emotional abuse acted out between intimates. This may include violence between husbands and wives, violence between boyfriends and girlfriends, violence within gay and lesbian relationships, and violence between family members (such as siblings, parents, etc.). IPV can include hitting, punching, slapping, pinching, shoving, and throwing objects at or near the victim, or threatening to do so. IPV also includes verbal and emotional abuse including name-calling, harassment, taunting, put-downs, and ridiculing, and sexual violence, such as forcing an intimate partner to engage in a sexual act without his or her consent.

The Centers for Disease Control and Prevention (CDC) estimates that one in three women (36 percent) and one in four men (29 percent) of the U.S. population report having been a victim of some form of IPV in 2010. One in four women (24 percent) and one in seven men (14 percent) have experienced severe IPV in 2010. Both men and women who were victims of IPV reported significantly higher rates of physical and |mental health problems than the general population. According to a CDC survey conducted in 2010, although both men and women are victims of IPV, women are victims far more often of multiple forms of violence, such as physical, sexual, and emotional violence, than men, who are far more often victims of solely physical violence. IPV has resulted in 1.3 million injuries each year, and 2,340 deaths in 2007, the majority of whom were women (Black et al., 2011).

> **Intimate partner violence results in 1.3 million injuries per year, and about 2,340 deaths in 2007, the majority of whom were women.**

Nearly 325,000 women are victims of domestic violence while pregnant, and research suggests that pregnancy can actually make women more vulnerable to abuse. Once considered a personal family matter, domestic violence in recent generations affects entire communities, both fiscally and socially. Women with a history of domestic violence report having significantly higher rates of physical health problems. Physical problems from assaults, partner rape, and the stress of living in a violent environment can lead to chronic pain, gynecological problems, HIV/AIDS, other sexually transmitted diseases, gastrointestinal problems, unwanted pregnancy, miscarriage, and premature births. The estimated health costs related to domestic violence is close to $6 million per year and $1.8 billion in lost productivity including lost time from work, unemployment, and increased dependence on public aid (CDC, 2003).

Domestic violence does not just affect the abused spouse. The children living in the home are victims as well, even if the violence is not aimed directly toward them. Boys who witness domestic violence are twice as likely to commit violence against their partners as adults (NCADV, 2007). IPV costs U.S. society approximately $5.8 million per year in lost revenue with victims of IPV missing collectively about 8 million days from work, and direct costs for mental and physical healthcare (CDC, 2003). Clearly, then, IPV is not a private family matter. The cost to society, both in injured members and in lost revenue, is far too high to ever allow this issue to be ignored again.

The Nature of Domestic Violence: The Cycle of Violence

Lenore Walker (1979) was the first to coin the phrase the *cycle of violence* to describe the pattern of interpersonal violence in intimate relationships. Most abusive relationships often begin in a *honeymoon-like state* with the abusers often telling their new partners that they are the only people in the world they can trust—the only ones who understand them. New partners are usually swept off their feet with compliments and many promises for a wonderful future. Once the abusers feel comfortable in the relationship, a dual process occurs. The abusers begin to feel vulnerable by recognizing their partner's power to hurt them deeply, and as familiarity in the relationship increases, the abusers often increase their sense of entitlement to have all their needs met.

Plagued with fears that they will be abandoned, taken advantage of, and humiliated (as many were in their childhoods), jealousy, possessiveness, and accusations begin. Emotional immaturity often prevents abusers from being able to separate their internal feelings from possible causes (i.e., are their feelings of jealousy caused by their own insecurities or caused by their partner's unfaithfulness?), thus a common assumption among batterers is that if they feel badly, their partners must be doing something to cause it.

In response to these threatening feelings of vulnerability and entitlement, and poised to be hurt once again, innocent partners often become the focus of the batterer's mistrust, fear, and ultimate rage. Abusers often misinterpret the intentions of their partners, mentally ticking off injustice after injustice. These types of negative misperceptions and misassumptions are prevalent and are rarely checked against fact.

Most partners of batterers will sense the increasing *tension* brought about by the abusers' underlying anger that is bubbling to the surface. Batterers might ask more questions, make sarcastic comments, ask why two cups are out rather than one, or question why the phone wasn't answered more quickly when they called. They will typically have a shorter fuse, becoming easily frustrated often without provocation. In response, most victims do their best to *walk on eggshells* to avoid an explosion. But no amount of running interference or offered reassurances will help because the process is an internal one, occurring within the mind of the abuser. In fact, most abusers have an actual *need* to be proven correct in their fear of being hurt and humiliated again because to a batterer, being too trusting is often synonymous with being an unsuspecting fool.

Eventually the *explosion* occurs despite all peacemaking efforts. Abusive rages can take on several forms including frightening bouts of screaming and yelling; intimidation; and physical abuse such as hitting, kicking, scratching, grabbing, slapping, and shoving. Attacks might also include throwing objects at or near the victim, punching walls, and making threats to harm either the person or the personal property of the victim.

Once batterers have experienced a violent rage, they are often temporarily relieved of their internal feelings of rage and in many respects take on the persona of a remorseful child seeking reassurance and approval. Batterers often *honeymoon* their partners and other family members who were victims of the abuse, promising never to repeat the abusive behavior. There is commonly a manipulative aspect to the batterer's professions of regret and apologies, with the extent of authentic remorse being somewhat questionable. One reason for this is that the batterer's apologies are often riddled with a series of "buts": "I'm sorry I hit you, *but* you know how I hate to be awakened early in the morning." "I'm sorry I shoved you, *but* you know I don't like you talking to other men." "I'm sorry I slapped you, *but* you know how stressed I get when work is so busy."

Rarely is the batterer's focus authentically placed on the pain and trauma caused to the partner or other family members. Rather, the honeymoon phase involves more of a panicked pleading, begging the victim not to leave, to forgive and forget, to move on quickly by minimizing the extent of the abuse. Statements intended to reframe the abuse, such as "I can't believe you think I shoved you! I clearly remember me reaching out to you and you jerking away and tripping," are common.

This can be an immensely confusing time for the victim, who usually knows instinctively that the batterer needs help, but any attempt to point out a pattern of abuse or to hold the batterer accountable (particularly after the batterer gets comfortable once again and stops apologizing) will hasten the tension-building phase, something the victim desperately wants to avoid. Attempts to demand authentic change in the abuser often result in the batterer accusing the victim of holding a grudge, being unforgiving, and punishing. Comments such as, "How dare you rub my face in this when I've already apologized . . . What do you want me to do? I've already said I'm sorry 100 times. Let's move on!" are common.

With the hope that the honeymoon phase might just last forever, victims often comply with the dangerous demands of the batterer to relinquish their own sense of reality and accept the reality of the batterer that the abuse was not that bad, that it will never happen again, and that it was a one-time event. Living in the here and now allows both the batterer and the victim to avoid seeing the pattern of abuse, which in some respects allows them both to avoid their fear of facing the truth and seriousness of the situation. But no matter how many promises the abusive partner makes or how desperately the victim wants to believe the abuse will never occur again, without intervention the cycle is destined to repeat itself.

Counseling Victims of Domestic Violence

WHOSE FAULT IS IT ANYWAY? ATTRIBUTING CAUSALITY OF ABUSE IN THE RELATIONSHIP   Counseling victims of domestic violence requires specialized training that focuses on the unique dynamics commonly at play in abusive relationships. Many of these dynamics relate to the cycle of violence discussed earlier, but many relate solely to the victim, including understanding common personality traits encountered in those who have a pattern of getting romantically involved with abusive partners, as well as traits commonly seen in individuals who will not leave or who continue to return to their abusive partners.

One significant element of counseling victims of domestic violence is assisting them in making decisions about their future that will not compromise their safety. Thus, although human service professionals may not actually tell the clients to leave an abusive relationship, they will often lead abused clients down this path, particularly if it is the only way to secure their safety and if the batterer has refused to enter into a structured treatment program.

Many victims of domestic violence have a locus of control that is far too internal. This means that they have a tendency to see themselves as responsible for more than they actually are and they do not necessarily recognize when their personal responsibility ends and when someone else's begins. In an unhealthy respect, this makes them a good "match" for a partner with an *external locus of control*. Those with an external locus of control have a tendency to see outside factors as responsible for the events in their lives. Batterers commonly have an external locus of control and blame their partners (as well as a host of other people and things) for their mistakes and failures. Those with a healthy locus of control will be able to recognize when something lies inside or outside their domain of responsibility. A healthy locus of control indicates that someone has good personal boundaries and will likely refuse to accept responsibility for something she knew was not her fault. But many victims of domestic violence do not have healthy personal boundaries and readily accept responsibility for virtually everything that is wrong in their relationship or with their partners. So, the batterer externalizes blame, and the victim internalizes blame.

A theory attempting to explain this core issue in domestic violence relationships focuses on *attribution theory,* specifically exploring how the victim attributes the partner's abusive behavior. If victims hold their partners at fault for the abusive behavior, attributing the abuse to personality factors such as an inability to manage anger, a refusal to take responsibility for their behavior, or a lack of empathy, then they will be more likely to leave the abusive relationship (Pape & Arias, 2000; Truman-Schram, Cann, Calhoun, & Vanwallendael, 2000). But victims who tend to attribute their partners' abusive behavior to situational or outside sources such as work stressors, family problems, or even alcoholism will have a greater likelihood of forgiving the batterer quickly and returning to the abusive relationship (Gordon, Burton, & Porter, 2004).

The human service professional can assist the victim in learning how to attribute causality of the abusive behavior to the batterer, incorporating an "even if" attitude: even if work is stressful, your mother is ill, you've had too much to drink, you've lost your job, money is tight, the kids are acting up, or you injured your knee, it's never okay to behave in an abusive manner. Victims of domestic violence also commonly need to develop more healthy personal boundaries so that they can understand what they are and are not responsible for in their relationships and with their abusive partners. For instance, the client might be responsible for responding to her husband's question in an irritable tone, but she is not responsible for her husband's choice to hit her in response; that was his choice, and it was unwarranted and an unreasonable response, one for which he was completely responsible.

A common clinical issue in helping someone develop new boundaries is the experience of unreasonable guilt. Many victims of domestic violence feel toxic guilt in response to setting limits with others, often believing that saying no to someone or upsetting another person is equivalent to being unkind. An emotionally healthy individual with good personal boundaries might feel badly when saying no to a request, or when firmly telling a partner that she is not responsible for his behavior, but she will not allow these bad feelings to influence what she knows to be true. In other words, she knows that despite feeling some guilt, she must honor her personal boundaries because to neglect them will negatively affect her self-esteem and self-respect. Yet victims of domestic violence will often allow their irrational guilt to determine their actions. If an action makes them feel guilty, they commonly assume that this action must be wrong.

Human service professionals can help clients see the irrationality of this way of thinking. Cognitive behavioral therapy (CBT) is a counseling technique commonly used to help victims of domestic violence recognize and change unhealthy relationship styles. Helping victims of domestic violence realize that feelings are not always the best indicators of appropriate action will assist them in setting better boundaries in their relationships and more efficiently recognizing the signs that a partner or potential partner is merely looking for a life scapegoat, rather than a life partner.

DOES SHE STAY OR DOES SHE GO?   One of the most frustrating aspects of counseling victims of domestic violence is the pattern of the victim returning to the abusive relationship despite intervention efforts and the risk of continued abuse. One theory that attempts to explain this dynamic is called the *social-exchange theory*. This theory posits that victims of domestic violence enter into a kind of cost-benefit analysis when attempting to make a decision about whether to stay or leave the abusive relationship. Is the cost more if the victims stay in the abusive relationship where they will be forced to endure more abuse? Or will the cost be higher if they leave, possibly facing economic insecurity, navigating the court system if a divorce is imminent, and managing work and family responsibilities alone? The *investment model of decision making* can be used when attempting to realistically weigh these pros and cons. This model involves the victim evaluating things such as her resources with and without the batterer, her ability to manage risk, and the risk involved in leaving, as well as estimating what will be gained or lost if she leaves the relationship (Rusbult & Martz, 1995).

For the objective observer the cost of staying means enduring abuse of increasing escalation and the cost of leaving may mean enduring financial hardship and other struggles relating to managing work and family alone. While the first option often results in worsening conditions, the latter option typically promises to improve with time. But victims of abuse often have a somewhat skewed perception of the risks of staying or leaving, using a positive bias when evaluating the cost-benefit analysis of staying—idealistically assuming that their partner will really change "this time," assuming that the abuse was "really not that bad," and overestimating their ability to rescue and compel change in their abusive partner. They may consider the difficulties they are bound to face the first few months on their own and assume that this transitional stage will last forever. They may use negative thinking, assuming that they will never get a job, will never be able to balance work and family, partly based on years of emotional abuse and partly based on the fear and low self-esteem that may have even been the prime motivators for getting into the unhealthy relationship in the first place.

Human service professionals can help victims of domestic violence more effectively process the pros and cons of leaving by helping them evaluate realistic risk factors and accurate scenarios. Counseling can also assist victims in learning how to manage risk more effectively without lapsing into negative thinking. In addition, practitioners can help the client "think outside of the box": exploring all alternatives and avoiding all-or-nothing thinking (I will be either financially secure or living on the streets, I will either be a part of an intact family or be constantly lonely and a social outcast). Encouraging the client to consider possibilities not previously acknowledged can help the client realize that she has far more control over her destiny than she might have previously thought. For instance, obtaining *factual* information about her financial situation, including learning laws related to an equitable division of property and the likely levels of child support and spousal maintenance, will assist victims of domestic violence in making good decisions that are based on fact, not fear.

Despite the specialized nature of working with victims of domestic violence, a generalist approach is most effective, often involving case management, court advocacy, individual counseling, group support, counseling children and adolescents, providing housing assistance, job coaching, and assistance with life skills. The human service professional working with victims of domestic violence must be familiar with contemporary theories of abuse, effective intervention strategies, common clinical disorders associated with being a survivor of domestic violence such as post-traumatic stress disorder (PTSD), domestic violence laws, the criminal justice process, and resources designed to meet the needs of victims and their children.

Human Systems

Understanding and Mastery of Human Systems: An understanding of capacities, limitations, and resiliency of human systems

Critical Thinking Question: In working with victims of domestic violence, a human service professional may find it deeply distressing when a client continues to return to an abusive partner. What generalist practice skills might a professional use to assist her/him in respecting the client's right to self-determination?

Domestic Violence Practice Settings

One of the most common practice settings where human service professionals work with victims of domestic violence is a *battered women's shelter*. Such shelters typically offer numerous services, including the following:

- A 24-hour hotline for immediate access to information and services
- Immediate safety shelters for domestic violence victims and their children
- Individual counseling for all victims
- Survivor support groups
- Court advocacy
- Children's programs
- Teen programs
- Information referral
- Medical advocates who provide on-site support at hospitals
- Immigrant programs (depending on the ethnic makeup of the community)

Although battered women's shelters often have a physical site where counseling and case management occur, their actual shelters are usually sprinkled throughout the community in confidential locations to ensure the safety of the victims utilizing shelter services. Shelters may include houses converted into shelters or even rented apartments located throughout a community. Victims of domestic violence and their children usually remain in a shelter for a time determined by their primary counselor, but the goal of shelter services focus on self-sufficiency, thus job placement, child care assistance, and transportation needs are also addressed.

Most shelters involve communal living, where residents share their living space with other victims. Residents are required to participate in group counseling sessions with other residents as well as assisting with the general functioning and maintenance of the shelter. Human service professionals are assigned to each shelter living space and facilitate in-house programs to maintain smooth functioning within the home, as well as among the residents. Most shelters institute rules such as a no alcohol or drugs policy and mandated maintenance of the confidentiality of the location of the shelter. Residents who release this information to their abusive partners will be asked to leave the program. Residents are also required to work on the meeting of program goals, and serious noncompliance may also be a reason to terminate services.

The Prosecution of Domestic Violence

In 1993 the federal government passed the Violence Against Women Act of 1994 (reauthorized in 2005 as the Violent Crime Control and Law Enforcement Act [Pub. L. No. 103-322]). The Violence Against Women Act established policies and mandates for how states were to handle domestic violence cases, such as encouraging mandatory arrests, encouraging interstate enforcement of domestic violence laws, and maintaining state databases on incidences of domestic violence. This act also provides for numerous grants for educational purposes (e.g., the education of police officers and judges), a domestic violence hotline, battered women's shelters, and to improve the safety of public areas such as public transportation and parks. Since the passage of the Violence Against Women Act incidents of domestic violence have been cut in half. Human service professionals are working alongside other advocates and pushing for more protection for immigrant women, as well as increased safety measures in the work place in future reauthorizations of Violence Against Women Act in order to address the growing problem of violence in these two arenas.

The Violence Against Women Act spurred several states to pass similar legislation, which continues to change the nature of domestic violence prosecutions. With regard to current policies regarding the prosecution of domestic violence, it is important to note that unlike a civil case, where a plaintiff brings an action and thus has the right to subsequently drop the case, in criminal cases the plaintiff is the state and the victims are witnesses. But in the past, prosecutors have allowed victims to drop a case (typically at the urgings of the batterer). Domestic violence legislation has for the most part put a stop to this practice. Instead, domestic violence is typically treated as any other crime where the victim is called as a witness and must appear at the trial to testify on behalf of the state.

This can create emotional tension for victims, who may initially want court involvement immediately after experiencing violence, but then want to resist any intervention when the honeymoon phase begins and renewed hope for authentic change seems possible. Counseling for the victim of abuse often focuses on the ways in which the victim can respond (often in counterintuitive ways) that will have the greatest likelihood of moving the batterer toward real change. As long as victims relinquish their own reality of the events and yield to the batterers' demands to forgive and forget without any real accountability, no real change will occur. Any effective counseling program must address the *denial, wishful thinking, indiscriminate forgiveness* (without accountability), and a desire to *protect* the batterer, as well as the *fear* of the future that many victims of domestic violence experience, which can prevent an honest and realistic appraisal of their abusive relationship.

Batterers Programs

It might be tempting to focus treatment efforts solely on the victims of abuse, leaving the perpetrators of abuse to fend for themselves. But if those who committed abuse were treated effectively, then domestic violence would no longer be a pressing social problem. It is also important to be aware that not all "batterers" are alike. In fact, although there are many batterers who are narcissistic with antisocial tendencies (sociopathy) and abuse their intimates with no remorse, there are also those who act out in anger but are truly remorseful, some who have never committed violence before but a combination of circumstances lowered their impulse control, some who are in reciprocally abusive relationships, and some who have been falsely accused.

It is vital that human service professionals take the time to understand the dynamics involved and not assume that if an accusation were made, it must be true. I have worked in domestic violence for years and worked with many authentic victims who had extremely abusive partners. Yet I will never forget the case involving a woman who presented with plausible stories of abuse at the hands of her husband, who was recently arrested for domestic violence. I was sold before having even met her husband, because my client's stories were convincing. Yet the criminal trial revealed that *she* had been emotionally abusive for years, and when he sought a divorce she threatened to seek revenge. She did so by causing self-injury and going upstairs privately to call the police. The tape of the 9-1-1 call was chilling as she screamed and cried while reporting the alleged abuse. If it had not been for the friend she told, who bravely testified at trial on behalf of the defense, her husband might have been convicted of a crime he did not commit, and she might have unfairly gained custody of their children because everyone, including me, was so quick to believe her simply because of her gender.

In the past the criminal justice system sought traditional forms of justice for those convicted of domestic violence, but this approach was often unsuccessful because judges were sometimes reluctant to break apart families, and more often victims of domestic violence were reluctant to testify against their partners or spouses, particularly if it meant a possibility of incarceration. Thus, several years ago domestic violence courts started mandating batterers to attend treatment programs often in lieu of jail.

Most batterers intervention programs are based upon the Duluth Model—a psychoeducational program drawn from feminist theory of domestic violence, which posits that domestic violence is caused by patriarchal ideology, and men's perception that they have the right to control their female partners. Many batterer intervention programs are also based upon group treatment using CBT and anger management training. Newer programs combined these models, based upon the premise that battering is a complex problem, thus a combination of psychoeducation, CBT, and anger management in a group setting will be most successful.

Programs range in duration from six weeks to one year and are often mandated by the court as a part of sentencing. Batterers are taught to respect personal boundaries, the difference between feelings and actions, the concept of personal rights and egalitarian relationships, and discover the dynamics of social learning theory including modeling so they can discover how their violent behavior is likely patterned after their parents or some other influential person in their lives. They also learn how to identify their personal triggers and learn strategies for managing their anger, including how to control impulses, and how to use "I" statements to avoid getting caught up in making accusations.

Most batterers' treatment programs have similar goals, including *increasing awareness of violent behavior* and *encouraging the batterer to take responsibility for violent behavior*. Common program philosophies include the following beliefs:

- Violence is an intentional act.
- Domestic violence uses physical force and intimidation as coercive methods to obtain and maintain control in the relationship.
- Using violence is a learned behavior and as such can be unlearned.

Many participants make authentic changes in group treatment not only because of the curriculum but also because of the built-in accountability that a group setting provides. Ironically it is the other group members who have been charged with domestic battery who often challenge those who refuse to engage or who consistently blame the victim. Unfortunately, at least an equal number of participants do not authentically change while in the program. Some batterers fail to complete the program, and others are reluctant to change because they actually love the rush and power they get from feeling intense anger (Pandya & Gingerich, 2002).

Whether batterer intervention programs actually work is a question that remains unanswered for the most part. A 2003 study commissioned by the U.S. Department of Justice (DOJ) found little support for the success of batterer intervention programs with regard to recidivism rates, or attitudes toward domestic violence. The only significant difference found was in the re-offense rates of men who completed programs 26 weeks or longer. Yet, while these men had significantly lower recidivism rates, their attitudes about domestic violence did not appear to change much. For instance, men in the experimental group (the batterer's intervention program) viewed their partners only slightly less responsible for the battering incident, than men in the control group. The study's authors cited numerous limitations of the study, which may have been responsible for the results, including a high drop-out rate, and questionable validity of the attitudinal surveys. Based upon these limitations, the authors recommended that batterer intervention programs be allowed to continue to evolve (since they are a relatively new tool in the fight against domestic violence), but in a manner that was responsive to the increased knowledge that is being gained about the nature of IPV, including common risk factors for becoming a batterer.

Sexual Assault

Another form of personal violence is the act of rape, or sexual assault. Sexual assault involves forcing some form of sexual act on another person without his or her consent.

Approximately one in five women in the United States have been raped sometime during their lifetime, and more than half of them were raped by intimate partners.

Determining the rate of sexual assault in the United States is difficult due to dramatic variations in the way sexual assault is defined. Although both men and women can be raped, women are victims of rape far more often than men. Approximately one in five women in the United States have been raped sometime during their lifetime, and more than half of them were raped by intimate partners (Black et al., 2011).

For the first time since 1927, the legal definition of forcible rape has been changed. According to the Uniform Crime Reports (UCR), the former definition was: "the carnal knowledge of a female, forcibly and against her will." That definition, unchanged since 1927, was outdated and narrow. It only included forcible male penile penetration of a female vagina. The new definition is: "[t]he penetration, no matter how slight, of the vagina or anus with any body part or object, or oral penetration by a sex organ of another person, without the consent of the victim." This is an important victory for advocates since this expanded definition now includes rape of both genders, rape with an object, and sexual acts with anyone who cannot give consent due to mental or physical disability.

Approximately 170,000 women, 12 years and older, and 15,000 men were raped or sexually assaulted in 2010. About 75 percent of all women who were raped were assaulted by perpetrators they knew, and about 25 percent were assaulted by strangers Black women are raped at a higher rate (relative to the population) than White or Hispanic women. Only half of all rapes and sexual assaults in 2010 were reported to police (Catalano, Smith, Snyder, & Rand, 2009; U.S. Department of Justice, 2011).

According to the CDC, rape and sexual assaults typically fall into four categories (Basile & Saltzman, 2002):

1. *Completed sexual acts* such as sexual penetration, but may also include any act of a sexual nature attempted or otherwise such as contact between a sexual organ and another part of the body
2. *Attempted sexual assault*
3. *Abusive sexual contact* such as intentional touching even through clothing
4. *Noncontact sexual abuse* such as intentional exposure and exhibitionism ("flashing") and voyeurism ("Peeping Tom")

Why People Commit Rape

Human service professionals who work with victims of sexual assault must understand the psychological dynamics of rape. One of the more common myths of why rape occurs includes blaming the victim by asserting that the victim wanted it, liked it, or in some way deserved the sexual assault because she provoked the assailant (by dressing or acting provocatively, etc.). Myths about rapists include assertions that only truly evil or insane men rape and that men just cannot control their sexual desires, and thus are not responsible for sexually assaulting women (Burt, 1991). The damage done by the proliferation of these rape myths is plentiful because they blame the victim while exonerating the perpetrator, which undermines societal prohibition against sexual violence.

In fact, a 1998 study at University of Mannheim in Germany (Bohner et al., 1998) found that such myths actually encourage sexual assault by giving rapists a way of rationalizing their antisocial behavior. In other words, although Western social customs may claim to abhor rape, popular rape myths provide rapists a way around such social mores by convincing themselves that the women in some way *asked for it* and that men simply *cannot control themselves,* thus they really haven't done anything wrong, or at least nothing that many other men don't do.

The Psychological Impact of Sexual Assault

The physical and psychological impact of sexual assault is serious and long-lasting and may include PTSD, depression, increased anxiety, fear of risk-taking, development of trust issues, increased physical problems including exposure to sexually transmitted diseases such as HIV/AIDS, chronic pelvic pain, gastrointestinal disorders, and unwanted pregnancy (CDC, 2005).

In 1975 Lynda Holmstrom and Ann Burgess coined the term *rape trauma syndrome* (RTS), a collection of emotions similar to PTSD, commonly experienced in response to being a survivor of a forced violent sexual assault. RTS includes an immediate phase where the survivor experiences both psychological and physical symptoms such as feeling extreme fear, consistent crying and sleep disturbances, and other reactions to the actual assault as well as the common fear of being killed during the assault. Survivors in subsequent phases of recovery experienced a variety of symptoms, including avoidance of social interaction, experiencing a loss of self-esteem, inappropriate guilt, and clinical depression. Many survivors deny the effects of the sexual assault because they do not want to be subject to the negative stigma associated with being a rape victim. In fact, one of the primary reasons most rape crisis advocates refer to clients as *survivors* rather than as *victims* is to reduce this stigma by focusing on the strength it takes to survive a sexual assault.

Male-on-Male Sexual Assault

Men are also victims of sexual assault, in the form of child sexual abuse, same-sex date rape, and male-on-male stranger rape. Research on male-on-male sexual assault is sparse with the exception of some early efforts to identify the nature and dynamics of male rape. The reason for the lack of studies in this area may be related to the belief that male rape is rare, at least outside prison walls. In fact, in many states, the legal definition of rape does not even account for men being victims.

Due to the stigma associated with being a victim of male-on-male sexual assault, most incidences of rape go unreported, thus it is impossible to know just how common this crime is. Even rapes that occur in prisons are often unreported not only because of the fear of retaliation but also because of the shame men feel in response to being victimized in this manner.

Treating men who have been sexually assaulted is similar in some respects to serving the female survivor population except that the shame men feel, although equal in intensity, tends to be more focused on their gender identity as males. Heterosexual men who were victims of rape reported questioning their sexual orientation because they were unable to fight off their attackers. Men also have a greater tendency to turn toward alcohol and drugs in response to the rape. Men also experience sexual dysfunction, problems

Hundreds of people take part in a candlelight march to call attention to violence against women and children during a "Take Back the Night" event.
Bettye Lane/Photo Researchers/Getty Images

getting close to people, particularly in intimate relationships, and as is the case with female victims, some male victims become sexually promiscuous (Mezey & King, 1989).

More studies need to be conducted on both female and male rape, particularly on the differing dynamics of sexual assault in minority populations. What research there is on ethnic minority populations seems to indicate that victims of sexual assault who are Caucasian and have higher levels of academic education tend to seek mental health counseling more often than victims of color or those with less education (Ullman & Brecklin, 2002; Vearnals & Campbell, 2001). This certainly has practical implications for human service professionals who through assessment or advocacy have the opportunity to reach out to victims or potential victims of sexual assault.

Common Practice Settings: Rape Crisis Centers

Human service professionals working in any practice setting will likely encounter a victim of sexual assault at some point in their careers. This might involve a recent victim seeking support services on the heels of an assault, but it is far more likely that rape victims will present for counseling at some point long after an assault, perhaps even years later, and might not even connect that the problems they are currently experiencing are with a past sexual assault.

Human service professionals who work directly with victims of sexual assault usually do so at a rape crisis center or sexual assault advocacy organization. Many states require that each county have at least one rape crisis center that offers a wide range of services including a 24-hour hotline, around-the-clock on-site advocacy during medical examinations and investigative interviews, and crisis counseling, as well as long-term individual and group counseling.

Many human service professionals who work with sexual assault victims receive from 40 to 50 hours of specialized training focusing on the history of the rape crisis movement, the nature of crisis counseling, the dynamics of RTS, rape myths, and the dangers of gender oppression. Training also includes information on normal child and adult developmental stages and how these stages are affected by sexual violence and trauma.

Victims of Violent Crime

Domestic violence and sexual assault are two types of violent crime that receive considerable attention within the human services field

> ### Human Services Delivery Systems
>
> Understanding and Mastery of Human Services Delivery Systems: Range of populations served and needs addressed by human services
>
> Critical Thinking Question: Given that sexual assault is one of the most underreported crimes in the United States, and that many victims do not seek help until months or years after being assaulted, how might human service professionals be proactive in getting services to more victims, sooner?

as well as within the public arena. There are other types of victimization that do not garner as much attention, but are also important. Every year millions of people in the United States become victims of a crime, many of which are violent crimes. In 2004, 24 million crimes were committed, 5.2 million of which were violent in nature (National Crime Victimization Survey, 2004). Although violent crime has been declining in recent years, the issue of victimization and the recognition and enforcement of victims' rights remains a relevant issue for human service professionals.

The victims' rights movement is a relatively new phenomenon having gained momentum in the 1980s when victims of crime came together along with advocates in the human services field to secure both a voice within the criminal justice community and some basic rights in the criminal justice system. Historically, victims of crime had virtually no rights in criminal proceedings because the U.S. criminal justice system is based on the presumption of innocence. Because defendants charged with a criminal offense are innocent until proven guilty, legally there can be no victims. If there are no victims prior to a defendant being convicted, then there are no rights to enforce. In addition, in criminal proceedings the case is considered an action committed against the state, thus other than being a witness, historically, victims of crime have had no special status. This logic, which is consistent with the U.S. criminal justice system, is completely backward for most victims and victim advocates.

> **Historically victims of crime had virtually no rights in criminal proceedings.**

The Victims' Bill of Rights

The victims' movement is based not on the desire to lessen the rights of criminal defendants, but rather on the desire to increase the rights of victims including being notified of court hearings, to appear at all legal proceedings, to make a statement at sentencing, and to be kept apprised of the incarceration status of the perpetrator.

Most victims and victim advocates state that a primary goal of the victims' movement is to ensure that crime victims have a voice within the community, specifically within the criminal justice system (Mika, Achilles, Halbert, Amstutz, & Zehr, 2004). How that voice gets heard is certainly up for debate. Whether through direct face-to-face meetings with criminal justice officials or through an active involvement in victim-sensitive training of police personnel, prosecutors, and judges, victims advocacy groups continue to work toward a system that sees victims as a central aspect of the criminal justice process (Quinn, 1998).

In response to the victims' movement and subsequent federal legislation (42 U.S.C. § 10606[b]), all states now have a Victims' Bill of Rights ensuring certain basic rights as well as protection. Although there is some variation from state to state, most states ensure that victims of violent crime be afforded the following rights:

- The right to be treated with dignity and fairness and with respect for the victim's dignity and privacy
- The right to be reasonably protected from the accused offender
- The right to be notified of court proceedings
- The right to be present at all public court proceedings related to the offense, unless the court determines that testimony by the victim would be materially affected if the victim heard other testimony at trial

- The right to confer with the attorney for the government in the case
- The right to restitution
- The right to information about the conviction, sentencing, imprisonment, and release of the offender (Victim's Rights Act of 1998).

Victim–Witness Assistance

In response to federal legislation and Victims' Bill of Rights state prosecution units within prosecutors' offices (state's attorney, district attorney, and attorney general offices) developed specialized units called Victim–Witness Assistance, designed to enforce victims' rights. Human service professionals work within these departments offering the following services:

- Crisis intervention counseling
- Referrals to coordinating human services agencies, such as rape crisis centers, battered women's shelters, and crime victim support groups
- Referrals to advocacy organizations such as Mothers against Drunk Driving (MADD), who have a presence in court to ensure enforcement of victims' rights
- Advocacy and accompaniment in court proceedings
- Special services or units for victims of domestic violence, child victims, older adults, and victims with disabilities
- Case status updates including notification of all public court proceedings
- Foreign language translation
- Assistance with obtaining compensation such as reimbursement for counseling and medical costs
- Assistance in preparation and writing of victim impact statements to be read by the victim at the sentencing hearing

Victim–witness advocates may have a master's degree in any of the applied social science disciplines (social work, psychology, general human services), but often work at the bachelor's level with some specialized training in the dynamics involved in violent crime victimization. Advocates must also be familiar with the inner workings of the criminal justice system because victims of violent crime often feel revictimized when they must endure the often confusing labyrinth of the prosecution system. The average person may not be familiar with the differing duties of a local police department and a state prosecuting office, nor may the average person know how a criminal case proceeds toward prosecution. Those individuals who have become victims of a crime must be quick studies so that they can be prepared for what is going to happen next. Victim–witness advocates can help crime victims understand the process of a criminal trial and the importance and value of each step within the prosecution process.

If a case goes to trial the victim–witness advocate will work closely with the victims to help prepare them for testifying. The clinical issues involved depend on the nature of the crime and victimization. For instance, if the defendant is the victim's spouse who is charged with domestic battery, the clinical issues will likely involve fear of retaliation and guilt in response to testifying against a spouse, particularly if there is a possibility that the defendant might have to serve time in jail or prison. If the defendant was charged with sexual assault, the victim will likely experience feelings of shame, embarrassment, and fear. A victim of home invasion might experience intense fear of retaliation once the defendant becomes aware of the victim's cooperation and testimony. In each instance the victim–witness advocate will work with community human services agencies and advocates to provide support and assistance to the victim in preparation for trial.

Once a defendant is found guilty, through either trial or a plea arrangement, a sentencing hearing is scheduled. In a sentencing hearing both sides have an opportunity to advocate for a sentence they believe is appropriate. It is the responsibility of the victim–witness advocate to assist victims in writing their victim impact statement, which will be read in open court before the judge, jury, and defendant. Although the statements are written in the words of the victim, they have a dual purpose—giving victims a voice in court and assisting the prosecutor obtain the desired sentence—thus it is important that victims receive guidance in the preparation for writing their statement. This also serves as another opportunity for victims to express and work through their pain, thus it is often an effective clinical tool.

Surviving Victims of Homicide

Some of the most emotionally intense and difficult cases for victim–witness advocates are homicide cases, particularly when the primary victim is a child. The victim–witness advocate must develop a high threshold for dealing with another's emotional pain because the pain of losing a loved one through violence is often unlike any other loss. Revictimization through the criminal justice process is almost a certainty as surviving victims of homicide are forced to balance their desire to represent their loved one in court by being present at all hearings with the trauma inherently involved in having to witness the gruesome details of the crime.

Research strongly suggests the importance of providing supportive counseling services and advocacy in the weeks immediately following the homicide. Surviving victims of intrafamilial homicides, where one family member kills another, are particularly prone to psychologically complex reactions involving both internal and external stressors. Most experts suggest the use of crisis counseling immediately following the crime that focuses on the concrete needs of the surviving victims. This approach is important in light of research, which suggests that surviving victims of homicide are mostly likely to utilize advocacy services during the initial crisis phase (Horne, 2003).

The needs of surviving victims of homicide are complex, particularly in the weeks and months after the murder. Surviving victims of homicide must cooperate with various law enforcement agencies and attend court proceedings at the same time that they must plan a funeral and contend with the effects and belongings of the murdered victim (which may include pets or even children in addition to physical belongings). This can be significantly overwhelming during a time when they are dealing with the paralyzing shock of losing a loved one in a sudden and violent manner.

Common Clinical Issues When Working with Victims of All Violent Crime

Regardless of the nature of the crime committed, victims of violent crime all have basic needs that need to be addressed by the human service professionals working with them in treatment (Courtois, 2004). These issues or treatment goals include the following:

1. Building formal and informal social support systems
2. Reinforcing ways to regain a sense of safety
3. Teaching victims how to manage their emotions, such as anger, sadness, and fear
4. Achieving physical and psychological stability
5. Building skills that will help victims regain a sense of personal power and control over their lives
6. Educating the client on the nature of the crime victimization so they know what to expect
7. Reconditioning victims to minimize negative triggering of the traumatic incident
8. Helping victims through the mourning process
9. Seeking resolution and closure, which leads to personal growth and allows the victim to regain the confidence and strength to trust people once again

By focusing on these core issues, as well as addressing the factors and needs specific to each type of crime victimization, the human service professional will be instrumental in fostering healing and growth in victims of crime so they can begin the process of seeing themselves no longer as victims but as true survivors.

Perpetrators of Crime

Forensic human service professionals working in the criminal justice arena often work with victims, but they may also work with offenders or perpetrators of crime. Direct practice with offenders might occur in an agency setting that offers mandated programs, such as batterers programs discussed earlier in this chapter, programs for alcoholics with drunken driving convictions, or group therapy for pedophiles. Many work within the criminal justice system in probation departments or juvenile justice programs, and many work in programs that facilitate outreach efforts focusing on gang members, recently released prisoners, or individuals who are at risk for continued criminal activity.

Gang Activity

Gangs consist of groups of individuals who actively participate in criminal activities on an organized or coordinated basis. Gang activity has become an increasingly severe problem in recent years, not only with regard to the number of gangs in operation within the United States (estimated to be somewhere between 700,000 and 800,000 nationwide), but also with regard to the type of violent activities in which many gang members participate. Gang activity remains primarily a big-city phenomenon, with some of the larger cities having more than 30 gangs operating at one time (National Youth Gang Center, 2005). Smaller towns and rural communities also experience gang problems, but these tend to be relatively sporadic with gangs that are loosely organized.

Gang members not only commit crimes such as theft and drug trafficking to support gang activity, but some of the most serious crimes committed by gang members involve turf wars where one gang is in conflict with another, leading to gang fights that often include both assaults and homicides. In some inner-city communities drive-by shootings are a way of life, and parents respond by keeping their young children off the streets and away from windows.

Most gang members are between the ages of 13 and 25, but some studies found gangs that have members as young as 10. Most gang members come from backgrounds of poverty and racial oppression, live in high-crime urban communities, and live in neighborhoods with high gang activity (Vigil, 2003). Although there has been a recent increase in female gang activity (Chesney-Lind, 1999), most gangs are still primarily comprised of males.

Risk Factors of Gang Involvement

There are several theories regarding why adolescents join gangs. Most sociological and anthropological theories focus on the sense of solidarity and feelings of belonging that gangs can provide disenfranchised youth. Identifying risk factors is important so that effective intervention strategies can be developed and put into action.

A comprehensive study facilitated by the DOJ evaluated the gang membership and backgrounds of over 800 gang members from 1985 to 2001 in an attempt to identify some of the reasons why adolescents join gangs. This study, referred to as the Seattle Social Development Project, confirmed that the majority of gang members are men (90 percent) and that gang members came from diverse ethnic backgrounds including Caucasian (European American), Asian, Latino, Native American, and African American, with African Americans having the highest rates of gang membership. Interestingly, the study found that the majority of gang members joined for only a short time, with 70 percent of youths belonging to a gang for less than a year (Hawkins et al., 2003).

The study identified multiple risk factors for gang membership, including living in high-crime neighborhoods, coming from a single-parent household, poverty, parents who approved of violence, poor academic performance, learning disabilities, little or no commitment to school, early drug and alcohol abuse, and associating with friends who commit delinquent acts. The study's authors recommended early prevention efforts that target youth with multiple risk factors. Programs need to focus on all aspects of the adolescent's life, including family dynamics, school involvement, peer group, and behavioral issues such as drug and alcohol abuse as well as any antisocial and delinquent behaviors.

What this study seems to underscore is that for youth with multiple risk factors gang membership may be less an option and more a way of life. Adolescents who are fortunate enough to have cohesive families, where high-functioning parents work hard to maintain structure, provide accountability, and keep teens engaged in positive activities, can often help adolescents avoid the temptation to join a gang. This is particularly true for black youth living in large urban areas (Walker-Barnes & Mason, 2001).

Adolescents without the benefit of such positive influences, including those who have neglectful and uninvolved parents, often face a reciprocal pull into gang life where they are targeted by existing gang members who recognize the existence of these risk factors, and the adolescents themselves are drawn to gang life because of the benefits gangs appear to provide such as a sense of belonging, a life of excitement, and the feeling of empowerment.

Human Services Practice Settings Focusing on Gang Involvement

Human service professionals who work with gang populations may do so on school campuses, in agencies that target at-risk youth, in faith-based outreach agencies, at police departments, or within the juvenile justice system. Most outreach programs target adolescents who live in large urban communities where gang activity is prolific and violent behavior a fact of life, especially those who come from single-parent homes, have poor academic histories, and have shown early signs of delinquent behaviors. Human service professionals also target social conditions on a macro level such as poverty, racism, and the lack of opportunities in urban communities, because these factors contribute to the development of gang activity.

Many human service programs that target at-risk adolescents operate after-school programs or evening community programs that give adolescents a place to go to socialize other than the streets. This is particularly important for youth who are in search of a sense of cohesion, security, and social belongingness, elements that might be missing from their home life. In light of the research indicating that most gang members have relatively loose, short-term affiliation with gangs, these types of programs have the potential of being successful in steering even active gang members away from gang life.

Finally, human service programs committed to reducing the gang problem must be willing to engage in active and aggressive outreach efforts, maintain a highly visible presence in the community, coordinate services with other gang intervention programs, and be willing to engage at-risk adolescents and their family on multiple levels.

Human Services in Prison Settings

The human services profession has a long history of association with the criminal justice system, most notably working in jails, prisons, government probation departments, police departments, and agencies offering services to recently released offenders. Human service professionals working within the criminal justice system may be employed as prison or correctional psychologists who conduct psychological evaluations on recently charged defendants or who provide assessment or counseling to offenders within the prison system. They may be licensed social workers who provide counseling and facilitate support groups focusing on various treatment issues designed to reduce *recidivism* (the process of relapsing into criminal behavior). They may be probation officers charged with the responsibility of coordinating treatment and supervising the offender's compliance with the conditions of probation (e.g., entering a drug treatment program, obtaining counseling, attending an anger management program, or completing community service), or they may be bachelor's level correctional treatment specialists or case managers who provide general counseling to the prison population, assisting them prepare for release and reentry into society.

Human service professionals may also work on a community level advocating for prison reform such as the development of mental health courts, substance abuse treatment programs in prisons, or increased mental health services for mentally ill prisoners. Thus, although this field of service is broad, the clinical issues are specialized, requiring training focusing on the common issues facing offenders both within prison and on release.

> **A key goal of the criminal justice system is to reduce recidivism; thus, "success" in terms of treatment is often focused on whether a prisoner once released reoffends and returns to prison.**

The U.S. prison system is plagued with violence including sexual assaults, drug problems, and mental illness. Human service professionals working within the area of corrections will likely encounter a wide range of issues that vary with the level of incarceration security, the gender and race of the prisoners, and the culture and climate of the specific prison. One of the chief problems affecting prisons across the country relates to the problem of overcrowding, with most state and federal prisons operating at either full or over capacity (Harrison & Beck, 2003). In an environment already wrought with tension, overcrowding can be the ingredient that leads to increased violence against both inmates and correctional staff.

The War on Drugs

Many people might be surprised to learn that violent crime in the United States has steadily declined since the early 1990s. Homicides, rapes, assaults, robberies, firearms-related crimes, and even violent juvenile crimes have all plummeted in recent years, yet the population in prisons and jails across the country has skyrocketed. In fact, the United States has the highest prison population of any country in the world (Walmsley, 2003). So what is to account for this seeming contradiction? Why, when virtually all forms of violent crime are on a downhill slide for many years, is the nation's prison system experiencing such a dramatic increase in population? Many social scientists agree that the primary reason for prison overcrowding relates to the U.S. War on Drugs.

In fact, approximately 55 percent of all federal prisoners are incarcerated for drug-related offenses (Harrison & Beck, 2003), and 80 percent of the increase in prisoners in the federal prison system between 1985 and 1995 is related to increased convictions of drug-related offenses (Bureau of Justice Statistics, 2004).

The U.S. war on drugs might seem like a good policy on the surface. Certainly no one would argue that the using and selling of illicit drugs is good for the American public. But many argue that the federal government's aggressive policies related to the prosecution and punishment of drug offenders unfairly targets poor, young ethnic minorities, many of whom are serving extremely long prison sentences due to minimum federal sentencing guidelines (sometimes 20 years to life), despite not committing any violent crime (Human Rights Watch, 2000a).

Human service professionals should be concerned about any governmental policy that either directly or indirectly targets a certain segment of the population. The war on drugs appears to do just this, evidenced by the significant overrepresentation of ethnic minorities, particularly African American men, within the federal and state prison system (Human Rights Watch, 2000b). Whether by design or not, one must ask why the U.S. government has not waged a "War on Domestic Violence" and a "War on Child Sexual Abuse," two social ills that have seriously negative consequences for U.S. society and that would target offenders across all socioeconomic levels and racial groups.

Human service professionals working within the U.S. criminal justice system must be aware of potentially unfair political policies to develop a truly objective perspective of social conditions leading to the overrepresentation of minorities in correctional facilities, the reasoning behind sentencing guidelines for various criminal offenses, even identifying social influences that tend to hold one behavior in a particular era as socially acceptable, only to criminalize it several decades later.

For instance, determining what drugs are socially acceptable and which ones are not is influenced by constantly shifting social mores. During the Prohibition era the use and sale of alcohol was considered criminal, yet today it is considered perfectly socially acceptable. Thus, there is a temporal aspect to the criminalization of certain behaviors, and it is vital that human service professionals recognize this dynamic.

> ## Human Systems
>
> Understanding and Mastery of Human Systems: Processes to effect social change through advocacy
>
> ___
>
> Critical Thinking Question: At first glance, the "War on Drugs" appears to be a beneficial social policy; however, on further inspection it turns out that this policy is having the effect of putting large numbers of young, ethnic minority men in prison for long periods of time for nonviolent drug offenses. How might a human service professional advocate on behalf of the populations disproportionately affected by the War on Drugs?

Clinical Issues in the Prison Population: The Role of the Human Service Professional

The issues confronting human service professionals working within the criminal justice system, particularly within a correctional facility, will vary depending on the gender, race, and type of crime committed by the defendant. A key goal of the criminal justice system is to reduce recidivism, thus "success" in terms of treatment is often focused on whether a prisoner once released reoffends and returns to prison.

MENTAL HEALTH PROGRAMS IN CORRECTIONAL FACILITIES   Behavioral programs within prisons can focus on many clinical issues, some related to criminal behavior and some related to other issues the inmates might be experiencing. Programs related to criminal behavior typically focus on issues such as drug abuse, sexual violence, domestic violence, anger management, and the development of social skills (for prisoners with antisocial tendencies). Programs designed to address psychosocial issues not directly related to criminal behavior typically focus on grief and separation issues, sexual abuse victimization (particularly for female inmates because a large proportion of the female inmate population has been the victim of sexual violence at some point in their lives), self-esteem, and issues related to the impact of being incarcerated.

PRISON AND PREGNANCY   Female inmates are often incarcerated for offenses related to drug addictions (writing bad checks, petty theft, prostitution, etc.), and those who are pregnant or parenting often have to rely on the county foster care system for the care of their children during their incarceration (Siefert & Pimlott, 2001). Human service professionals working in a correctional facility will likely encounter women (particularly women of color) who are grieving over the loss of their children or are anticipating their loss once they give birth. One of the roles of human service professionals is to work with outside agencies that can arrange to transport children to see their incarcerated mothers to maintain the mother–child bond. Parenting issues are often explored as well as the impact of drug abuse during pregnancy, with the goal of maintaining close family ties and reducing the incidence of prenatal damage and infant mortality related to drug use during pregnancy.

Some prisons have grant-funded programs that provide intensive prenatal care, nutrition counseling, substance abuse treatment, and individual and group counseling. One such program is called the Women and Infants at Risk (WIAR), which helps mothers break intergenerational cycles of abuse, giving infants the best start in life possible. This is particularly important in light of how the "cards" are already stacked against infants who are born behind prison walls (Siefert & Pimlott, 2001).

SEXUALLY TRANSMITTED DISEASES AND AIDS Another significant issue often confronting both inmates and human service professionals involves the high rate of infectious diseases that exists within the prison population, made worse by the ongoing problem of sexual assaults. Diseases such as hepatitis B and hepatitis C are prevalent in some prisons, and HIV/AIDS remains a serious concern among prisoners and correctional staff alike. A 2002 report by the National Commission on Correctional (NCCHC) indicated that the incidence of AIDS in the U.S. prison population is five times that of the general population, and the primary method of transmission is sexual assault (Robertson, 2003).

The fear of being raped is the number one fear among men serving time in prison, and although no one is certain of the exact number of male-on-male sexual assaults within the prison system, it is estimated that between 7 and 12 percent of the male prison population have been a victim of sexual assault while incarcerated, although the actual number is presumed to be much higher (Human Rights Watch, 2001), with many prisoners suffering multiple rapes throughout their incarceration. This issue is of such significant concern that in 2003, President George W. Bush signed an act appropriating $13 million to fund rape prevention programs within the prison system (Robertson, 2003).

Barriers to Treatment

One complaint among mental health providers in correctional settings is the underfunding and understaffing of mental health programs often experienced in many jails and prisons across the country. Developing effective and comprehensive mental health services within correctional facilities is an important aspect of efforts to reduce recidivism rates among the prison population, but the U.S. criminal justice system is punitive in nature and not based on a rehabilitation model; thus mental health programs are often not a priority within the criminal justice system, evidenced by a consistent lack of funding, understaffing, and limited outreach.

Human Services Delivery Systems

Understanding and Mastery of Human Services Delivery Systems: Range and characteristics of human services delivery systems and organizations

Critical Thinking Question: Providing human services to incarcerated individuals is a distinctly different experience from working with clients in contexts other than the criminal justice system, and even different from working with those who are court-mandated to attend treatment but who are not in jail. How might a human service professional adapt her/his methods of service delivery to best meet the special needs of incarcerated clients?

Yet even in prisons that have sufficient mental health services, barriers still exist that often prevent prisoners from accessing these services. A 2004 study surveying prisoner attitudes about mental health services identified several perceived barriers to service, including being uncertain how or when to access counseling, a belief that mental health services are for "crazy" people, the lack of confidentiality involved in the counseling relationship with a fear that the information shared would later be used against them, a fear that other prisoners would believe they were a snitch, a belief that people should deal with their own problems, a preference for talking with friends and family rather than a professional counselor, and having had a past bad experience with counseling (Morgan, Rozycki, & Wilson, 2004).

Human service professionals need to be aware of these common perceptions held by prisoners so that strategies can be designed to overcome both real and perceptual barriers to seeking mental health counseling. Although many of these negative perceptions held are common among the general population as well, many are related to being in custodial care where prisoners' personal rights are extremely limited by necessity.

Concluding Thoughts on Forensic Human Services

Working within the criminal justice system offers rich opportunities for human service professionals at all education levels. The opportunity to interact with several other advocacy organizations and to coordinate services with agencies offering complementary services provides the human services professional with a broad range of professional experiences. Human service professionals provide counseling, case management, and advocacy to both victims and offenders, thus making a difference in the lives of the members of society most in need.

Victims of violent crime such as domestic violence, sexual assault, and other violent crimes need advocacy and counseling to turn tragedy into triumph and powerlessness into empowerment. Human service professionals are on the front lines of bringing issues formerly kept in the dark out into the open, removing stigmas, and creating change that makes survivors out of victims.

Criminal activity and subsequent incarceration leaves long-lasting scars on the families of offenders, often plunging them into a cycle of poverty and social isolation. This process significantly increases the likelihood of creating an intergenerational pattern of incarceration, thus some of the most important work that forensic human service professionals do involves working with the family members of prisoners, particularly children who not only feel abandoned by their incarcerated parents but often are forced to enter the foster care system if no family members are available to care for them.

Rehabilitation offers the most hope of lowering recidivism rates among the prison population, yet a correctional philosophy that incorporates rehabilitation is controversial because in the eyes of many in the general public, counseling and other mental health programs feel too much like a luxury, not deserved by those who have committed crimes. Yet not only are prisoners not a homogeneous group (i.e., many prisoners have been incarcerated for relatively minor offenses), but those who have committed the most serious offenses are in many cases those who need mental health services the most. Unfortunately, mental health programs are often the first to be cut from state and federal budgets because on the whole the prisoner population does not garner much sympathy within the general public. For this reason it is imperative that human service professionals advocate for the basic rights and needs of prisoners, as they do with all vulnerable populations.

The following questions will test your knowledge of the content found within this chapter.

1. Human service professionals who work in practice settings dealing with domestic violence, sexual assault, gang activity, and criminal justice agencies such as police departments, probation, state and county prosecutor office, and within correctional facilities are considered:
   a. criminal justice social workers
   b. forensic human service providers
   c. criminal justice human service workers
   d. Both A and C

2. Domestic violence includes
   a. violence between heterosexual intimate partners
   b. violence between same sex partners
   c. violence between siblings
   d. All of the above

3. Approximately _____ percent of sexual assault victims knew their assailant.
   a. 12
   b. 32
   c. 70
   d. 55

4. Myths about rapists include assertions that
   a. only truly evil or insane men commit rape
   b. men just cannot control their sexual desires
   c. all men rape women
   d. Both A and B

5. Most victims and victim advocates state that a primary goal of the victim's movement is to ensure that:
   a. crime victims have a voice within the community, specifically within the criminal justice system
   b. the rights of defendants are minimized
   c. defendants charged with violent crimes are not released on bond
   d. All of the above

6. Risk factors of gang membership include all but the following:
   a. living in high crime, impoverished neighborhoods
   b. coming from a single-parent household
   c. having significant health problems early in life
   d. poor academic performance and/or learning disabilities

7. What is the "War on Drugs"? Has this government policy and approach to drug enforcement been successful in stemming the drug trade? Why or why not? What have social advocates cited as complaints about this set of policies?

8. Describe the roles and functions of human service providers working within prison settings. Provide some key demographic information of inmates, including female inmates. Include ways in which recidivism rates can be lowered.

Suggested Readings

Lord, J. H. (1990). *No time for goodbyes: Coping with sorrow, anger and injustice after a tragic death.* Ventura, CA: Pathfinder Publishing.

Internet Resources

American Civil Liberties Union: http://www.aclu.org
Family Violence Prevention Fund: http://endabuse.org
Legal Services for Prisoners with Children: http://prisonerswithchildren.org/index.htm

National Center for Victims of Crime: http://www.ncvc.org/ncvc/Main.aspx
National Coalition against Domestic Violence: http://www.ncadv.org

National Organization for Victim Assistance: http://www.trynova.org

Office for Victims of Crime: http://www.ojp.usdoj.gov/ovc

Prisoner Policy Initiative: http://www.prisonpolicy.org/index.html

Rape, Abuse & Incest National Network (RAINN): http://www.rainn.org

YWCA: http://www.ywca.org

References

Basile, K. C., & Saltzman, L. E. (2002). *Sexual violence surveillance: Uniform definitions and recommended data elements* (Version 1.0). Atlanta, GA: Centers for Disease Control and Prevention, National Center for Injury Prevention and Control. Retrieved September 14, 2005, from http://www.cdc.gov/ncipc/pub-res/sv_surveillance/sv.htm

Black, M. C., Basile, K. C., Breiding, M. J., Smith, S. G., Walters, M. L., Merrick, M. T., Chen, J., & Stevens, M. R. (2011). *The National Intimate Partner and Sexual Violence Survey (NISVS): 2010 Summary Report.* Atlanta, GA: National Center for Injury Prevention and Control, Centers for Disease Control and Prevention.

Bohner, G., Reinhard, M., Rutz, S., Sturm, S., Kerschbaum, B., & Effler, B. (1998). Rape myths as neutralizing cognitions: Evidence for a causal impact of anti-victim attitudes on men's self-reported likelihood of raping. *European Journal Social Psychology, 28,* 257–268.

Bureau of Justice Statistics. (2004). *Crime in the United States, annual, uniform crime reports.* Washington, DC: U.S. Department of Justice, Office of Justice Programs. Retrieved November 4, 2005, from http://www.ojp.usdoj.gov/bjs/glance/tables/drugtab.htm

Burt, M. R. (1991). *Rape myths and acquaintance rape.* In A. Parrot & L. Bechhofer (Eds.), *Acquaintance rape: The hidden crime* (pp. 26–40). New York: Wiley.

Catalano, S., Smith, E., Snyder, H., & Rand, M. (2009, September). Female victims of violence. (NCJ 228356). Bureau of Justice Statistics Selected Findings. Retrieved from http://www.ojp.usdoj.gov/bjs/pub/pdf/fvv.pdf

Centers for Disease Control and Prevention (CDC). Costs of intimate partner violence against women in the United States. Atlanta (GA): CDC, National Center for Injury Prevention and Control; 2003.

Centers for Disease Control and Prevention. (2005). *Sexual violence: Fact sheet.* Atlanta, GA: National Center for Injury Prevention and Control. Retrieved September 15, 2006, from http://www.cdc.gov/ncipc/factsheets/svfacts.htm

Chesney-Lind, M. (1999). Challenging girls' invisibility in juvenile court. *Annals of the American Academy of Political and Social Science, 564,* 185–202.

Courtois, C. (2004). Complex trauma, complex reactions: Assessment and treatment. *Psychotherapy: Theory, Research, Practice, Training, 41*(4), 412–445.

Gordon, K. C., Burton, S., & Porter, L. (2004). Predicting the intentions of women in domestic violence shelters to return to partners: Does forgiveness play a role? *Journal of Family Psychology, 18*(2), 331–338.

Harrison, P. M., & Beck, A. J. (2003). *U.S. Department of Justice, Bureau of Justice Statistics, Prisoners in 2002.* Washington, DC: U.S. Department of Justice.

Hawkins, J. D., Smith, B. H., Hill, K. G., Kosterman, R., Catalano, R. F., & Abbott, R. D. (2003). Understanding and preventing crime and violence: Findings from the Seattle Social Development Project. In T. P. Thornberry & M. D. Krohn (Eds.), *Taking stock of delinquency: An overview of findings from contemporary longitudinal studies* (pp. 255–312). New York: Plenum.

Holmstrom, L. L., & Burgess, A. W. (1975). Assessing trauma in the rape victim. *American Journal of Nursing, 75*(8), 1288–1291.

Horne, C. (2003). Families of homicide victims: Service utilization patterns of extra- and intrafamilial homicide survivors. *Journal of Family Violence, 18*(2), 75–81.

Human Rights Watch. (2000a). *Key recommendations from punishment and prejudice: Racial disparities in the war on drugs.* Retrieved November 4, 2005, from http://www.hrw.org/campaigns/drugs/war/key-reco.htm

Human Rights Watch. (2000b). *Punishment and prejudice: Racial disparities in the war on drugs, 12*(2). Retrieved November 4, 2005, from http://hrw.org/reports/2000/usa/index.htm#TopOfPage

Human Rights Watch. (2001). *No escape: Male rape in U.S. prisons.* Retrieved September 27, 2005, from http://www.hrw.org/reports/2001/prison/report.html

Mezey, G., & King, M. (1989). The effects of sexual assault on men: A survey of 22 victims. *Psychological Medicine, 19,* 205–209.

Mika, H., Achilles, M., Halbert, E., Amstutz, L., & Zehr, H. (2004). Listening to victims—a critique of restorative justice policy and practice in the United States. *Federal Probation, 68*(1), 32–39.

Morgan , R. D. Rozycki , A. T. Wilson , S. (2004). Inmate perceptions of mental health services. Professional Psychology: Research and Practice, 35, 389–396.

National Coalition Against Domestic Violence [NCADV] (2007). Domestic violence facts. Retrieved August 17, 2012 from http://www.ncadv.org/files/DomesticViolenceFactSheet(National).pdf.

National Youth Gang Center. (2005). *Highlights of the 2002–2003 national youth gang surveys.* Washington, DC: U.S. Department of Justice, Office of Justice Programs, Office of Juvenile Justice and Delinquency Prevention.

Pandya, V., & Gingerich, W. J. (2002). Group therapy intervention for male batterers: A microethnographic study. *Health & Social Work, 27*(1), 47–55.

Pape, K. T., & Arias, I. (2000). The role of attributions in battered women's intentions to permanently end their violent relationships. *Cognitive Therapy and Research, 24,* 201–214.

Planty, M. and Truman, J.L. 2012. Criminal Victimization, 2011. Bureau of Justice Statistics Bulletin. Washington D.C.: Bureau of Justice Statistics. Retrieved October 2012 from: http://www .bjs.gov/content/pub/pdf/cv11.pdf.

Quinn, T. (1998). *An interview with former visiting fellow of NIJ, Thomas Quinn*. Washington, DC: The National Institute of Justice Journal, Office of Justice Programs, U.S. Department of Justice.

Robertson, J. E. (2003). Rape among incarcerated men: Sex, coercion and STDs. *AIDS Patient Care and STDs, 17*(8), 423–430.

Rusbult, C. E., & Martz, J. M. (1995). Remaining in an abusive relationship: An investment model analysis of nonvoluntary dependence. *Personality and Social Psychology Bulletin, 21*, 558–571.

Siefert, K., & Pimlott, S. (2001). Involving pregnancy outcome during imprisonment: A model residential care program. *Social Work, 42*(2), 125–134.

Truman-Schram, D. M., Cann, A., Calhoun, L., & Vanwallendael, L. (2000). Leaving an abusive dating relationship: An investment model comparison of women who stay versus women who leave. *Journal of Social and Clinical Psychology, 19*, 161–183.

U.S. Department of Justice. (2011). *National crime victimization survey*. Washington, DC: Office of Justice Programs, Bureau of Justice Statistics. Retrieved September 15, 2012, from http://bjs. ojp.usdoj.gov/content/pub/pdf/cv10.pdf

Vearnals, S., & Campbell, T. (2001). Male victims of male sexual assault: A review of psychological consequences and treatment. *Sexual and Relationship Therapy, 16*(3), 279–286.

Victim's Rights Act of 1998, 42 U.S.C. § 10606(b) (West 1993).

Vigil, J. M. (2003). Urban violence and street gangs. *Annual Review Anthropology, 32*, 225–242.

Violent Crime Control and Law Enforcement Act of 1994, Pub. L. No. 103-322, Title IV, § 40001 *et seq.*, 108 Stat. 1902 (1994).

Walker-Barnes, C. J. & Mason, C. A. (2001). Ethnic differences in the effect of parenting upon gang involvement and gang delinquency: A longitudinal, HLM perspective. Child Development, 72, 1814–1831.

Walmsley, R. (2003). *World prison population list* (5th ed.). London: Home Office Research, Development and Statistics Directorate.

CHAPTER 15

Macro Practice and International Human Services

UNHCR/J. Wreford

When students consider entering the field of human services, they often do so because they want to help people meet their basic needs by counseling them, helping them obtain much-needed services, and teaching them to learn new ways of meeting their needs in the future. In other words, most students think of direct clinical practice with individuals and families when considering a career in the human services profession. But many times the "personal troubles" a client is encountering are being caused by some external source—an injustice that is structural or systemic such as the school system that offers no bus service and therefore inadvertently contributes to low-income students' truancy rates, or a government social welfare policy that inadvertently punishes single mothers who work part-time by cutting their benefits, or a "three-strikes" law that sends a young man to jail for 25 years for a third, yet relatively minor, offense. How does the human service professional combat harmful policies that punish when they should reward or unfair legislation that hurts certain segments of the population?

The human services profession is grounded in the notion that people are a part of larger systems and to truly understand the individual one must understand the broader system this individual is operating within. The discussion of Bowen's Family Systems Theory in Chapter 4 is a good place to start in understanding how systems work, noting that there is a reciprocal dynamic involving both the individual and the system, where each has an impact on the other. Hence, an individual can receive years of counseling, but until structural deficiencies are addressed, they will continue to experience difficulty in some manner.

Learning Objectives

- Recognize current and historic disenfranchised populations and understand societal conditions and dynamics that render groups vulnerable to abuse and exploitation
- Understand the various aspects of macro practice such as community development, community organization, and policy practice
- Become familiar with the nature of globalization and its affect on the human service profession
- Identify major human rights violations such as crimes against women and children, indigenous populations, labor violations, the effects of civil war, and genocide
- Identify some of the ways in which the international community responds to global crises and international human rights violations

Since when do you have to agree with people to defend them from injustice?
—Lillian Hellman

> **The human services profession is grounded in the notion that people are a part of larger systems and to truly understand the individual one must understand the broader system this individual is operating within.**

It is important, then, for human service professionals to recognize that people can be helped by approaching problems on various levels. By way of comparison, if as a human service professional you were committed to eradicating violence within society, you might choose to work with victims of domestic violence in the hope that counseling them might help your clients recognize the signs of abuse and avoid engaging in abusive relationships in the future. This approach would involve *micro practice*—practice with individuals. You might also decide to facilitate treatment groups for batterers, believing that the greatest likelihood of change can be accomplished by addressing the perpetrators of violence in a group setting where each group member can learn from others. This approach would involve *mezzo practice*—practice with groups.

But, if you decided to address the problem of violence by working with an entire community, locally, nationally, or perhaps even globally, by creating a new program in your agency, by conducting a public awareness campaign to educate the population about the prevalence of violence, or by lobbying for the passage of antiviolence legislation, then you would be conducting *macro practice*—practice with communities and organizations.

Macro practice involves addressing and confronting social issues that can act as a barrier to getting one's basic needs met on an organizational level by creating structural change through social action. The most basic themes involved in macro practice include advocating for *social and economic justice* and *human rights* for all members of society to end human oppression and exploitation (Weil, 1996). There are several ways *social change* is accomplished through macro practice, including *program development, community development* through *community organizing, policy practice,* and *international* or *global advocacy.*

Thus, although direct clinical practice is important, working with entire systems to promote positive structural change on all fronts is equally important. Some human service professionals work solely in macro practice in administrative positions or policy practice conducting no direct practice whatsoever, but a great many human service professionals who are involved in micro practice are also involved in macro practice on at least some level. For instance, when I worked as a victim advocate for a local state's attorney's office, I counseled victims of violent crime. But I also served on a domestic violence advisory coalition that evaluated community concerns and interagency coordination.

Why Macro Practice?

Human service professionals might ask themselves why they should be concerned about what is happening to people in an entire community, in a different part of the country, or in a completely different part of the world. But a foundational value of the human services profession is a commitment to social justice and human rights achieved through social action and social change. This is particularly relevant to human service professionals living in the United States in light of the fact that many clients in need of human services assistance have emigrated from countries where they were victims of oppression and human rights violations. This requires an understanding on the part of the human service professional of the wide range of global abuses related to social injustice and human rights abuses, as well as recognizing how these abuses have implications on direct practice with individual clients.

Human service professionals must also be aware of the history of social injustices and human rights abuses that have occurred within U.S. borders as well as develop an awareness of what groups are most likely to be targets of discrimination and oppression. For instance, Calkin (2000) discussed the abuse and oppression of minorities and the poor within the U.S. criminal justice system and the importance of human service professionals accepting a call to social action:

> Moment by moment in the practice process, there are opportunities to recognize and support, or to ignore, the power that people bring or could bring to their lives and communities. There are opportunities to act respectfully toward someone for whom that is so uncommon, or not to—and to acknowledge when we really can't understand, to acknowledge the errors of sensitivity we make so often. Human services organizations and professionals can easily be seduced into colluding with violations of human rights, ranging from disrespect toward people already struggling with mental illness or substance abuse to acceptance or resignation in the face of deprivations of basic human rights. (p. 2)

This foundational commitment to social justice is so integral to the human services profession that the professional obligation to social action is reflected in the ethical principles of the discipline. For instance, the National Organization for Human Services (NOHS) (1996) ethical standards reference the human service professionals' responsibility to society, which includes remaining aware of social issues that impact communities, and initiating social action when necessary by advocating for social change. The National Association of Social Workers (NASW) (1999) ethical standards go one step further by expanding the social workers' responsibility to the international level stating that "[s]ocial workers should promote the general welfare of society, from local to global levels, and the development of people, their communities, and their environments" (p. 26).

Unfortunately, the human services profession has gradually moved away from its original call to community action, turning instead to a model of individualized care (Mizrahi, 2001). This is likely due to an increased focus on the increasing popularity of individual psychotherapies within all the mental health professions in the 20th century. This doesn't mean that macro practice or social advocacy has ceased. Rather, as those in the human services fields have pulled away from community work, other disciplines have moved in to fill the vacuum, such as urban and public planners and those in the political sciences. This pattern has resulted in the human services profession often being out of the loop of community building and organizing efforts (Johnson, 2004). Concerns have also been expressed regarding the trend of neglecting the subject of macro and community practice in human services and social work educational programs, thus compounding the tendency for human service professionals to avoid macro practice because many recent graduates feel ill equipped to enter into social advocacy or policy practice on an organizational level (Polack, 2004).

This movement away from macro practice is apparently an international trend as well because studies generated outside the United States have made some similar observations. For instance, Weiss (2003) cited examples of how many human service professionals in Israel do not feel competent addressing social issues on a community or global level because the majority of their training focused on practice with individual clients. Weiss encourages those in the human services professions both in Israel and abroad to reengage in policy-related activities and social advocacy on a macro level.

The reality is that social issues such as poverty and human exploitation must be addressed through advocacy efforts for social change on a macro level as well as a micro level to create much-needed structural changes. Influencing changes in social policy that affects public aid (such as welfare reform legislation), mental healthcare (such as mental health parity laws), and even domestic violence issues (such as policies that mandate cooperation between criminal justice agencies and battered women's shelters) are an integral aspect of human services that directly affect clients' daily lives.

At-risk and Oppressed Populations

Before beginning any discussion on social advocacy efforts it is important to identify populations that are often the target of social injustice, oppression, and human rights violations. It is challenging to comprise a comprehensive list of at-risk populations because there is some shifting in oppressed people from era to era. For instance, Chapter 5 discussed how children although still quite vulnerable are no longer considered an oppressed group in the same way that they were around the turn of the century when poverty and harsh economic conditions led to thousands of children flooding the streets of New York, leading to a significant reduction in sympathy toward orphaned children.

In essence, an at-risk population can include any group of individuals who are vulnerable to exploitation due to lifestyle, lack of political power, lack of financial resources, and lack of societal advocacy and support. Currently, at-risk and oppressed populations include ethnic minorities, immigrants (particularly those who do not speak English), indigenous people, older adults, women, children in foster care, prisoners, the poor, the homeless, single parents, lesbians, gays, bisexual transgendered individuals, members of a religious minority, and the physically and intellectually disabled. In addition, in many regions of the world certain groups of individuals are selected and oppressed due to their ethnic background, religious heritage, and caste (their level of status within society, which in many regions of the world is a level one is born into), and although these individuals may not be in the minority as far as numbers, they typically have little to no political power and are subject to mistreatment and exploitation.

> An at-risk population can include any group of individuals who are vulnerable to exploitation due to lifestyle, lack of political power, lack of financial resources, and lack of societal advocacy and support.

At-risk populations often share unique characteristics not shared by others within a particular culture (within mainstream population and/or those in the majority) (Brownridge, 2009), and it is this uniqueness that can often increase their risk of oppression, discrimination, injustice, and exploitation. At-risk populations are thus *at greater risk* of experiencing a variety of social problems than other populations within the mainstream of society, which undoubtedly then affect the broader population (even if those in power do not believe so).

Vulnerability increases with what is called *intersectionality*—where an individual possesses more than one social and cultural vulnerable characteristic, leading to increased risk of disadvantage. The concept of intersectionality was originally applied to race and gender; the concept is now applied to a variety of marginalizing categories in addition to gender and race, such as level of disability, sexuality, socioeconomic status, social class, immigration status, nationality, and family status (Knudsen, 2005; Meyer, 2002; Samuels, 2008). An example of intersectionality of vulnerability would be an African American older lesbian who is economically disadvantaged, physically disabled, and struggling with homelessness. This profile reveals a woman who experiences multiple forms of vulnerability to injustice on a variety of levels, likely needing various types of advocacy (Martin, in press).

Social forces can combine as well, increasing the risk of discrimination, prejudice, oppression, and injustice. For instance, social conditions such as white privilege (advantage experienced by Caucasians to varying degrees), nativism (a bias against foreign-born residents or those who are perceived as threats to a country's nationalism), xenophobia (an irrational fear of immigrants and foreigners), and other forms of prejudice often combine to increase a group's vulnerability to oppression, marginalization, and exploitation (Martin, in press). Within the human services field there is a recognition that at-risk populations often need advocacy because many of the challenges that lie before them are created within society through policies, laws, and attitudes that create an "uneven playing field", where some groups enjoy greater access to benefits (often referred to as "privilege") whereas other groups are systematically excluded from such societal benefits.

Human Systems

Understanding and Mastery of Human Systems: An understanding of capacities, limitations, and resiliency of human systems

Critical Thinking Question: "Intersectionality" refers to the combined influence on an individual of two or more characteristics that place her/him in an "at-risk" population: for example, being Native American AND being female. How can human service professionals use their understanding of the concept of intersectionality to guide their treatment of, and advocacy for, such clients?

A Human Rights Framework: Inalienable Rights for All Human Beings

Before human service professionals can effectively engage in work on a macro level, whether doing community organizing or more direct social justice advocacy on behalf of at-risk and oppressed populations, they must first become aware of what a just society looks like. What is an ideal society? At the root of any discussion of an ideal society is the assumption that all human beings have inalienable rights simply because they are human. Yet, history is replete with examples of egregious human rights violations, often waged in the belief that such actions are justified on some level. Slavery, a caste system that deems one group of people more worthy than another, a patriarchal system that subjugates females within society, the genocide or "ethnic cleansing" of a particular cultural group, and the sale and exploitation of women and children are all examples of the gross mistreatment of individuals, often because there is some defining characteristic about these individuals that makes them different from another group. Such differences are often used to justify their mistreatment, where members of a more powerful group place themselves above the members of a more vulnerable group. Members of a just society recognize that no one group should have oppressive power over another, and that all human beings have basic rights that must be protected. Since some groups of individuals are more vulnerable than others, human service professionals working in macro practice, particularly on an international level, take responsibility for being the voice of the voiceless (Martin, in press).

In the next few sections I will explore some ways in which human service professionals engage in practice on a macro level, including community development, community organizing, and policy practice. These areas of macro practice are quite general, and you'll likely notice that there is quite a bit of overlap between each of these areas, but gaining at least a cursory understanding of the different types of macro practice is important so that you can better understand how human service professionals' work goes from identifying social problems within society to finding ways of effectively addressing them.

Mobilizing for Change: Shared Goals of Effective Macro Practice Techniques

Macro practice is a multidisciplinary field shared by those in the human services, social sciences, political sciences, and urban planning disciplines. Within the general field of macro practice, models have been developed to frame the various ways of approaching social concerns on a broad level. Although there is a very broad range of theories and models of macro or community practice, most models have at their core the basic goal of societally based social transformation where a community on any level (local, national, or global) incorporates values that reflect the human dignity and worth of *all* its members.

Within most macro practice models empowerment strategies are used that focus on social and economic development, creating liaisons between community members and community organizations, political and social action, which will likely involve advocating for policy changes that address injustices and inequalities within society (Netting, Kettner, & McMurtry, 2009). Various aspects of macro practice will vary depending on the area of concern and the vulnerable population being targeted, but virtually all models of macro practice include a focus on community development, which can refer to the development of a geographic community, such as a neighborhood or city, or a community of individuals, such as women, immigrants, or children.

Common Aspects of Macro Practice

COMMUNITY DEVELOPMENT   Community development dates back to the settlement house movement when Jane Addams and her colleagues worked with politicians, various community organizations, political activists, and community members to create a better community for all members. Addams was personally concerned with child labor, compulsory education, rights of immigrants, and voting rights for woman (women's suffrage). By engaging residents, community leaders, local politicians, and other community organizations, Addams was able to develop a sense of community cohesion, which resulted in several laws being passed that benefited the members of her community, including those who resided in the settlement houses.

Community development in Addams's day is similar in many respects to today, where effective community building depends on the participation of community organizations and community members working together to address issues that are of concern to the entire community (Austin, 2005). The actual issues involved could be anything from addressing crime in the community to educational concerns such as low state test scores, developing an after-school program to combat juvenile delinquency, bringing new businesses to the community to create jobs for community members, or rallying community leaders to develop more open spaces, including parks in densely populated neighborhoods.

A community development approach is empowering because the mutual collaboration of several agencies and area organizations provides support for community members in ways not possible through human service agencies alone. Another empowering aspect of community development is that the collaboration process can create a sense of collective self-sufficiency that often leads to civic pride for community members. In fact, effective community development is based on the conviction that any community is capable of mobilizing "economic, social, and political resources to support families" (Austin, 2005, p. 109).

There are several necessary components of successful community development, including diversity among group members, a sense of shared values among members, positive and collaborative teamwork, good communication, equal participation of all team members, and a good network of connections outside the community (Gardener, 1994). Good community development also depends on the ability to secure enough funding to support group members' activities and efforts. Good networking skills are also essential as are good technology skills because so much of networking in contemporary society is accomplished through email and other technological means (Austin, 2005; Weil, 1996).

COMMUNITY ORGANIZING   Community development depends on the efforts of community organizing efforts, which in turn depends on the efforts of community organizers. The first step in community organizing is to create a consensus on what the community needs, in particular what negative issues the community is facing or areas of needed improvement. Once community members agree on the problems to be addressed, community organizers set about to recruit members to join in the effort to create change. It is important to once again note that the term *community* does not necessarily refer to a geographic community, but might also refer to a community of people, such as women, victims of domestic violence, prisoners, or foster care children.

Community organizers can be professional policy makers or licensed social workers, or they can be individual people with a particular passion and calling for social action. A schoolteacher who gets a group of his students together to remove graffiti from public buildings is a community organizer. The single mother of three who organizes a voluntary after-school tutoring program for the kids in her neighborhood is a community organizer. The father of a child victim of sexual abuse who organizes a campaign to increase prison time for sexual offenders is a community organizer. The licensed social worker whose agency is hired to canvas a neighborhood in an antidrug educational campaign is a community organizer.

Community organizing efforts usually begin around a problem or concern of many people in a community. Once a problem has been identified, community organizers must conduct research to define the issues, understanding how the problem or issue developed and what if any forces exist to keep the problem in place. For instance, the community activist who is organizing efforts to increase the labor rights of undocumented immigrants will likely encounter opposition from factory owners who benefit by paying untaxed low wages to undocumented workers. Thoroughly researching this issue will enable community organizers to identify constituents in the community who will support their cause as well as those who will oppose it. Research will also enable community organizers to identify additional harm done by unfair labor practices not initially identified that might increase the strength of any collating forces.

Once the problem has been identified and research has been conducted, a plan of action must be determined based on the research conducted. Community organizers might decide to picket factories where they perceive abuse of undocumented workers; they might decide to distribute press releases and have a press conference to gain media involvement, organize a work walkout, or conduct a letter-writing campaign to local political leaders. Successful community organizers also organize fund-raising efforts to support their social activism. Sources of fund-raising can include a number of strategies including a direct request for donations, auctions, fund-raising dinners, membership fees, or government grants.

POLICY PRACTICE   Policy practice is a more narrow form of community practice where the human service professional works within the political system to influence government policy and legislation on a local, state, federal, or even global level. The form that policy practice takes depends in large part on the issues at hand, but certain activities in policy practice are consistent despite the issue. This is a relatively new field within human services, with few researchers focusing on policy practice prior to the 1980s. It remains an often neglected area of practice, both within human services and social work education and within human services practice setting. One reason for this may be that effective policy practice relies on a broad range of skills that reaches far beyond the clinical realm (Rocha & Johnson, 1997).

Policy activities center on either reforming current social policy or initiating the development of new policy that addresses the needs of the underserved and marginalized members of society with the primary goal of social justice through social action and advocacy. Policy practice is based on the belief that many problems in society, such as poverty, are structural in nature and can be addressed through making structural changes within society (Weiss, 2003).

Although various approaches to policy practice have been defined within academic literature, Iatridis (1995) has defined several skills necessary for effectively integrating social policy practice into direct service or micro practice. The first skill involves the human service professionals' ability to understand the nature of social policy, including what it is, how it is developed, its influences and effect on society, as well as how social welfare policies are most often implemented. The second skill involves the ability and willingness to view direct practice from a systems perspective, where individual practice is seen as a part of a greater whole. In other words, human service professionals engaged in policy practice must be able to link issues confronted in direct service to structural problems in society (i.e., institutionalized racism, laws that oppress certain groups) by using a P-I-E paradigm (Person-in-Environment), a concept addressed throughout this text relating to the importance of viewing social issues such as poverty on a societal as well as an individual level. Another equally important skill involves the human service professionals' commitment to improving social justice within society by working toward a more equitable distribution of the community's resources.

Those who engage in policy analysis research various social issues in an attempt to determine the short- and long-term effect of new policies and legislation. Policy activists and analysts might focus their attention broadly on social injustices in general, or they may focus on more narrow issues such as the quality of mental health delivery systems, or the focus may be extremely narrow such as the social injustices confronted by those seeking mental healthcare. Human service professionals engaging in policy practice must be able to identify key trends and issues, as well as become familiar with legislation or pending legislation that will affect the area of concern. Let's assume you are involved in policy practice working for an agency concerned with the older adult population. The federal administration's policies regarding Social Security funding would be a matter of great concern to you. Yet if you were involved with policy practice advocating for the rights of the children of undocumented immigrants, you'd be very concerned about possible legislation that would prohibit these children from attending public school. Regardless of the area of concern, policy analysts must be able to identify the "ripple effect" of new policies and legislation to identify their potential harm or benefit to their target population as well as the entire community.

Human Services Delivery Systems

Understanding and Mastery of Human Services Delivery Systems: Skills to effect and influence social policy

Critical Thinking Question: How does a human service professional engaged in policy-related advocacy use the Person-in-Environment (PIE) perspective to guide her/his formulation of social problems and their solutions?

The Global Community: International Human Services

The world is getting smaller, not in terms of population, of course, but in terms of globalization—the increase in international connectedness among all countries and, consequently, all people. No longer are countries completely isolated either in their financial economy or political climate. In the world's new globalization, each country is connected to every other country through increased ease in communication, the development of a global economy (international financial interdependence, mutual trade, and financial influence), and increased international migration, combining to create a situation where the political state of one country influences the economic and political climate of another (Ahmadi, 2003).

Although many consider the term *globalization* to refer solely to matters of economics where businesses can sell goods and trade services as if there were no geographic borders, it also reflects the increased awareness, communication, and cooperation among social advocates. In fact, social reform on a global level is more possible now than ever before. Consider the impact the Internet has had on the exchange of information between relatively remote communities and on regions wrought with oppression. Although limits can be placed on information exchange, the Internet has made global awareness of social issues as easy as pressing a few buttons. Of course that is a somewhat simplistic statement, but the importance of the Internet cannot be underscored both in regard to direct communication and in regard to global awareness of social issues through website publication. For instance, Amnesty International (www.amnesty.org) includes a comprehensive list of human rights abuses and concerns occurring throughout the world. Within this website, individuals can obtain detailed information on the types of abuses currently occurring throughout the world, as well as instructions on how to take steps to assist in the global campaign to stop such oppression and abuse.

This increased ease in global communication has meant that human service professionals in one part of the world can quickly communicate with human service professionals in another part of the world, sharing valuable information and coordinating efforts and services. In fact, there are several international organizations that exist for this very purpose. The International Federation of Social Workers (IFSW) is an international organization founded in 1956 that works with other international human services and human rights organizations to encourage international cooperation and communication among human service professionals around the globe. The IFSW has members from 80 different countries throughout the world, including countries in Africa, Asia, Europe, Latin America, and North America.

The International Association of Schools of Social Work (IASSW) is a support organization and information clearinghouse that works to "develop and promote excellence in social work education, research and scholarship globally in order to enhance human well being" (www.iassw-aiets.org). The IASSW also supports an exchange of information and expertise between social work educational programs.

The International Council on Social Welfare (ICSW) is an independent organization founded in 1928 in Paris, which is committed to social development and works with the United Nations (UN) on matters related to social development, social welfare, and social justice throughout the world. The work of the ICSW is an excellent example of community development at work using networking and international liaisons with other organizations to achieve its goals. The ICSW mission captures the way in which macro practice occurs through a comprehensive network of agencies and organizations on all levels of society to achieve the global mission of eliminating social injustice (refer to paragraph 3 at http://icsw.org/intro/missione.htm).

Even professional counselors whose training has traditionally leaned more in the direction of clinical practice have recently been encouraged to venture into global matter by advocating for social justice. Chi-Ying Chung (2005) made several recommendations to professional counselors to get involved in international human rights work, suggesting that they apply their training in multicultural counseling and competencies to the international arena to combat human rights abuses.

Although the human services profession exists worldwide, and concerns about specific social issues such as violence and children's rights are shared among all countries, the nature of the social issues and the function and role of the human service professional will vary depending on the political and economic conditions unique to each country. Human service professionals around the globe have many shared values but have differences in values as well. For instance, in the United States, self-determination is very highly valued in all the human services, particularly the social work profession, but not only is self-determination not considered a core value of the profession in other countries, in Asia, Africa, and even Denmark the concept of self-determination is considered either unimportant or dangerous as it detracts from the value of community and cooperation (Weiss, 2005).

Overall, though, human service professionals in virtually every country place a high value on the protection of human rights, social justice, and the end to human oppression in whatever form it might be taking within that particular region. For instance, a primary concern of the human service professionals in South Africa relates to issues of race emanating from its former system of apartheid. School social workers are commonly used to teach positive race relations among the students in South African public schools. Race issues take on a different form in the United States related to its history of slavery and mass immigration.

HIV/AIDS Pandemic

AIDS, a life-threatening disease found disproportionately in sub-Saharan Africa, has had a devastating effect on families, particularly children. The life expectancy in many African countries has dropped from 61 to 35 years of age, and has had a profound effect on children and the quality of their childhoods. For instance, as of 2007, of the approximately 17 million children estimated to have been orphaned by the AIDS epidemic, approximately 15 million live in Sub-Saharan Africa UNAIDS, 2008; UNICEF, 2012). This represents an increase over prior years despite the fact that adult HIV-infection rates have declined in recent years, and use of antiviral medications have become increasingly available, particularly in several sub-Saharan African countries (UNAIDS, 2008). In Zimbabwe alone United Nations Children's Fund (UNICEF, 2004) estimates that 30 percent of all children have been orphaned due to AIDS. Many developing countries have neither the funding nor the capacity to place child welfare issues as a priority (Dhlembeu & Mayanga, 2006). Women bear the primary burden of this disease with regard to both stigma and the brunt of caregiving, despite the fact that they are being infected at far higher rates than men (Joint United Nations Programme on HIV/AIDS, 2004).

Human service professionals working in the highest risk countries in sub-Saharan Africa, including Ethiopia, Nigeria, South Africa, Zambia, and Zimbabwe, must contend with the devastating impact of the HIV, including the very complicated and far-reaching implications of so many children being orphaned as a result of the death of one or both of their parents due to AIDS. This situation is further complicated by the fact that many of the child welfare agencies in these countries (if they even exist) are ill equipped to handle the vast number of orphans, many of whom are not being well cared for and may be infected with the HIV virus as well.

In many countries in Africa as well as other regions, traditional beliefs and stigmas exist which are counterproductive to HIV/AIDS treatment protocol compliance. But even in situations where a country is highly compliant with international health-care protocols, such as the case of Rwanda, the management of the AIDS pandemic is extremely complex and presents numerous challenges to human service professionals. For instance, in Rwanda, thousands of women were infected by HIV/AIDS by Hutu, the genocidal government's Interahamwe militia who raped the majority of women during the genocide. Those women who were not then cut down by machetes, learned months or years later that they were infected with HIV (Des Forges, 1999). Thus in the Rwandan context, an entirely new generation of orphans was created due to conditions directly linked to the 1994 genocide. Further, many of these orphans are HIV-positive as well. The agency WeActx in Kigali works with HIV-infected women and their children, providing them with both healthcare and trauma services. The director of this agency recently shared that a significant concern among the youth population being served by this agency relates not only to their daily provision and education needs, but also to the common refusal of many of the youth to adhere to the AIDS treatment protocol because they are in "denial" that they have this disease. Their HIV status is yet another ongoing reminder of the genocide, which has affected and will continue to affect the Rwandan population, particularly Tutsi survivors, for generations to come (it is important to note that many Hutu women were raped during the genocide as well, and were also infected with HIV, which is why the WeActx agency does not restrict its services to only Tutsi genocide survivors, but to Hutu women as well).

Several human services agencies exist solely to care for these orphaned children. Other agencies focus their efforts on education and testing. This public health crisis has far-reaching implications that must be addressed internationally if there is going to be any real remedy that will positively affect the lives of those infected and those at risk of infection.

Crimes Against Women and Children

Crimes against women and children are of concern to countries throughout the world, and human service professionals, including social workers, psychologists, and professional counselors as well as human rights workers are involved in advocacy, counseling, and political activism on all levels to create international awareness and social action to put a stop to atrocities such as government-sanctioned honor killings, punitive sexual assaults, exploitation and harassment, and discrimination that strips women and children of their basic human rights.

FEMALE GENITAL MUTILATION   Another issue often confronting human service professionals in all of Africa involves female genital mutilation (FGM), or "female circumcision," where historical tradition and tribal culture prescribes that a girl's external genitalia, typically including her labia and clitoris, be cut away in a rite of passage ceremony celebrating her entry into her womanhood. The most serious type of FGM is Type 3, which includes the cutting away of the labia minora and the sewing together of the labia majora (the outer vaginal lips), which then creates a seal with only a small opening for the passing of menstrual blood and urine. The vaginal seal is intended to keep the women in the tribe from having sexual relations before marriage. It is literally torn open during the woman's first sexual encounter with her husband, which not only causes extreme pain, but also has serious health consequences such as bleeding and possible infection. In some cultures the torn pieces of labia are actually sewn together again if the woman becomes pregnant and are then torn open again during childbirth.

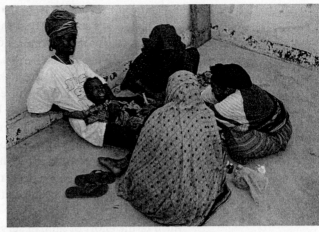

Young girl endures female genital mutilation in Somalia. (Source: http:// www.global-sisterhood-network.org/content/view/1470/59/) Jean-Marc Bouju/Impact/HIP/The Image Works

It is estimated that nearly 100 to 130 million girls have undergone FGM, which can cause serious health risks including lifelong pain, infertility, and death (World Health Organization, 1998). FGM is rarely performed by a physician, but is frequently conducted by a village leader with no pain medication. Girls are often tied down and subjected to this surgery, which is intended to ensure chastity and purity. There has been a recent backlash among women in some African countries who are discouraging FGM in their communities, although this practice is still quite prevalent in many rural regions. Human service professionals are conducting educational campaigns to influence local leaders who have the power to discourage this practice, as well as influencing many Western countries to add those escaping FGM to qualify for refugee status.

HUMAN SEX TRAFFICKING   Human service professionals in many Asian countries must contend with numerous human rights violations, the most prevalent and disturbing of which includes the human trafficking of women and children for the purposes of slavery, forced marriage, and the sex trade. For instance, according to the Human Rights Watch (HRW, 2002), approximately 10,000 women and girls are "recruited" from Burma to brothels in Thailand each year. The most recent U.S. Department of State (2010) Trafficking in Persons report states that government corruption and the involvement of public officials in the human trafficking trade makes matters even more challenging for human rights workers who are attempting to achieve social justice for these women and children.

As of 2011 there were approximately 12.3 million individuals who were victims of human trafficking worldwide, the majority of whom were young females, the majority

As of 2011 there were approximately 12.3 million individuals who were victims of human trafficking worldwide, the majority of whom were young females trafficked for sexual purposes.

of whom were trafficked for sexual purposes (U.S. Department of State, 2012). In fact, young girls are the most sought after targets of large criminal organizations that are in the business of trafficking human beings. Although people can be sold for various reasons, including forced servitude and child labor, the majority of human trafficking involves forced sexual slavery, where young women and girls are forced to become prostitutes. Girls are sold into sex slavery by family members in need of money, are kidnapped, or are lured into the sex trade with promises of modeling contracts or domestic work in other countries. Many of these girls are kept in inhumane environments where they are forced to have sex with up to 10 men a day. Many contract HIV/AIDS and are cast out onto the street once they become too sick with AIDS to be useful (U.S. Department of State, 2012).

Much of the effort of human service professionals in countries with high rates of human trafficking, including India, Burma, Thailand, and Sri Lanka, is focused on rescuing these women and children and ensuring that they are delivered to safe communities where they will not be exploited again. Complicating intervention strategies is the fact that many government officials in these Asian countries either look the other way when confronted with the illegal sex trade or openly contribute to it by protecting criminal organizations responsible for human trafficking. Human rights organizations have reported that many police officers, members of the military, and other government officials in Thailand often arrest victims who attempt to escape, putting them in prison on charges of prostitution, a clear act of retaliation, rather than helping them to escape (HRW, 2004).

STREET CHILDREN   Human service professionals in Central and South American as well as Eastern European countries must contend with the significant problem of thousands of homeless street children roaming the streets in search of food and shelter. The problem of street children is growing around the globe, leading several human rights organizations to call human service professionals to action. Street children are

A child from the "Untouchable" caste in India begging for food
Xander Martin

sometimes orphans, but are often children who have parents but have left home due to poverty or lack of supervision. In many Eastern European countries, including Romania, the problem of street children is a direct result of political policies resulting from families having a large number of children with the promise of government provisions, only to be left in terribly vulnerable positions when these governments failed, leaving parents with no means for providing for their exceptionally large families. Street children are at risk of abuses by older children as well as police and government officials who often physically abuse children as young as 5 years (HRW, 2002). Children have even been murdered by the police with no official response. Drug abuse is also rampant within the street children population, who often sniff glue to keep warm and to abate hunger pains.

Human service professionals have organized agencies that reach out to these children by finding homes for them, either with religious organizations or through international adoption. International human services agencies work with local agencies to bolster aid efforts, including lobbying government officials to address this issue by funding child welfare efforts.

CHILD LABOR AND ECONOMIC INJUSTICE   Child labor is a social justice issue across the globe, but is a particular concern in Asian, African, and Latin American countries, where children as young as 4 years are required to work up to 12 hours per day in jobs that put them in both physical and psychological danger. Child labor abuses include children in India who plunge their hands into boiling water while making silk thread and children as young as 4 years in Asia who are tied to rug looms for many hours a day and forced to make rugs.

Of the 120 million children forced into full-time labor, 61 percent reside in Asia, 32 percent in Africa, and 7 percent in Latin America (HRW, 2004). International human rights organizations such as HRW, Amnesty International, and UNICEF work diligently to protect children's rights, including lobbying of international policies and legislation that protect children as well as funding human rights efforts in specific countries allowing for intervention at the local level. But the problem of child labor, particularly in sweatshops in the Global South (Central and South America, Southeast Asia, India, and the Southern region of Africa), remain a serious problem impacting the entire world both socially and economically.

For instance, Polack (2004) discussed the impact of hundreds of billions of dollars in loans made to countries in the Global South by countries in the North (England, Spain, France, the United States, etc.). Polack argued that the cumulative impact of these loans to some of the poorest countries in the world has been devastating to the poorest members of these countries because these loans (1) financed large-scale projects, such as hydroelectric plants, that either benefited the North or displaced literally millions of people, pushing them even further into poverty, (2) financed military armaments for government regimes that oppressed the countries' most vulnerable and poorest residents, or (3) lined the pockets of corrupt leaders of many countries in the Global South, resulting in increased oppression of the country's least-privileged members.

Very little if any of this loan money has benefited the majority of the citizens of these countries; rather, it has harmed them and in fact continues to harm them by increasing the poverty within these already devastatingly poor regions. In an attempt to repay this debt many countries of the Global South exploit their own

Teenage boys working in the ship breaking yards in Bangladesh
Xander Martin

workers to make loan payments. For example, countries in South America have sold sections of rain forest formerly farmed by local residents to Northern timber companies, and other countries have been forced to privatize and then sell utility services formerly provided by the government, resulting in dramatic increases in the cost of utilities. These developments have resulted in many Northern companies making millions of dollars literally at the expense of the poorest residents of these debt-ridden countries.

One of the most devastating impacts of what has now evolved into trillions of dollars of debt for these Southern countries is the evolution of the sweatshop industry, large-scale factories that develop goods exported to the North. Some of the poorest people in the world, including children, work in sweatshops throughout Asia, India, and Southern Africa, where horrific abuses abound. This occurs legally in many of these countries because in a desperate attempt to attract export contracts, many countries in Asia, including India, created "free-trade" agreements or free-trade zones for Western corporations, allowing them to circumvent local trade regulations, such as minimum wage, working hour limits, and child labor laws, if they would open factories in their impoverished countries.

Polack (2004) suggests that literally every major retail supplier in the United States benefits from these sweatshop conditions such as extremely low wages, extremely poor working conditions, physical and sexual exploitation without retribution, excessively long working hours (sometimes in excess of 12 hours per day with no days off for weeks at a time), and severe retribution such as immediate termination for complaints or requests for better working conditions. Child labor is the norm in these sweatshops with most sweatshop owners preferring adolescent girls as employees because they tend to be more compliant and are more easily exploited.

Although local and international human rights advocates work diligently to change these working conditions, at the root of the problem of child exploitation is economic injustice rooted in generations of intercountry exploitation. Thus, there is significant complexity not easily confronted without government involvement, which is often slow in coming when large corporations are making millions of dollars with the system as it currently operates. For instance, as labor unions have become the norm in the United States, many companies such as Nike and Wal-Mart moved their factories to Asia and Central and South America, where millions of dollars can be saved in wages and benefits cuts (National Labor Committee, n.d.). Addressing the issue of child labor and economic injustice will take the lobbying efforts of many international human rights organizations working with the media to create public awareness where buying power is often the only tool powerful enough to influence sweatshop owners and large retail establishments.

CASE STUDY 15.1

Testimony of Mahamuda Akter, MNC Garment Factory, September 2002

My name is Mahamuda Akter. I am 18 years old. I've only had the chance to go through fifth grade. I was 13 when I began working in the garment factories. For the last two years I have been working at the MNC factory in the Chittagong Export Processing Zone, where we sew clothing for Wal-Mart. I am a sewing operator.

Until September 5, we were working on Ozark Trail shirts. Before that—for six or seven months—we worked constantly on Sportrax athletic clothing. Now we are sewing Faded Glory shorts. Depending upon the type of garment we are working on, my job is to join the collar, or to sew either the pocket or the hem of the sleeves. Attaching the collars is very complicated since you must match the patterns of the fabric. The supervisors scream at us to do 40 pieces an hour. But it's impossible. Working as fast as we can, I can only finish 30 collars in an hour.

The supervisors tell us we have to meet Wal-Mart's target. There is constant pressure on us to work faster. They beat us. They slap our faces or slap us on the back of the head. They grab us by the hair and jerk our heads. They push and shove us.

I was beaten several times in August and September. My supervisor, who is a man, slapped my face and cursed at me that I was a son of a bitch and that my parents were whores. They use vulgar and filthy words, they made me cry. Many of us girls cry, but they make you keep working.

I work on Line "D." In July, the supervisors kicked one of the girls on our line, yelling that she had made a mistake. They threw her against the wall and her mouth was bleeding. They took her to the office and fired her that afternoon.

Another thing they do as punishment is to make a girl stand on a bench in front of all the other workers, forcing her to hold her ears and pull them down. It's a shameful insult. They do this especially to the young girls and it makes them feel terrible.

There are 4,000 workers in our factory. Eighty-five percent of us are women. We have lots of helpers who are 10 to 12 years old.

Our regular work schedule is from 7:30 A.M. to 10:00 P.M. But they often force us to work until 3:00 A.M. In August, I had to work 13 nights till 3:00 A.M. In other sections it was even worse, and they had to work 20–25 nights to 3:00 in the morning. We work seven days a week. In August we had just one day off. For the year, I think I got a total of 15 days off.

When we work through to 3:00 A.M., we get three breaks, a half hour for lunch from 1:00 to 1:30 P.M.; 10 minutes from 7:00 to 7:10 P.M., and an hour off for supper from 11:00 to midnight. After the 3:00 A.M. shift, we sleep in the factory. It is so crowded that we sleep sitting on our benches slumped over our sewing machines. There is no place to even lie down on the floor. At 5:00 A.M. they ring a loud bell to wake everyone up, so we can get ready to start work again. We wash our faces, use the bathroom, eat something and go back to work. Sometimes we are forced to do these 19½-hour shifts three days in a row.

We are exhausted. Many times the workers faint. The supervisors throw water on their faces and they have to get back to work. They also play loud music to keep us awake.

I earn 2,100 taka a month, which is $35.60. I'm told this comes to 17 cents an hour.

We are not allowed to talk at work, and if we are caught we are punished. You need permission to use the bathroom. When we work until 3:00 in the morning for example, we can use the bathroom just three times in the entire shift.

We have a daycare center at the factory, but it is a joke. It is just for show to the buyers. It is never really used.

We are not allowed sick days, or national holidays, or any vacation.

They also cheat us on our overtime wages. They keep two sets of time cards. The phony one is for Wal-Mart. It says that we work just from 7:30 A.M. to 6:30 P.M., in other words, that we work two hours of overtime a day. It also says that we receive every Friday off. That's a lie.

None of us have ever heard of the Wal-Mart Code of Conduct. Before the Wal-Mart buyers come to the factory, the factory is always cleaned. The supervisors tell us to lie if

Understanding and Mastery of Human Services Delivery Systems: International and global influences on service delivery

Critical Thinking Question: The ethical codes of organizations such as the NASW call on human service professionals to advocate for social and economic justice around the world. What responsibilities do human service workers in the United States have to children like Mahamuda Akter?

the buyers ever question us—we are supposed to say that we work just to 6:30 P.M. and that we have one day off a week. The buyers always walk around with the manager. Everyone is so frightened, no one dares complain. Sometimes the buyers ask us to smile and they take a picture. They usually come around 1:00 or 2:00 in the afternoon. They never come at 10 P.M. or 3:00 A.M.

I live in one room with three other girls who are coworkers. We must pay 1,150 taka rent each month. We cannot even afford a fan or a TV. We share one water pump, an outhouse, and one gas stove with 20 other people.

Every day we eat rice, rice with lentils or with mashed potatoes. Sometimes we have an egg at night. I'm always hungry. I weigh 79 pounds. Maybe once in a month we can eat beef.

We work so hard, but it is not right that they mistreat us so and pay us so very little.

I am afraid of getting old. Living and working like this, by the time you are 20 you are already old, and your health is failing. When you reach 30, they fire you. It is not just. I have no savings. I have nothing.

I would like a better life for myself and the other girls.

Source: Institute for Global Labour & Human Rights (formerly National Labour Committee)

Indigenous populations are often forced to live in the midst of environmental degradation
Xander Martin

Indigenous People

Protecting the rights of indigenous people is a common concern of human service professionals practicing in countries such as the United States, Australia, and many Central and South American countries. Indigenous populations are often forced to engage in harsh and dangerous labor practices, such as working in fields sprayed with insecticides, transporting supplies on their person, or begging, in order to survive.

The human rights issues pertaining to indigenous peoples of Australia, primarily comprised of Aborigines, are similar in nature to those in the United States, where the historic immigration of Europeans displaced the indigenous tribal communities. In addition, both countries engaged in an official campaign of discrimination and cultural annihilation as indigenous tribes were forced off their lands and onto restricted areas, where they were unable to practice traditional methods of self-support. Both Native Americans in the United States and Aborigines in Australia were subject to the mass forced removal of children, who were mandated to attend schools where they were forced to abandon their cultural heritage and native language.

The 36-year civil war in Guatemala, which ended in 1996, involved what many human rights organizations consider the genocide of indigenous populations, or what is commonly referred to as the "disappearance" of indigenous populations. The UN Truth and Reconciliation Committee estimates that up to 200,000 people were killed by government forces (HRW, 2008).

In response to the intergenerational trauma that has resulted from physical and cultural genocide, many indigenous people have experienced a decimation of their population as well as extreme poverty, forced migration, and marginalization often manifesting in physical and mental health problems. Human service professionals work with indigenous people in reconciliation efforts to restore them to a level of self-sufficiency and cultural pride. Several movements are underway within indigenous tribal communities intended to move them toward wholeness and a life without substance abuse, depression, and the brokenness in families that has so often been the result of social ills.

One program within a Native American community was developed by a tribal member who suffered from alcoholism for years and who received inspiration and input from tribal elders who shared wisdom regarding traditional cultural laws for authentic change. The four laws of change became known as the *Healing Forest Model,* which is based on the philosophy of the Medicine Wheel, a Native American concept that addresses the interconnectedness of everything in life. According to the teachings of the Medicine Wheel, the pain of one person creates pain for the entire community, thus there are no individual issues or concerns. This community concept of healing is very consistent with a model of macro practice, which posits that there are no such things as individual problems but instead people make up communities and therefore all individual problems become community problems. This philosophy may be counterintuitive to North Americans, who as a society place an exceedingly high value on individuality, oftentimes at the cost of community. Yet many believe that the key to reclaiming physical and mental health in indigenous culture is through such a community practice approach (Coyhis & Simonelli, 2005).

Refugees

According to the Office of the United Nations High Commissioner for Refugees (UNHCR) there are approximately 42 million displaced people who have been forcibly removed from their homes and communities due to civil war, conflict, political and cultural persecution, natural disaster, ethnic cleansing, and genocide.

The Immigration and Nationality Act defines "refugee" as:

(A) any person who is outside any country of such person's nationality or, in the case of a person having no nationality, is outside any country in which such person last habitually resided, and who is unable or unwilling to return to, and is unable or unwilling to avail himself or herself of the protection of, that country because of persecution or a well-founded fear of persecution on account of race, religion, nationality, membership in a particular social group, or political opinion, or (B) in such circumstances as the President after appropriate consultation (as defined in section 207(e) of this Act) may specify, any person who is within the country of such person's nationality or, in the case of a person having no nationality, within the country in which such person is habitually residing, and who is persecuted or who has a well-founded fear of persecution on account of race, religion, nationality, membership in a particular social group, or political opinion. (Sec. 101(a)(42))

Individuals may become refugees through a variety of circumstances. In the last two decades there have been between 17 and 33 armed civil conflicts at any one time, leading to civil unrest and instability in several developing countries. In the midst of a civil war innocent civilians are often forced to flee in search of safety, a phenomenon referred to as *forced migration.* If civilians flee but do not cross international boundaries, they are referred to as internally displaced persons (IDPs), but if they are forced to flee into another country, then they often receive the legal designation of refugee. Refugees may live in secret, in a country with closed borders, thus are considered by the host country as illegal immigrants. Life as an illegal immigrant is lived on the fringes, in constant fear of detection, detainment, and repatriation. In other situations, refugees are warehoused in refugee settlements or camps. Most refugee camps are managed by the UNHCR, and despite such management, they remain a place of great risk and despair. In many refugee camps refugees are not allowed to leave and are often considered a serious risk to the host country. Most refugee camps are established in "border" regions and may remain in close proximity to the war that caused the displacement in the first place. The majority of refugees in protracted situations develop a sense of significant despair as their situation lingers on for generations, as with the Burundi, who have been in refugee camps in Tanzania since the early 1970s. Those refugees fortunate enough to be selected for resettlement in the United States often face years of challenges as they struggle to survive in a complex society, often underemployed and socially isolated (Hollenbach, 2008; Loescher, Milner, & Troeller, 2008). Human service professionals often work with refugees in a variety of practice settings, including refugee resettlement agencies (contracted with the U.S. Department of State), schools, and mental health agencies. Macro practice involves advocacy and policy practice effecting changes in policies that create additional challenges to an already immensely vulnerable and traumatized population.

Refugee communities, also referred to as diaspora, should not be considered powerless victims without personal agency though as many come together to form quite powerful lobby groups advocating for their agendas both within their host countries as well as in home country affairs. In fact, recent research has shown that a country in postconflict is at a significantly higher risk of renewed conflict if there is a related diaspora that is politically active and advocating against the home country government (Collier & Hoeffler, 2000; Lyons, 2007). Thus it is vital that human services workers working with diaspora groups be aware of the sociopolitical dynamics related to the history of conflict in the refugees' country of origin so that they can assist the diaspora members to engage in ways that will support peace processes, and not exacerbate old and existing conflicts.

Lesbian, Gay, Bisexual, and Transgendered Rights

Individuals who have nontraditional sexual orientations, including lesbian women, gay men, bisexual men and women, and transgendered individuals (those who have undergone surgery to physically become the opposite gender) have long been the victims of abuse, discrimination, and at the very least a tremendous amount of misunderstanding. *Homophobia* is defined as irrational fear of homosexuals or of homosexual behavior. Lesbian, gay, bisexual, and transgendered (LGBT) individuals are subjected to homophobic sentiments and outright discrimination and violence in all parts of the world. Until recently the majority opinion of those in Western culture was that LGBT individuals were either morally perverse or mentally ill. In fact, it wasn't until 1987 that all references to homosexuality were completely removed from the *Diagnostic and Statistical Manual of Mental Disorders.*

Acts of harassment and violence against LGBT individuals based on their sexual orientation are prevalent all over the world, causing significant distress, depression, and even suicidal ideation (Huebner, Rebchook, & Kegeles, 2004). LGBT youth are at risk of discrimination in school and community settings in both the United States and the United Kingdom, although many school districts now use policies designed to protect adolescents whose sexual orientation are known to others in the school or community (Ryan & Rivers, 2003). LGBT individuals are commonly the victims of direct or subtle discriminatory practices, verbally abused and harassed, and the victims of violence, sometimes even murder, solely because of their sexual orientation.

Although abuse and discrimination against LGBT individuals is assumed to be far worse in developing countries, this is not always the case. In many regions of the world the line between heterosexuality and homosexuality is quite porous, particularly compared to Western cultural norms. This contention is based on the practice of male-on-male sexual activity commonly practiced in many parts of the world when one or both men are married. For instance, in Bangladesh married men often frequent male prostitutes but do not necessarily consider themselves homosexual. They are rarely victims of harassment or abuse because they do not violate gender stereotypes, which essentially means that men continue to act like men and women continue to act like women (Dowsett, 2003). The relevance of this is that in many parts of the world violence against LGBT is based more on behavior that is contrary to traditional gender stereotypes than it is on their sexual activities.

Yet in many regions of the world homosexual behavior is considered a criminal act punishable by anything from a prison sentence to death. Homosexuality is considered illegal in South Africa, and LGBT individuals are often the victims of human rights abuses, including punitive rapes. In addition, they are often unjustly blamed for the HIV/AIDS crisis currently occurring in Africa (Graziano, 2004). LGBT individuals in Saudi Arabia are subject to public floggings and imprisonment for even suspected homosexual behavior. In Egypt vice officers travel through towns in vans arresting in excess of 100 men at a time for suspected

Rally against California Proposition 8 barring gay marriage in New York City 2008.
Tricia Serfas

homosexuality. Many of these men were arrested because they knew what the word *gay* meant, a North American word assumed to be known only by homosexual men. Men arrested on suspected homosexuality are then subject to severe beatings until they agree to sign arrest papers admitting to their homosexuality. Signing these papers means a lifetime of certain harassment and refusing to sign them means certain death. In Jamaica LGBT individuals are often the target of horrible human rights abuse, oftentimes fueled by the police who often invite bystanders to attack men suspected of homosexual behavior. One incident reported to a human rights organization involved a man suspected of being gay who was attacked by police and ultimately beaten and stabbed to death in the middle of the street by bystanders who joined in the beating. Police in Jamaica also commonly stop individuals suspected of being LGBT on the streets, searching them looking for any sign of homosexual activity such as condoms or lubricants. If these items are found, the men are often beaten and arrested (HRW, 2005).

Several countries in Africa, including Uganda and Nigeria, are currently considering antihomosexuality laws that would make homosexual activity illegal and punishable by brutal penalties, including death. What is particularly disturbing about this recent antihomosexuality trend in Eastern Africa are reports that some U.S. Evangelical leaders are behind the effort to criminalize homosexuality, based upon a belief that the "homosexual agenda" threatens the traditional family (Gettleman, 2010). Human rights organizations have expressed outrage in response to the reported link between antigay legislation in Africa and the U.S. Evangelical church for a variety of reasons, chief among them the potential for dictatorships with poor human rights records to use such legislation to silence (either through long-term incarceration or death) anyone who opposes their autocratic rule (HRW, 2009). One might question whether any such organized efforts emanating from any developed country is a form of neocolonialization, reflecting significant ignorance of the history of the region as well as paternalistic attitudes common during colonial rule of African countries. Regardless, such misplaced advocacy has a great possibility of significantly increasing human rights abuses against an already marginalized population.

> **Several countries in Africa, including Uganda and Nigeria, are currently considering antihomosexuality laws that would make homosexual activity illegal and punishable by brutal penalties, including death.**

Human service professionals and human rights workers around the globe are working tirelessly to reduce crimes against LGBT individuals through the passage of policies and legislation designed not only to protect individuals whose sexual orientation is not traditional, but also to decriminalize homosexual behavior in all countries. The recent passage of the Matthew Shepard & James Byrd Jr. Hate Crimes Prevention Act (P.L. 111-84) in the United States, signed into law in October 2009 by President Obama, makes it a federal crime to assault individuals because of their sexual orientation, gender, or gender identity. The passage of this highly contested legislation has been lauded by civil rights organizations as a significant step forward in this fight for equality and protection of the LGBT population (Human Rights Campaign, 2009).

What might be one of the most important issues to consider is that regardless of whether one considers homosexuality a lifestyle choice, a genetically predetermined orientation, a nontraditional sexual orientation no better or worse than heterosexuality, or an act of perversion and immorality, violence against someone based on their sexual orientation is never permissible under any conditions, thus even those human service professionals who because of religious faith or cultural tradition believe that heterosexuality is the only physically and psychologically healthy lifestyle, should be called to action to ensure that *all* individuals, despite their sexual orientation, are treated with compassion and dignity.

Torture and Abuse

Countries in Eastern Europe as well as countries in Northern and Western Africa are overwhelmed with the repercussions of war and genocide where human service professionals and human rights workers deal with numerous human atrocities such as torture, war crimes, and the crisis of thousands of refugees. But the problem of abuse and torture is truly worldwide, and as much as members of industrialized countries would like to believe that human torture is a problem known only to lesser developed countries, the physical and sexual torture of the Iraqi detainees at Abu Ghraib prison is a clear reminder that human torture occurs on all soils at the hands of people from the most "civilized" of countries.

Countries in the midst of war are particularly vulnerable to human rights abuses involving torture because war seems to have a diminishing effect on human compassion and empathy. Human torture and abuse can include anything from random physical abuse to the systematic abuse and even murder of groups of people common in genocide, prisoner of war camps, and refugee camps. Many of the abuses documented in the Taliban-ruled Afghanistan included sexual assault, government-sanctioned gang rapes of women who brought disgrace on their countrymen, and physical torture such as the cutting off of limbs for minor infractions (U.S. Department of State, 2001).

Most if not all victims of wartime atrocities such as rape and torture, many of whom are being revictimized in refugee camps, suffer from post-traumatic stress disorder (PTSD) and other psychiatric conditions related to grief and loss. Human service professionals work with victims of torture on all fronts—some within refugee camps, and some in other countries who have accepted victims on refugee status. The psychological issues involved are vast and in addition to the disorders mentioned earlier include depression, anxiety, and adjustment disorders. Most human service professionals in developing countries and former Soviet bloc countries are employed by the government and deliver broad-ranging services on a community level, focusing on the manifestation of a history of war, as well as the ramifications of transitioning from a communist society to a democracy. For instance, a relatively significant portion of human services in Croatia is focused on postwar issues as well as the care of Bosnian refugees and other war victims, focusing on trauma recovery and helping victims to manage the comprehensive impact of war on the individual and families (Knežević & Butler, 2003).

Throughout the Bush/Cheney administration several advocacy organizations, including Amnesty International, HRW, and the International Red Cross, cited numerous egregious examples of torturing prisoners suspected of involvement in the September 11, 2001, terrorist attacks, or of being a supporter of "enemy combatants." Both former president George W. Bush and former vice president Cheney defended their policy of using "enhanced" interrogation techniques, denying that such practices constituted a violation of the Geneva Convention, a collection of international humanitarian laws that among other remedies provides parameters on how prisoners of war are to be treated.

In 2006 the HRW submitted a report to the Human Rights Committee detailing numerous human rights violations occurring under the Bush/Cheney administration in violation of International Covenant on Civil and Political Rights (ICCPR), including the secret and indefinite detention of prisoners at Guantanamo Bay and at undisclosed locations abroad. According to the report, most of these prisoners have not been charged with any crimes and have thus been denied due process. Other human rights violations include the use of torture as an interrogation technique, such as sleep deprivation, isolation, sexual humiliation, and water boarding (which gives the subject the sensation of drowning). Federal legislation that was enacted in 2005 supported the use of information obtained from torture and also "precludes detainees at Guantanamo Bay from bringing any future challenge to their ongoing detention or conditions of confinement before the courts" (HRW, 2006, p. 7), including torture, and cruel inhuman and degrading treatment. The following case studies were included in an HRW report submitted to the United Nations Human Rights Committee:

> Consider the cases of Kahled el-Masri and Maher Arar. El-Masri, a German citizen, states that he was seized in Macedonia in December 2003 and eventually transferred to a CIA-run prison in Afghanistan where he was beaten and held incommunicado for several months. In May 2004, he was flown to Albania, deposited on an abandoned road, and eventually made his way back to Germany. El-Masri states that one of the detaining officials admitted that his arrest and detention was a mistake. El-Masri filed a suit in U.S. federal court against the former CIA Director George Tenet and the corporations and individuals allegedly involved in his rendition. He alleged violations of his due process rights and the international prohibitions against arbitrary detention and cruel, inhuman, or degrading treatment. The U.S. government, however, moved to dismiss, arguing that discovery in the case would require revealing "state secrets." Despite the fact that the case had been widely reported in the U.S. and international media. [*sic*]The court agreed and on February 16, 2006, dismissed the case. El-Masri plans to appeal the ruling. If he loses, he will have no avenue for seeking relief and compensation for the 5-month period of physical and psychological abuse. Maher Arar, a Canadian citizen, was detained by the United States in September 2002. U.S. immigration authorities held him for two weeks, during which time he was unable to challenge either his detention or imminent transfer to a country likely to torture him. Relying on diplomatic assurances from Syria, the United States then flew Arar to Jordan, where he was driven across the border to Syria and detained there for ten months. Arar reports that he was beaten by security officers in Jordan and tortured repeatedly, often with cables and electrical cords, during his confinement in a Syrian prison. Arar sued former Attorney General John Ashcroft and others involved in his detention and rendition for compensation for the physical and psychological harm suffered in Syria. The United States asserted a national security

privilege. The district court agreed and dismissed the case, reasoning that it could not second-guess the government's claims that the need for secrecy was paramount and that discovery about what happened in the case could have negative impacts on foreign relations and national security. Arar, like el-Masri, is denied a remedy, even though the facts of his case, like in the el-Masri case, are widely reported. In both cases, the U.S. government has shut down any inquiry into practices that appear to violate international prohibitions on *non-refoulement* and use of torture and cruel, inhuman, and degrading treatment. Violations of non-derogable rights cannot and should not be justified or shielded from review on grounds of national security. (OHCHR, 2006, p. 10)

Some of the most egregious policies have been passed during times of crisis when people are scared and willing to sacrifice civil and human rights for the sake of security. Yet as human service professionals we must advocate for human rights in all situations, and resist the temptation to dehumanize any group, which tends to make it far easier to justify such horrendous mistreatment.

Human Services Delivery Systems

Understanding and Mastery of Human Services Delivery Systems: Range of populations served and needs addressed by human services

Critical Thinking Question: There are tens of thousands of torture survivors living in the United States, many of them from Latin America, Africa, Eastern Europe, and the Middle East. In addition to having been tortured, they struggle with issues common to immigrants: adjustment to a new culture, language barriers, loss of family and community support networks, and lack of employment. How can human service professionals best serve the numerous and interconnected needs of this vulnerable population?

Genocide and Rape as a Weapon of War

The 1948 UN Convention on the Prevention and Punishment of Genocide defines genocide as any act committed with the intention to destroy, in whole or in part, a national ethnic, racial or religious group: killing members of the group; causing serious bodily or mental harm to members of the group; deliberately inflicting on the group conditions of life calculated to bring about its physical destruction in whole or in part, imposing measures intended to prevent births within the group, and forcibly transferring children of the group to another group (UN General Assembly, 1948).

Genocides most typically occur within a broader armed civil or international conflict, thus determining whether civilian deaths as a result of a conflict rise to the level of genocide is somewhat political in nature, as is determining that massacres or crimes against humanity do not rise to the level of genocide. Such a determination can be made by any country that is a signatory of the Genocide Convention, as well as by the General Assembly of the United Nations. Yet it is important to note that just because an incident of civilian killings is not deemed "genocide" by the international community does not mean that genocide has

Tutsi Genocide survivor Yvette Nyombayire Rugasaguhunga washing her grandmother, Tereza Kamagaju's, a genocide victim's bones, in a post-genocide ritual honoring the dead
Yvette Nyombayire Rugasaguhunga

not occurred, as there may be political reasons why the United Nations does not level charges of genocide against a particular government.

There have been several genocides in the world's recent history, each one seemingly more gruesome than the next. The U.S. genocide of Native Americans during the 1700s through the 1800s and Turkey's genocide of the Armenians in 1917 are examples of genocides that have never been officially recognized by the international community. More recent genocides include the Nazi Holocaust against the Jews in Europe during World War II, the Serbian genocide against the Bosnians in 1992 through 1994, and the Rwandan genocide in 1994 where approximately 800,000 to 1,000,000 Tutsis were macheted to death by government-sponsored Hutu militia. Each of these genocides also involved *rape as a weapon of war*—the raping of women of the targeted ethnic or religious group for the purposes of either humiliating the targeted group, or impregnating the women forcing them to have children of another ethnic/religious group. For instance, in Rwanda, the Habyarimana government's armed forces, government-sponsored militia groups called Interahamwe, and Hutu civilians not only used machetes to kill and maim hundreds of thousands of Tutsis, but they also subjected hundreds of thousands of Tutsi women to sexual violence with the goal of impregnating them as well as infecting them with HIV (Buss, 2009; Cohen et al., 2009; Des Forges, 1999; HRW, 1996).

Rape as a weapon of war is a systematic tactic used in armed conflict targeting the civilian population (primarily women and girls) involving sexual violence in an officially orchestrated manner and as a purposeful policy to humiliate, intimidate, and instill fear in a community or ethnic group (Buss, 2009; HRW, 1996). Thus, as articulated by Buss (2009), rape during wartime is not a by-product of armed conflict, but an instrument of it. In June of 2008 the United Nations Security Council passed Resolution 1820, which recognizes rape as a weapon of war and establishes a commitment to addressing sexual violence in conflict, including punishing perpetrators (UN Security Council, 2008). This resolution became an important part of convictions by international criminal tribunals in response to genocides in former Yugoslavia (the ICTY), Rwanda (the ICTR), and in the United Nations–backed Special Court for Sierra Leone (SCSL) (UNDPKO, 2010).

Macro Practice in Action

Local advocacy organizations such as the YWCA (Young Women's Christian Association) lobby for governmental policies and laws that protect victims of crime, including sexual assault. Mothers against Drunk Driving (MADD) has been instrumental in lowering the legal alcohol limit for driving to 0.08 from 0.10, as well as establishing stiffer penalties for alcohol-related crashes. Amnesty International advocates for human rights and social justice for oppressed individuals around the world, releasing annual reports of human rights violations within each country. The passage of one domestic violence law can protect thousands of women. An antidrug educational campaign can convince thousands of adolescents to stay off drugs. One press release can lead to a boycott that can increase wages for thousands of young women in sweatshops in India. Direct practice with individuals can change the lives of a few people, but macro practice can change the lives of an entire community or a whole country. The power of macro practice should serve as an impetus for all human service professionals to consider embracing macro practice on some level, whether that means conducting voter registration drives in politically underserved areas, conducting a letter-writing campaign in support of legislation designed to protect a vulnerable population, or working on behalf of an international human rights organization that works tirelessly on behalf of exploited children, abused women, or traumatized refugees. Such positions offer significant rewards to those human service professionals willing to develop multidisciplinary expertise through education and experience that when combined with the networking power of other organizations can create positive change for all members of society.

Supporters of same-sex marriage organized a very successful and well-attended series of rallies held across the United States in response to the passage of an amendment to the California Constitution that defined a valid marriage as being between a man and a woman. The legislation was placed on the ballot after the California courts legalized gay marriage. The LGBT community and their many supporters flooded the streets in cities across the nation demanding equal rights under the U.S. Constitution.

An example of a grassroots organization that is working to end FGM in Eastern Africa is Termination of FGM, a project of the Loreto Sisters of Eastern Africa Province, located in Kenya. The project was started by Sr. Dr. Ephigenia Gachiri, a member of the Kikuyu tribe, who lives in a convent in Nairobi. Gachiri grew up with FGM as a part of her culture and didn't realize the very serious ramifications of the ritual until she had the opportunity to attend a UN convention on women's rights and heard a presentation on the grave consequences of FGM. She states that she made a decision after this conference to spend the rest of her career fighting FGM in her native country of Kenya. Sr. Ephigenia conducts educational seminars with village elders, tribal leaders, as well as school-aged children in order to confront dangerous long-standing myths, such as the belief that women who are not circumcised will become promiscuous, even potentially entering the life of prostitution. Sr. Ephigenia has developed alternate rites of passage based upon Christian beliefs which she advocates should replace FGM as a rite of passage into adulthood. In order to facilitate the replacement rite of passage, Sr. Ephigenia and her colleagues conduct training seminars in schools across Kenya where girls and boys engage in educational activities culminating in the alternate rite of passage ceremony where they and their families commit to not allowing the girls in the family to undergo FGM. Sr. Ephigenia describes the serious ramifications of this choice since in many tribes, including the Maasai tribe, a girl who is uncircumcised is not only unable to marry, but will often be completely shunned from her community—barred from engaging in communal meals, and even barred from collecting water at the same time as the other women. Sr. Ephigenia credits her success to the fact that she is not perceived as an outsider among her neighboring villages, thus she has greater legitimacy and credibility than outsiders from Western countries would likely have.

Social Action Effecting Social Change

One of the most dramatic forms of social change occurred during the 2008 presidential campaign when millions of Americans, many of whom had not been previously politically active, including many disenfranchised groups, advocated for now President Barack Obama, the country's first African American president. President Obama's message of real change for the country—one that promised for human rights and a renewed commitment to social justice led to a grassroots movement that many believe was something this country has never seen in previous elections. Political affiliations aside, what is important for our purposes is the recognition that virtually all people have the power to affect social change on a broad scale when they are motivated and well organized.

It is sometimes easy to see all of the problems in our world and respond with a feeling of futility, yet what many human service professionals soon realize is that making the world a better place is possible, particularly for those with a passion for meeting the needs of the most vulnerable members of society in a way that reflects empathy, compassion, justice, and respect for human dignity.

The following questions will test your knowledge of the content found within this chapter.

1. Macro practice involves addressing and confronting social issues that can act as a barrier to optimal functioning by working
   a. on an organizational level by creating structural change through social action
   b. with families to create change on a systemic level
   c. with groups to create change on a systemic level
   d. with individuals to create changes within society

2. According to Calkin (2000), human service organizations and professionals can easily be "seduced" into colluding with violations of human rights, including:
   a. disrespect toward people already struggling with mental illness or substance abuse
   b. acceptance or resignation in the face of deprivations of basic human rights
   c. actively advocating for the oppression of marginalized populations
   d. All of the above

3. The first step in community organizing is to:
   a. create a consensus on what the community needs
   b. develop steps in developing new policies
   c. create a consensus on intervention strategies
   d. create an intervention strategy addressing negative areas impacting the community

4. Of the 120 million children forced into full-time labor, the majority reside in:
   a. Africa
   b. Asia
   c. Latin America
   d. the Middle East

5. Rape as a weapon of war
   a. is a systematic tactic used in armed conflict targeting the civilian population (primarily women and girls)
   b. involves sexual violence during war time that is officially orchestrated
   c. is a purposeful policy to humiliate, intimidate, and instill fear in a community or ethnic group
   d. All of the above

6. Refugees who are housed in camps for long periods of time are referred to as
   a. long-term diaspora groups
   b. extended refugee problems
   c. protracted refugee situations
   d. nondurable refugee situations

7. Describe the treatment many LGBTQs experience worldwide and current advocacy efforts on a local and global level.

8. Describe dynamics associated with human sex trafficking including a description of those most vulnerable to being trafficked, underlying reasons for why trafficking occurs and current efforts to stop the practice of human sex trafficking.

Suggested Readings

Hokenstad, M. C., & Midgley, J. (Eds.). (2004). *Lessons from abroad: Adapting international social welfare innovations.* Washington, DC: NASW Press.

Langer, L. L. (1991). *Holocaust testimonies: The ruins of memory.* New Haven, CT: Yale University Press.

Rosenfeld, L. B., Caye, J. S., Ayalon, O., & Lahad, M. (2004). *When their worlds fall apart.* Washington, DC: NASW Press.

Van Soest, D. (1997). *The global crisis violence: Common problems, universal causes, shared solutions.* Washington, DC: NASW Press.

Internet Resources

American Indian Movement (AIM): http://www.aimovement.org

American Red Cross: http://www.redcross.org

AmeriCares—Humanitarian Lifeline to the World: http://www.americares.org

Amnesty International: http://www.amnesty.org

Anti-Defamation League: http://www.adl.org

Anti-Slavery: http://www.antislavery.org

AntiRacismNet: http://www.antiracismnet.org/main.html

Cultural Survival: http://www.culuralsurvival.org

Doctors on Call: http://www.docs.org

Doctors without Borders: http://www.doctorswithoutborders.com

Human Rights Watch: http://www.hrw.org

International Federation of Social Workers: http://www.ifsw.org

References

Ahmadi, N. (2003). Globalisation of consciousness and new challenges for international social work. *International Journal of Social Welfare, 12,* 14–23.

Austin, S. (2005). Community-building principles: Implications for professional development. *Child Welfare, 84*(2), 105–122.

Brownridge, D. (2009). *Violence against women: Vulnerable Populations.* New York: Routledge.

Buss, D. E. (2009). Rethinking "Rape as a Weapon of War". *Feminist Legal Studies, 17*(2), 145–163.

Calkin, C. (2000, June). Welfare reform. *Peace and Social Justice: A Newsletter of the NASW Committee for Peace and Social Justice, 1*(1). Retrieved September 17, 2005, from http://www.naswdc.org/practice/peace/psj0101.pdf

Chi-Ying Chung, R. (2005). Women, human rights & counseling: Crossing international borders. *Journal of Counseling and Development, 83,* 262–268.

Cohen, M. H., Fabri, M., Cai, X., Shi, Q., Hoover, D. R., Binagwaho, A., & … Anastos, K. (2009). Prevalence and predictors of post-traumatic stress disorder and depression in HIV-infected and at-risk Rwandan women. *Journal of Women's Health* (15409996), 18(11), 1783–1791. doi:10.1089/jwh.2009.1367

Collier, P. & Hoeffler, A. (2000), Economic causes of civil conflict and their implications on policy. In C. Crocker, F. Hampson, and P. Aall (Eds.), *Leashing the dogs of civil war: conflict management in a divided world* (pp. 197–218).

Coyhis, D., & Simonelli, R. (2005). Rebuilding Native American communities. *Child Welfare, 84*(2), 323–336.

Des Forges, A. (1999). Leave none to tell the story. New York: Human Rights Watch. Available online at http://www.hrw.org/legacy/reports/1999/rwanda/rwanda0399.htm

Dhlembeu, N., & Mayanga, N. (2006). Responding to orphans and other vulnerable children's crisis: Development of Zimbabwe's national plan of action. *Journal of Social Development in Africa, 21*(1), 5–49.

Dowsett, G. W. (2003). HIV/AIDS and homophobia: Subtle hatreds, severe consequences and the question of origins. *Culture, Health & Sexuality, 5*(2), 121–136.

Gardener, J. W. (1994). *Building community for leadership training programs.* Washington, DC: Independent Sector.

Gettleman, J. (2010, January 4). *Americans' role seen in Uganda anti-gay push.* New York Times. Retrieved January 11, 2010, from http://www.nytimes.com/2010/01/04/world/africa/04uganda.html

Graziano, K. J. (2004). Oppression and resiliency in a post-apartheid South Africa: Unheard voices of black gay men and lesbians. *Cultural Diversity and Ethnic Minority Psychology, 10*(3), 302–316.

Hollenbach, D. (2008). *Refugee rights: Ethics, advocacy and Africa.* Washington, DC: Georgetown University Press.

Huebner, D. M., Rebchook, M., & Kegeles, S. M. (2004). Experiences of harassment, discrimination, and physical violence among young gay and bisexual men human rights watch. *American Journal of Public Health, 94*(7), 1200–1203.

Human Rights Watch. (2002). *Burmese women and girls trafficked to Thailand.* The Human Rights Watch Report on Women's Human Rights. Retrieved September 30, 2005, from http://www.hrw.org/about/projects/womrep/General-123.htm#P1937_535306

Human Rights Watch. (2004). *All Jamaicans are threatened by a culture of homophobia.* Retrieved September 30, 2005, from http://hrw.org/english/docs/2004/11/23/jamaic9716.htm

Human Rights Watch. (1996). Shattered lives: Sexual violence during the Rwandan genocide and its aftermath. New York: Human Rights Watch.

Human Rights Campaign. (2009). *President Obama signs hate crimes legislation into law.* Retrieved January 11, 2010, from http://www.hrc.org/13699.htm

Human Rights Watch. (2005). *Saudi Arabia: Men "behaving like women" face flogging: Sentences imposed for alleged homosexual conduct violate basic rights.* Retrieved September 30, 2005, from http://hrw.org/english/docs/2005/04/07/saudia10434.htm

Human Rights Watch. (2008). *Guatemala: World Report 2009.* Retrieved January 10, 2009, from http://www.hrw.org/en/node/79213

Human Rights Watch. (2009). *Uganda: "anti-homosexuality" bill threatens liberties and human rights defenders proposed provisions illegal, ominous, and unnecessary.* Retrieved January 10, 2009, from http://www.hrw.org/en/news/2009/10/15/uganda-anti-homosexuality-bill-threatens-liberties-and-human-rights-defenders

Iatridis, D. (1995). Policy practice. In R. L. Edwards (Ed.), *Encyclopedia of social work* (19th ed., pp. 1855–1866). Washington, DC: NASW Press.

International Council on Social Welfare (n.d.). What is our mission? Available online at: http://www.icsw.org/intro/missione.htm

Johnson, A. (2004). Social work is standing on the legacy of Jane Addams: But are we sitting on the sidelines? *Social Work, 49*(2), 319–322.

Joint United Nations Programme on HIV/AIDS. (2004). *Report on the global AIDS epidemic.* Geneva: UNAIDS.

Knežević, M., & Butler, L. (2003). Public perceptions of social workers and social work in the Republic of Croatia. *International Journal of Social Welfare, 12*, 50–60.

Knudsen, S. (2005). Intersectionality: A theoretical inspiration in the analysis of minority cultures and identities in textbooks. Presented at the Eighth International Conference on Learning and Educational Media "Caught in the Web or Lost in the Textbook?" *IUFM DE CAEN (France) 26–29 Octobre 2005.* Article accessed on March 25, 2008, at: http://www.caen.iufm.fr/colloque_iartem/pdf/knudsen.pdf

Lyons, T. (2007), Conflict-generated diasporas and transnational politics in Ethiopia. *Conflict, Security, and Development, 7,* 4: 529–549.

Loescher, L., Milner, J., & Troeller, G. (2008). *Protracted refugee situations: Political, human rights and security implications.* New York: United Nations University Press.

Meyer, B. 2002. Extraordinary stories: Disability, queerness, and feminism. *NORA 3,* 168–173.

Mizrahi, T. (2001). The status of community organization in 2001: Community practice context, complexities, contradictions, and contributions. *Research on Social Work Practice, 11,* 176–189.

National Association of Social Workers. (1999). *Code of ethics of the National Association of Social Workers.* Washington, DC: Author.

National Labor Committee. (n.d.). *Working conditions in China.* Retrieved December 21, 2005 from http://www.nlcnet.org/campaigns/archive/report00/introduction.shtml

National Organization for Human Services. (1996). *Ethical standards of human service professionals.* Washington, DC: Author.

Netting, E., Kettner, P., & McMurtry, S. (2009). *Social work macro practice.* Boston: Pearson Education.

Polack, R. (2004). Social justice and the global economy: New challenges for social work in the 21st century. *Social Work, 49*(2), 281–290.

Rocha, C., & Johnson, A. (1997). Teaching family policy through a policy framework. *Journal of Social Work Education, 33*(3), 433–444.

Ryan, C., & Rivers, I. (2003). Lesbian, gay, bisexual and transgender youth: Victimization and its correlates in the U.S. and U.K. *Culture, Health and Sexuality, 5*(2), 103–119.

Samuels, G. M., & Ross-Sheriff, F. (2008). Identity, oppression, and power: Feminisms and intersectionality theory. *Affilia, 23,* 5–9.

UNAIDS. (2008). 2008 Report on the global AIDS epidemic. Geneva: UNAIDS. Available at: www.unaids.org.

United Nations Children's Fund [UNICEF], U.S. Agency for International Development. (2004). *Children on the Brink 2004: A Joint Report of New Orphan Estimates and a Framework for Action.* The Joint United Nations Programme on HIV/AIDS. New York: United Nations Children's Fund.

Office of the High Commissioner for Human Rights. (2006). *Human Rights Watch supplemental submission to the Human Rights Committee during its consideration of the second and third periodic reports of the United States.* Retrieved December 9, 2010, from http://www2.ohchr.org/english/bodies/hrc/docs/ngos/HRW.pdf

United Nations, Department of Peacekeeping Operations [UN-DPKO] (2010). Review of the sexual violence elements of the judgments of the international criminal tribunal for the former Yugoslavia, the International Criminal Tribunal for Rwanda, and the special court for Sierra Leone in the light of Security Council Resolution 1820. Available online at: http://www.unrol.org/files/32914_Review%20of%20the%20Sexual%20Violence%20Elements%20in%20the%20Light%20of%20the%20Security-Council%20resolution%201820.pdf

U.N. Security Council, 5916th Meeting. "Resolution 1820 [Sexual Violence as a War Tactic]" 19 June 2008, pp. 51–52. In *Resolutions and Decisions of the Security Council 2008* (S/RES/1820). Official Record. New York, 2008.

UN General Assembly, Prevention and punishment of the crime of genocide, 9 December 1948, A/RES/260, available at: http://www.unhcr.org/refworld/docid/3b00f0873.html [accessed 5 November 2012]

USCIS Immigration and Nationality Act 101(a) 41 http://www.uscis.gov/ilink/docView/SLB/HTML/SLB/0-0-0-1/0-0-0-29/0-0-0-101/0-0-0-195.html

U.S. Department of State. (2012). *Trafficking in persons report.* Washington, DC: U.S. Government Printing Office. Retrieved October 12, 2012, from http://www.state.gov/j/tip/rls/tiprpt/2012/

U.S. Department of State. (2001). *Afghanistan: Country Reports on Human Rights Practices, Bureau of Democracy, Human Rights, and Labor.* Retrieved November 7, 2009, from http://www.state.gov/g/drl/rls/hrrpt/2000/sa/721.htm

Weil, M. O. (1996). Community building: Building community practice. *Social Work, 41*(5), 481–499.

Weiss, I. (2003). Social work students and social change: On the link between views on poverty, social work goals and policy practice. *International Journal of Social Welfare, 12,* 132–141.

Weiss, I. (2005). Is there a global common core to social work? A cross-national comparative study of BSW graduate students. *Social Work, 50*(2), 102–110.

World Health Organization. (1998). *Female genital mutilation—an overview.* Geneva: Author.

Epilogue
The Future of Human Services in an Ever-Changing World

The human services profession exists to assist people meet their basic needs. One of its strengths is its multidisciplinary approach wherein individuals with education and training in various disciplines—including human services, social work, and counseling—work side by side, addressing the barriers to self-sufficiency and optimal living. Unlike many other mental health disciplines, human service professionals are true generalists, and their specializations are less often focused on particular psychological disorders, and more often focused on a particular social problem, such as interfamily violence or child welfare.

The passion to create meaningful change in the lives of others creates a drive in many human service professionals that may compensate for the relatively low pay and often less-than-ideal working conditions (although it would be incorrect to assume that just because one wants to enter the human services field, he or she cannot earn a decent living). Nonetheless, it is this drive and passion that pushes so many individuals forward in a career that does not have particularly high status, but affords the unique experience of having the power to make a significant difference in the lives of others by reminding people of their worth, holding the hand of the dying, reminding a grieving child that there is still hope, or standing with victims of violence who are facing their attackers in court. This is an empowering career, one that changes with every new client.

Human services is a unique career in that it can often lead to other opportunities including a career in academia, writing, public speaking, policy analysis, or international human rights work. Even a career track that leads to clinical private practice can remain exciting and varied if the human service professional remains committed to social justice and advocacy.

Avoiding Professional Burnout

As wonderful as this career is, it is also wrought with stress, crisis, and a significant potential to "burn out" quickly. There are many ways to avoid burnout, and several of these ways involve developing mental paradigms that help professionals avoid becoming overinvolved in the lives of their clients. One paradigm that benefits many human service professionals is to recognize that their clients are on a journey—on *their own* journey—and the role of the human service professional is to assist the client on a small portion of this journey. Many human service professionals experience professional burnout because they take too much responsibility for the lives of their clients. Understanding that clients are on their own journey and trusting that the human service professional is one of many mentors, counselors, or guides who will come along in their clients' lives puts the clinician–client relationship into healthy perspective.

Another paradigm that can be useful in helping human service professionals to avoid burnout is to make a commitment to never work harder than your clients. Human service professionals typically enter the helping profession because they care about people and want to help them have better lives. It is easy for human service professionals to fall into the trap of overworking for a client whom they so much want to help. But one must ask whether doing too much for a client is actually helpful. Or could it be harmful to clients who may already feel powerless and unable to take the steps necessary to make positive changes in their lives? This does not mean that it is inappropriate for a human service professional to help an overwhelmed client make a telephone call or that it is "enabling" for the counselor to make initial contact to a referral. But whenever I begin to feel overwhelmed working with certain clients, the first question I ask is whether I am working harder than they are. If the answer is "yes," then I need to step back and give my clients the room to decide whether or not to take the necessary, albeit often difficult, steps to create positive change in their lives. If my clients choose not to exert the necessary energy, then, as saddened as this might make me, I must accept my clients' inaction as a choice to remain in whatever situation they are in.

Human Services and Technology

Technology has changed (and continues to change) the world; the human services profession has been slow in making use of technological changes. Reasons for this include the lack of security in e-mail communication, which has an impact on confidentiality. E-mail communication between practitioner and practitioner discussing clients or e-mail communication between practitioner and client may expose a human services agency to legal liability if privacy cannot be guaranteed. Another reason for human services agencies' general reluctance to become more technologically based relates to the costs associated with purchasing and maintaining computer systems.

Despite these concerns, the Internet can be a wonderful resource for human service professionals searching for appropriate referrals for clients. Most counties have websites that include comprehensive information about available services. Many human services organizations, government assistance programs, and various grant-giving agencies not only have invaluable information on their websites, but also allow applicants to apply for services online, expediting the application process.

The Internet can be tremendously useful for human service professionals who want to coordinate services with other professionals or obtain information on a particular issue. Technology is also being used to facilitate various types of testing, including personality and career assessments, ADHD (attention deficit/hyperactivity disorder) evaluation, and adaptive functioning evaluations. Advocacy efforts have been made easier through the Internet: Legislation can be researched online and a virtual letter-writing campaign can be conducted in minutes.

Despite the concerns about privacy and confidentially, technology can serve both human service professionals and clients. The Internet can be empowering for clients, enabling them to be more self-sufficient in finding resources, including housing, job opportunities, and child care. In addition, there are resources for homebound individuals who might not be able to benefit from an on-site support group but can garner some of the same benefits from online support groups or bulletin boards.

The Effect of the Economic Crisis and Changes in the Political Landscape on Human Service Practice

The human services field is expected to continue to grow in the coming decades. There are various reasons for this, including the increasing complexity of society that results in numerous challenges for families. As the challenges facing societies increase, human services agencies will continue to be a valuable resource providing services for a broad range of clients. Whether working in schools, hospitals, criminal justice agencies, or the government, human service professionals serve those individuals who do not have the resources to meet their most basic needs.

The economic crisis that began in 2007 resulted in numerous employment layoffs, home loan foreclosures, and a significant increase in the economic vulnerability of many people living in the United States. Any one type of vulnerability within the lives of individuals will no doubt increase their vulnerability in all areas their lives, thus increasing the incidence of all of the social problems explored in this book, including domestic violence, child abuse, homelessness, mental health issues, and physical illness, which in turn will increase the need for human service professionals across the wide range of populations and practice settings. Unfortunately, this increased need exists in the face of significant fiscal cuts on local and national levels, most of which affect the funding of social service programs. The long-term effect of this economic crisis on the human services field remains to be seen, but those in the human services field have a lot to be optimistic about in light of the Obama administration's stated commitment to social justice in policies affecting the country's most vulnerable members.

Globalization

Our world is shrinking due to a variety of domains becoming "globalized," which is having a dramatic impact on the world and how it functions. The globalization of world market economies means that if one country sinks into a recession, it will likely take the rest of the world with it. If civil war rages in a far-off country, the ripple effect will be felt worldwide, whether through forced migration and refugee flow or international community involvement. The globalization of communication means that we can switch on our television sets, or our laptops, and know instantly what is happening thousands of miles away. We can Skype friends and family across the globe, text for free using our "smart phones" attached to a wireless connection, and make connections with old friends from elementary school and new friends in foreign countries using social media sites such as Facebook, Tumblr, or Twitter. These are exciting times for communication, but such rapid technological developments create both positive and negative consequences. Migrants can remain connected to home on a daily basis (good), and wage "virtual war" against their homeland governments using the Internet (bad). Child pornography is rampant online (bad), but law enforcement can use virtual online stings to catch consumers (good). The human services profession has no doubt been affected by the globalization of technology because our clients have.

Those in the human services field are committed to addressing problems in society, often before those within society are prepared to admit that such problems even exist. Human service professionals are consistently on the frontlines of social problems, creating change in the lives of individuals and communities, and globally. Society is constantly evolving, which creates the sometimes negative by-products of conflict, complexity, and challenges for many. It is for this reason that human service professionals will always be needed to recognize and confront human problems, helping society's most vulnerable members meet their basic emotional, physical, and spiritual needs.

Index